Understanding Advanced Physical Inorganic Chemistry

The Learner's Approach

Revised Edition

Understanding Advanced Physical Inorganic Chemistry

The Learner's Approach

Jeanne Tan

BSc (Hons), PGDE (Sec), MEd (LST)

Kim Seng Chan

BSc (Hons), PhD, PGDE (Sec), MEd, MA (Ed Mgt), MEd (G Ed), MEd (Dev Psy)

World Scientific

NEW JERSEY • LONDON • SINGAPORE • BEIJING • SHANGHAI • HONG KONG • TAIPEI • CHENNAI • TOKYO

Published by

WS Education, an imprint of

World Scientific Publishing Co. Pte. Ltd.

5 Toh Tuck Link, Singapore 596224

USA office: 27 Warren Street, Suite 401-402, Hackensack, NJ 07601

UK office: 57 Shelton Street, Covent Garden, London WC2H 9HE

British Library Cataloguing-in-Publication Data
A catalogue record for this book is available from the British Library.

UNDERSTANDING ADVANCED PHYSICAL INORGANIC CHEMISTRY
The Learner's Approach
Revised Edition

ISBN 978-981-4733-95-3 (pbk)

Typeset by Stallion Press
Email: enquiries@stallionpress.com

Printed in Singapore

Preface

The primary purpose of this book is to support students in developing a more holistic and integrative understanding of physical inorganic chemistry at the advanced level not only to meet the requirements of the national examination, but to really appreciate the centrality and relevancy of chemistry as a science subject. Hopefully with this book, students find the subject fascinating and exciting enough to want to continue to study chemistry after they have left school.

Our students usually find the learning of chemistry very difficult and lack intellectual stimulation. Thus, because of this, their learning is usually disjointed and mostly fact driven. The impetus behind this book was to translate what we have learnt from our students into a book that is pedagogically driven to develop their critical understanding of chemistry. This book should prove not only useful to students attempting to gain meaningful understanding of advanced chemistry, but also an asset to science teachers who are fresh into teaching service, to make the learning of chemistry of their students more engaging and mind stimulating. There are certainly pragmatic reasons for experienced science teachers to share the relevance of this book amongst the teaching community.

The main feature of this book is that we have incorporated various think-aloud questions at specific points in the content. Through this Socratic questioning, we want to challenge students to be analytical, to evaluate the chemical facts that are presented and to find a synthesis among facts and ideas being presented among disparate chapters. Ultimately, we hope that they develop their scientific thinking, gain better understanding of the nature of science and a more meaningful understanding of fundamental concepts, and appreciate that, although science evolves amongst human social-cultural activity, it is nonetheless an extremely powerful tool to develop knowledge and advance technology. The other feature that we are proud of, as it is really useful to students for integrative understanding, is the incorporation of questions at the end of each chapter. These integrated questions will prove useful to students in helping them to revise fundamental concepts learnt from previous chapters as well as see the importance and relevancy in the application to their current learning. This book offers a vision of understanding chemistry meaningfully and fundamentally from the learners' approach.

This revised edition has been updated to meet the minimum requirements of the new Singapore GCE A level syllabus that would be implemented in the year 2016. Nevertheless, this book is also highly relevant to students who are studying chemistry for other examination boards. In addition, the authors have also included more Q&A to help students better understand and appreciate the chemical concepts that they are mastering.

Acknowledgements

We would like to express our sincere thanks to the staff at World Scientific Publishing Co. Pte. Ltd. for the care and attention which they have given to this book, particularly our editors Lim Sook Cheng and Sandhya Devi, our editorial assistant Chow Meng Wai and Stallion Press.

Special appreciation goes to Ms Ek Soo Ben (Principal of Victoria Junior College), Mrs Foo Chui Hoon, Mrs Toh Chin Ling and Mrs Ting Hsiao Shan for their unwavering support to Kim Seng Chan.

Special thanks go to all our students who have made our teaching of chemistry fruitful and interesting. We have learnt a lot from them just as they have learnt some good chemistry from us.

Finally, we thank our families for their wholehearted support and understanding throughout the period of writing this book. We would like to share with all the passionate learners of chemistry two important quotes from the Analects of Confucius:

學而時習之，不亦悅乎？(Isn't it a pleasure to learn and practice what is learned time and again?)

學而不思則罔，思而不學則殆 (Learning without thinking leads to confusion, thinking without learning results in wasted effort)

Contents

CHAPTER 1

Atomic Structure and the Periodic Table

1.1 The Subatomic Particles of Matter

The three fundamental subatomic particles of matter are **protons**, **neutrons** and **electrons**.

The table below shows the properties of these subatomic particles.

	Proton	Neutron	Electron
Actual mass	1.673×10^{-27} kg	1.675×10^{-27} kg	9.109×10^{-31} kg
Relative mass	1	1	1/1840
Charge	$+1.602 \times 10^{-19}$ C	0	-1.602×10^{-19} C
Relative charge	+1	0	−1
Location within atom	in the nucleus	in the nucleus	around the nucleus

Protons and neutrons are collectively known as nucleons. The nucleons reside in the small nucleus of the atom whereas the electrons revolve around it in a "vast" empty space. The size of an atom is easily more than 10,000 times that of the nucleus. Yet, it is the nucleus that accounts for most of the mass of an atom since the mass of the electron is negligible as compared to the mass of the protons and neutrons.

An atom is electrically neutral and contains equal numbers of electrons and protons.

1.1.1 *Behaviour in an electric field/magnetic field*

The behaviour of particles in an electric field **depends on their mass and charge**. Like charges **repel** while unlike charges **attract**. Hence, a beam of charged particles passing through an electric field is **deflected**.

Since neutrons are neutral, they are not deflected but travel in a straight path perpendicular to the direction of the electric field.

When a beam of protons passes through an electric field, the positively charged particles are deflected towards the negative plate.

As for a beam of negatively charged electrons, these are deflected towards the positive plate and to a **greater extent** since they are much lighter than protons.

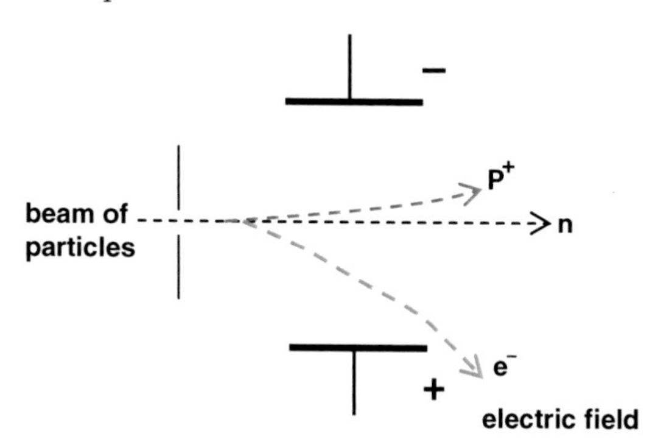

The amount of deviation from its original direction of movement is known as the angle of deflection. The angle of deflection θ can be expressed by:

$$\text{angle of deflection} \propto \frac{\text{charge}}{\text{mass}}.$$

Another way to measure the amount of deviation is to use the radius of deflection:

$$\text{radius of deflection} \propto \frac{\text{mass}}{\text{charge}}.$$

This is possible if we imagine that after deflection the particle moves in a circular path. Hence, the factors affecting the radius of deflection are reciprocal to that for the angle of deflection.

Q: Why is the angle of deflection directly proportional to the charge of the particle?

A: The greater the charge of the particle, the greater is the attractive force exerted on it from the oppositely charged plate, and the greater is the deviation from its original direction of motion.

Q: Why is the angle of deflection inversely proportional to the mass of the particle?

A: If two particles are moving at the same speed but one is more massive than the other, the heavier particle has a greater kinetic energy. Thus, more energy must be exerted on the heavier particle to cause it to deflect. Since the applied electric field is exerting the same amount of force on these two different particles with different masses, the heavier particle is deflected to a lesser extent.

The figure below shows how a beam containing protons ($^1_1H^+$), deuterons ($^2_1H^+$) and protiums (^1_1H) is affected when passing through an electric field.

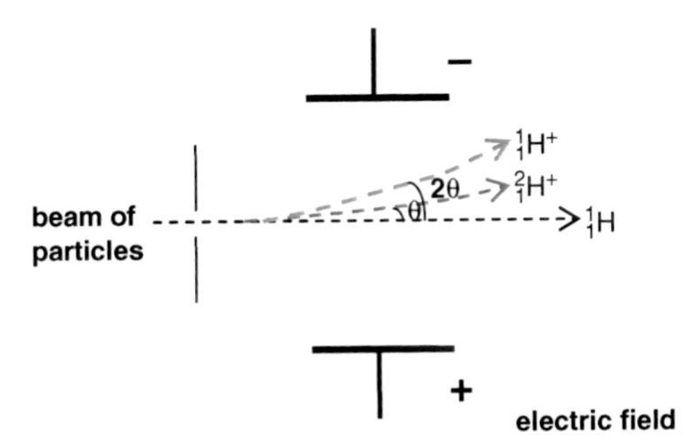

Being positively charged, both the protons and deuterons will deflect towards the negatively charged plate.

However, the angle of deflection for the deuterons will be half that for the protons since a deuteron has twice the mass of a proton.

$^1H^+$ has a charge/mass ratio of 1/1 while $^2H^+$ has a charge/mass ratio of 1/2.

As for the neutral hydrogen atoms, they will not be deflected but travel in a straight path perpendicular to the electric field.

1.1.2 *Isotopes*

Isotopes are atoms of the **same element** that have **different masses** due to the presence of **different numbers of neutrons**. An example is the deuterium (^2_1H) and the protium (^1_1H) isotopes. The latter is loosely called the hydrogen atom.

Isotopes usually have the same chemical properties since they have the same number of protons and hence the same number of electrons. However, their physical properties differ since they have different numbers of neutrons and hence different masses.

For instance, molecules of $^{35}Cl_2$ and $^{37}Cl_2$ undergo the same chemical reactions but pure $^{35}Cl_2$ will have a lower density, melting point and boiling point than $^{37}Cl_2$ (refer to Chap. 2 for an explanation of these differences).

Q: Why do isotopes react similarly?
A: This is because in a chemical reaction, it is the electrons that are transferred between different atoms; atoms either gain, lose or share electrons. The nucleus remains intact.

To distinguish among the isotopes present, a classification system has been devised. In this system, the nuclide of an element is represented by three notations:

$$^A_ZX$$

- The atomic symbol (X) represents each element in the periodic table.
- **Atomic number/Proton number** ($\boldsymbol{Z}$) is the number of protons in the nucleus.
- **Mass number/Nucleon number** ($\boldsymbol{A}$) is the sum of protons and neutrons in the nucleus.

Q: What is the meaning of this symbol: $_8O^{2-}$?
A: This represents an oxide ion. The subscript on the left is the atomic number whereas the superscript on the right indicates the extra electrons oxygen has acquired. For this $_8O^{2-}$ species,

the number of protons (8) is not equal to the number of electrons (10). The number of electrons is the same as the number of protons for a neutral atom only.

Q: What is the meaning of this symbol: ${O_2}^{2-}$?

A: This represents a peroxide ion. The subscript on the bottom right indicates that there are two oxygen atoms in this species **covalently bonded** to each other. The superscript indicates the extra electrons this species has acquired.

1.1.3 *Relative masses of an element*

The mass of an atom is very small, ranging from 10^{-24} to 10^{-22} grams. Chemists use a relative atomic mass scale to compare the masses of different atoms. Initially, protium (^1H) was used as the atom for comparison as its nucleus only consists of one proton.

In 1961, the carbon-12 atom was adopted by the International Union of Pure and Applied Chemistry (IUPAC) as the reference standard for relative atomic masses. On the carbon-12 scale, an atom of the isotope ^{12}C is assigned a mass value of 12 atomic mass units (a.m.u.). The relative masses of all other atoms are obtained by comparison with 1/12 of the mass of a carbon-12 atom. Thus, the relative atomic mass of an atom is a dimensionless quantity (without physical units).

Term	Definition	Remarks
Relative isotopic mass	mass of one atom of the *isotope* of an element relative to 1/12 of the mass of one atom of ^{12}C	• Dimensionless. • It is actually similar to mass number since the bulk mass of an atom results from its massive nucleus. The electrons have negligible mass. • It is a whole number. • It can be determined using mass spectrometry (Refer to *Understanding Advanced Organic and Analytical Chemistry* by KS Chan & J Tan).

(*Continued*)

(*Continued*)

Term	Definition	Remarks
Relative atomic mass (A_r)	*average* mass of one atom of the element relative to 1/12 of the mass of one atom of ^{12}C	• Dimensionless. • This quantity has taken into consideration the composition of various isotopes present in an element. It is usually not a whole number.
Relative molecular mass (M_r)	*average* mass of one molecule of the substance relative to 1/12 of the mass of one atom of ^{12}C	• Dimensionless. • It is equal to the sum of the A_r of the atoms shown in the molecular formula.
Relative formula mass (M_r)	*average* mass of one formula unit of the substance relative to 1/12 of the mass of one atom of ^{12}C	• Dimensionless. • A formula unit is the smallest collection of atoms from which the formula of an ionic compound can be established. It is equal to the sum of the A_r of the atoms shown in the formula unit.

Example 1.1: Determine the A_r of chlorine given that there exist two isotopes, ^{35}Cl and ^{37}Cl, with percentage isotopic abundance 75% and 25%, respectively.

Solution: Since A_r is a weighted average, one has to take into consideration the relative amount of each isotope. The following shows how to calculate a weighted average:

$$A_r \text{ of chlorine} = \frac{75(35) + 25(37)}{(75 + 25)} = 35.5.$$

Q: It seems that the relative isotopic mass is a whole number. Is it really so?

A: The word "relative" means that the value obtained is measured with respect to something else; hence, this number has no physical unit. In this case, the relative isotopic mass of an isotope is the mass measured with respect to 1/12 of the mass of a ^{12}C atom, which has a value of 1 unit ($1/12 \times 12$). As the mass of an atom arises mainly from the nucleus and since the total number of nucleons in ^{35}Cl is 35, the relative isotopic mass seems to have a value of 35 too. But in reality, the actual relative isotopic mass is less than the mass of all nucleons added up. This phenomenon is known as mass defect. The difference in the masses, which is less than 1% and arise because part of the mass has been converted into binding energy according to $E = mc^2$, is necessary to hold the nucleons together in the nucleus.

1.2 Orbitals and Quantum Numbers

1.2.1 *The nature of electron*

Electrons are negatively charged particles and thus they cannot crowd together at the same point in space. They need to spread themselves out to minimise interelectronic repulsion (like charges repel) by moving about. As a result, the electrons in an atom revolve round the nucleus, occupying space from near the nucleus to much further away. In effect, the large size of an atom arises solely because of the huge space occupied by the electrons.

When scientists studied the atoms of various types of elements, they found out that each of the electrons in an atom occupied a specific region in space known as an **atomic orbital**. The atomic orbital is a region of space from near the nucleus to a certain distance away. In the atomic orbital, one has an approximately 90 percent chance of finding the electron there. Do not forget that an electron is not stationary but constantly moving around!

Through some mathematical calculations using Quantum Mechanics, scientists created a scientific model to give a unique identity to each and every single electron in the atom. This identity is characterised by four numbers known as quantum numbers, namely, the Principal Quantum Number (n), the Angular Momentum Quantum Number (l), the Magnetic Quantum Number (m_l), and the Spin Quantum Number (m_s). No two electrons in the same atom have exactly the same set of four quantum numbers.

Due to this unique occupancy in space, each electron has a specific amount of energy, and we say that the energy is quantised. Thus, the four quantum numbers can be used to represent a specific energy level that an electron might have.

The importance of the principal quantum number is that it indicates the **average distance** of the orbitals from the nucleus, and hence the amount of attractive force the nucleus is exerting on the electron residing in the atom. The value of n takes on integers starting from 1, 2, 3 and so forth.

An electron in the $n = 1$ principal quantum shell is the closest to the nucleus, and is thus most strongly attracted and has the least energy. In other words, **the greater the value of n, the further the electrons are from the nucleus, and the greater is the energy of the electrons**.

Q: Why does the energy of an electron increase as we move away from the nucleus?

A: By convention, when an electron is "free", that is, not subjected to any other electrostatic interactive forces (attractive or repulsive), it has an assigned energy value of zero. This is when the electron is "infinitely far" from the nucleus. But now if you want to bring an electron from $n = 1$ principal quantum shell to infinity, you have to do work against the attractive force of the nucleus; you have to "break" the "bond" between the electron and the nucleus. Breaking bonds requires energy. The energy that you expend while doing work is gained by this electron (energy is conserved from the Law of Conservation of Energy) and hence its energy increases. When an electron moves from infinity and is attracted by the nucleus, a "bond" is formed, and energy will be released. The following diagram helps to explain.

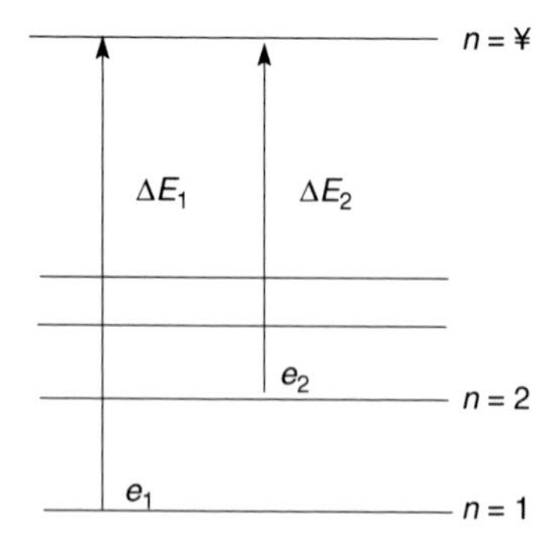

At an infinite distance from the nucleus, the energy of a free electron is zero, i.e., $E_{\text{infinity}} = 0$.

$n = 2$ has an energy level E_2, which by convention is a negative value. For example, $-300\,\text{kJ mol}^{-1}$.

$n = 1$ has an energy level E_1, which by convention is a negative value. For example, $-500\,\text{kJ mol}^{-1}$.

Energy of Electron 1, $e_1 <$ Energy of Electron 2, e_2.
Energy to remove e_1, $\Delta E_1 = E_{\text{final}} - E_{\text{initial}} = 0 - E_1 = -E_1$.
Energy to remove e_2, $\Delta E_2 = E_{\text{final}} - E_{\text{initial}} = 0 - E_2 = -E_2$.

> Since $\Delta E_2 < \Delta E_1$, we say that more energy is absorbed by Electron 1 than by Electron 2 in order to reach the same infinite distance from the nucleus. Therefore, Electron 1 must be at a lower energy level than Electron 2. Such a consideration is important in order to be in line with the concept that when energy is being absorbed, it is a positive quantity. This corresponds with work being done against an opposing force and in this case, bond breaking. In contrast, the release of energy is a negative quantity, corresponding to the formation of a bond. This also explains why, by convention, scientists assign a negative sign to the energy possessed by an electron in an atom.

Each principal quantum shell comprises of subshell(s). A subshell is a group of orbitals with the **same energy level but different orientation in space**. These subshells are represented by the letters s, p, d, and f. Each of these subshells contains specific numbers of such orbitals, each of which can only accommodate a maximum of two electrons.

- The s subshell comprises ONE s-orbital; the p subshell comprises THREE p-orbitals; the d subshell comprises FIVE d-orbitals; and the f subshell comprises SEVEN f-orbitals.
- The number and type of subshells possible is in turn limited by the principal quantum shell number. When $n = 1$, only the $1s$ subshell is possible. When $n = 2$, there are two types of subshells, namely the $2s$ and $2p$ subshells. When $n = 3$, there are the $3s$, $3p$ and $3d$ subshells.

When we piece all the information together, we end up with a pattern as shown in Table 1.1.

Table 1.1 Pattern of Shells

Shell/Principal Quantum No. (n)	Type of Subshells	Number of Orbitals	Number of Electrons	Maximum number of Electrons
1	$1s$	1	2	2
2	$2s$	1	2	8
	$2p$	3	6	
3	$3s$	1	2	
	$3p$	3	6	18
	$3d$	5	10	
4	$4s$	1	2	
	$4p$	3	6	
	$4d$	5	10	32
	$4f$	7	14	

From Table 1.1, it can be shown that in the nth principal quantum shell, there are n subshells consisting of n^2 orbitals with a maximum number of $2n^2$ electrons.

The diagram below is a schematic representation of the energy levels, indicating the relative energies of the various orbitals. Generally, for the **same value of n**, the relative energies of the orbitals increase in the following order: $s < p < d < f$.

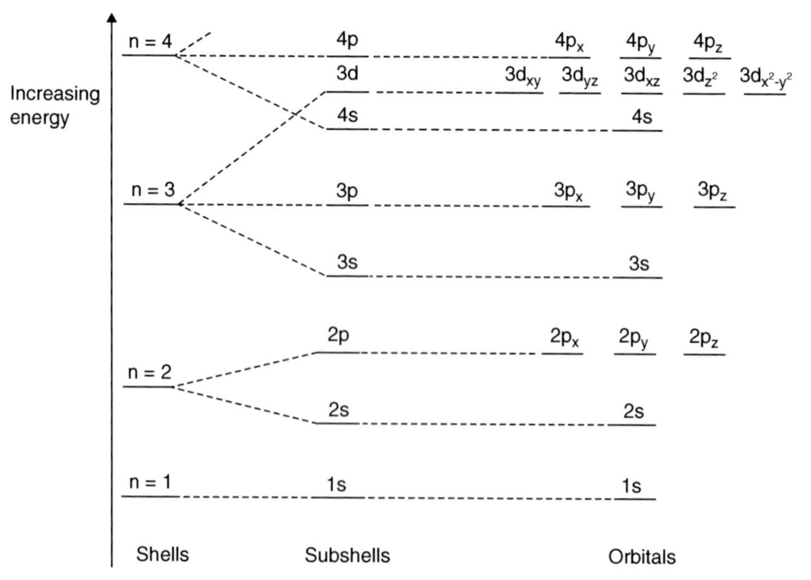

Q: Why does the $4s$ subshell have a lower energy than the $3d$ subshell given that the $n = 4$ principal quantum shell should have a higher energy than the $n = 3$ principal quantum shell?

A: $n = 4$ **should** have a higher energy than $n = 3$. But the s subshell has a relatively lower energy than the d subshell for the same principal quantum shell. These two factors counteract each other. As the increased in energy of the principal quantum shell is not exactly matched by the lowered in energy of the subshell, coincidentally, it results in the $4s$ subshell having a lower energy than the $3d$ subshell of different principal quantum numbers. The same explanation accounts for the relative energies of the $5s$ and $4d$ subshells.

Q: Does that mean that the $4s$ subshell is now closer to the nucleus than the $3d$ subshell?

A: No. On average, the $n = 4$ principal quantum shell is still further away from the nucleus than $n = 3$. So, although the energy of the $4s$ subshell is lower than the $3d$ subshell, it does not imply that the distances have changed. Remember that a principal quantum shell is actually a band, not a single discrete line.

1.2.2 *Shapes of orbitals*

Instead of representing orbitals as an electron map showing the electron density, we make use of geometrical shapes to denote the orbitals.

- s orbitals:
  - These are spherical, where the spherical surface indicates the region where there is a 90% probability of locating an electron.

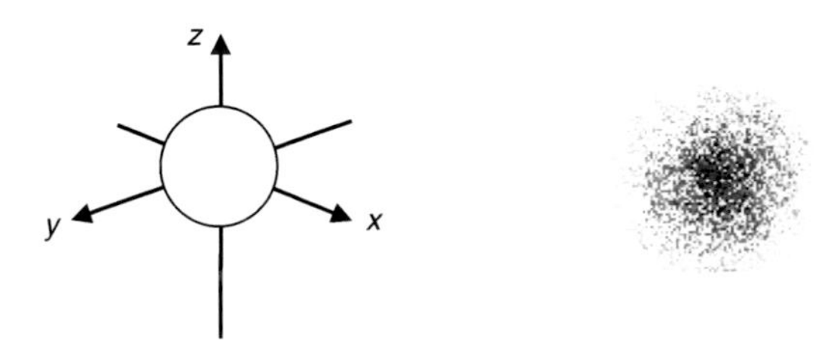

- p orbitals:
  - These are shaped like "dumb-bells".

- A set of three degenerate (same energy) *p* orbitals make up the *p* subshell.
- These three orbitals differ only in their spatial arrangements, being mutually perpendicular to each other as shown in the diagram.
- Each *p* orbital is approximately one-third the size of the *s* orbital in the same principal quantum shell, which means that three *p* orbitals have the same volume as an *s* orbital, of the same principal quantum number.

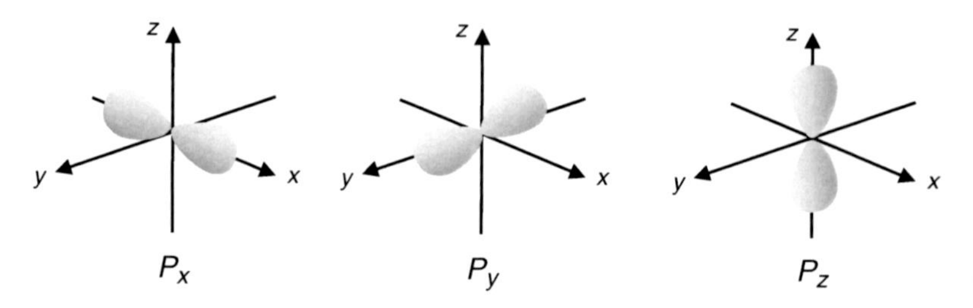

- *d* orbitals:
  - These are shaped like a "four-petal flower".
  - A set of five degenerate *d* orbitals make up a *d* subshell. Each *d* orbital is about 20% the size of an *s* orbital.

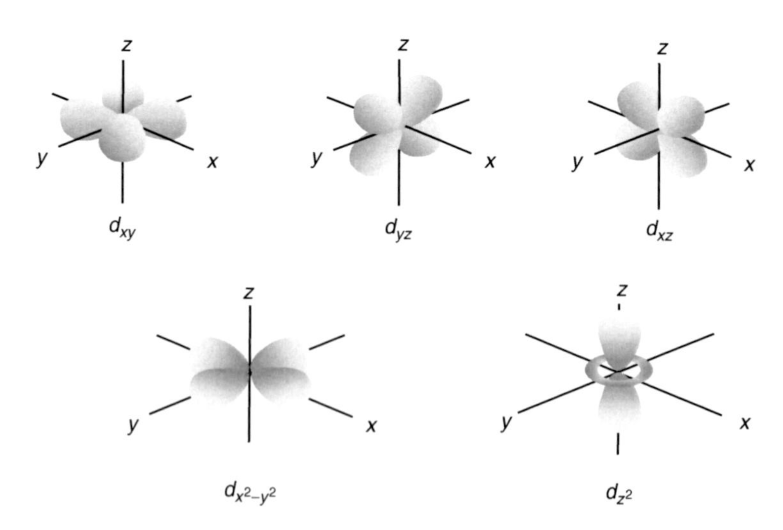

For a particular type of orbital, when the **value of *n* increases**, the **orbitals become more diffused**. What do we mean by this?

Taking the s orbitals for instance, all s orbitals are spherical but they get larger as the value of n increases. With a $2s$ orbital larger than a $1s$ orbital, the probability of finding an electron in a $2s$ orbital is spread across a greater region, and thus this electron is less strongly attracted by the nucleus as compared to one in the $1s$ orbital. As we view the electron as a negatively charged cloud, this means a lower charge density and hence a more diffuse orbital.

Thus far, there are four important points to understand:

(i) The greater the principal quantum number, the higher the energy of the electron, and hence the lesser the amount of energy required to remove it (a process known as ionisation).

(ii) Different subshells have different energy levels, i.e., $s < p < d < f$. Hence, more energy is required to remove an electron from the s subshell which is at a lower energy level (slightly closer to the nucleus) than from the p subshell.

(iii) Within the same subshell, the orbitals have the same energy. We say that they are degenerate.

(iv) For increasing n values, orbitals of the same subshell type occupy a bigger region of space, and are thus more diffuse. This is important as we will use it for understanding the strength of the covalent bond in Chap. 2.

1.3 Electronic Configurations

The electronic configuration represents the electronic structure of an atom. It refers to the way in which the electrons are distributed among the various orbitals.

There are two common notations for writing electronic configurations:

(i) *spdf* notation (e.g., for $_7$N, it is written as $1s^2 2s^2 2p^3$) or

(ii) orbital-as-box diagram, e.g., for $_7$N, orbitals are represented by boxes and each electron is represented by a half-headed arrow:

⥮	⥮	↿ ↿ ↿
1s	2s	2p

Q: Why do we need to know the electronic configuration of an element?

A: Knowing the correct electronic configuration enables us to know which electron is to be removed and which orbital it resides in. This is important as the removal of different electrons from different orbitals requires different amounts of energy.

1.3.1 *Rules used in working out electronic configuration*

(i) Place electrons into orbitals, starting with those of the lowest energy (nearest to the nucleus) and then working outwards. This is known as the Aufbau Principle.

As shown in the diagram below, the order progresses as $1s2s2p3s3p4s3d4p$, and so forth.

When the electrons in an atom occupy orbitals of lowest available energy, the atom is said to be in its *ground* state.

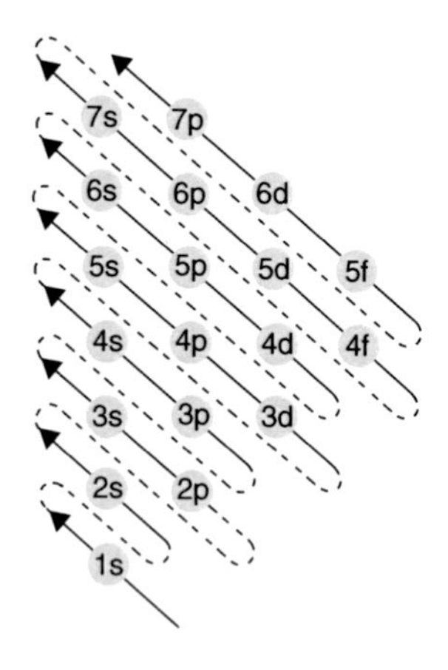

When an electron is promoted to an orbital of a higher energy level, the atom is unstable and is said to be in an *excited* state.

(ii) Each orbital can only contain a maximum of two electrons of opposite spins. This is known as the Pauli Exclusion Principle.

⇅ ☑ ⇈ ☒

Q: What is electron spin?

A: You can imagine an electron like Earth, rotating on a particular axis.

Q: Why can't two electrons in the same orbital have the same spin?

A: When an electron spins, this spinning charged particle creates a magnetic field. If two electrons spin in the same direction, the magnetic field created would be repulsive and the energy level of these two electrons would be higher as compared to if they spin in opposite directions to create an attractive magnetic field. An analogy can be used here: picture a spinning electron as moving in one particular direction. Two such spinning electrons will be moving in opposite directions and the chances of them "meeting" will be lower. This results in lesser inter-electronic repulsion.

(iii) When filling a set of degenerate orbitals (orbitals in the same subshell and of the same energy), ensure that each orbital is half full before completely filling any one. This is known as Hund's Rule of Multiplicity.

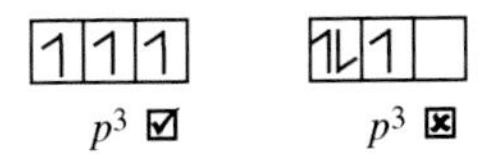

p^3 ☑ p^3 ☒

Q: Why do we need to first place electrons in empty orbitals of the same subshell before pairing them in an orbital?

A: Electrons repel each other. By occupying different orbitals, the electrons remain as far apart as possible from one another, thus minimising electron–electron repulsion. Note that each orbital represents a particular region of space, and hence two different orbitals are two different regions of space separated from each other. Take for instance the three *p* orbitals in a *p* subshell: each is oriented perpendicularly from one another, occupying different regions in space.

Example 1.2: State the electronic configuration of an $_8O$ atom. Sketch, on the given axes, the shapes of all the orbitals that are occupied by electrons.

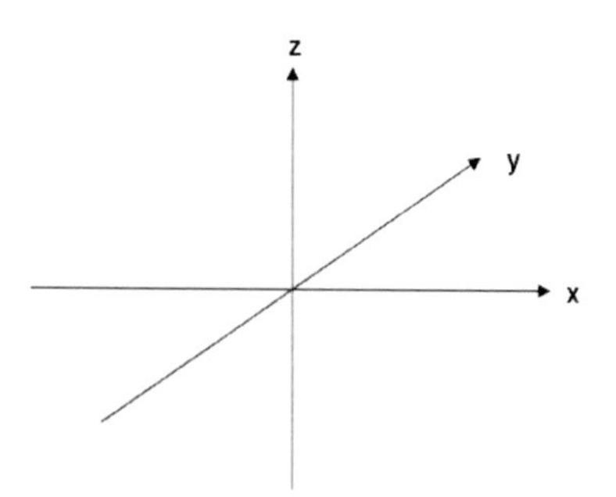

Solution: The electronic configuration of $_{8}$O atom: $1s^22s^22p^4$

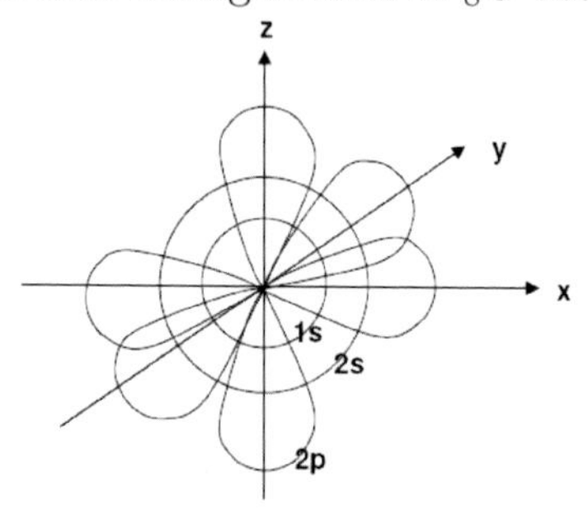

Example 1.3: State the electronic configuration of the $_{25}$Mn atom. Sketch an energy level diagram to illustrate its electronic configuration.

Solution: The electronic configuration of $_{25}$Mn: $1s^22s^22p^63s^23p^6$ $3d^54s^2$

energy
4s
3d
3p
3s
2p
2s
1s

1.3.2 *Electronic configuration of transition elements*

Q: Why is the electronic configuration of $_{25}$Mn not $1s^22s^22p^63s^2$ $3p^63d^7$?

A: The 4s orbital is filled first before the 3d orbitals. This is because the 4s orbital has a lower energy level as compared to the 3d orbitals.

Q: If 4s orbital is filled before the 3d orbital, why is the electronic configuration of $_{25}$Mn not written as $1s^22s^22p^63s^23p^64s^23d^5$?

A: The electronic configuration is always written in the order of increasing Principal Quantum Number. This also indicates the order of increasing energy level of the various subshells.

Example 1.4: State the electronic configuration of $_{25}Mn^{2+}$. How many unpaired electrons does it have?

Solution: $_{25}Mn^{2+}$ has the electronic configuration of $1s^22s^22p^63s^2$ $3p^63d^5$ and thus it has five unpaired electrons.

It is wrong to write it as $1s^22s^22p^63s^23p^63d^34s^2$. ⊠

When an atom of a transition element undergoes ionisation, forming a cation, the 4s electrons are removed first before the 3d electrons. This is because the 3d subshell is closer to the nucleus. Once the 3d subshell is occupied by electrons, they repel the 4s electrons further from the nucleus and cause the latter to be at a higher energy level.

Exercise: Write down the electronic configurations of the following species:

(a) $_{15}P^{3-}$, (b) $_{19}K^{+}$, (c) $_{22}Ti^{2+}$, and (d) $_{24}$Cr.

Solution:

(a) $_{15}P^{3-}$: $1s^22s^22p^63s^23p^6$,
(b) $_{19}K^{+}$: $1s^22s^22p^63s^23p^6$,
(c) $_{22}Ti^{2+}$: $1s^22s^22p^63s^23p^63d^2$, and
(d) $_{24}$Cr: $1s^22s^22p^63s^23p^63d^54s^1$.

Approach:

(i) First, write down the electronic configurations for the neutral atom.

(ii) Remember to add electrons to the $4s$ orbital before the $3d$ orbitals.
(iii) Remove or add the relevant number of electrons to the electronic configuration of the neutral atom to get that of the cation or anion.
(iv) Remember to remove electrons from the orbitals with the highest energy. Thus, remove electrons from the $4s$ orbital before the $3d$ orbitals.

$_{15}P^{3-}$ and $_{19}K^{+}$ are an example of a pair of isoelectronic species.

Isoelectronic: species containing the same number of electrons
Isotopic: species containing the same number of protons
Isotonic: species containing the same number of neutrons

Q: Do isoelectronic species have the same electronic configuration?
A: Not necessary. Take for instance, Fe^{2+} (24 electrons) and Cr (24 electrons) are isoelectronic but their electronic configuration are not the same.

1.3.3 *Anomalous electronic configurations*

Apparently, electronic configurations with *half-filled* or *fully filled* 3d subshells are unusually stable due to the symmetrical distribution of charge around the nucleus. The electronic configuration of $_{24}Cr$ is $1s^22s^22p^63s^23p^63d^54s^1$ and that of $_{29}Cu$ is $1s^22s^22p^63s^23p^63d^{10}4s^1$.

More detailed explanation about the writing of electronic configurations for transition elements is covered in Chap. 11.

Q: Why is symmetrical distribution of similar charge preferred?
A: If similar charges are distributed symmetrically, this means that all charges are spread out evenly and as far apart as possible. Such a situation results in a similar amount of electrostatic repulsion at each point in space. Consequently, such a state has lower energy as compared to a state of asymmetrical distribution.

1.4 Ionisation Energies

Ionisation involves the removal of electron(s), forming a cation.

As electrons, being negatively charged, are naturally attracted to the positively charged nucleus, their removal requires energy. Thus, ionisation energies are positive values, indicating energy is absorbed during ionisation.

- **First ionisation energy** (1st I.E.) is the energy required to **remove one mole of electrons** from **one mole of gaseous atoms** in the ground state to form one mole of gaseous **singly charged cations**. For example,

$$\mathrm{Ca(g)} \rightarrow \mathrm{Ca^{+}(g)} + \mathrm{e^{-}}, \quad \Delta H = \mathrm{1st\,I.E.} = +590\,\mathrm{kJ\,mol^{-1}}.$$

Q: Why must the atom be in the gaseous state?
A: When we carry out ionisation, the species must be gaseous atoms. In the gaseous state, atoms have very minimal interactions with one another. Thus, the energy input would solely be responsible for removing the electron and not for overcoming other types of bond. Remember that the gaseous state symbol is very important here.

- The second ionisation energy (2nd I.E.) is the energy required to **remove one mole of electrons** from **one mole of gaseous singly charged cations** to form one mole of gaseous **doubly charged cations.**

$$\mathrm{Ca^{+}(g)} \rightarrow \mathrm{Ca^{2+}(g)} + \mathrm{e^{-}}, \quad \Delta H = \mathrm{2nd\,I.E.} = +1150\,\mathrm{kJ\,mol^{-1}}.$$

1.4.1 *Factors influencing the magnitude of ionisation energies*

The magnitude of ionisation energy serves as a measure of the strength of attraction between the positively charged nucleus (due to the protons present) and the valence electron that is to be removed.

The ionisation energy of an atom or ion is mainly influenced by the following factors:

(i) the effective nuclear charge acting on the electron,
(ii) interelectronic repulsion between electrons in the same orbital,
(iii) the orbital where the electron comes from,

(iv) the distance between the nucleus and the electron to be removed, and

(v) the charge of the cation.

Q: What is a valence electron?

A: *Valence* refers to the *outermost.* Thus, a valence electron "sits" in the outermost principal quantum shell and is furthest from the nucleus. The principal quantum number for the valence shell corresponds to the period number of the element. All other principal quantum shells of electrons before the valence principal quantum shell are known as the inner core electrons.

1.4.1.1 *The effective nuclear charge acting on the electron*

An electron in the atom faces two main electrostatic forces, namely, the **attractive force exerted by the nucleus and the repulsive force of electrons closer than itself to the nucleus**. This repulsive force is termed "**screening effect**" or "**shielding effect**" as it seems to "block" the electron from being attracted by the nucleus. The repulsive force caused by electrons from the same principal quantum shell is usually ignored as its effect is quite minimal; electrons in the same shell exert a negligible shielding effect on one another since they hardly come between each other and the nucleus.

Thus, one can see that the net attractive force on a valence electron is actually less than the number of protons in the nucleus. This net attractive force is termed the effective nuclear charge (ENC) and it can be approximated by:

$$\text{ENC} \approx \text{Nuclear charge} - \text{Number of inner core electrons.}$$

For example:

Element	Nuclear charge (No. of Protons)	No. of Inner Shell Electrons	Effective (Net) Nuclear Charge
Mg	+12	10	+2
P	+15	10	+5
Cl	+17	10	+7

The greater the ENC, the greater the amount of energy needed to remove the electron(s), i.e., ionisation energy will increase, which means that a valence electron is easier to be removed than an inner core electron. And very importantly, **ENC increases across a period of elements**.

Q: What is the ENC on *each* of the valence electrons of an oxygen atom?

A: The electronic configuration of an O atom is $1s^22s^22p^4$. The nuclear charge consists of eight protons and the number of inner core electrons is two (since there is only one Principal Quantum Shell of electrons before the $n = 2$ valence shell); therefore, the ENC is ≈ 6.

Q: Does that mean that each of the six valence electrons is attracted by 1/6 of the ENC, which is one proton?

A: No. Each of the six valence electrons is attracted by six protons. This is because the six valence electrons are moving around the nucleus within the same distance from the nucleus and the nucleus is considered a point charge. Therefore, the ENC is the same on each of the valence electrons.

Q: Does that mean that the ENC on each of the two $1s$ electrons is equivalent to eight protons?

A: Yes. There are no other inner core electrons before the $n = 1$ Principal Quantum Shell; therefore, there is no shielding effect. In fact, if there are electrons in subsequent Principal Quantum Shells after $n = 1$, these electrons would "push" the electrons in the $n = 1$ Principal Quantum Shell nearer to the nucleus. The ENC on each of the electrons in the $n = 1$ subshell is the same, i.e., equivalent to eight protons.

1.4.1.2 *Interelectronic repulsion between electrons in the same orbital*

Ionisation energy can be lowered if the electron is being removed from an orbital that contains another electron as compared to the case where the orbital contains a singly unpaired electron.

For instance, the 1st I.E. of oxygen (1310 kJ mol^{-1}) is lower than that of nitrogen (1400 kJ mol^{-1}) because the interelectronic repulsion makes the removal of one of the paired electrons easier.

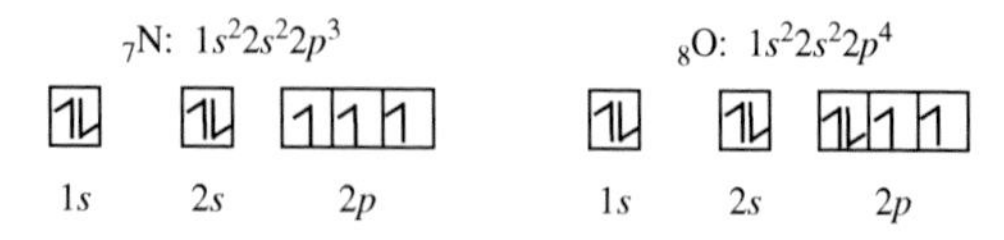

This interelectronic repulsion factor can be used to account for the lower than expected I.E. of a species with the ns^2np^4 valence shell electronic configuration compared to another with the ns^2np^3 valence shell electronic configuration.

However, it cannot be used to account for the difference in I.E. between atoms or ions of the ns^1 versus the ns^2 valence shell electronic configuration. For instance, beryllium ($1s^2 2s^2$) has higher 1st I.E. than lithium ($1s^2 2s^1$) because of its higher effective nuclear charge. Although the interelectronic repulsion between the $2s$ electrons of Be does make the removal of one of the paired electrons easier than the removal of the valence electron in Li, the ENC effect far outweighs the interelectronic repulsion effect!

Q: The ENC of an O atom is greater than the N atom. So shouldn't this cause the O atom to have a higher 1st I.E.?

A: Yes, the ENC of the O atom is greater than the N atom and this should cause it to have a higher 1st I.E. But experimentally, it has been found that the O atom has a lower 1st I.E than what was expected. There must be some other factors that have yet to be considered. The only reason we can use to explain the data collected would be to employ the interelectronic repulsion factor. As shown here, there is the interplay of two opposing factors but it seems that interelectronic repulsion is a more dominant factor than ENC.

Q: Why is the domineering effect of the interelectronic repulsion over the ENC factor observed in the oxygen versus nitrogen case but not in the beryllium-lithium scenario?

A: Remember that we mentioned before that a p orbital is one-third the size of an s orbital from the same principal quantum shell? Thus, because of this difference in size, interelectronic repulsion is more prominent in a p orbital than in the much bigger s orbital.

1.4.1.3 *The orbital where the electron comes from*

The orbital where the electron comes from also affects the ionisation energy.

For instance, the 1st I.E. of boron ($799\,\text{kJ mol}^{-1}$) is lower than that of beryllium ($900\,\text{kJ mol}^{-1}$) because the electron that is being removed from the boron atom comes from a $2p$ subshell that is at a higher energy level than the $2s$ subshell which requires less energy to be removed.

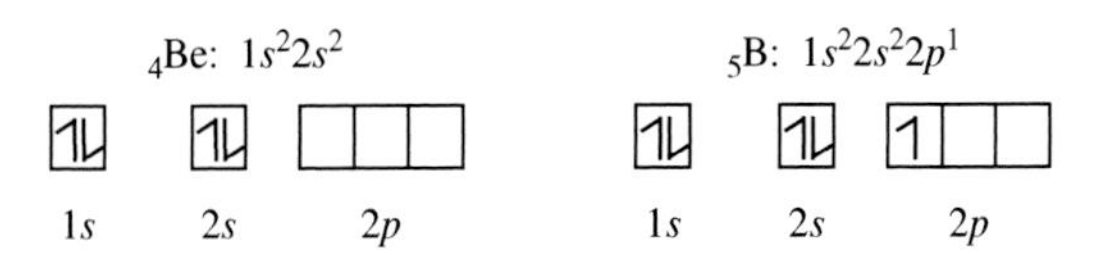

This factor can be used to account for the lower than expected I.E. of a species with the ns^2np^1 valence shell electronic configuration compared to another with the ns^2 valence shell electronic configuration.

Q: The ENC of a B atom is higher than that of a Be atom but there is interelectronic repulsion faced by the electrons in the $2s$ subshell of Be. Are these two factors less dominant than the "difference in energy level" factor?

A: Yes. Based on the ENC factor, B is expected to have a higher 1st 1.E. than Be. But this does not agree with experimental data. If we are to consider the interelectronic repulsion factor, Be should have a lower 1st I.E., but observed data prove otherwise. Hence, the only factor that can be used to account for the observed experimental data is that the $2p$ electron is at a higher energy level than the $2s$ electron.

1.4.1.4 *Distance between the nucleus and the electron to be removed*

The further an electron is from the nucleus, the weaker is its attraction to the positively charged nucleus. Consequently, it is easier to be removed and less energy is required to do so. This factor is used to account for the trend in I.E. down a group of elements.

Thus, we expect less energy is required to remove an electron from a higher principal quantum shell (n increases down a group) as it is further away from the nucleus.

Q: Why is the attractive force weaker the further an electron is from the nucleus?

A: The electrostatic force (F) between the nucleus and electron can be approximated to $F \propto 1/r^2$, where r is the distance of separation between the charges. Thus, as r increases, the strength of the electrostatic force decreases drastically.

1.4.1.5 *The charge of the cation*

The removal of an electron from a +2 cation is greater than from a +1 cation, and so forth for the same element. For instance, the 2nd I.E. of Ca is greater than its 1st I.E.. This is because when we remove the second electron from a $Ca^+(g)$ species, the total number of electrons in $Ca^+(g)$ is fewer than in Ca(g). But both Ca(g) and $Ca^+(g)$ still have the same nuclear charge. The interelectronic repulsion of the electron cloud in $Ca^+(g)$ is less than that of Ca(g), resulting in a stronger net attractive force for the electrons in $Ca^+(g)$, and thus higher I.E. The same reason applies for the increase in consecutive ionisation energies of a given element.

For example:

$$\mathrm{Na(g) \rightarrow Na^+(g) + e^-} \qquad \text{1st I.E.} = +494\,\mathrm{kJ\,mol^{-1}}$$

$$\mathrm{Na^+(g) \rightarrow Na^{2+}(g) + e^-} \qquad \text{2nd I.E.} = +4560\,\mathrm{kJ\,mol^{-1}}$$

$$\mathrm{Na^{2+}(g) \rightarrow Na^{3+}(g) + e^-} \qquad \text{3rd I.E.} = +6940\,\mathrm{kJ\,mol^{-1}}$$

$$\mathrm{Na^{3+}(g) \rightarrow Na^{4+}(g) + e^-} \qquad \text{4th I.E.} = +9540\,\mathrm{kJ\,mol^{-1}}.$$

Thus, we can make use of successive I.E. values to deduce information regarding the number of valence electrons an element has. This number of valence electrons corresponds to the group number of the element in the periodic table.

Example: The first four successive I.E.s (in $\mathrm{kJ\,mol^{-1}}$) of element **W** are as follows: 577; 1820; 2740; 11,600.

Determine which group of the periodic table element **W** belongs to.

Solution: W belongs to Group 13.

A big jump in energy is seen between the 3rd and 4th I.E. This implies that the fourth electron to be removed belongs to a principal quantum shell that has a lower energy than the one that the valence electrons occupied. Therefore, there are only three electrons in the valence shell of **W**.

Approach:

(i) Tabulate the differences between successive ionisation energies.
(ii) To determine the group to which **W** belongs, we must determine the number of valence electrons. The point where a big jump in I.E. occurs indicates the removal of successive electrons from an inner principal quantum shell.

Exercise: The first seven successive I.E.s (in kJ mol^{-1}) of element **X** are as follows: 1010; 1905; 2910; 4965; 6275; 21,270; 25,430.

Which group of the periodic table does element **X** belong to?

(**Answer:** Element **X** belongs to Group 15.)

1.5 Periodic Table: Trend in Ionisation Energy

We will now apply the factors that influence ionisation energies to account for Periodic Table trends across the period and down the group.

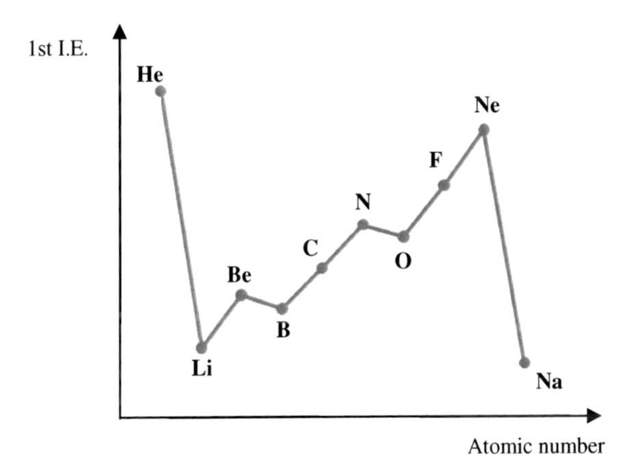

The graph shows the trend in 1st I.E. across Period 2.

- In general, 1st I.E. increases across a period.
- From Li to Ne, nuclear charge increases but the shielding effect is relatively constant since the number of inner core electrons is the same across a period.

(i) effective nuclear charge acting on the electron

Overall, effective nuclear charge increases, resulting in electrons being pulled closer towards the nucleus. It becomes increasingly difficult to remove the valence electron, i.e., more energy is required, and thus 1st I.E. increases.

If we go by the general trend of increasing I.E. across the period, we would expect the 1st. I.E. of B to be higher than that of Be, and likewise, the 1st I.E. of O to be higher than that of N. However, experimental data concluded that this is not the case.

- B has a lower 1st I.E. than Be since less energy is needed to remove its $2p$ electron, which is at a higher energy level (further away from the nucleus) than the $2s$ electron of Be.

(ii) orbital where the electron comes from

- O has a lower 1st I.E. than N as less energy is required to remove one of the paired $2p$ electrons since there is interelectronic repulsion between these electrons in the same orbital.

(iii) Interelectronic repulsion between electrons in the same orbital

- In general, ionisation energy decreases down any group.
- Na has a lower 1st I.E. than Li as less energy is required to remove a $3s$ electron in Na, which is further away from the nucleus than a $2s$ electron in Li.

(iv) the distance between the nucleus and the electron to be removed.

As we move down a group, nuclear charge increases but electrons are being added to a higher principal quantum shell which is further away from the nucleus. This results in the valence electrons experiencing weaker attractive force, and they are consequently easier to be removed.

What is wrong with the following explanation?
"Across a period, the atomic radius decreases so the distance between the nucleus and the valence electron decreases. The electron to be removed is thus more tightly held and it is increasingly more difficult to remove it. Hence, 1st I.E. increases".

Answer:
Both atomic radius and ionisation energy are phenomena that are accounted for by a similar fundamental reason, which is the ENC factor. Thus, you cannot use one property to explain the other.

In fact, we will apply the same ENC factor that influences ionisation energies to account for trends in atomic radii across the period. As for down the group, we would use the distance from the nucleus factor to account for both the trend of decreasing I.E. and increasing in atomic sizes.

1.6 Periodic Table: Trend in Atomic Radii

Atoms have no definite outer boundary and thus no definite circumference unlike a solid spherical object such as a bowling ball. Nonetheless, the size of an atom can be estimated by using the atomic radius which describes the distance from the nucleus to the outermost energy level.

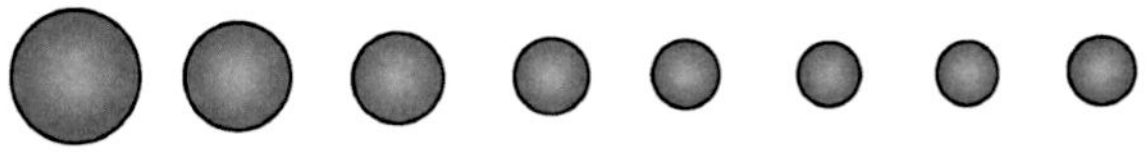

Period 2	Li	Be	B	C	N	O	F	Ne
Atomic Radii (pm)	152	112	88	77	70	66	64	70

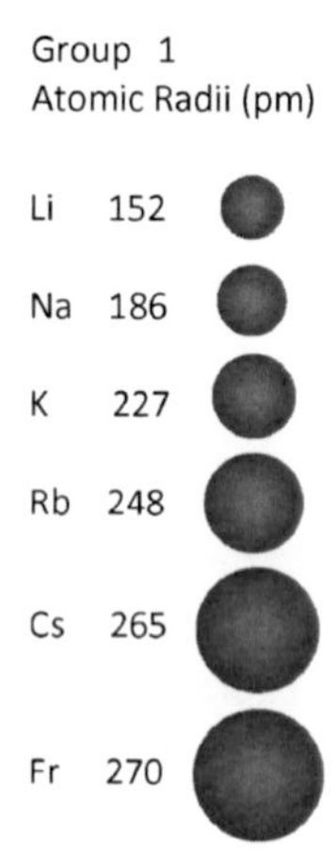

- In general, atomic radius decreases across a period.
- From Li to Ne, nuclear charge increases but the shielding effect is relatively constant since the number of inner core electrons is the same across a period. Overall, effective nuclear charge increases, resulting in electrons being pulled closer towards the nucleus. Thus, atomic radius decreases.
- In general, atomic radius increases down a group.
- As we move down a group, nuclear charge increases but electrons are being added to a higher principal quantum shell which is further away from the nucleus. This results in the valence electrons experiencing weaker attractive force, which gives rise to greater atomic radius.

Thus far, there are five important points to understand:

(i) I.E. **generally** increases across a period because of greater ENC. This is due to the increase in nuclear charge but a relatively constant shielding effect as the number of inner core electrons is the same across a period.

(ii) Elements from the **same period** may have different I.E. for the following reasons:

  - ENC factor; e.g., a nitrogen atom has a higher I.E. than a carbon atom because a nitrogen atom has greater ENC.

- Interelectronic repulsion factor; e.g., an oxygen atom has a lower I.E. than a nitrogen atom due to interelectronic repulsion between its paired electrons in the same orbital.
- A *p* electron has higher energy than an *s* electron; e.g., a boron atom has lower I.E. than a beryllium atom.

(iii) I.E. decreases down a group because, although nuclear charge increases, electrons are being added to a higher principal quantum shell which is further away from the nucleus. This results in the valence electrons experiencing weaker attractive force, and are consequently easier to be removed.

(iv) Successive I.E.s of an element increases because the interelectronic repulsion, due to a decrease in the number of electrons, decreases and since nuclear charge remains the same, net attractive force increases.

(v) Both I.E. and atomic radius across a period can be accounted for by the ENC factor. But down a group, the factor that is responsible is due to the increase in the distance from the nucleus.

My Tutorial (Chapter 1)

1. Naturally occurring boron consists of two isotopes, ^{10}B and ^{11}B, having abundances of 19.7% and 80.3%, respectively.

   (a) Explain the terms *isotope* and *relative isotopic mass*.

   (b) Calculate the relative atomic mass of naturally occurring boron.

2. (a) (i) Give the electronic configuration of an atom of the isotope of calcium, $^{45}_{20}Ca$.

   (ii) Give the names and numbers of each type of nucleon present in the isotope.

   (iii) State one reason why the information in (a)(i) is usually more useful to chemists than that in (a)(ii).

   (b) $^{45}_{20}Ca$ is radioactive, decaying by giving off electrons. State and explain what would happen to the electrons if they were passed through an electric field.

(c) (i) For the following species, sketch a graph of ionic size against the respective atomic numbers:

$$Na^+, O^{2-}, S^{2-}, P^{3-}, F^-, Cl^-.$$

(ii) Explain the trends shown in the graph sketched in part (c)(i).

3. Given the first four ionization energies, in kJ mol^{-1}, of gallium and cobalt:

	1st	2nd	3rd	4th
Gallium	577	1980	2960	6190
Cobalt	757	1640	3230	5100

(a) Provide an explanation for the increases in the successive ionization energies.

(b) With reference to the electronic configuration, explain why a discontinuity is present in gallium but not in cobalt.

(c) Naturally occurring gallium consists of two isotopes, ^{69}Ga and ^{71}Ga. Calculate the percentage abundance of each isotope.

4. Given the 1st I.E.s, in kJ mol^{-1}, of the elements lithium to neon:

Li	Be	B	C	N	O	F	Ne
519	900	799	1090	1400	1310	1680	2080

(a) Give an equation representing the 1st I.E. of oxygen.

(b) (i) Explain why the 1st I.E. generally increases across the period.

(ii) Explain the irregularities in trend of the 1st I.E. of B and O.

(c) An element W has successive ionization energies as follows:

$$786; 1580; 3230; 4360; 16,000; 20,000; 23,600; 29,100\,\text{kJ mol}^{-1}.$$

(i) Provide with an explanation which group in the periodic table element W belongs to.

(ii) Give the valence electronic configuration of an atom of W.

(iii) Suggest formulas for TWO chlorides of W.

5. The ionization energies of successive elements in the periodic table are shown in the next page:

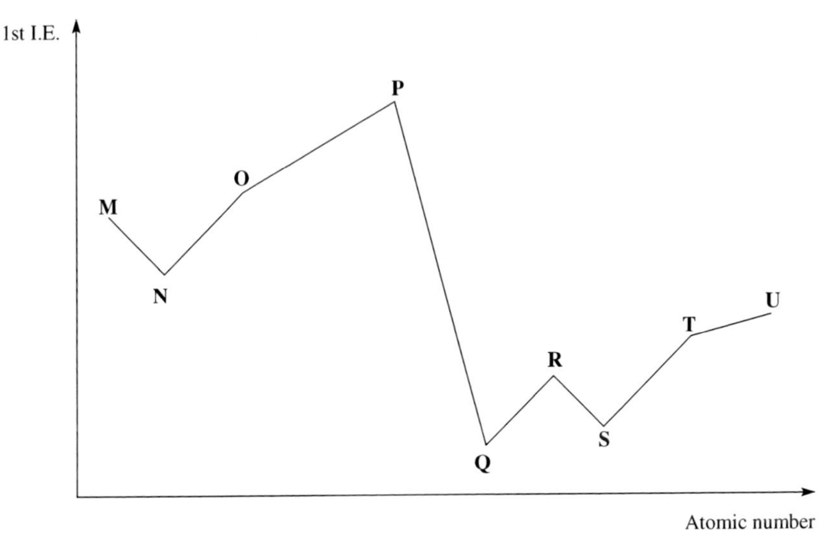

(a) Which two elements are in the same group?
(b) Which element has the largest atomic radius?
(c) Explain why the ionization energy of

(i) N is smaller than that of M;
(ii) O is greater than that of N;
(iii) P is greater than that of O;
(iv) P is greater than that of N;
(v) P is greater than that of Q;
(vi) R is greater than that of Q;
(vii) T is smaller than that of U.

CHAPTER 2

Chemical Bonding

Chemical bonds are electrostatic forces of attraction (positive charge attracting negative charge) that bind particles together to form matter. When different types of particles interact electrostatically, different types of chemical bonds are formed. There are four different types of conventional chemical bonds, namely, metallic, ionic, covalent and intermolecular forces.

2.1 Metallic Bonding

Within a metal, atoms partially lose their loosely bound valence electrons. These electrons are mobile and delocalised, not belonging to any one single atom and yet not completely lost from the lattice.

A metal can thus be viewed as a rigid lattice (Fig. 2.1) of positive ions surrounded by a sea of delocalised electrons. What holds the lattice together is the strong metallic bonding: the electrostatic attraction between the positive ions and the delocalised valence electrons.

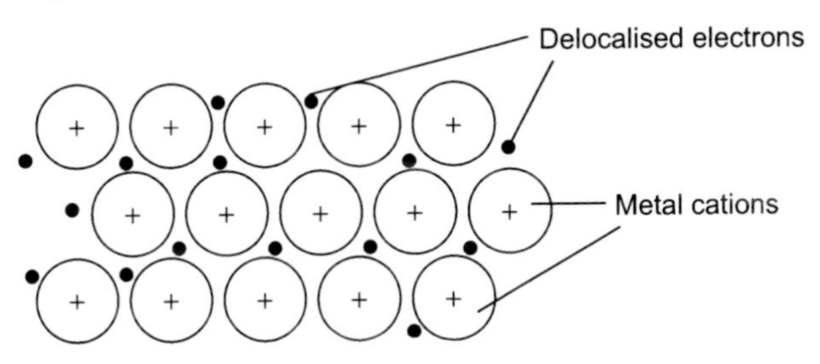

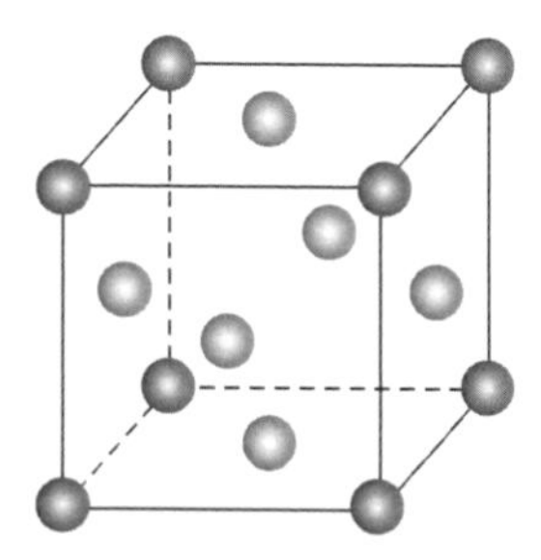

Fig. 2.1. The copper atoms occupy the corners and the centre of each face of a cube.

Metallic bonds are strong and non-directional (so we do not talk about shape here).

Since metallic bonding is the result of the interaction between the delocalised electrons and the positive ion, the strength of a metallic bond depends on:

- The number of valence electrons available for bonding. (This factor is useful in explaining the increase in metallic bond strength across a period of metals).
- The size of the metal cation. (This factor is useful in explaining the decrease in metallic bond strength down a group of metals).

2.1.1 *Physical properties of metals*

2.1.1.1 *Metals are malleable and ductile*

These properties are related to the ability of the cations to move over one another without the breaking of metallic bonds (Fig. 2.2).

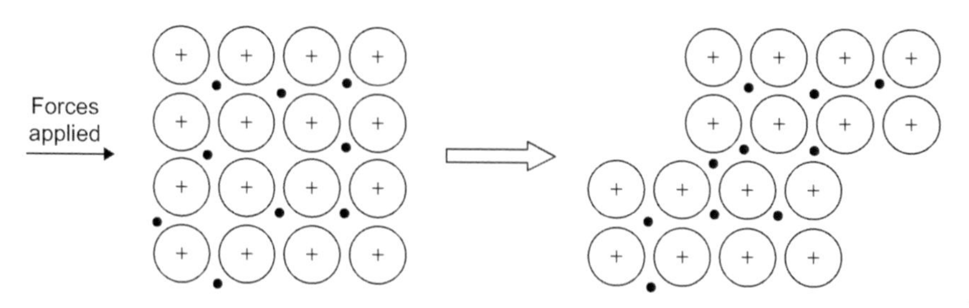

Fig. 2.2. The arrangement of metal atoms (a) before and (b) after "slip".

2.1.1.2 *High melting and boiling points*

These properties are related to the strong metallic bonding.

For instance, Mg and Al have high melting points as a great amount of energy is needed to overcome the strong metallic bonds between the positively charged ions and the "sea of delocalised valence electrons".

The higher melting point of Al (m.p. 660°C) compared to Mg (m.p. 650°C) is due to a **greater number of delocalised electrons** available for metallic bonding.

2.1.1.3 *Good thermal and electrical conductivity*

These properties are related to the presence of delocalised electrons in the metallic lattice.

For instance, Na, Mg and Al are good conductors of electricity as there exist delocalised electrons that can act as mobile charge carriers when a potential difference is applied. Electrical conductivity increases from Na to Al due to a **greater number of delocalised electrons** in the metallic structure.

2.2 Ionic Bonding (or Electrovalent Bonding)

Ionic compounds are generally formed between metals and non-metals. The metal atoms lose electrons from the outermost principal quantum shell, forming positively charged ions (cations) and the non-metal atoms receive these electrons, place them in the outermost principal quantum shell, forming negatively-charged ions (anions). The resultant electrostatic attraction between the cations and anions is known as the ionic bond. (*This is another case of positive charge attracting negative charge*!)

Q: Why are electrons removed from or placed in the outermost principal quantum shell?

A: Electrons from the outermost principal quantum shell are the easiest to remove as compared to the inner core electrons as they are furthest away from the nucleus and less strongly attracted.

In addition, the outermost shell is not fully filled; thus, it can accept additional electrons.

Q: How does the cation actually attract the anion? Is it the "+" of the cation attracting the "–" of the anion?

A: When we actually look at a cation or anion, it just looks like a spherical ball, like a neutral atom. You will not see a "+" sign on the cation nor a "–" sign on the anion. When we talk about the ionic bond, it is actually the attractive force between the nucleus of the cation and the electron clouds of the anion, and *vice versa*. At the same time, there is repulsion between the electron clouds of the cation and anion, and also between the two nuclei. If the attractive force is greater than the repulsive force, there is a net attractive force that exists, and this net attractive force is the ionic bond that holds the two species together. For simplicity, we just say a cation attracts an anion.

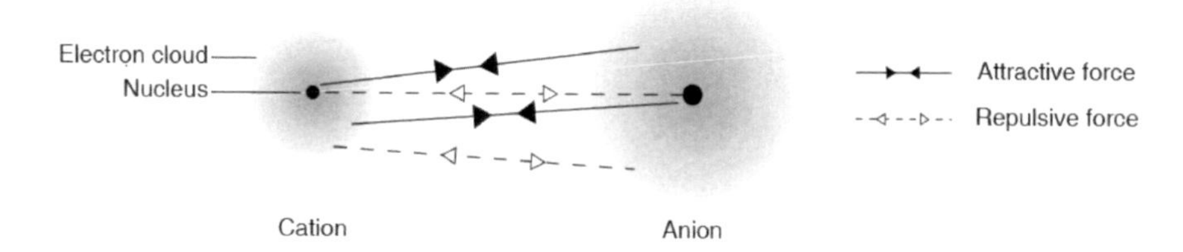

Ionic compounds have crystalline lattices wherein the ions are arranged in an orderly manner and held in fixed positions by the strong ionic bonds. In the lattice of NaCl(s), each Na^+ ion is surrounded by six Cl^- ions and each Cl^- ion, in turn, is surrounded by six Na^+ ions as nearest neighbours (Fig. 2.3). For metal complexes or crystal structures, the number of nearest neighbours (atoms, ions or molecules) around a central atom or ion is termed the coordination number. When used for describing simple ionic structures, the coordination number indicates the number of nearest oppositely charged ions around the ion of concern. Hence, both Na^+ and Cl^- have coordination numbers of six, that is, each ion is surrounded by six nearest ions that are oppositely charged. Another common lattice structure is exemplified by CsCl(s) wherein Cs^+ and Cl^- have coordination numbers of eight; each Cs^+ ion is surrounded by eight Cl^- ions and each Cl^- ion is surrounded by eight Cs^+ ions (Fig. 2.3).

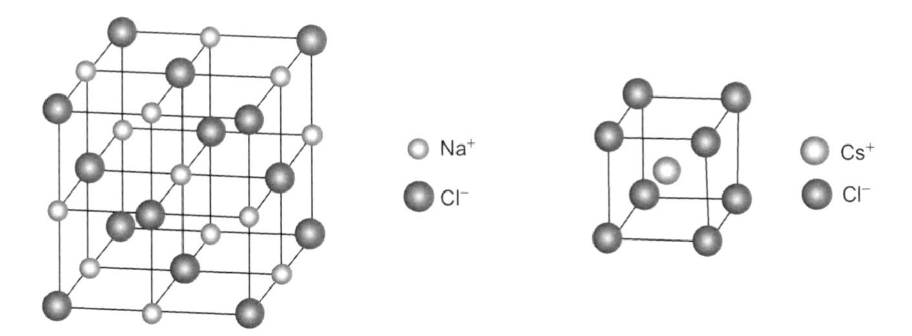

Fig. 2.3. Lattice structure of NaCl (left) and CsCl (right).

It must be noted that ionic bonds are not singular bonds that hold together just one cation and one anion. Ionic bonds are actually non-directional. Take NaCl(s) for example: each Na^+ ion is equally attracted to the neighbouring Cl^- ions without preference for any Cl^- ion in particular.

Q: How do we determine the number of electrons to be gained or lost in forming ionic compounds?

A: We make use of the Octet Rule, which is an *observational* conclusion that "Atoms tend to lose, gain or share electrons until they are surrounded by eight valence electrons."

The Octet Rule stems from the fact that all noble gases (except helium) have eight valence electrons. They have very stable electronic arrangements, as evidenced by their high ionisation energy, low affinity for additional electrons, and general lack of reactivity. However, one must note that there are many exceptions to the Octet Rule, which we will see in later sections. The Octet Rule should not be regarded as a driving force in explaining why ionic compounds are formed.

Apart from transition metals and Group 13 elements, Group 1 and Group 2 metals will lose the number of electrons corresponding to their group number. Group 13 elements would lose three electrons

$(13 - 10 = 3)$. The non-metals will tend to gain the required number of electrons to make a total of eight electrons in the valence shell.

Take for instance NaCl(s), which is formed from its elements in the following reaction:

$$2\text{Na(s)} + \text{Cl}_2\text{(g)} \longrightarrow 2\text{NaCl(s)}.$$

In this reaction, the Na atom loses its one valence electron to form Na^+, which has a stable octet configuration:

$$\begin{array}{ccc} \text{Na} & \longrightarrow & \text{Na}^+ + \text{e}^- \\ 1s^2 2s^2 2p^6 3s^1 & & 1s^2 2s^2 2p^6 \end{array}$$

The Cl atom gains one electron to form Cl^-, which also has a stable octet configuration:

$$\begin{array}{ccc} \text{Cl} + \text{e}^- & \longrightarrow & \text{Cl}^- \\ 1s^2 2s^2 2p^6 3s^2 3p^5 & & 1s^2 2s^2 2p^6 3s^2 3p^6 \end{array}$$

We can depict the electron transfer using the following dot-and-cross diagram:

Na× + :Cl· ⟶ [Na]$^+$ + [:Cl:×]$^-$

x = valence electron of Na
• = valence electron of Cl

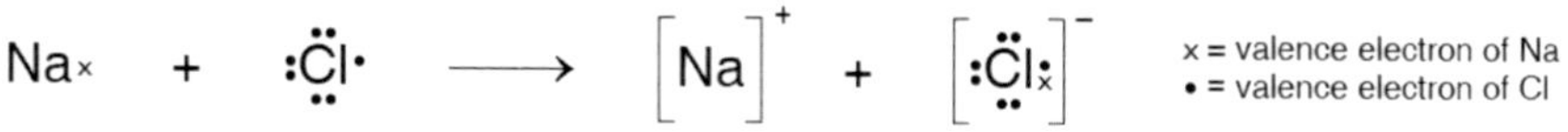

There are other changes that occur in the reaction such as breaking of the Cl–Cl bond, but these will be covered in depth in Chap. 4 on Chemical Thermodynamics.

2.2.1 *Guidelines for drawing dot-and-cross diagrams*

These diagrams represent the valence shell electrons as dots or crosses. Inner-shell electrons are not depicted since it is only the valence electrons that are responsible for bonding.

Let us examine how the dot-and-cross diagram for the following reaction is drawn:

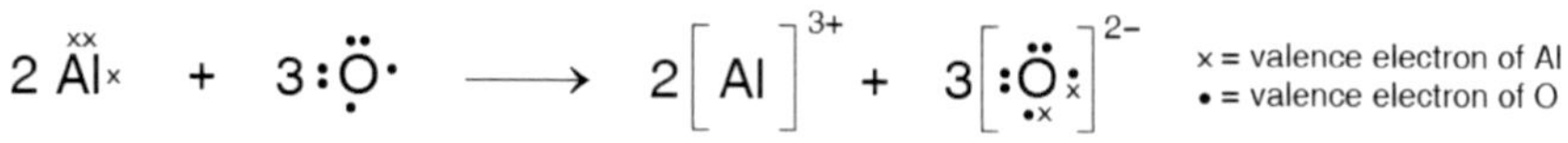

2 Al× + 3 :O· ⟶ 2 [Al]$^{3+}$ + 3 [:O:×]$^{2-}$

x = valence electron of Al
• = valence electron of O

- First of all, draw the valence electrons around each atomic symbol for the atoms Al and O. The number of valence electrons is essentially the group number on the periodic table. Use "×" and "·" to differentiate the electrons from each element.
- Take note that the metal atom loses all its valence electrons easily to form a cation. Thus, the dot-and-cross diagram for the cation should show no valence electrons but the atomic symbol enclosed in square brackets with a superscript that denotes the charge of the cation (its value corresponding to the number of electrons lost).
- The non-metal atom "seeks" to form a stable outer octet configuration. Thus, in the case of O, it will accept two electrons. For the anion, there should be eight valence electrons depicted and the electrons that are gained should be differentiated from the original valence electrons that the non-metal atom has. The anion thus has a charge that corresponds to the number of electrons gained.
- The number of electrons lost by a metal atom need not necessarily be equal to that gained by a non-metal atom. To ensure electrical neutrality, you need to balance the charges by adding the required coefficients (on the left-hand side of the symbols for atoms or ions).

Example 2.1: Give the dot-and-cross diagrams for (i) CaF_2, (ii) Na_2S and (iii) Mg_3N_2.

Solution:

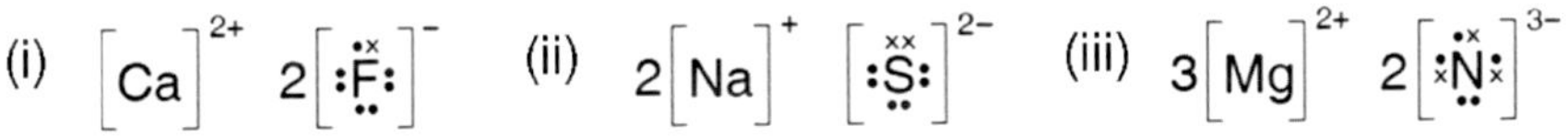

(i) $[Ca]^{2+}$ $2[:F:]^{-}$ (ii) $2[Na]^{+}$ $[:S:]^{2-}$ (iii) $3[Mg]^{2+}$ $2[:N:]^{3-}$

Q: In the dot-and-cross diagram for Na_2S, the two "electrons gained" are paired together on S whereas in the case of Mg_3N_2, the three "electrons gained" are each paired with one of the original valence electron belonging to N. Is it important to distinguish between the types of "pairing" when drawing dot-and-cross diagrams?

A: It does not matter how pairs of both source of electrons are drawn and in which combinations since all electrons are indistinguishable, i.e., the following dot-and-cross diagram for Mg_3N_2 is also acceptable:

$$3[\text{Mg}]^{2+} \quad 2[\text{N}]^{3-}$$

2.2.2 *Physical properties of ionic compounds*

2.2.2.1 *High melting and boiling points*

Ionic compounds have high melting and boiling points, indicating that the ionic bonds are strong.

For melting to occur, the crystal lattice has to be broken down so that the ions are free to move about. Therefore, a great amount of energy, in the form of heat, is required to overcome the strong electrostatic attraction between the oppositely charged ions (i.e., ionic bonds).

Likewise, when molten ionic compound undergoes boiling, energy is needed to overcome the ionic bonds between the mobile ions so that they can break away from one another and have greater freedom of movement.

For a good indication of the strength of the ionic bond, we can make use of an energy term known as the lattice energy (L.E.) shown in Eq. (2.1):

$$\text{L.E.} \propto \frac{q_+ \times q_-}{r_+ + r_-}, \tag{2.1}$$

where q_+ = charge on cation, q_- = charge on anion, r_+ = radius of cation, r_- = radius of anion.

Lattice energy is defined as the **energy released** when **one mole** of a pure solid ionic compound is formed from its **constituent gaseous ions** under standard conditions (of 1 bar and 298 K).

For example, Eq. (2.2) depicts L.E. for magnesium oxide:

$$Mg^{2+}(g) + O^{2-}(g) \longrightarrow MgO(s). \tag{2.2}$$

Since energy is released when a bond is formed, **lattice energy is always a negative value (exothermic)**. A greater magnitude of L.E. indicates a stronger ionic bond.

Q: Why must the formation of the ionic compound start from "its constituent gaseous ions"?

A: When a bond is formed, energy is evolved. In the gaseous state, the cations and anions have minimal interaction and when they form the ionic solid, the particles are in closest proximity and they attract one another. This is very similar to what we have discussed in Chap. 1 on Atomic Structure. The closer the electron is to the nucleus, the stronger is the attractive force faced by the electron. Therefore, the energy that is evolved is a very good indication of the strength of the bond that is formed.

Hence, the magnitude of L.E. depends on two factors:

- Charges on the ions: The higher the charge, the greater is the attraction between the oppositely charged ions, which then leads to a more exothermic L.E.
- Ionic radius of the ions: The smaller the ions, the closer they are and the greater the attraction, which leads to a more exothermic L.E.

Example 2.2: Explain why MgO (m.p. 2850°C) has a higher melting point than Na_2O (m.p. 1130°C).

Solution: Both Na_2O and MgO have giant ionic structures with strong electrostatic forces of attraction between the oppositely charged ions in the ionic lattice. Mg^{2+} has a greater charge and smaller ionic radius than Na^+. These two factors result in a more exothermic lattice energy for MgO, which indicates stronger ionic bonding. This accounts for its higher melting point compared to Na_2O.

Common misconception: It is WRONG to say that "since there are two Na^+ in the formula Na_2O, the "number" of ionic bonds in Na_2O is double that of MgO".☒

The ionic bond is non-directional, and when we talk about an ionic bond, we are just referring to the attractive force between oppositely charged ions.

The formula unit, written as Na_2O in this case, simply indicates the numbers of cations and anions that are present to give an electrically neutral ionic compound. It gives us the simplest ratio of both types of ions in the giant ionic lattice since there are countless such ions, depending on the crystal size. The only valuable information that we can derive from these numbers would be to infer the charges on the cation and anion.

Example 2.3: Explain the following:

(i) The melting point of KCl is lower than NaCl.
(ii) The melting point of NaCl is lower than NaF.
(iii) The melting point of KCl is lower than K_2S.

Solution:

(i) The cationic size of K^+ is greater than Na^+, and therefore the ionic bond strength of NaCl is stronger than KCl.
(ii) The anionic size of Cl^- is greater than F^-, and therefore the ionic bond strength of NaF is stronger than NaCl.
(iii) The anionic charge of S^{2-} is higher than Cl^-, resulting in the ionic bond strength of K_2S being stronger than KCl.

Q: What kind of explanation should we offer if we were not told which of the K_2S or KF has a higher melting point?

A: Well, the ionic bond strength depends on the factors of S^{2-} versus F^-. Based on the charge factor, S^{2-} has a higher charge than F^-. Thus, this factor alone should cause the melting point of K_2S to be higher than KF. But based on the size factor, F^- is smaller in size than S^{2-}. Thus, the ionic bond in KF should be stronger than K_2S. And hence, the melting point of KF should be higher than K_2S. So, to know which is the more domineering factor, we really have to look at the actual experimental data and then conclude accordingly.

2.2.2.2 *Ionic compounds are hard and brittle*

Ionic compounds are hard because of the strong ionic bonds present within the compound. When you apply pressure or force on a crystal of NaCl, it is difficult to separate the ions.

Yet they are brittle. As the particles are arranged in an orderly manner, the ionic bonds are easily broken if a force is applied specifically along certain "cleavage" planes in the crystal lattice. The layers of particles will slide past each other and this will result in repulsion between like-charged particles, causing the crystal to break apart (Fig. 2.4).

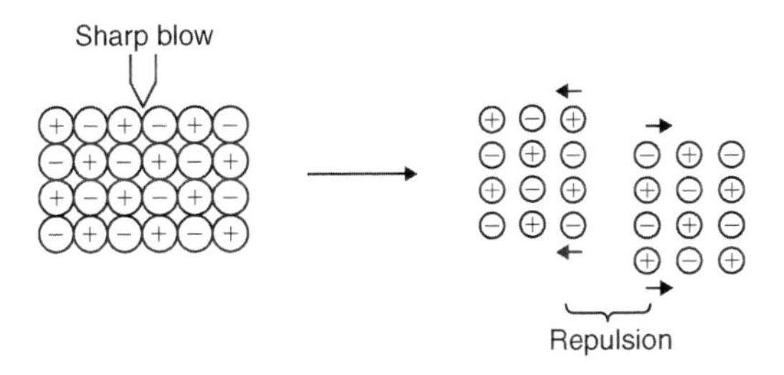

Fig. 2.4. Cleavage of crystal lattice of an ionic compound.

2.2.2.3 *Good conductors of electricity*

Ionic compounds are conductors of electricity in the molten (liquid) state or when dissolved in water. In these cases, there are mobile ions to act as charge carriers.

They do not conduct electricity in the solid state as the ions are held in fixed positions in the crystal lattice and thus their lack of mobility means that they cannot act as charge carriers.

Q: So, when we have a solution that contained dissolved ions, does that mean that the compound that dissolved in it is an ionic compound?

A: Not necessary! Some students thought that if a solution contains ions, the compound that dissolved in it must be ionic in nature. There are some covalent compounds that contain molecule, for examples HCl and NH_3, but when dissolved in a solvent such as water, undergoes ionization to produce ions.

2.2.2.4 *Generally soluble in polar solvents*

Ionic compounds are generally soluble in polar solvents such as water. Not all ionic compounds are soluble in water as they have a wide range of solubilities that depend on factors such as lattice energy and hydration energy (see Chap. 4 on Chemical Thermodynamics).

Hydration describes the exothermic process whereby an ion is attracted by layers of polar water molecules via the electrostatic interaction known as the ion–dipole interaction. If molecules other than water molecules are involved, the process is known as solvation. The stronger the ion–dipole interactions, the more energy will be evolved. (Bonds form, energy evolves!) The *magnitude* of the lattice energy reflects the amount of energy that is required to break up the lattice and separate the ions. If a particular ionic compound can dissolve in water, it goes with the saying that the energy released during hydration is able to compensate for the smaller amount of energy that is required to break up the lattice, and *vice versa.*

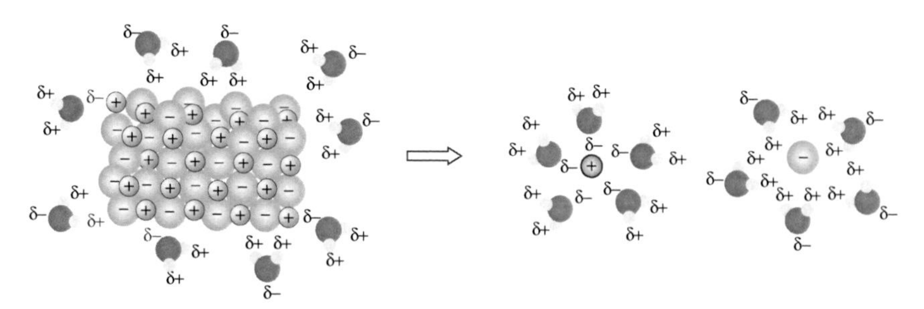

Ionic compounds are insoluble in non-polar solvents such as hexane.

Q: Why does an ionic compound dissolve in one solvent but not in another?

A: For an ionic compound to dissolve, that is, to have the ions separated from one another, energy that is released from the solvation process must be sufficient to offset the energy needed to overcome the strong ionic bonds and hence break up the crystal lattice.

To ensure substantial energy released from solvation, there must be favourable interactions between the ions and the solvent molecules that are stronger than, or at the very least, of similar magnitude to, those between the ions themselves. This is the driving force for the ions to be separated. Likewise, to have solvent molecules attracted to the ions, the interactions between them must be stronger than, or similar to, those between the solvent molecules. Such an explanation is the reasoning behind the saying "like dissolves like".

Q: How does an ionic compound dissolve in water?
A: We can imagine the process of an ionic solid dissolving to be like peeling an onion. The positive end of the dipole of a water molecule is attracted to the anions on the surface of the ionic solid, and the negative end of the dipole is attracted towards the cations. The formation of the various ion–dipole interactions releases energy. The energy released is transferred to the cations and anions, which increases the vibrational energy of these ions. As more ion–dipole interactions occur, releasing more energy, the greater amount of vibrational energy enables the ions to be freed from the lattice.

2.3 Covalent Bonding

Covalent compounds are usually formed between non-metals.

A covalent bond is formed by the **sharing of electrons** between **two atoms**. The shared electrons are **localised between the two nuclei**, in contrast to a non-bonding electron, which moves three-dimensionally around its own nucleus.

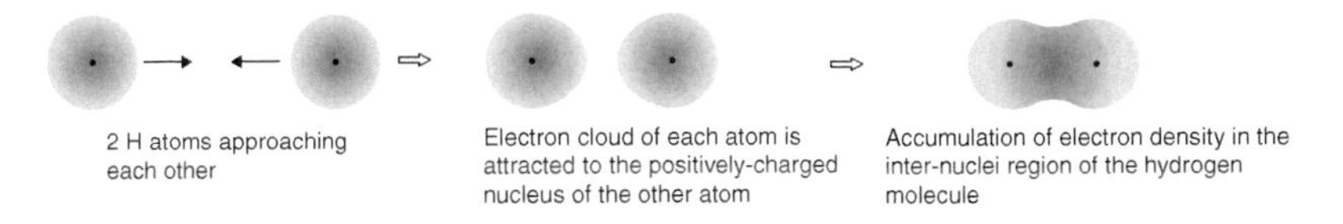

The **covalent bond is the electrostatic attraction between the localised shared electrons and the two positively charged**

nuclei. *(This is yet again another case of positive charge attracting negative charge! This is what "bonding" is all about!).*

Q: Why are covalent compounds usually formed between non-metals whereas ionic compounds are usually formed between a metal and a non-metal?

A: From Chap. 1 on Atomic Structure, we learnt that effective nuclear charge increases across the period from left to right. This means that the electron clouds of the elements are more strongly attracted by the nucleus as we move across the period.

Elements on the left are metals and they are characterised by the relative ease with which they lose electrons as compared to elements on the right of the period. So when a metal and a non-metal meet, you have one that "wants to give" and another that is "willing to accept" electrons. A great fit! But what happens when two non-metals meet? Well, neither of them wants to give, so they "compromise" by sharing, resulting in an accumulation of electron density in the inter-nuclei region.

Q: Will there not be repulsion when the two nuclei come close to each other?

A: Yes, there will be repulsion between the two nuclei and even repulsion between the non-bonding electron clouds of the two atoms when a covalent bond is formed. But if these two repulsive forces are weaker than the attractive forces the two nuclei have for the "shared" electrons (which is the electron cloud that is accumulated in the inter-nuclei region), then you will have a net attractive force that holds these two atoms together. This is, in fact, a covalent bond!

Q: So does this mean that as long as the reaction is between a metal and a non-metal, ionic bonds must be formed? Whereas, if it is between two non-metals, then covalent bonds must form?

A: Not necessarily! Metals reacting with non-metals to give ionic compounds is a very useful GUIDELINE, but it is not always true. For e.g., Al is a metal and Cl is a non-metal but $AlCl_3$ is a covalent compound (refer to Section 2.4.1 for more details). Ultimately, what makes a reaction occur is determined by whether the *energy change* is favourable (see Chap. 4 on Chemical

Thermodynamics) and whether there is *sufficient energy* for the particles to react (see Chap. 5 on Reaction Kinetics).

2.3.1 *Covalent bond formation*

To form a covalent bond, the **valence orbitals** of the two bonding atoms must overlap for the sharing of electrons to occur. There are two types of orbital overlap which result in two different types of bond:

- Head-on overlap of orbitals forms a sigma bond.
- Side-on overlap of orbitals forms a pi bond.

Q: Why do we need to bring in the concept of "orbitals" here?
A: In Chap. 1 on Atomic Structure, we have learnt that electrons can be perceived as residing in regions of space known as orbitals. Orbitals come in different shapes and sizes. Thus, we can think of a covalent bond forming when two partially filled valence atomic orbitals overlap with each other. We focus only on the valence orbitals because they are in the outermost principal quantum shell, which is only partially filled up. The inner core shells are completely filled with the respective number of electrons and therefore cannot accommodate any extra electrons.

Q: Why are there different types of orbital overlap?
A: Orbitals come in different shapes and when they overlap with each other, geometrical considerations result in either a head-on or side-on overlap. Such overlap of orbitals resulting in an **accumulation of electron density in the inter-nuclei region** is known as **effective overlap of orbitals**. We shall look at this in the following section.

2.3.1.1 *Formation of sigma bond (σ bond)*

A sigma bond is formed when two orbitals overlap head-on.

It can occur between two *s* orbitals as in the case of H–H covalent bond formation:

H ↑ + ↓ H ⟶ H ↑↓ H ⟶ H ↑↓ H

1*s* orbital 1*s* orbital

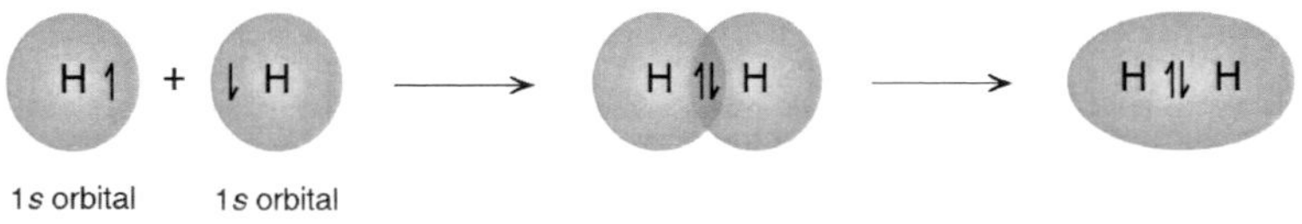

It can also be formed by the overlap of an *s* orbital and a *p* orbital as in the case of H–F covalent bond formation:

1*s* orbital 2*p* orbital

Furthermore, it can be formed by the overlap of two *p* orbitals as in the case of Cl–Cl covalent bond formation:

3*p* orbital 3*p* orbital

2.3.1.2 *Formation of pi bond (π bond)*

After a sigma bond is formed between two atoms, it is possible to have the valence *p* orbitals of these atoms overlap in a side-on manner, if required. This side-on overlap of orbitals is known as the pi bond.

It usually occurs when two *p* orbitals overlap side-on.

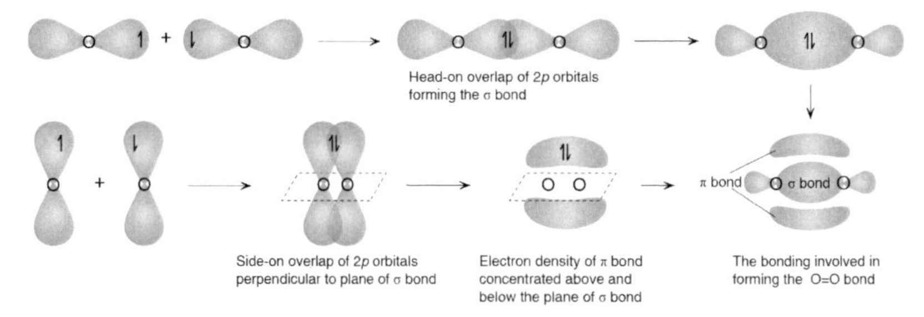

A pi bond is weaker than a sigma bond.

Q: Can a pi bond be formed without the presence of a sigma bond between two atoms?

A: Absolutely not. If given the options of forming a sigma bond or a pi bond, the two atoms "prefer" to form a sigma bond as it is more stable and stronger than the pi bond. This is because if you look at the sigma bond and compare it with the pi bond, the electron density accumulated in the inter-nuclei region is greater than that in the pi bond. We thus say that the sigma bond results from **more effective overlap** of atomic orbitals as compared to the pi bond. This greater accumulation of electron density minimises inter-nuclei repulsion to a greater extent than in the case of the pi bond.

Q: Can there be two sigma bonds formed between the two atoms?

A: Absolutely not. The formation of two sigma bonds will result in too much accumulation of electron density within the inter-nuclei region. The inter-electronic repulsion will be too great to be contained. In addition, if you look at the electronic configuration of the two atoms involved in covalent bond formation, you will not be able to find, on the same atom, two valence atomic orbitals with the correct orientation in space, to form two sigma bonds with the other atom. Thus, the different ways of forming the sigma bond and the pi bond result in spreading out the electron density in the inter-nuclei region with minimal inter-electronic repulsion but maximum bond strength!

2.3.2 *Dative covalent bond (coordinate bond)*

Usually, a single covalent bond consists of each bonding atom contributing one electron. It is also possible to have a single covalent bond formed whereby the pair of sharing electrons come only from one atom, known as the donor atom. This type of bond is known as a dative covalent bond or coordinate bond.

There is one donor atom with a lone pair of electrons to donate and a receiver atom which must have a low-lying vacant orbital to receive the pair of electrons.

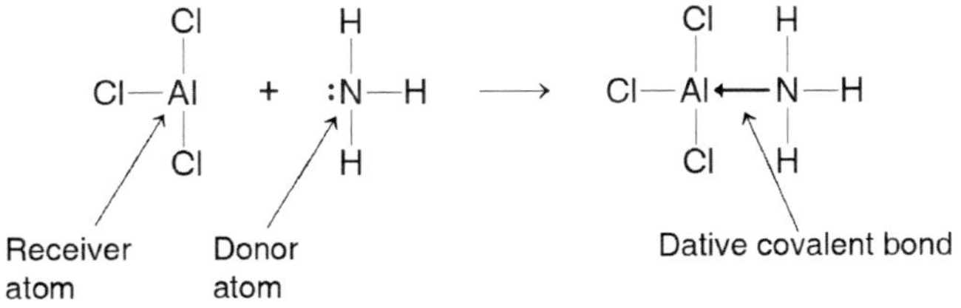

Cl Cl H Cl H
Cl—Al + :N—H ⟶ Cl—Al←—N—H
Cl H Cl H
Receiver atom
Donor atom
Dative covalent bond

Once a dative covalent bond is formed, it is no different to any other single covalent bond.

The dative covalent bond is represented by an arrow "→" that shows the direction of electron flow from donor atom to receiver atom.

Other examples include the formation of NH_4^+, BF_3–NH_3 adduct and Al_2Cl_6 dimer:

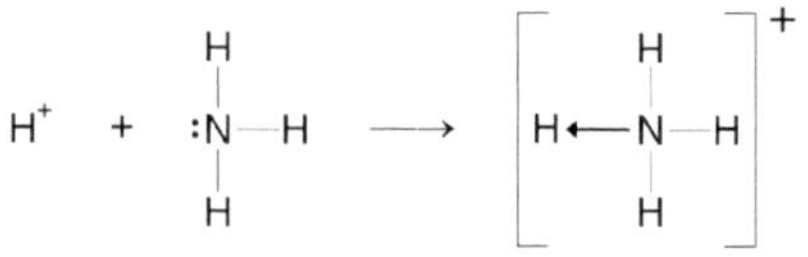

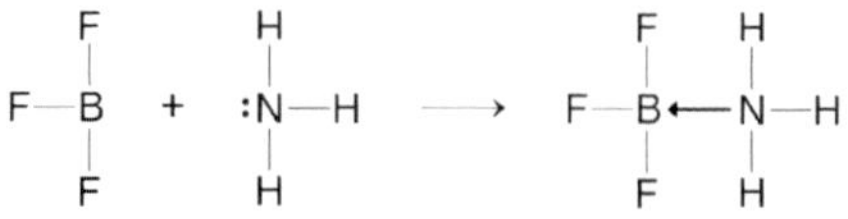

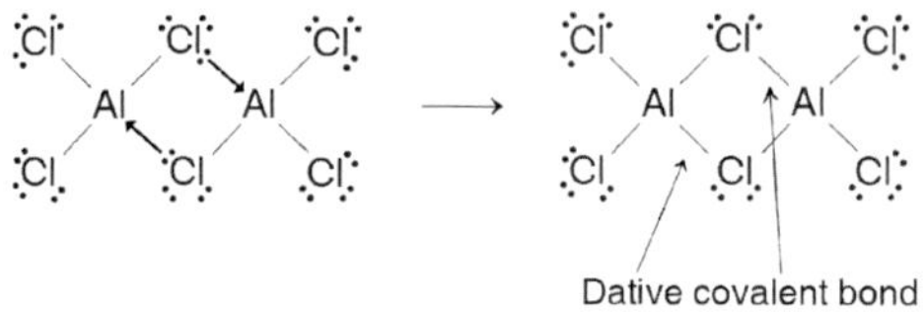

$$H^+ + :NH_3 \longrightarrow [H \leftarrow NH_3]^+$$

$$F_3B + :NH_3 \longrightarrow F_3B \leftarrow NH_3$$

$$2\,AlCl_3 \longrightarrow Al_2Cl_6$$

Dative covalent bond

Q: Is there really no difference between a single covalent bond and a dative covalent bond?

A: In terms of the concept of sharing, you cannot differentiate between a single covalent bond and a dative covalent bond as the electrons are within the inter-nuclei region. But in reality, you will not expect these two types of covalent bond to be of the same strength as the donor atom will not be so "altruistic" as to fully donate the two electrons for sharing. The electron density will always be more lopsided towards the donor atom. This means that a dative covalent bond is actually polar in nature. This is why some books use the following symbol to denote a dative covalent bond:

$$\overset{+}{A}\!-\!\!-\!\!\overset{-}{B}$$

The positive sign on atom A indicates that it is the donor atom, which loses an electron, hence the positive charge. The negative charge on B indicates that it is the acceptor atom, which gains an electron.

2.3.3 *Factors affecting strength of covalent bond*

Bond energy and bond lengths are two **indicators** of the strength of covalent bonds.

Bond energy (BE) is the average amount of energy required to break **one mole** of a particular bond in any compound in the gas phase. A larger bond energy implies a stronger covalent bond.

Bond length is defined as the distance between the nuclei of two bonded atoms. A shorter bond length implies a stronger covalent bond.

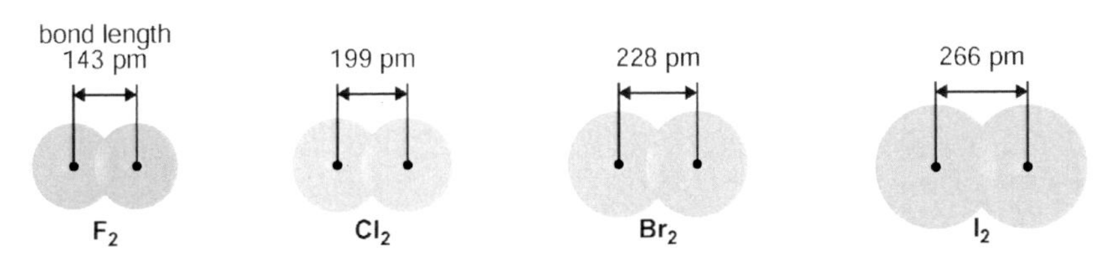

The strength of a covalent bond depends on:

- Effectiveness of orbital overlap in bond formation.

Example 2.4: Account for the trend of bond energies of chlorine to iodine: BE(Cl–Cl): 244 kJ mol^{-1}; BE(Br–Br): 193 kJ mol^{-1}; BE(I–I): 151 kJ mol^{-1}.

Solution: The orbitals used for covalent bonding increase in size from Cl to Br to I. The **larger** orbitals are more **diffuse** and when the larger orbitals overlap each other, the accumulation of electron density within the inter-nuclei region is lower. Thus, the **orbital overlap is less effective** and this accounts for the weaker bond strength that is reflected in the bond energies.

Common mistake: It is WRONG to cite the bond energy or bond length as the reason to account for the strength of a covalent bond.☒

The effectiveness of orbital overlap factor is usually used to account for differences in bond strengths for atoms that come from the same group, e.g., H–Cl, H–Br, H–I or C–Cl, C–Br, C–I, etc.

- Difference in electronegativity of bonding atoms.

Q: What is electronegativity?

A: Electronegativity refers to the ability of an atom to polarise the electron cloud that is shared between the two atoms. The higher the effective nuclear charge (ENC), the higher the electronegativity value. Thus, we expect electronegativity to increase across a period (ENC increases across period) and decrease down a group (because the valence shell is further away from the nucleus and thus electrons residing in the valence shell are less strongly attracted).

Example 2.5: Why is the bond energy of H–S greater than that of H–P?

Solution: S is more electronegative than P. This results in polarisation of the shared electron density, creating a dipole. The additional attractive force within the dipole strengthens the covalent bond. (A dipole is created when there is a separation of charges in a bond).

The difference in electronegativity of bonding atoms factor is usually used to account for differences in bond strengths for atoms that come from the same period, e.g., N–H versus O–H, C–C versus C–N, etc.

- Number of bonds between the bonding atoms, e.g., single, double, triple bond, etc.

As the number of shared electrons increases within the inter-nuclei region for multiple bonds, the attractive force for these electrons increases. Thus, the covalent bond strength increases from single to double to triple.

Q: If the covalent bond strength increases with the number of shared electrons, then why does quadruple bond not exist?

A: An increase in the number of shared electrons does increase the amount of "glue" within the inter-nuclei region. But at the same time, the amount of inter-electronic repulsion between this greater number of electrons also increases. This opposes the "glue" effect. For quadruple bonds, the inter-electronic repulsion among the eight electrons outweighs the "glue" effect, and therefore cannot exist.

Example 2.6: Why is N_2 less reactive (or more inert) than F_2?

Solution: More energy is needed to overcome the stronger triple bond in N_2 than the single bond in F_2. Hence, N_2 is less reactive than F_2. Thus, BE(N≡N):994 kJ mol^{-1}; BE(F–F): 158 kJ mol^{-1}.

2.3.4 *Shapes of molecules — The VSEPR model*

Unlike both metallic and ionic bonds, covalent bonds are directional, leading to various possible types of molecular shapes. This is because covalent bond formation involves the overlap of atomic orbitals and each atomic orbital has a particular size, shape and orientation in space.

The Valence Shell Electron Pair Repulsion (VSEPR) model is but one useful model that is used to predict the shapes of molecules.

The shape of a molecule can be predicted by using the two main principles under the VSEPR model:

- Electron pairs, both bond pairs (BP) and lone pairs (LP), around a central atom arrange themselves as far apart as possible to minimise inter-electronic repulsion.
- The strength of electrostatic repulsion decreases in the order: LP–LP > LP–BP > BP–BP. Lone pairs repel more than bond pairs. This is because a lone pair on an atom is non-bonding and thus the electron density pretty much resides close to this atom as compared to the electron density of a bond pair, which is spread out between two nuclei.

The first principle helps to predict the shapes of molecules. The second helps in estimating the bond angles in molecules.

2.3.4.1 *Predicting shapes of molecules using the VSEPR model*

Approach:

(i) Draw the dot-and-cross diagram of the molecule in question.
(ii) Count the regions of electron densities around the central atom. This will give you the electron-pair geometry (refer to Table 2.1).

Table 2.1 Shapes of Molecules and Polyatomic Ions (VSEPR theory)

Number of Electron Pairs	Electron–Pair Geometry	No. of Lone Pairs	Molecular Geometry	Examples
2	Linear	0	Linear (180°)	Cl–Be–Cl, O=C=O
3	Trigonal planar	0	Trigonal planar (120°)	BF_3
		1	Bent	SO_2
4	Tetrahedral	0	Tetrahedral (109.5°)	CH_4
		1	Trigonal pyramidal	NH_3
		2	Bent	H_2O
5	Trigonal bipyramidal	0	Trigonal bipyramidal (120°)	PCl_5
		1	Distorted tetrahedral (or seesaw or sawhorse)	SF_4
		2	T-shaped	ClF_3
		3	Linear	$[IF_2]^-$
6	Octahedral	0	Octahedral (90°)	SF_6
		1	Square pyramidal	BrF_5
		2	Square planar	XeF_4

A region of electron density (electron pairs) encompasses the shared electrons between the central atom and an atom bonded to it. Thus, single, double, or triple bonds, or a lone pair of electrons are each considered as one region of electron density.

(iii) Next, count the number of lone pairs (non-bonding electrons) around the central atom.

(iv) The molecular geometry (shape) is deduced without taking into consideration the presence of the lone pairs of electrons (refer to Table 2.1).

Example: To determine the shape of ClF_3 molecule using the VSEPR theory.

(i)

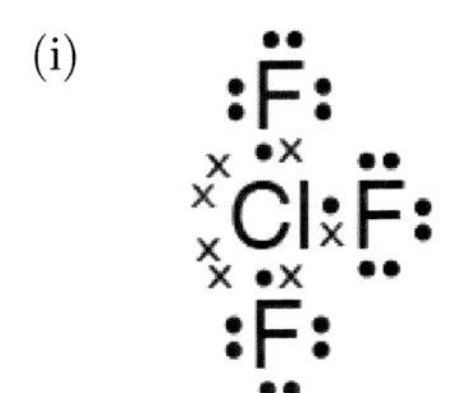

(ii) Around the central Cl atom, number of electron pairs = 5. Therefore, electron-pair geometry is trigonal bipyramidal.

(iii) Number of lone pairs: 2.

(iv) Molecular shape of ClF_3 is T-shaped.

Q: What is the difference between "electron-pair geometry" and "molecular geometry"?

A: Electron-pair geometry gives the arrangement of electron densities which is inclusive of lone pairs of electrons around the central atom. On the other hand, the molecular geometry depicts the arrangement of the atomic nuclei around the central atom. One first has to deduce the electron-pair geometry, followed by the molecular geometry.

Q: What is meant by the "arrangement of the atomic nuclei"? Why is it so?

A: In an atom, the nucleus is more massive than the electron cloud, and is thus more detectable. Remember Rutherford's scattering experiment?

Table 2.1 may look intimidating with its myriad possible shapes: a molecular species can be linear in shape with two regions of electron density and no lone pair OR five regions of electron density with three lone pairs OR even with six regions of electron densities and four lone pairs!

First, determine the number of electron densities by drawing the dot-and-cross diagram to give you the electron-pair geometry. Then, count the number of lone pairs and the remaining bonding electron densities will give you the molecular geometry.

Q: Why is it that ClF_3 is T-shaped with the lone pair of electrons placed on the equatorial position and not on the axial position?

A: If you placed the lone pair of electron densities on the axial position, there would be more 90° lone pair–bond pair repulsion than in the equatorial position; these close-range repulsive forces are significant and need to be minimised.

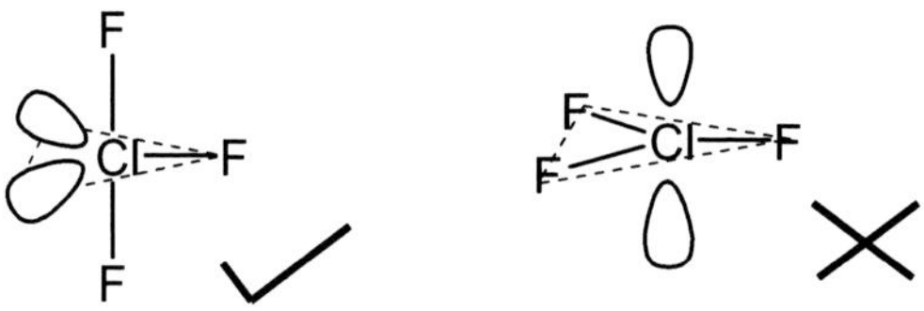

Q: Does the lone pair of electrons on the peripheral atoms of ClF_3, i.e. those on the F atoms, affects the electron-pair geometry or even the overall shape of the molecule?

A: In VSEPR theory, we ignore this effect and do not take into consideration the effect that the lone pair of electrons on the peripheral atoms have on the electron-pair geometry about the central atom. At most, the lone pair of electrons on the peripheral atoms would affect the bond strength that the peripheral atom has with the central atom through inter-electronic repulsion between lone pair of electrons on the two different atoms, for example, the F_2 molecule (refer to Chapter 10 on Group 17 chemistry).

Q: So, does it mean that the more electronegative F atoms of ClF_3 would not distort the electron cloud of the lone pair of electrons on the central atom?

A: Yes, indeed. The more electronegative atom among the two atoms that are involved in covalent bond formation would at most withdraw the shared electron density only. The lone pair of electrons is too far away to be distorted by the other bonding atom. Thus, we would ignore the effect the more electronegative atom has on the lone pair of electrons on the other atom.

2.3.4.2 *Guidelines for drawing dot-and-cross diagrams*

- Try to identify the central atom and draw the valence electrons around the central atom using either a dot or cross symbol. Note: The number of valence electrons an atom has is equivalent to its Group number in the Periodic Table for Groups 1 and 2. For Groups 13 to 18, one need to subtract a factor of 10 units from the Group number in order to deduce the number of valence electrons.
- Draw the peripheral atoms around the central atom. Take note of the number of electrons each peripheral atom requires in order to achieve octet configuration.
- Decide on the types of bond (dative, single, double or triple) a peripheral atom needs to form with the central atom in order to achieve octet configuration.
- Always ensure that the peripheral atom achieves octet configuration. If in order to fulfil this criterion, the central atom needs to expand octet, then allow it. This is viable for an element that belongs to Period 3 and beyond. Elements from Period 2 cannot expand beyond the octet configuration!
- If the species is negatively charged, this means that there are extra electrons on the species; these electrons should go to the central atom provided that the peripheral atoms have already achieved octet configuration. If not, the extra electrons should go to the peripheral atoms to make them achieve octet configuration. This is especially true if there are oxygen atoms surrounding the central atom due to the high electronegativity nature of the O atom.
- If the species is positively charged, the electrons removed (that is why it is positively charged) should come from the central atom because the peripheral atoms have already been made to achieve octet configuration.

Example 2.7: Give the dot-and-cross diagrams for (i) CO, (ii) NH_3, (iii) N_2O, (iv) NO_2^+ and (v) CO_3^{2-}.

Solution:

(i)  C O (ii) 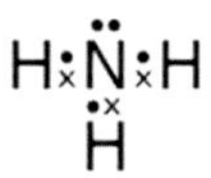 H N H, H (iii)  N N O

(iv) 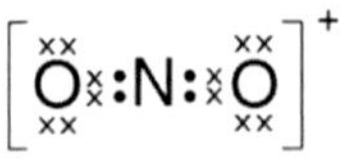 $[O\,N\,O]^+$ (v) 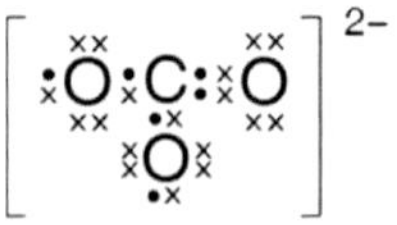 $[O\,C\,O, O]^{2-}$

2.3.4.3 *Exceptions to the octet rule*

- Species with an odd number of valence electrons, e.g., NO:

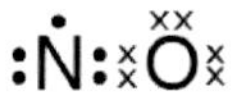

N O

- Electron deficient compounds, e.g., BF_3:

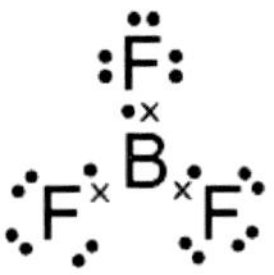

F, B, F, F

- Species with expanded valence shells. (Note: The central atom is from Period 3 and beyond.) For example, SF_6 and PCl_5:

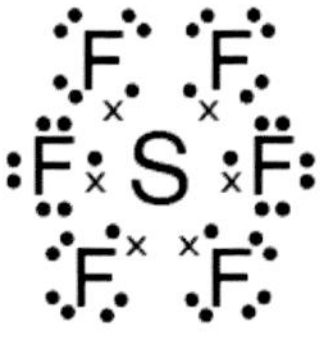

F F F S F F F   Cl Cl P Cl Cl Cl

Exercise: State the shapes of the following species by using the VSEPR theory: (i) BF_3, (ii) XeF_4, (iii) O_3, (iv) HCN, (v) ICl_4^+, (vi) SO_4^{2-}, (vii) SbF_5^{2-}, (viii) PF_5 and (ix) N_2H_4.

Solution: (i) Trigonal planar, (ii) square planar, (iii) bent, (iv) linear, (v) seesaw/distorted tetrahedral, (vi) tetrahedral, (vii) square pyramidal, (viii) trigonal bipyramidal and (ix) trigonal pyramidal with respect to each N atom (both N are considered central atoms).

Example 2.8: Predict the shapes of CCl_4 and H_3O^+ and state the bond angles in these species.

Solution: There are **four regions of electron densities** around the C atom in the CCl_4 molecule and the O atom in the H_3O^+ ion. These electron pairs must be directed in a **tetrahedral** manner (i.e., to be as far apart as possible) in order to minimise electrostatic repulsion.

Since these are **all bond pairs** in the case of CCl_4, it is **tetrahedral** in shape and **all** bond angles are **109.5°**.

As for H_3O^+, there are **three bond pairs** and **one lone pair**, and thus H_3O^+ is **trigonal pyramidal** in shape. Since **lone pair–bond pair repulsion is greater than bond pair–bond pair repulsion**, the bond pairs in H_3O^+ are forced closer together, resulting in a bond angle **less than 109.5°**, i.e., ~107°.

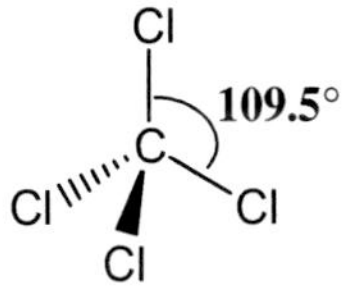

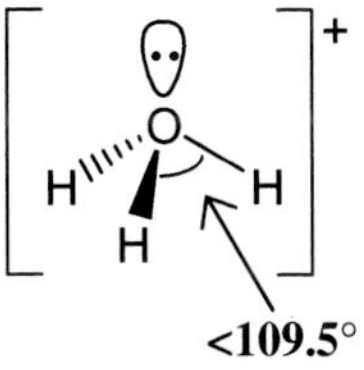

Cl C Cl Cl Cl 109.5°

H O H H + <109.5°

Example 2.9: Account for the difference in bond angles in CH_4, NH_3 and H_2O.

Solution: There are **four regions of electron densities** around the central atom in each molecule, i.e., C in CH_4, N in NH_3 and O in H_2O.

These electron pairs must be directed in a **tetrahedral** manner (i.e., to be as far apart as possible) in order to minimise electrostatic repulsion.

Since these are **all bond pairs** in the case of CH_4, it is **tetrahedral** in shape and **all** bond angles are **109.5°**.

As for NH_3, there are **three bond pairs** and **one lone pair**, and thus it is **trigonal pyramidal** in shape. Since **lone pair–bond pair repulsion > bond pair–bond pair repulsion**, the bond pairs in NH_3 are forced closer together, resulting in a bond angle of ~107°.

As for H_2O, there are **two bond pairs** and **two lone pairs**, and thus it is **bent** in shape. Since **lone pair–lone pair repulsion > lone pair–bond pair repulsion > bond pair–bond pair repulsion**, the bond pairs in H_2O are forced much closer together, resulting in a smaller bond angle of ~105°. Take note that, generally for the presence of every lone pair of electrons, the bond angle decreases by 2°.

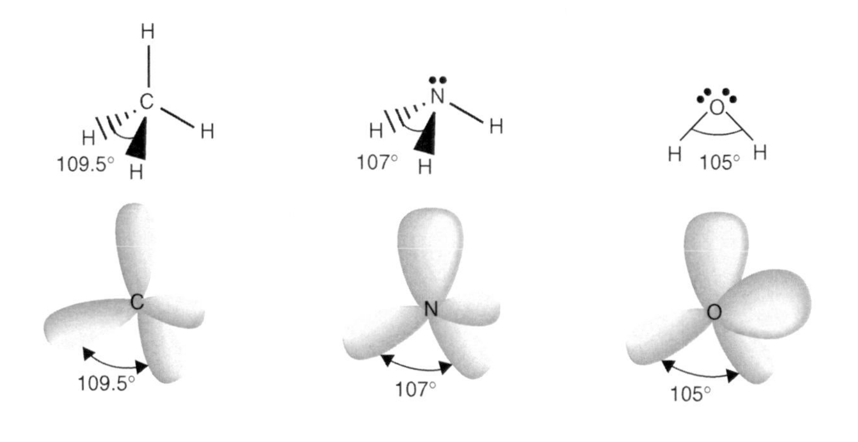

Q: Does the lone pair of electrons on the peripheral atoms affect the bond angle around the central atom?

A: The VSEPR theory does not take into consideration the impact of the lone pair of electrons of the peripheral atoms on the bond angle around the central atom as it is too far away. Similarly, the electronegativity of the peripheral atoms would only affect the shared electrons between the peripheral and central atoms and not the lone pair of electrons on the central atom.

2.3.5 *Using the Hybridisation Model to understand shape*

Q: Why must we learn about the Hybridisation Model?

A: An electron resides in an orbital and each orbital comes with a particular shape and orientation in space. Consider, for instance, a CH_4 molecule. The carbon atom is covalently bonded to four hydrogen atoms. We can simply unpair the pair of electrons in the 2*s* orbital and promote one to the 2*p* subshell. This does not really require too much energy and thus can be easily achieved. We now have four unpaired electrons sitting in four orbitals (the 2*s* orbital and three 2*p* orbitals).

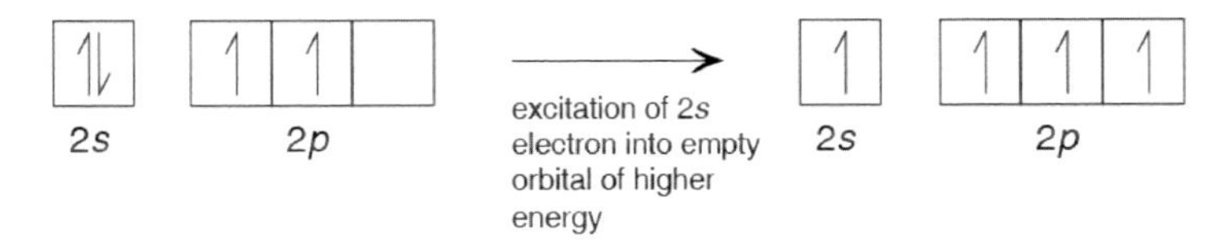

If these four orbitals are now overlapped with the 1*s* orbital of four hydrogen atoms, we are going to get four C–H bonds. However, since the three 2*p* orbitals that carbon has used in bond formation are perpendicular to one another, we are going to get three C–H bonds that are at 90° to one another. In addition, not all of the four C–H bonds are going to be of equal length as different orbitals have been used by the carbon atom for bond formation.

A problem arises because, experimentally, it has been found that all the bond angles in a CH_4 molecule are the same, at 109.5°. And all the C–H bonds are of the same strength and bond length. So, we have used the wrong approach to try to understand a CH_4 molecule!

We need a different model to explain what we have observed experimentally. Thus, the Hybridisation Model was formulated mathematically to help us. Hybridisation is essentially the mathematical mixing of atomic orbitals to create new orbitals (known as hybrid orbitals) that are then used for bonding. Combinations of different numbers of atomic orbitals give rise to different types of hybrid orbitals.

Remember that the model is "man-made"; nature does not understand what hybridisation is. So let us now look at the various types of hybridisation and understand its purpose and usage, especially for organic molecules.

2.3.5.1 sp^3 *hybridisation*

Let us consider how hybridisation can be used to account for both the bonding and shape of methane, CH_4.

Consider the valence orbitals of C:

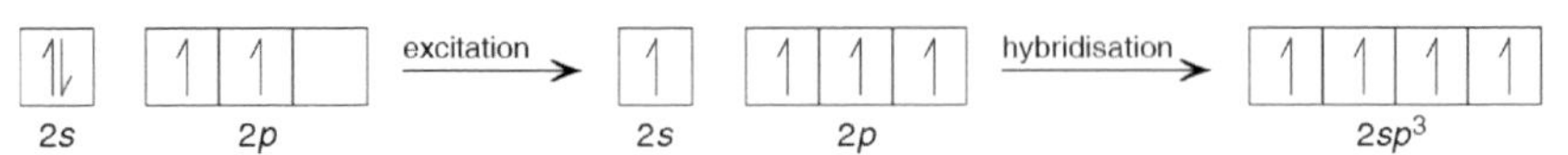

Four atomic orbitals (one 2s and three 2p orbitals) give rise to **four** equivalent sp^3 hybrid orbitals which are orientated 109.5° apart, at the corners of a tetrahedron.

In forming methane, each of the sp^3 hybrid orbitals of the C atom overlaps head-on with the 1s orbital of the H atom forming four σ bonds and resulting in the tetrahedral geometry of the molecule.

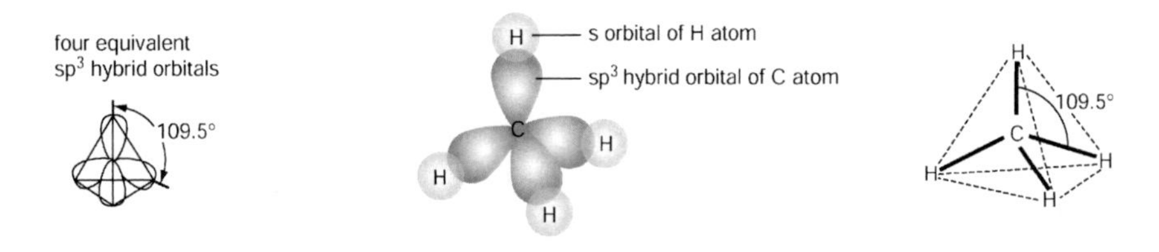

2.3.5.2 sp^2 *hybridisation*

Consider how hybridisation can be used to account for both the bonding and shape of boron trifluoride, BF_3.

Consider the valence orbitals of B:

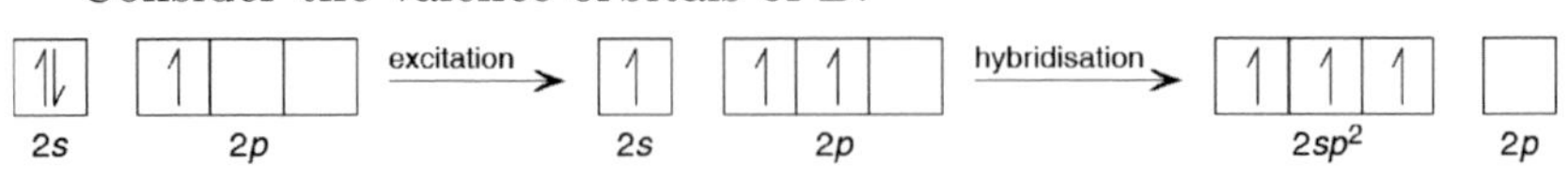

Three atomic orbitals (one 2s and two 2p orbitals) give rise to **three** equivalent sp^2 hybrid orbitals which are orientated 120° apart in a triangular plane.

In forming BF_3, each of the sp^2 hybrid orbitals of the B atom overlaps head-on with the 2*p* orbital of the F atom forming three σ bonds and resulting in the trigonal planar geometry of the molecule.

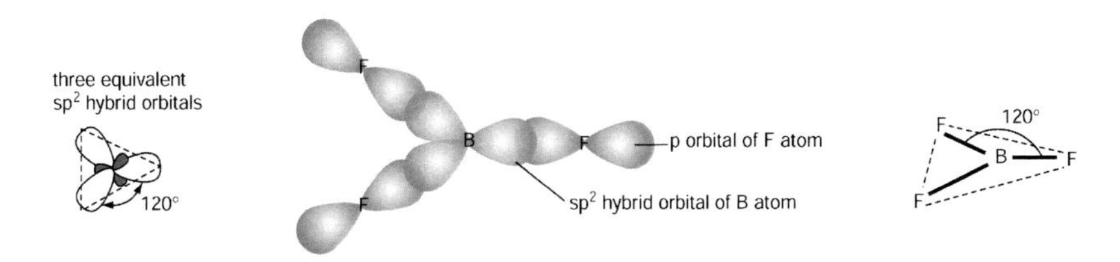

Consider how it can be used to account for both the bonding and shape of ethene, C_2H_4.

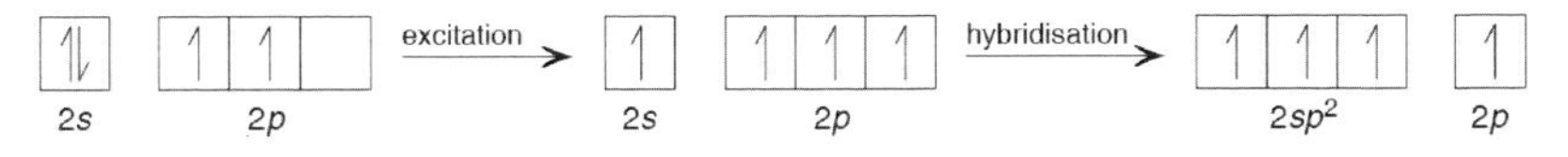

Three atomic orbitals (one 2*s* and two 2*p* orbitals) give rise to **three** equivalent sp^2 hybrid orbitals which are orientated 120° apart in a triangular plane. The unhybridised *p* orbital lies perpendicular to this plane.

In forming ethene, two of the sp^2 hybrid orbitals of each C atom overlaps head-on with the 1*s* orbital of the H atom forming the four C–H σ bonds. The remaining sp^2 hybrid orbitals of the C atoms overlap with each other to form the C–C σ bond. The unhybridised *p* orbitals are thus brought closer to each other and they overlap side-on to form the pi bond.

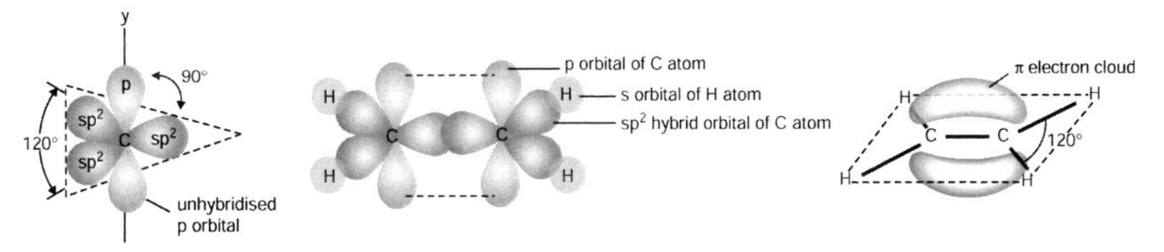

2.3.5.3 *sp hybridisation*

Consider how hybridisation can be used to account for both the bonding and shape of beryllium chloride, $BeCl_2$.

Consider the valence orbitals of Be:

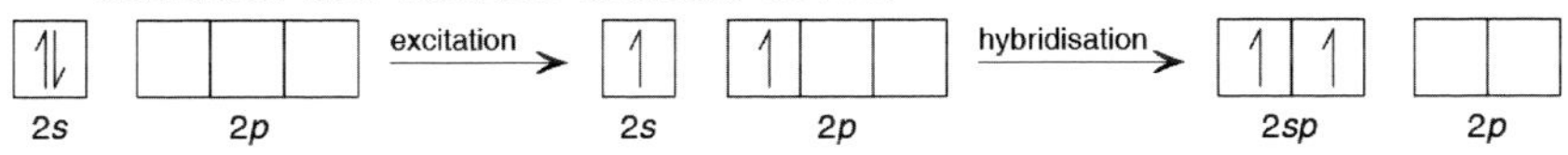

Q: Why is the energy of the hybrid orbitals lower than the unhybridised p orbital?

A: A hybrid orbital, in these cases here, make use of the s and p orbitals. An s orbital has lower energy level than a p orbital. Hence, the hybrid orbital that is formed would have lower energy level than the unhybridised p orbital but higher energy than the unhybridised s orbital.

Q: So, does it mean that covalent bond that is formed using an sp^2 hybridised orbital is stronger than using an sp^3 hybridised orbital?

A: Yes, absolutely. An sp^2 hybridised orbital has higher s character (33%), which means that the electron that resides in this sp^2 hybridised orbital is closer to the nucleus than in the sp^3 hybridised (25% s character) orbital (i.e. why sp^2 hybridised orbital has lower energy level than sp^3 hybridised orbital). Similarly, bond formed using an sp hybridised orbital is stronger than sp^2 or sp^3 hybridised orbital.

Two atomic orbitals (one $2s$ and one $2p$ orbitals) give rise to **two** equivalent sp hybrid orbitals which are orientated 180° apart.

In forming $BeCl_2$, each of the sp hybrid orbitals of the Be atom overlaps head-on with the $3p$ orbital of Cl atom forming two σ bonds and resulting in the linear geometry of the molecule.

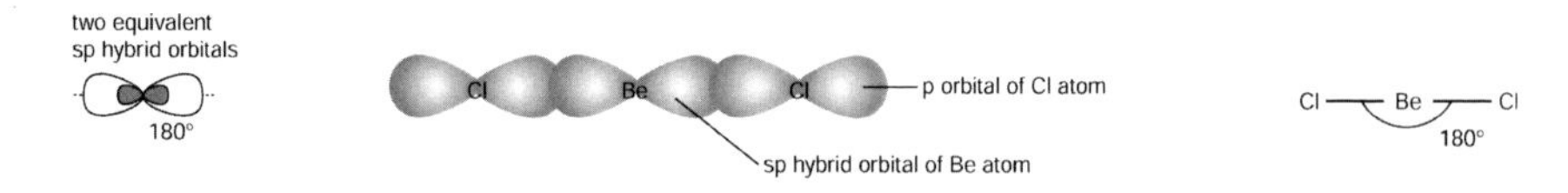

Consider how hybridisation can be used to account for both the bonding and shape of ethyne, C_2H_2.

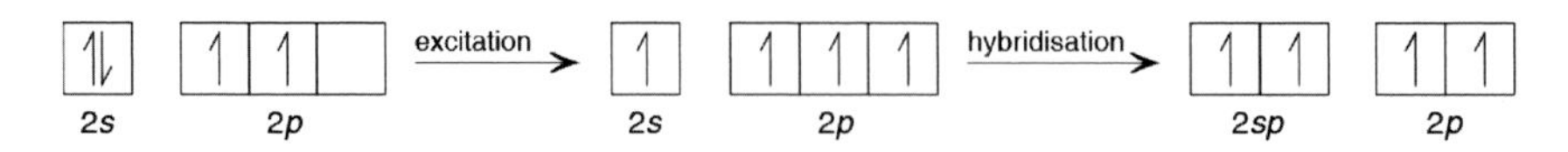

In forming ethyne, one of the sp hybrid orbitals of each C atom overlaps head-on with the $1s$ orbital of the H atom, forming the two C–H σ bonds. The remaining sp hybrid orbitals of the C atoms overlap with each other to form the C–C σ bond. There are two unhybridised p orbitals on each C atom that lie perpendicular to the plane containing the hybrid orbitals. These overlap side-on to form two pi bonds.

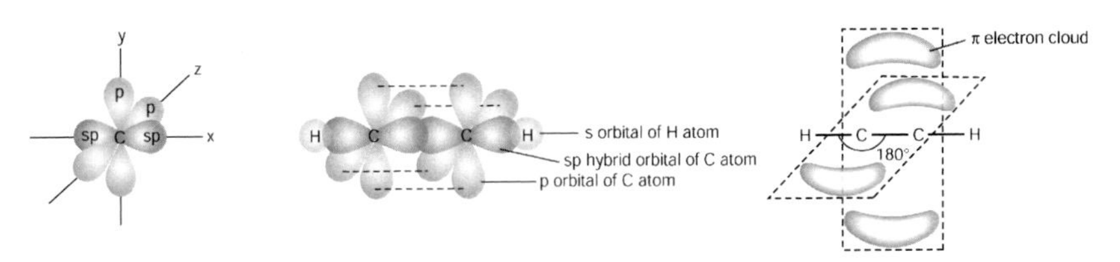

Q: Why does PCl_5 exist, but not NCl_5?

A: NF_5 does not exist since its formation requires five N–F bonds, which means five unpaired electrons in the N atom are required for bonding. Since the $2p$ electrons are already unpaired, one of the $2s$ electrons would have to be promoted to a **low-lying vacant orbital**. The next available empty orbital is the $3s$ orbital, which has a much **higher energy** than the $2s$ orbital. Since too much energy is involved in promoting the electron, the N atom cannot expand its octet to accommodate all the bonding electrons. Thus, NF_5 does not exist. On the other hand, PF_5 exists because the P atom can expand its octet by using its low-lying vacant $3d$ orbitals to accommodate the bonding electrons.

In short, to determine the hybridization state of an atom, we can make use of the following guidelines:

- Four single bonds *OR* four sigma bonds *OR* four regions of electron densities $\Rightarrow$ atom is sp^3 hybridized with tetrahedral electron-pair geometry;
- Two single bonds and one double bond *OR* three sigma bonds and one pi bond *OR* three regions of electron densities $\Rightarrow$ atom is sp^2 hybridized with trigonal planar electron-pair geometry;
- One single bond and one triple bond *OR* two double bonds *OR* two sigma bonds and two pi bonds *OR* two regions of electron densities $\Rightarrow$ atom is sp hybridized with linear electron-pair geometry.

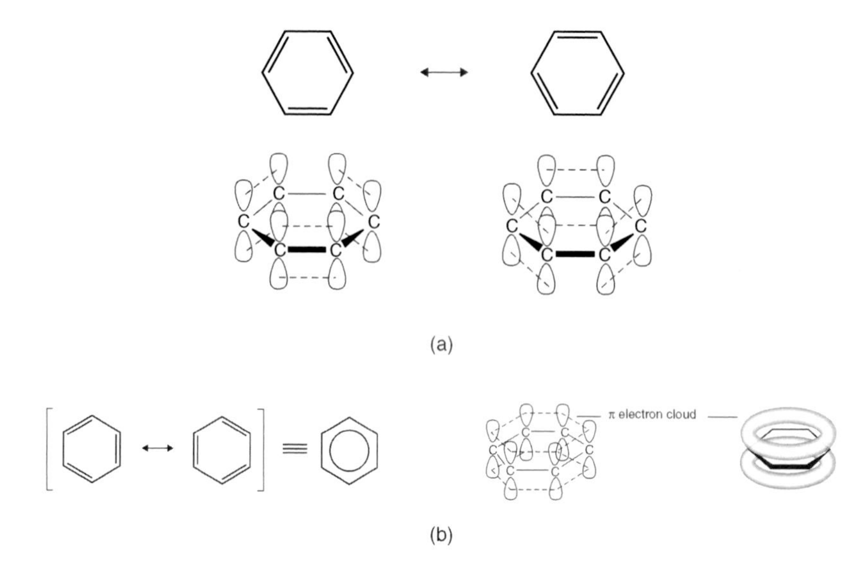

Fig. 2.5. (a) Canonical forms of benzene: Localised structures of benzene. (b) Resonance hybrid: Delocalised structure of benzene.

2.3.6 *Delocalised bonding/resonance*

The bonding in some molecules and ions cannot be adequately explained using one dot-and-cross diagram or Lewis structure.

Take for instance the organic molecule benzene, C_6H_6. If we are to draw either one of the localised structures using the dot-and-cross method (shown in Fig. 2.5a), we are expected to have two types of bond length, C–C single and C=C double bonds.

However, experimental data show that all carbon–carbon bond lengths in benzene are identical and the bond length is in between that of the C–C single and C=C double bond lengths.

This evidence implies that there is delocalisation of π electrons in the benzene molecule, giving rise to the actual structure, known as the resonance hybrid, which is described by an average of the two resonance forms (canonical forms) that are equivalent. Note that resonance DOES NOT mean that the molecule constantly flips from one structure to the other (Fig. 2.5).

Other examples include the following polyatomic anions:

Carbonate ion, CO_3^{2-}

Nitrate(V) ion, NO_3^-

Q: Does the occurrence of delocalised electrons explain why graphite is a conductor of electricity ONLY in a particular direction?

A: Indeed. The continuous overlapping of the unused *p* orbitals (each *p* orbital contains an electron) of the carbon atoms in the graphite structure allows the delocalised electrons to move along the plane BUT not perpendicular to the plane.

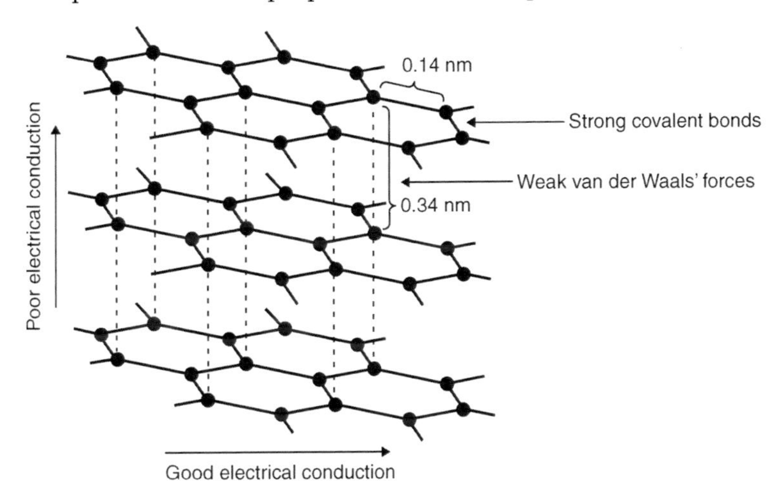

2.4 Intermediate Bond Types

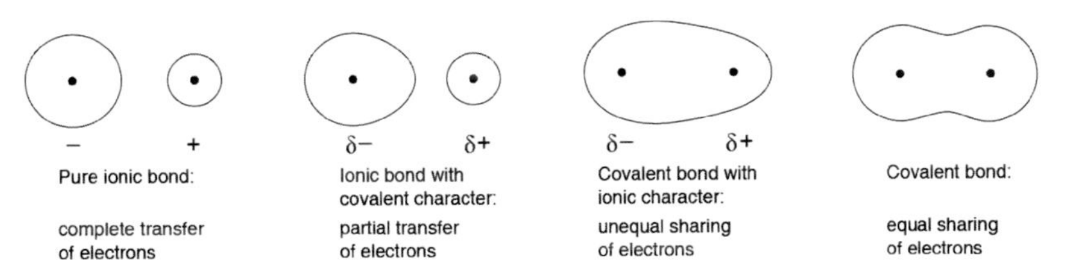

H_2 can be considered a molecule with a pure covalent bond, i.e., equal sharing of electrons between the bonding atoms. In the case of HCl, there is unequal sharing of electrons because of a difference in electronegativity of the bonding atoms. This gives rise to a bond which has a permanent dipole (otherwise known as a covalent bond with ionic character).

At the other end of the continuum of bond types, we have the pure ionic bond as a result of complete electron transfer. But experimental data have shown that even for some supposedly ionic substances, there is some degree of electron sharing between the oppositely charged ions.

One example would be aluminium chloride. We would expect aluminium chloride to be an ionic compound just like aluminium fluoride. However, the extent of electron sharing between the ions is so great that aluminium chloride actually exhibits characteristics of covalent compounds. This specific example clearly indicates that the notion "metals and non-metals combine to form ionic compounds" is not completely correct in all instances! There are numerous other examples, but we will focus instead on the explanation for such a phenomenon.

2.4.1 *Covalent character in ionic bonds*

For electron sharing to occur between anion and cation, there must be a distortion (polarisation) of the electron cloud of the anion by the positively charged cation. Electron density is drawn into the region between the two nuclei, resulting in partial sharing of electrons. When this happens, the ionic bond is said to exhibit some covalent character.

Greater polarisation imparts a greater degree of covalency in the compound.

The extent of polarisation therefore depends on:

- Polarising power of the cation
  A general indicator of the polarising power of a cation is a term known as charge density. The charge density can be approximated as

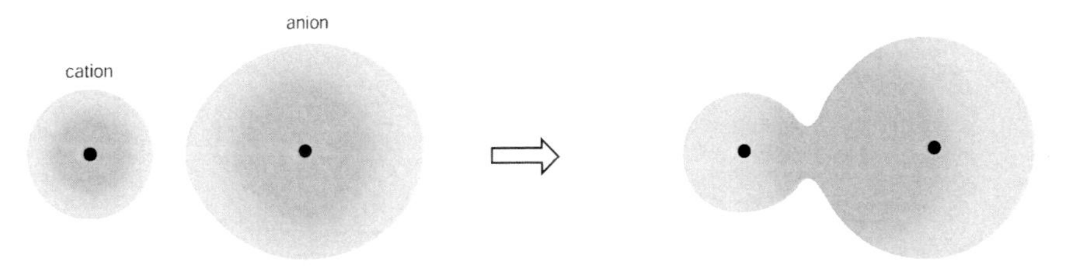

Fig. 2.6. Polarisation of an anion's electron cloud by a cation.

being proportional to the ratio of the cationic charge over the cationic radius, i.e., charge density $\propto (q_+/r_+)$. The more "concentrated" the positive charge, the more "powerful is its attractive power".

Therefore, a cation of higher charge density has greater polarising power.

Example 2.10: Why is $AlCl_3$ covalent and NaCl ionic?

Solution: Due to its smaller ionic size and higher charge, the Al^{3+} ion has higher charge density and hence greater polarising power than Na^+. The electron cloud of the Cl^- ion is therefore distorted to such a great extent that it leads to sharing of electrons. Hence, $AlCl_3$ is covalent.

- Polarisability of the anion
  There are two general indicators of the polarisability of an anion: the size and the charge of an anion.
  - The greater the anionic size, the further the electron cloud is from the nucleus, which means it will be less tightly attracted to the nucleus and hence more easily distorted.
  - The higher the anionic charge, the greater the number of extra electrons the anion has as compared to its parent atom, and hence the greater the possibility of the distortion of the electron cloud.

Example 2.11: Why is $AlCl_3$ covalent and AlF_3 ionic?

Solution: Due to its larger ionic size, the Cl^- ion is highly polarisable compared to the F^- ion.

The electron cloud of the Cl^- ion is more easily distorted by the cation, resulting in an accumulation of electron density in the inter-nuclei region, causing the formation of the covalent bond.

For the compound AlF_3, even though the charge density of Al^{3+} is high, the polarisability of F^- is not particularly great. For that reason, AlF_3 is primarily ionic.

Note that if the charge density of the cation and the polarisability of the anion are high, the compound is likely to be covalent in nature. This explanation also accounts for compounds such as PbO_2 being covalent and not ionic, although Pb is a metal. Can you imagine if PbO_2 were ionic, how large the charge density of a Pb^{4+} ion would be? Not to mention that it is energetically demanding to remove four electrons from a Pb atom!

The concept of the polarising power of cations is useful in explaining certain other properties of elements and their compounds. You have seen how it helps to account for the covalent character of $AlCl_3$. In subsequent chapters, this same idea helps to address questions such as the following:

Q: Why is NaCl neutral in water but a solution of $AlCl_3$ is acidic? (Refer to Chaps. 7 and 9 on Ionic Equilibria, Periodicity.)

Q: Why is it more difficult to decompose $BaCO_3$ than $MgCO_3$? (Refer to Chap. 10 on Chemistry of Groups 2 and 17.)

2.4.2 *Ionic character in covalent bonds*

Electronegativity denotes the ability of an atom in a molecule to attract electrons to itself. When the bonding atoms have the same electronegativity, the bonding electrons are shared equally between the two nuclei, forming a pure covalent bond. This normally happens when the bonding atoms are identical.

As long as the bonding atoms are different, there exists a difference in their relative electronegativity values. The bonding electrons will be more strongly attracted to the more electronegative atom, giving rise to a permanent separation of charges (dipole) in the bond; the presence of ionic character in the covalent bond. Such covalent bonds with unequal sharing of electrons are termed polar bonds. For example, HF in the following:

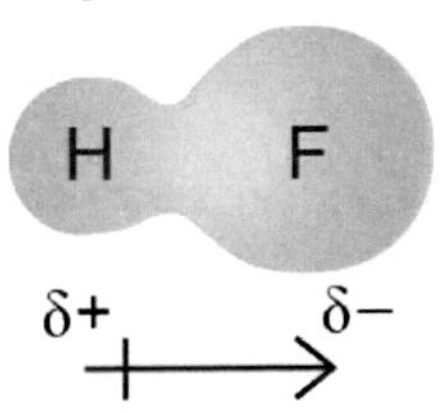

A greater difference in electronegativity implies a greater polarity in the covalent bond.

Partial charges are used to illustrate the relative ability of the bonding atoms to attract the shared electrons. The more electronegative atom, i.e., having a stronger hold on the shared electrons, is considered to be "electron rich" and gains a partial negative charge ($\delta-$). The other bonding atom, which is less electronegative, partially loses its hold on the electrons and acquires a partial positive charge ($\delta+$) — an indication of its being "electron deficient".

Example 2.12: Which is more polar, the C–Cl bond or the C–I bond?

Solution: The C–Cl bond is more polar as the difference in the electronegativity values of C and Cl is greater than that between C and I. Therefore, the shared electron cloud in C–Cl is more distorted.

Polar covalent bonds are generally stronger than non-polar ones, as evidenced in the differences in bond energies of H–H and H–O bonds: $\mathrm{BE(H-H)} = 436\,\mathrm{kJ\,mol^{-1}}$ and $\mathrm{BE(H-O)} = 460\,\mathrm{kJ\,mol^{-1}}$.

Q: An O atom has a greater effective nuclear charge (ENC) than an H atom. Is it not because of this that the bonding electrons are more strongly attracted, leading to a stronger bond?

A: Yes. It may be because of the ENC factor which gives rise to a more electronegative O atom and hence a more polar bond. It all boils down to the level of analysis that is being called for, i.e., the context wherein the problem is to be analysed. You can use either the ENC concept or the electronegativity concept to explain the above differences in bond energies.

Q: Is there any advantage in knowing if ionic character is present in a covalent bond?

A: A resounding yes! An atom that is relatively more electron rich (or electron deficient) can "attract" an attack by a complementary species that is electron deficient (or electron rich). This is how chemical reactions can come about. You will learn more about this in Organic and Inorganic Chemistry.

Q: Can the concepts "ionic character in a covalent bond" and "covalent character in an ionic bond" be viewed interchangeably?

A: No. For instance, you cannot say that there is covalent character in the ionic bond for HCl because there is no such thing as an ionic H^+Cl^- species. Neither can you say that there is ionic character in the covalent bond for MgO because there is no such thing as covalent Mg–O whereby the Mg atom would achieve octet configuration by sharing electrons with an O atom. Whether it is "ionic character in a covalent bond" or "covalent character in an ionic bond" depends on which end of the continuum you have started with.

2.5 Physical Properties of Covalent Compounds

2.5.1 *Properties of giant covalent compounds*

Some non-metallic substances such as diamond and silicon have high melting and boiling points, which means that the attractive forces binding the particles together are very strong. For melting to occur, a great amount of energy in the form of heat is required to overcome the strong covalent bonds between the atoms.

In fact, diamond, an allotrope of carbon, is made up of a large network of carbon atoms each covalently bonded to four other carbon atoms as shown in Fig. 2.7. Silicon has a similar structure to diamond

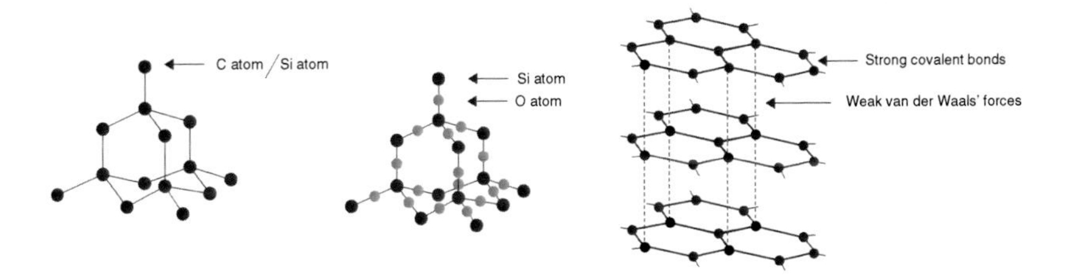

Fig. 2.7. Lattice structures of diamond and Si (left), SiO_2 (middle) and graphite (right).

whereas the macromolecular structure of silicon dioxide (SiO_2) consists of each Si atom covalently bonded to four oxygen atoms and each oxygen atom is in turn covalently bonded to two Si atoms.

Example 2.13: Account for the higher melting point of diamond (3350°C) compared to silicon (1410°C).

Solution: The orbitals used for forming C–C covalent bonds are smaller and less diffuse than those used for forming Si–Si covalent bonds. There is thus more effective overlap of the orbitals of C and this accounts for the stronger C–C bond, which requires more energy to be cleaved. Hence, diamond has a higher melting point than silicon.

Unlike diamond, graphite, another allotrope of carbon, has a giant layered structure. Within each layer, the carbon atoms are each covalently bonded to three other carbon atoms. There is an unpaired electron "sitting" in a p orbital of each carbon atom. All these p orbitals overlap side-on, resulting in a continuous network of delocalised electrons. These electrons are free to move throughout the layer, giving rise to the uni-directional electrical conductivity of graphite along the layers but not between layers.

Q: Why is the bond length between two C atoms within a layer shorter than the bond length between two C atoms from adjacent layers? Does this mean that the bond strengths are different? Are they not all C–C covalent bonds?

A: Indeed, the longer bond length between C atoms in adjacent layers indicates that the bonding between these atoms is weaker than those between C atoms within the same layer. The forces of attraction holding these layers together are not covalent bonds. Rather, they belong to a class of intermolecular forces of attraction known as the instantaneous dipole–induced dipole interactions.

In fact, it is this class of intermolecular attractive forces that accounts for the much lower melting and boiling points of simple covalent molecules such as H_2O and NH_3 in contrast to those of diamond and graphite. These latter molecules, along with Si and SiO_2, are termed macromolecules, or giant covalent molecules.

Thus, note that covalent compounds can exist as either simple discrete molecules or macromolecules.

2.5.2 *Properties of simple covalent compounds*

- Simple covalent compounds, such as CO_2 and H_2O, have low melting and boiling points.

Melting and boiling are phase changes. When a covalent compound undergoes a phase change, it still retains its molecular character. What differs is the extent of interactions among its molecules and also their degree of freedom of motion.

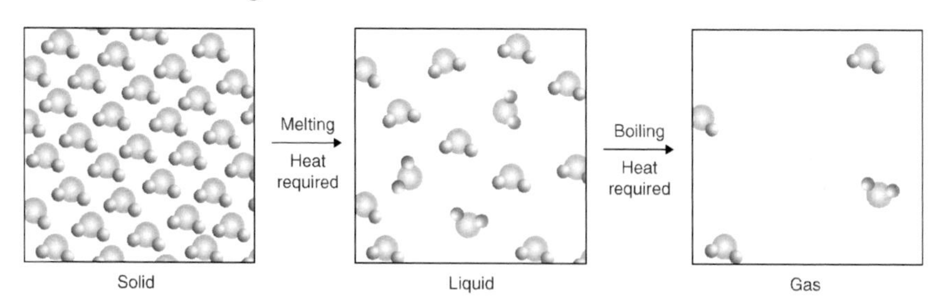

For melting to occur, the crystal lattice has to be broken down so that the discrete molecules are free to move about. Therefore, energy (in the form of heat) is required to overcome the intermolecular forces of attraction among the molecules. Likewise, for boiling to occur, energy is needed to overcome the intermolecular attractions between the mobile molecules so that they can break away from one another and have greater freedom of movement.

- Most covalent compounds are non-conductors of electricity (an exception is the macromolecule graphite).
- Covalent compounds readily dissolve in organic solvents, such as benzene and tetrachloromethane, rather than in polar solvents like water.

Some physical properties of simple covalent compounds are often determined by the type and strength of attractive forces between the particles. In the next section, we will cover the various types of intermolecular forces of attraction.

2.6 Intermolecular Forces of Attraction

Covalent bonds hold atoms in a molecule together. But what helps to keep the millions of water molecules intact together in a cube of ice? In this section, we will cover various types of intermolecular forces of attraction (yet another type of electrostatic interaction) that hold molecules together and differentiate these from the **intra**molecular covalent bonds that bind atoms within a molecule.

Depending on the type of molecules, one of the following intermolecular forces of attraction, or combinations of these, could exist between molecules:

- van der Waals' forces of attraction
  - Instantaneous dipole–induced dipole interactions (id–id)
  - Permanent dipole–permanent dipole interactions (pd–pd)
- Hydrogen bonding

It must be noted that these intermolecular attractions are also known as intermolecular bonding! They are not the same as conventional bonds such as ionic, covalent and metallic bonds. The attractive forces between molecules are very much weaker than the conventional bonds.

2.6.1 *Instantaneous dipole–induced dipole (id–id) interactions*

Id–id interactions exist for all types of molecules, but they may be used to account for the observed physical properties of matter.

As electrons are always in random motion, at any point in time, there is an uneven distribution of electrons in a molecule. This separation of charges creates an instantaneous dipole in the molecule.

The instantaneous dipole in one molecule can induce the formation of dipoles in nearby unpolarised molecules. As a result, a weak electrostatic attraction forms between these dipoles.

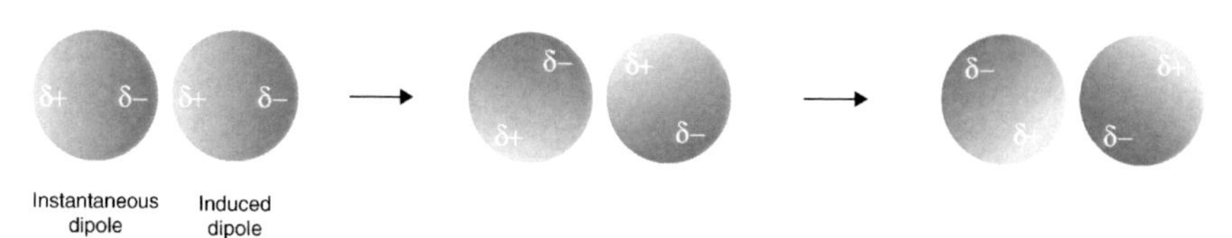

For the very fact that electrons are constantly moving, these dipoles are short-lived. However, new dipoles are soon formed again and the process repeats itself so that on average, there are "permanent" forces of attraction between these molecules, which give rise to their observed physical properties such as melting point, boiling point, non-ideal behaviour, and so forth.

The strength of id–id interactions depends on:

- The number of electrons in the molecule: the greater the number of electrons (i.e., the bigger the electron cloud), the more polarisable the electron cloud, and the stronger the id–id interactions.
- The surface area for contact of the molecule: the greater the surface area of contact possible between the molecules, the greater the extent of id–id interactions.

Example 2.14: Account for the higher boiling point of Br_2 (b.p. 59°C) compared to Cl_2 (b.p. –34°C).

Solution: Since both compounds have simple molecular structures, boiling involves breaking of intermolecular forces. The two compounds are non-polar in nature; thus, the molecules are attracted to one another via id–id interactions. Their boiling point is determined by the strength of the id–id interactions, which increases with an increase in the number of electrons per molecule. Since Br_2 has a greater number of electrons per molecule than Cl_2, it has a higher boiling point than Cl_2.

Q: The bond energy of Br–Br ($193\,kJ\,mol^{-1}$) is smaller than the bond energy of Cl–Cl ($244\,kJ\,mol^{-1}$). Should Br_2 not have a lower boiling point since it is easier to cleave the Br–Br covalent bonds?

A: It is a common misconception to think that it is the intramolecular bonds, i.e., covalent bonds in the molecule, that have to be broken for a phase change to occur for simple covalent compounds. For melting or boiling to occur, it is the intermolecular forces between molecules that need to be overcome; the molecules themselves remain intact.

Example 2.15: Account for the lower boiling point of dimethylpropane compared to pentane.

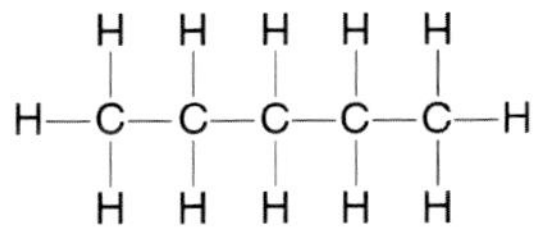

Pentane
(b.p. 36 °C)

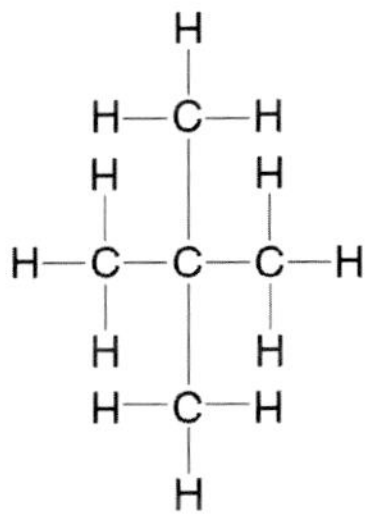

Dimethylpropane
(b.p. 10 °C)

Approach: Pentane and dimethylpropane are what we call a pair of constitutional/structural isomers that have the same molecular formula (C_5H_{12}) but different arrangements of bonding atoms. Since the two isomers have the same number of electrons, we cannot use this factor to account for differences in boiling points. Instead, we need to look at the second factor, which relates the surface area of molecules to the extensiveness of id–id interactions. If we are to look at the space-filling model of these molecules, we find that the surface area for contact of the pentane molecule is greater than that of the seemingly spherical dimethylpropane molecule.

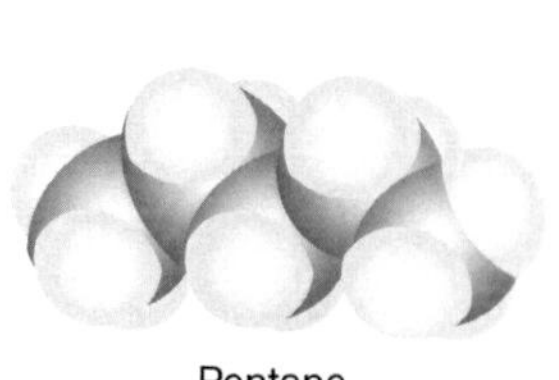

Pentane
(b.p. 36 °C)

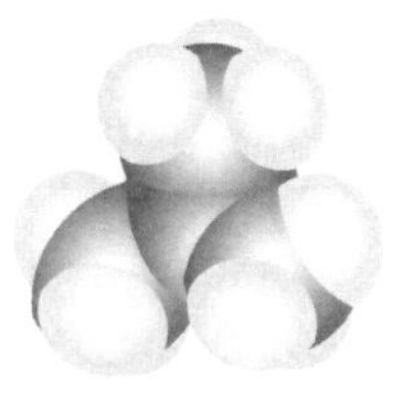

Dimethylpropane
(b.p. 10 °C)

Solution: Both dimethylpropane and pentane exist as simple discrete molecules. Since pentane is more linear than the spherical dimethylpropane, it has a greater surface area of contact which leads to more extensive id–id interactions between its molecules than for dimethylpropane molecules. Thus, more energy is needed to overcome the stronger id–id interactions between pentane molecules, thereby accounting for its higher boiling point.

Q: CH_4 has a higher boiling point than N_2. How do we account for this? An N_2 molecule has greater number of electrons than a CH_4 molecule. Thus, should the id–id interactions not be stronger between N_2 molecules than between CH_4 molecules and hence N_2 should have a higher boiling point?

A: Based on the "number of electrons" factor, there are stronger id–id interactions between N_2 molecules BUT the surface area of CH_4, being greater than that of N_2, gives rise to more extensive id–id interactions. The latter factor is more dominant than the first.

2.6.2 *Permanent dipole–permanent dipole (pd–pd) interactions*

Pd–pd interactions exist for polar molecules only. As there is an uneven distribution of electrons in polar bonds, permanent separation of charges (dipole) is found within polar molecules.

The permanent dipoles in neighbouring polar molecules attract each other. As a result, a weak electrostatic attraction forms between these dipoles, known as pd–pd interactions.

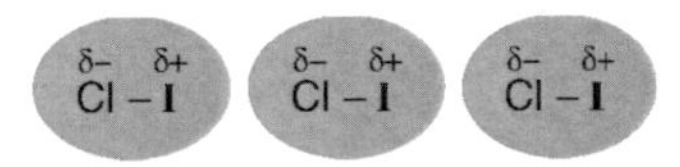

The strength of pd–pd interactions depends on the magnitude of the molecule's net dipole moment. The greater the magnitude of the dipole moment, the more polar is the bond.

The magnitude of the dipole, in turn, depends on the magnitude of the electronegativity difference between the bonding atoms.

The greater the difference in electronegativity, the greater is the dipole moment, and the more polar is the covalent bond.

Dipole moment is a vector quantity that has both magnitude and direction. It is represented by the symbol $\longmapsto$, which depicts the $\delta+$ end pointing towards the $\delta-$ end of the dipole.

To determine whether a molecule is polar and hence if pd–pd interactions exist between molecules of its own kind, two criteria must be fulfilled:

- There must be polar bonds within the molecule.
- There must be a net dipole moment for the molecule.

2.6.2.1 *How to determine whether a bond is polar*

As long as the bonding atoms are different, they will have different electronegativity, which leads to a dipole moment in the bond, i.e., the bond is polar.

2.6.2.2 *How to determine whether a molecule is polar*

Other than diatomic molecules such as Cl_2 and HBr, most other molecules will contain more than one bond. Thus, there are two possible scenarios:

- The dipole moments of the individual polar bonds cancel one another out, leaving a molecule with no net dipole moment.

- There is a collective effect of the individual polar bonds, giving rise to a net dipole moment and hence a polar molecule.

How do we know if there is cancellation of the individual dipoles? Here, we need to bring in the relative spatial orientation of the bonds, how they are arranged in relation to one another. Based on certain geometries, the dipoles, treated as vectors, may cancel out one another. For instance, the bonds between the central C atom and each of the O atoms in a CO_2 molecule are polar. In the linear molecule, these dipole moments have the same magnitude, but since they are pointing in opposite directions, there is cancellation of dipoles that renders the molecule non-polar.

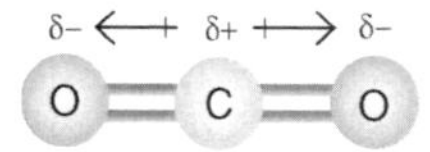

You will be happy to note that there is no complex mathematics needed to calculate whether there is a resultant dipole moment or not. A simple yet effective way is to note the following conditions:

- As long as a molecule has one of the standard shapes shown below, with identical bonding atoms and no lone pairs on the central atom, then the **molecule has no net dipole moment.**

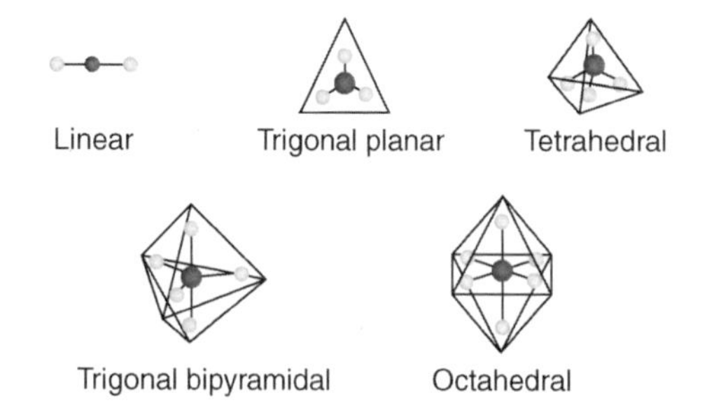

Both CH_4 and CCl_4 are tetrahedral with the central C atom bonded to identical peripheral atoms. They are non-polar molecules even though they contain polar C–H and C–Cl bonds. There is no net dipole moment since the individual dipole moments cancel one another out.

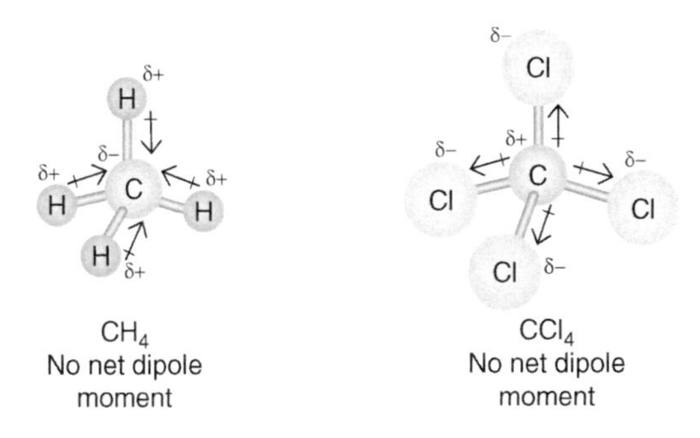

- If the molecule contains only one lone pair on the central atom, then the **molecule has a net dipole moment**, e.g., NF_3.

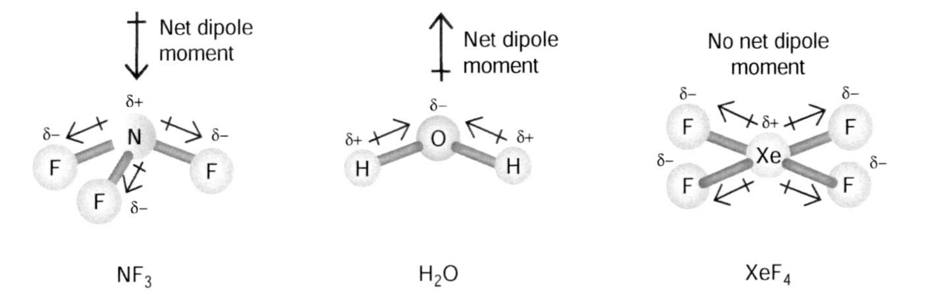

However, a molecule with more than one lone pair on the central atom may or may not be polar. H_2O is a polar molecule but XeF_4 is non-polar even though both central atoms, in each molecule, contain two lone pairs.

In summary:

- The presence of polar bonds need not equate to the molecule being polar.
- A molecule is considered polar if it possesses a net dipole moment.

2.6.3 *Hydrogen bonding*

Hydrogen bonding is present between molecules that have one highly electronegative atom — F, O, or N — covalently bonded to an H atom.

X, being more electronegative than H, attracts the bonding electrons in the H–X bond closer to itself. As a result, it is more "electron rich" and gains a partial negative charge ($\delta-$). The H atom, on the other hand, acquires a partial positive charge ($\delta+$).

Being "electron deficient", this H is strongly attracted to the lone pair of electrons on a highly electronegative atom (electron-rich region) in other molecules. This electrostatic attraction between the H atom and the lone pair of electrons on the highly electronegative atom, F, O or N, is known as hydrogen bonding.

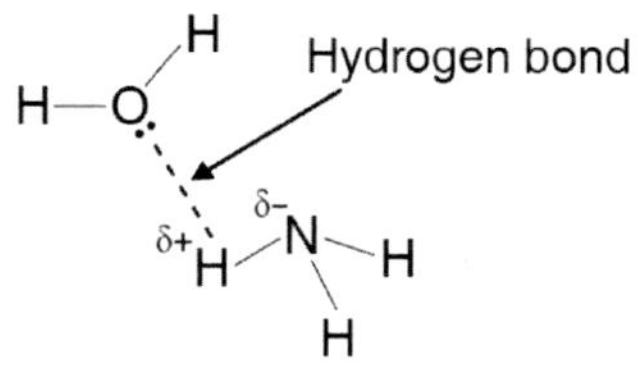

Q: This means O—H---Cl—H is not considered hydrogen bonding?
A: No, indeed not. So make sure you know your **definition** well.
Q: How about O—H------N—H?
A: Yes, this is a type of hydrogen bonding.

Hydrogen bonding can be regarded as a more specific type of pd–pd interaction that is applicable to certain polar molecules only. But note that the terms "hydrogen bonding" and "pd–pd interactions" are non-interchangeable. Examples of compounds that exhibit hydrogen bonding include H_2O, HF, NH_3 and organic compounds such as alcohols, carboxylic acids, amines and amides.

The strength of a hydrogen bond depends on:

- Dipole moment of the H–X bond, where X is O, F, or N:
  F–H- - -F–H > O–H- - -O–H > N–H- - -N–H
- Ease of donation of a lone pair on Y, where Y is O, F, or N:
  N–H- - -N–H > O–H- - -O–H > F–H- - -F–H

Overall, hydrogen bond strength is in the following order:
F–H- - -F–H > O–H- - -O–H > N–H- - -N–H

Q: Does that mean that HF has the highest boiling point since its hydrogen bonding is the strongest?

A: No. Water has the highest boiling point because it can form **more extensive hydrogen bonding** than the rest. It can actually form two hydrogen bonds per water molecule whereas both NH_3 and HF can only form one hydrogen bond per molecule.

This explanation also accounts for why ethanol has a lower boiling point than water.

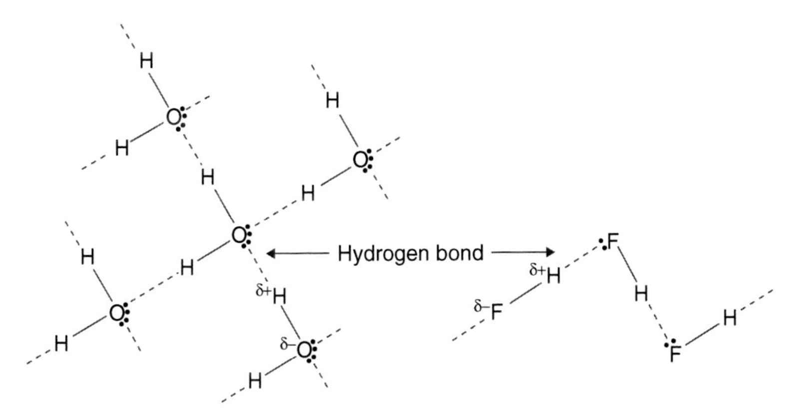

2.6.3.1 *Importance of hydrogen bonding*

Hydrogen bonding helps to explain the following observations:

- Dimerisation of carboxylic acids.

Carboxylic acid molecules dimerise, through hydrogen bonding, in non-polar solvents. Such dimerisation occurs in the vapour phase (just above boiling point) and in non-polar solvents such as benzene.

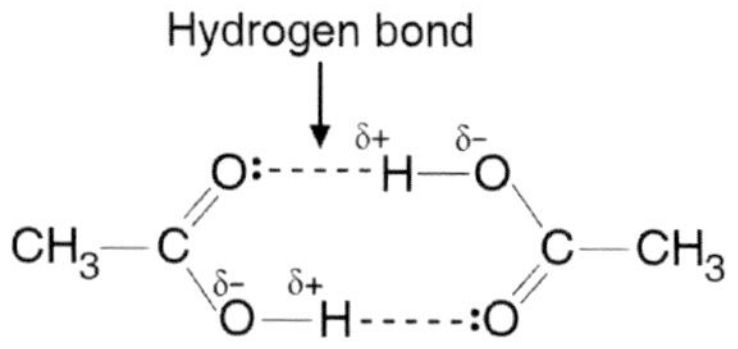

Hydrogen bond

δ+ δ−

CH_3—C(=O:)...H—O—C—CH_3

δ− δ+

Dimerisation of these acids does not occur in aqueous solutions. Instead, they form hydrogen bonds with the abundant water molecules. Some degree of dissociation also occurs for these weak acids in aqueous solutions.

- High solubility of such substances as ammonia and short-chain alcohols (e.g., methanol, ethanol) in water. Solubility of alcohol in water decreases with increasing carbon chain due to the hydrophobic nature of the carbon chain (R).

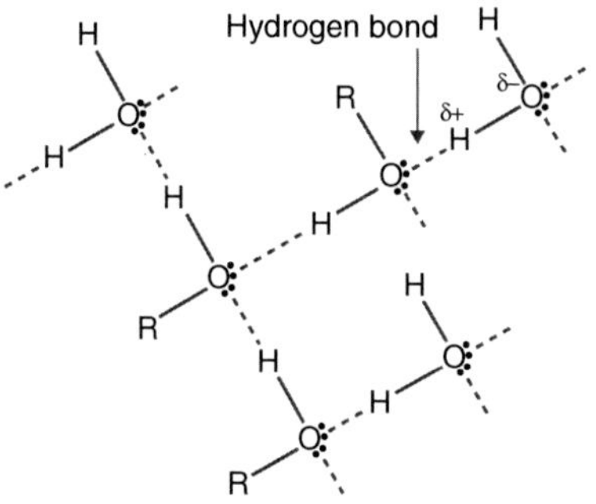

Hydrogen bond

- **Inter**molecular hydrogen bonding versus **intra**molecular hydrogen bonding.

The boiling point of 2-nitrophenol (214°C) is lower than that of its isomer, 4-nitrophenol (259°C). The presence of **intra**molecular hydrogen bonding limits the number of sites available for **inter**molecular hydrogen bonding. This causes 2-nitrophenol to have a lower boiling point and also solubility.

Intramolecular hydrogen bond

Intermolecular hydrogen bond

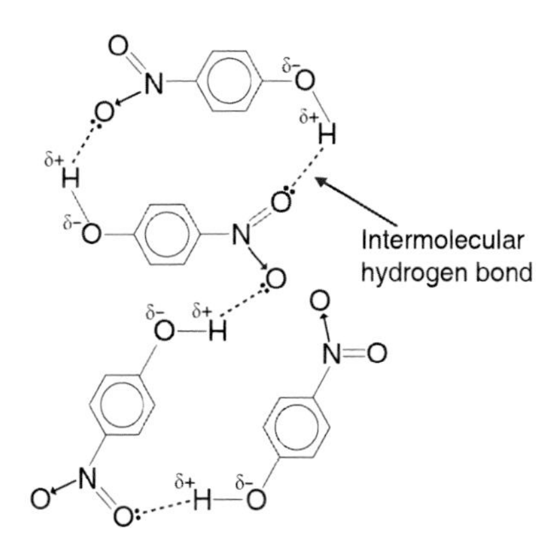

(a) Less extensive intermolecular hydrogen bonding in 2-nitrophenol

(b) More extensive intermolecular hydrogen bonding in 4-nitrophenol

2-nitrophenol is less soluble in water than 4-nitrophenol because the presence of **intra**molecular hydrogen bonding within 2-nitrophenol limits the number of sites available for **inter**molecular hydrogen bonding with water molecules.

- Relative densities of ice and liquid water.

Ice is less dense than water.

Hydrogen bonding gives rise to an open structure, causing the ice to have a lower density than water. Hence, ponds and lakes freeze from the surface downwards. The layer of ice helps to insulate the water underneath from further heat loss. Living things can survive in ponds and rivers even when temperature falls below freezing.

Hydrogen bonding in ice

Hydrogen bonding in water

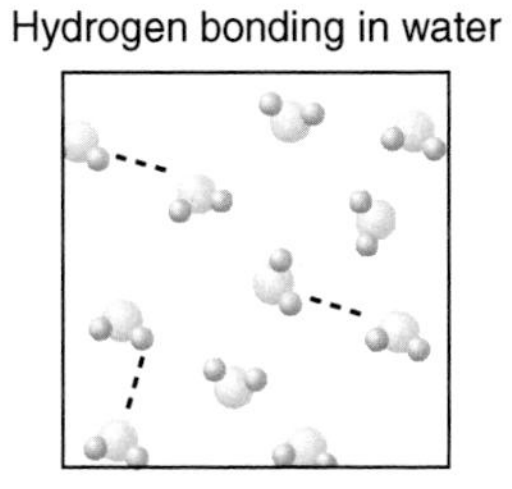

- Stabilisation of structure of proteins.

For instance, the presence of hydrogen bonding helps to stabilise the secondary structures of proteins.

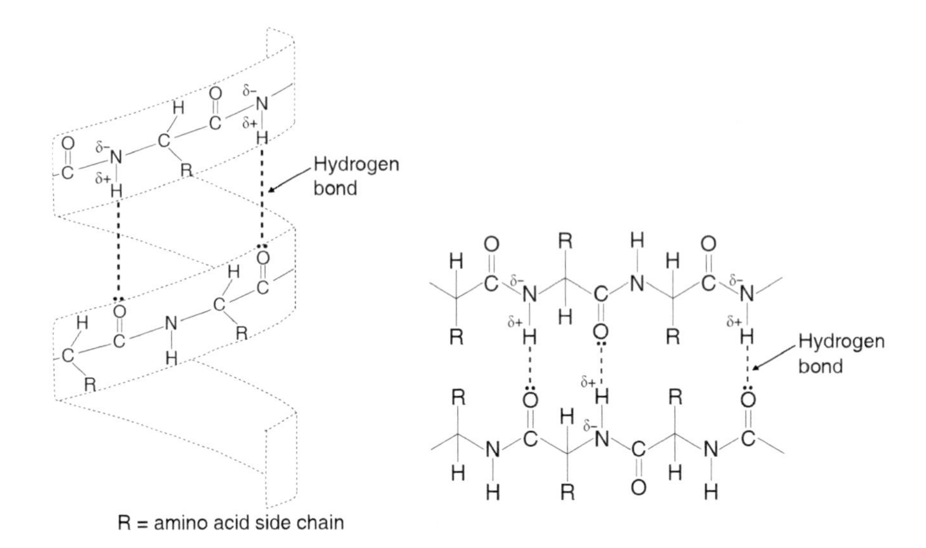

(a) The alpha helix structure. (b) The beta pleated sheet structure.

2.7 Summary of Chemical Bonding: Putting it all Together

In relating the type of structure, bonding or intermolecular forces of attraction to the physical properties of substances, it is useful to start with the general trends:

- The metallic bond, ionic bond and covalent bond are all strong bonds. These are much stronger than intermolecular forces of attraction between discrete molecules.
- The
  order of decreasing strength of intermolecular forces of attraction:
  Hydrogen bonding >> pd–pd interactions > id–id interactions
  (strongest) (weakest)

Example 2.16: Account for the higher melting point of Si (1410°C) compared to Mg (650°C).

Solution: Mg has a giant metallic lattice structure whereas Si has a giant molecular structure. More energy is required to overcome the covalent Si–Si bonds that are stronger than the metallic bonds between Mg^{2+} ions and the sea of delocalised valence electrons. Thus, the melting point of Si is higher than that of Mg.

Example 2.17: Account for the higher melting point of AlF_3 (1290°C) compared to NF_3 (–207°C).

Solution: AlF_3 has a giant ionic lattice structure whereas NF_3 has a simple molecular structure. More energy is required to overcome the strong ionic bonds between the oppositely charged ions than the weaker van der Waals' forces of attraction between NF_3 molecules.

Example 2.18: Account for the lower boiling point of ethanal compared to ethanol.

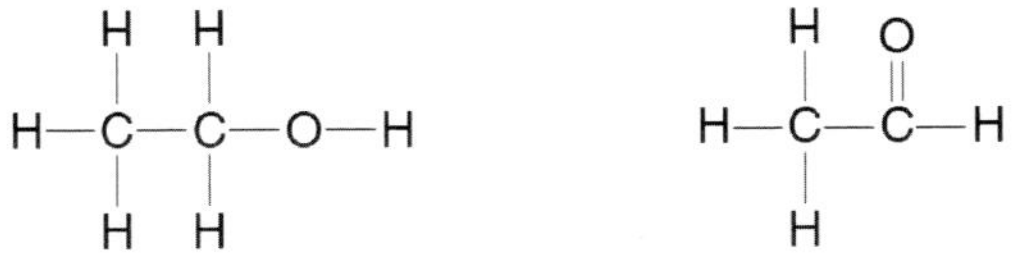

Ethanol
(b.p. 78.4 °C)

Ethanal
(b.p. 20.2 °C)

Solution: Both ethanol and ethanal are polar molecules with simple molecular structures. The lower boiling point of ethanal as compared to ethanol indicates that the pd–pd interactions are weaker than hydrogen bonding. Thus, less energy is required to overcome the attractive forces between ethanal molecules compared to the hydrogen bonding between ethanol molecules.

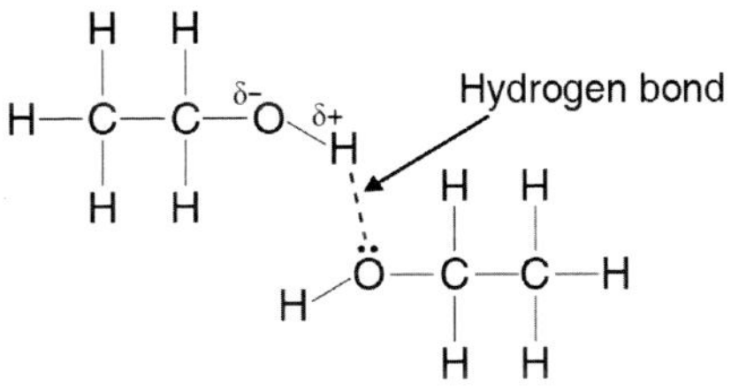

Hydrogen bond

However, there are times that the above trends do not help to account for physical data given. In such cases, you have to analyse the data carefully to conclude what the appropriate explanation is for the trend given. Table 2.2 is a summary of the different types of bonding and structures for various compounds.

For instance, the physical states of iodine (a solid) and water (a liquid) at room temperature, indicate that the intermolecular forces

Table 2.2

Type of Compound	Type of Structure and Bonding	Physical Properties
Giant ionic compound, e.g., NaCl, AlF_3	• Giant ionic crystal lattice structure • Strong electrostatic forces between oppositely charged ions	• High m.p. and b.p. • Hard and brittle • Conducts electricity only in the molten state and in aqueous solution • Soluble in water
Giant metallic compound, e.g., Fe, Cu	• Giant metallic lattice structure • Strong electrostatic forces between cations and sea of delocalised valence electrons	• High m.p. and b.p. • Good conductor of electricity and heat • Malleable and ductile
Giant molecular compound, e.g., C, Si, SiO_2	• Giant molecular structure • Strong covalent bonds between atoms throughout the entire lattice	• High m.p. and b.p. • Hard solid • Insulator of heat and electricity • Insoluble in water
Simple molecular compound, e.g., I_2, $AlCl_3$, S_8	• Simple molecular structure • Strong covalent bonds between atoms within the discrete molecule • Weak van der Waals' forces of attraction between molecules	• Low m.p. and b.p. • Soft solid • Insulator of heat and electricity • Insoluble in water • Soluble in non-polar solvent

of attraction in iodine are much stronger compared to the hydrogen bonding present among water molecules.

Although there are only id–id interactions among the iodine molecules, these interactions, albeit weaker, are so extensive to the extent that they are stronger than the hydrogen bonding present between water molecules (think about a Velcro strap: a hook and loop pair is rather weak, but when you have a patch of hooks and loops, you get a strong fastener).

My Tutorial (Chapter 2)

1. (a) Solids with crystalline structures may be classified as *molecular*, *giant covalent* or *ionic*. With reference to each crystalline structure, explain how *low melting point*, *high melting point* and *non-conduction of electricity* are related to bonding and structure.

   (b) Explain why the boiling points of the noble gases increase from helium to xenon.

2. (a) Boron trihydride, BH_3, reacts with trimethylamine, $(CH_3)_3N$, to form a 1:1 adduct.

   (i) Using BH_3 and $(CH_3)_3N$ as examples, explain how the shapes of simple molecules can be determined by simply considering the repulsions between electron pairs.

   (ii) Name the type of bond that is formed between BF_3 and $(CH_3)_3N$ in the adduct.

   (b) Boron and carbon are adjacent to each other in the periodic table. There are some B–N compounds known to be *isoelectronic* with C–C compounds. An example would be boron nitride, an electrical insulator that has a planar hexagonal layered structure of alternating boron and nitrogen atoms, similar to graphite.

   (i) Explain the meaning of the term *isoelectronic.*

   (ii) Give the type of bonding which is present *within* the layers.

   (iii) Give the type of interaction *between* the layers.

(iv) Give a possible use in which this compound could replace the industrial usage of the corresponding carbon compound.

(v) When heated under high pressure, this form of boron nitride is converted into another form which is an extremely hard solid. Suggest the type of structure adopted by this new material.

3. Sulfur reacts with nitrogen to yield nitrogen disulfide (NS_2), an electronically symmetrical gas phase radical which is highly unstable. This radical undergoes further reaction to give a cyclic molecule, dinitrogen disulfide (N_2S_2). The N_2S_2 is explosive in nature, and can also be heated to give the polymeric $(SN)_x$, which is metallic in nature. At very low temperatures (0.33 K), the polymeric compound is a superconductor.

   (a) With the aid of a dot-and-cross diagram, deduce the shape of NS_2 and N_2S_2.
   (b) Give a possible explanation why both NS_2 and N_2S_2 are highly unstable.
   (c) Give a possible structure of the polymeric $(SN)_x$, depicting its unidirectional electrical conducting property.
   (d) Give a possible reason why the polymeric compound becomes superconducting at low temperatures.
   (e) Give a reason why it is possible for such a polymeric compound to form between sulfur and nitrogen.

4. This question is concerned with the shapes of molecules and the forces between them.

   (a) Depict, with clear diagrams, the shapes of the following molecules: (i) CH_4; (ii) H_2O.
   (b) Explain the main features of the theory that is used to predict the shapes of these molecules.
   (c) The boiling points and molar masses of some first row hydrides are tabulated below.

Substance	Boiling point (K)	Molar mass ($g\,mol^{-1}$)
CH_4	109	16
NH_3	240	17
H_2O	373	18

(i) Explain the difference in boiling points between NH_3 and CH_4 in terms of structure and bonding.
(ii) Why does H_2O have a higher boiling point than NH_3?

(d) 1,2-dihydroxybenzene (compound A) has a boiling point of 518 K but its isomer 1,4-dihydroxybenzene (compound B) has a boiling point of 558 K.

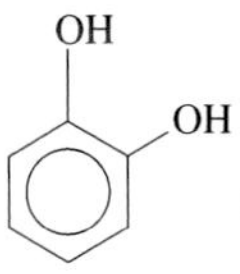

OH OH (A) OH OH (B)

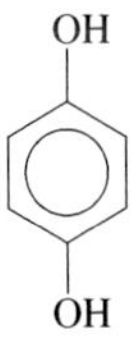

Give one reason why 1,4-dihydroxybenzene has a higher boiling point.

5. (a) Molecular crystals are composed of simple discrete molecules held together in a regular array.

(i) With references to bromine, Br_2, and bromomethane, $CH_3Br(s)$, which are crystalline solids at low temperature, describe and explain the types of intermolecular forces for holding the molecules together in such crystals.
(ii) Give a reason for one physical property associated with molecular crystals.

(b) In contrast, substances such as silicon dioxide and sodium chloride consist of giant lattice crystal structures, which do not contain simple discrete molecules. For each of these substances:

(i) describe their crystalline structures,
(ii) state the type of chemical bonding in the crystals, and
(iii) explain why silicon dioxide adopts one type of bonding and sodium chloride another.

CHAPTER 3

Ideal Gas and Gas Laws

Gases do not have a fixed volume and shape. Gases are compressible. Gases have these properties because they are made up of particles that are separated by large distances, and these particles are constantly moving around. The large separation is due to the weak intermolecular forces between the particles and such separation between the particles can be varied because the intermolecular forces can be overcome with ease. For example, a fixed amount of gas when placed in different containers of different volumes or shapes would be able to take up the space of the container easily. When pressure is applied to a volume of gas, its volume decreases. This decrease in volume is due to the increase in proximity between the particles, that is, a decrease in the distance of separation.

The constant movement of gaseous particles results in the particles possessing a large amount of translational kinetic energy (K.E.). The amount of translational kinetic energy is temperature dependent, i.e., average K.E. is directly proportional to temperature. These constant movements in the container result in constant collisions with the walls of the container, giving rise to the phenomenon known as pressure, which is basically the average force exerted per unit area when the gaseous particles collide. In addition, this contact with the wall of the container is also a means whereby gaseous particles exchange energy from the external environment. But because of the large separation between gaseous particles, energy transfer

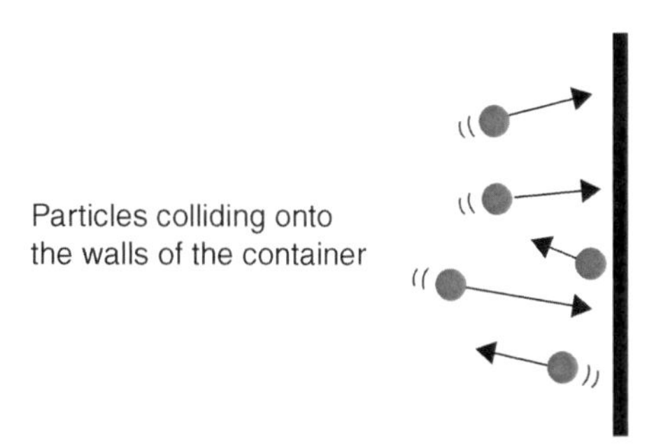

Fig. 3.1. Particles colliding with the walls of a container.

between these particles is not as fast as that between particles in liquids and solids. Therefore, gases are poor heat conductors, which also makes them good insulators.

3.1 Gas Laws

To gain a better understanding of the variables that affect gases, gas laws have been established that relate macroscopic properties of gases such as volume (V), temperature (T) and pressure (p).

In this chapter, we will cover the following important gas laws and see how these laws are applied in real-life situations:

- Boyle's Law (discovered in 1662)
- Charles' Law (discovered in 1787)
- Gay-Lussac's law (discovered in 1809)
- Avogadro's Law (discovered in 1811)
- The Ideal Gas Law
- Dalton's Law of Partial Pressure

3.1.1 *Boyle's law*

For a fixed mass of gas at constant temperature, its pressure is inversely proportional to its volume, i.e.,

$$p \propto \frac{1}{V}.$$

With Boyle's Law, we can deduce the effect of changing pressure on the volume of this fixed mass of gas at constant temperature and *vice versa* by using the equation:

$$p_1V_1 = p_2V_2.$$

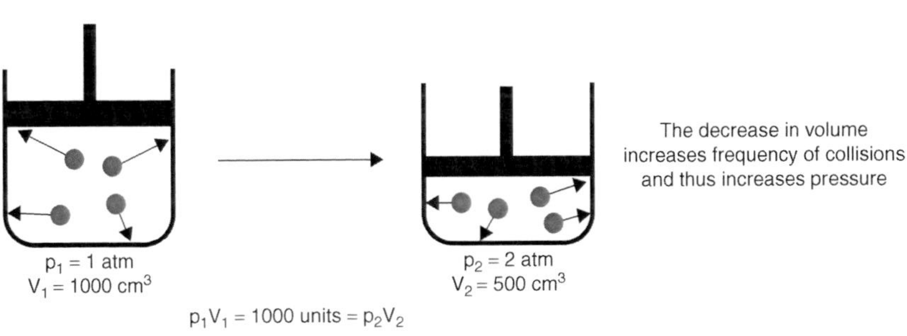

Fig. 3.2.

Real-life scenario: One of the ways to rescue a person who is choking is to perform abdominal thrusts. The theory behind this action is that by exerting pressure on the diaphragm, we are decreasing the volume inside the lungs, which then increases the pressure inside the lung, forcing out the foreign object lodged in the airway.

3.1.2 *Charles' law*

For a fixed mass of gas at constant pressure, its volume is directly proportional to its temperature, i.e.,

$$V \propto T.$$

With Charles' Law, we can deduce the effect of changing temperature on the volume of a fixed mass of gas at constant pressure and *vice versa* by using the equation:

$$\frac{V_1}{T_1} = \frac{V_2}{T_2}.$$

Real-life scenario: Have you ever experienced holding a balloon while walking in the hot afternoon sun and it suddenly burst? You looked around for what could have pricked the balloon, but there was nothing in sight. . . it just mysteriously popped? What you actually witnessed was an application of Charles' Law. The balloon was at constant pressure. In the hot sun, its temperature rose and according to Charles' Law, its volume expanded, in this case, to the extent that the latex vessel could no longer contain the heated gases and it burst.

3.1.3 *Gay-Lussac's law*

For a fixed mass of gas at constant volume, its pressure is directly proportional to its temperature, i.e.,

$$p \propto T.$$

With Gay-Lussac's Law, we can deduce the effect of changing temperature on the pressure of this fixed mass of gas at constant volume and *vice versa* by using the equation:

$$\frac{p_1}{T_1} = \frac{p_2}{T_2}.$$

Real-life scenario: If someone were to put a bag of potato chips into a freezer, thinking this would preserve their flavour, he would be in for a disappointment. This is because in order to maintain freshness, preservatives are added. These preservatives need a constant internal environment which is provided by a constant internal pressure. Placing the bag in the freezer would cause a reduction in pressure, as per Gay-Lussac's Law, and this would actually foil the original plan to maintain freshness.

3.1.4 *Avogadro's law*

Avogadro's hypothesis states that at constant temperature and pressure, gases of the same volume contain the same number of particles.

It follows from Avogadro's hypothesis that one mole of **any gas** occupies the same volume as one mole of any other gas at the same pressure and temperature.

Q: Why should the same amount of any type of gas have the same volume at the same temperature and pressure? Don't different gases have molecules of different sizes?

A: The size of a gas particle (atom or molecule) is insignificant as compared to the volume of the container that it occupies. Yes, at the molecular level, different types of gas particles have different sizes. However, the difference in size is still insignificant when compared to the volume of the container. (We will later see that the size factor is only significant at high pressure.) This is also why when you increase the number of gas particles in a system without allowing the volume to change, the container can still accommodate the greater number of gas particles. There is simply plenty of space in between the gaseous particles for other gaseous particles to come in!

From Avogadro's hypothesis stems Avogadro's Law, which states that for a gas at constant pressure and temperature, its volume is directly proportional to the number of moles of the gas, i.e.,

$$V \propto n.$$

Real-life scenario: If someone inflates a balloon, the size of the balloon increases because more gaseous particles are being introduced into the balloon at a constant external atmospheric pressure and temperature.

Now we have a useful stoichiometric equivalence between volume of gas and the amount of gas in terms of moles. However, take note that this stoichiometric equivalence is only applicable in cases of constant temperature and pressure. The following example should demonstrate this.

3.1.4.1 *Problems involving the use of the relationship $V \propto n$*

Example 3.1: $20\,\text{cm}^3$ of gaseous ethane C_2H_6 is burnt in $80\,\text{cm}^3$ of oxygen.

(a) What is the volume of the residual gaseous mixture at the end of the reaction?

(b) The residual gaseous mixture is treated with NaOH(aq). What is the volume of gas remaining after the treatment?

Assume all volumes are measured at room temperature and pressure.

Approach for part (a): The residual gaseous mixture can include gaseous products formed and also unreacted gaseous reactants. Thus, we need to find out if there are any excess reactants.

Hydrocarbons are covalent compounds containing only carbon and hydrogen atoms. Their complete combustion produces only two products, water and carbon dioxide.

The general equation for the complete combustion of hydrocarbons is

$$C_xH_y + \left(x + \frac{y}{4}\right)O_2 \longrightarrow xCO_2 + \frac{y}{2}H_2O.$$

Solution for part (a): Write out the balanced equation for the combustion of the hydrocarbon ethane:

$$C_2H_6(g) + \frac{7}{2}O_2(g) \longrightarrow 2CO_2(g) + 3H_2O(1).$$

Since under constant temperature and pressure, $V \propto n$, we can establish the following stoichiometric equivalencies between volume and amount of gas:

	$C_2H_6(g)$	$+\frac{7}{2}O_2(g)$	$\longrightarrow 2CO_2(g)$	$+3H_2O(l)$
Mole ratio/mol	1	7/2	2	3
Volume ratio/cm^3	1	7/2	2	3

For 20 cm^3 of C_2H_6, the volume of O_2 needed for complete combustion $= 20 \times \frac{7}{2} = 70$ cm^3.

However, with 80 cm^3 of O_2 introduced, there is an excess of unreacted O_2 of 10 cm^3.

$$\begin{aligned}\text{Volume of } CO_2 \text{ produced} &= 2 \times \text{volume of } C_2H_6 \text{ reacted}\\ &= 2 \times 20\\ &= 40\,\text{cm}^3.\end{aligned}$$

$$\begin{aligned}\text{Volume of residual gaseous mixture} &= \text{Volume of } CO_2 \text{ produced} \\ &\quad + \text{volume of unreacted } O_2 \\ &= 40 + 10 \\ &= 50\,\text{cm}^3.\end{aligned}$$

Approach for part (b): When a gaseous mixture is treated with NaOH(aq), acidic gases present will react with the alkali and be removed. The remaining gases, after the alkali treatment, are those that do not react with it.

Likewise, if a gaseous mixture is treated with HCl(aq), basic gases such as NH_3(g) will be neutralised by the acid and thus removed from the gaseous mixture.

In this problem, CO_2(g) is the acidic gas that will react with NaOH(aq):

$$2\text{NaOH(aq)} + \text{CO}_2\text{(g)} \longrightarrow \text{Na}_2\text{CO}_3\text{(aq)} + \text{H}_2\text{O(l)}.$$

This reaction is very similar to $Ca(OH)_2$(aq) + CO_2(g) $\longrightarrow$ $CaCO_3$(s) + H_2O(l), except that the calcium carbonate formed is insoluble in water. In essence, the important reaction is just between the carbon dioxide and hydroxide ions.

Solution for part (b): Volume of gas that remained after the alkali treatment

$$\begin{aligned}&= \text{Volume of residual gaseous mixture} - \text{volume of } CO_2 \\ &= 50 - 40 \\ &= 10\,\text{cm}^3.\end{aligned}$$

Note that for this type of problem, the product H_2O is taken to have reverted to the liquid state and is thus not part of the gaseous mixture.

Example 3.2: $15\,\text{cm}^3$ of a gaseous hydrocarbon is reacted with $140\,\text{cm}^3$ of oxygen which is in excess. After the reaction, the residual gas mixture is found to occupy $110\,\text{cm}^3$ when measured under room conditions. When this mixture is treated with KOH(aq), the volume decreases to $50\,\text{cm}^3$. What is the molecular formula of the hydrocarbon?

Solution: The residual gas mixture comprises both unreacted O_2 and the product, CO_2.

Since CO_2 reacts with KOH(aq),

$$\text{Volume of } 50\,\text{cm}^3 = \text{Volume of unreacted } O_2.$$

Thus,

$$\text{Volume of } O_2 \text{ reacted} = 140 - 50 = 90\,\text{cm}^3$$

and

$$\text{Volume of } CO_2 \text{ produced} = 110 - 50 = 60\,\text{cm}^3.$$

$$C_xH_y + \left(x + \frac{y}{4}\right)O_2 \longrightarrow xCO_2 + \frac{y}{2}H_2O$$

	C_xH_y	O_2	CO_2
Volume ratio/cm^3	15	90	60
Mole ratio/mol	1	6	4

Hence, $x = 4$ and $(x + y/4) = 6 \Rightarrow y = 8$.

The molecular formula of the hydrocarbon is C_4H_8.

Example 3.3: $20\,\text{cm}^3$ of a hydrocarbon is reacted with excess oxygen. At the end of the reaction, there is a reduction in volume of $50\,\text{cm}^3$. When the gaseous mixture is treated with $Ca(OH)_2$(aq), there is a further reduction in volume of $60\,\text{cm}^3$. What is the molecular formula of the hydrocarbon? Assume all volumes are measured at room temperature and pressure.

Solution: CO_2(g) is the acidic gas that reacts with $Ca(OH)_2$(aq). Thus, its volume is $60\,\text{cm}^3$.

Since $20\,\text{cm}^3$ of the hydrocarbon reacts with excess O_2 to produce $60\,\text{cm}^3$ CO_2, the value of $x = 3$.

$$C_xH_y + \left(x + \frac{y}{4}\right)O_2 \longrightarrow xCO_2 + \frac{y}{2}H_2O$$

	C_xH_y	O_2	CO_2
Volume ratio/cm^3	20	?	60
Mole ratio/mol	1		3

$$\begin{aligned}\text{Contraction of } 50\,\text{cm}^3 &= \text{Volume of hydrocarbon} + \text{volume of O}_2 \\ &\quad \text{reacted} \\ &\quad - \text{volume of CO}_2 \\ &= 20 + V_{O_2} - 60.\end{aligned}$$

Thus,

$$V_{O_2} = 90\,\text{cm}^3.$$

Therefore,

$$\begin{aligned}(x + y/4) &= 90/20 \\ 3 + y/4 &= 9/2 \\ y &= 6.\end{aligned}$$

The molecular formula of the hydrocarbon is C_3H_6.

3.1.5 *The ideal gas law*

If you have noticed thus far that these gas laws seem to be interrelated, you are right! Combining Boyle's Law, Charles' Law and Gay-Lussac's Law gives us the Combined Gas Law [Eq. (3.1)], which allows us to manipulate two variables and see the effect on a third variable:

$$\frac{p_1V_1}{T_1} = \frac{p_2V_2}{T_2}. \tag{3.1}$$

If we are to include Avogadro's Law into Eq. (3.1), we will arrive at a single equation which allows us to predict what happens to a gas when not just two but three or more variables are changed at the same time. This equation is called the Ideal Gas Law, as given in Eq. (3.2):

$$pV = nRT. \tag{3.2}$$

The terms and units in Eq. (3.2) are:

- temperature (T), expressed in kelvins
- amount (n), of gas expressed in moles
- pressure (p), commonly expressed in Pascal (Pa) or Nm^{-2} or atmospheric pressure (atm) or bar

- volume (V), commonly expressed in dm^3 or m^3
- R, the molar gas constant.

Depending on the units chosen for pressure and volume, the molar gas constant R will have different values. R is frequently expressed as either 8.31 J K^{-1} mol^{-1} (S.I. unit) or 0.083 bar dm^3 K^{-1} mol^{-1}. Thus, there are two sets of units that you should be acquainted with for the two values of R:

(i) When R = 8.31 J K^{-1} mol^{-1}. (ii) When R = 0.083 bar dm^3 K^{-1} mol^{-1}.

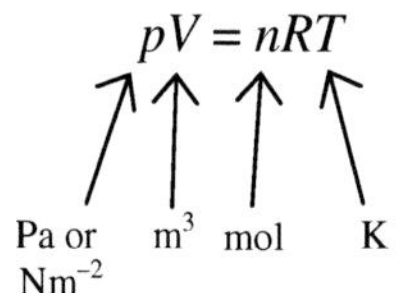

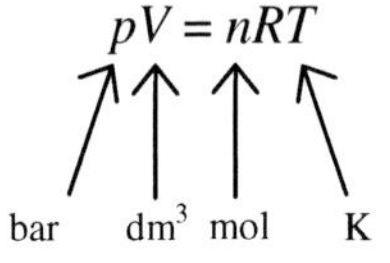

You will find the following conversion data useful in calculations:

- 1 bar = 100,000 Pa or 100 kPa
- 1 dm^3 = 1000 cm^3
- 1 dm^3 = 10^{-3} m^3
- 1 cm^3 = 10^{-6} m^3

The volume occupied by one mole of gas is known as the *molar* volume and it is dependent on physical conditions. In this book, you will come across two common sets of conditions: (i) standard temperature and pressure (denoted as s.t.p.) and (ii) room temperature and pressure (denoted as r.t.p.).

Using the Ideal Gas Equation, we can derive the molar volume of an ideal gas in these conditions:

- At s.t.p. (where T = 273 K and p = 100 kPa or 1 bar), the molar volume of an ideal gas is 22.7 dm^3 mol^{-1}.
- At r.t.p. (where T = 298 K and p = 100 kPa or 1 bar), the molar volume of an ideal gas is 24.8 dm^3 mol^{-1}.

Example 3.4:

$$MgCO_3(s) + 2HCl(aq) \longrightarrow MgCl_2(aq) + CO_2(g) + H_2O(l).$$

What is the mass of magnesium carbonate required to produce $20.0\,dm^3$ of dry carbon dioxide at s.t.p.?

Solution: Assume 1 mole of gas occupies $22.7\,dm^3$ at s.t.p.
Amount of CO_2 to be produced $= \frac{20}{22.7} = 0.8811\,mol$.
Amount of $MgCO_3$ needed $=$ Amount of $CO_2 = 0.8811\,mol$.
Molar mass of $MgCO_3 = 24.0 + 12.0 + 3(16.0) = 84.0\,g\,mol^{-1}$.
Mass of $MgCO_3$ required $= 0.8811 \times 84.0 = 74.0\,g$ (3 s.f.).

Example 3.5: 0.5 moles of N_2 and 0.2 moles of CO_2 are introduced into a $5\,dm^3$ evacuated vessel at 23°C. Assuming no reaction between the two gases, what is the total pressure exerted on the vessel?

Solution: Data provided:
$V = 5\,dm^3 = 5 \times 10^{-3}\,m^3$.
$T = 23°C = 273 + 23 = 296\,K$.
$n =$ total amount of gas $= 0.5 + 0.2 = 0.7$ moles.
Substitute these values into the Ideal Gas Equation:
$pV = nRT$
$p = \frac{0.7 \times 8.31 \times 296}{5 \times 10^{-3}} = 3.44 \times 10^5\,Pa$.

3.1.6 *Further manipulations of the ideal gas equation*

The Ideal Gas Equation can be manipulated to determine properties such as concentration, mass, molar mass (or relative molecular mass) and density of a gas.

- To determine concentration of a gas, the Ideal Gas Equation is rearranged as follows:

$$p = \frac{n}{V} \times RT,$$

where $n/V =$ concentration.

- To determine mass (m), molar mass (M) or relative molecular mass (M_r) of a gas, the Ideal Gas Equation is modified as follows:

$$pV = \frac{m}{M} \times RT,$$

where m/M = amount of gas in moles.

M_r has the same numerical value as molar mass. The only difference is that M_r is a dimensionless quantity while molar mass has units of g mol^{-1}. The mass calculated is in grams.

- To determine the density of a gas, the Ideal Gas Equation is modified as follows:

$$pV = \frac{m}{M} \times RT$$

$$p = \frac{m}{V} \times \frac{RT}{M}$$

$$p = \rho \times \frac{RT}{M},$$

where m/V = density of gas (denoted by Greek letter Rho, ρ).

Example 3.6: A gas in a 900 cm^3 vessel is found to have a mass of 2.90 g at 30°C and 100 kPa. What is the relative molecular mass of the gas?

Solution: Assuming that the gas behaves ideally, we can make use of the following equation to solve for M_r:

$$pV = nRT$$

$$pV = \frac{m}{M} \times RT. \tag{3.3}$$

Data provided:
$p = 100\,\text{kPa} = 100{,}000\,\text{Pa}$ (Note: remember to convert kPa to Pa).
$V = 900\,\text{cm}^3 = 9 \times 10^{-4}\,\text{m}^3$.
$T = 30°\text{C} = 273 + 30 = 303\,\text{K}$.
$m = 2.90\,\text{g}$.

Substitute these values into Eq. (3.3):
$pV = \frac{m}{M} \times RT$
$100{,}000 \times 9 \times 10^{-4} = \frac{2.90}{M} \times 8.31 \times 303$

$M = 81.1\,\text{g}\,\text{mol}^{-1}$.
Thus, M_r of the gas is 81.1.

Exercises

1. A gas is found to have a density of $2.65\,\text{g}\,\text{dm}^{-3}$ at 24°C and 102 kPa. Determine the relative molecular mass (M_r) of the gas. (**Answer: 64.1.**)
2. A sample of $NH_3(g)$ has a volume of $36\,\text{dm}^3$ at 293 K and 98 kPa. What is the mass of $NH_3(g)$ in the sample? (**Answer: 24.6 g.**)

3.1.7 *Dalton's Law of Partial Pressure*

Dalton's Law of Partial Pressure states that the total pressure of a mixture of gases is simply the sum of the partial pressures of the individual gases at the same temperature. The assumption herein is that the gases do not react with one another.

For a mixture which consists of gas X, Y and Z, Dalton's Law is represented as

$$p_{\text{total}} = p_x + p_y + p_z,$$

where p_{total} = total pressure of the gaseous mixture, and p_x, p_y and p_z = partial pressure of gas X, Y and Z respectively.

Q: What is partial pressure?
A: Partial pressure of a gas is essentially the pressure the gas will exert under the same conditions of temperature and volume if it alone was to occupy the container.

Partial pressure can be calculated in the following ways:

- When the volume, temperature and amount of gas is given, use the Ideal Gas Equation (i.e., $pV = nRT$) to solve for p.
- Make use of the relationship between partial pressure, mole fraction and total pressure:

$$p_y = \chi_y \times p_{\text{total}}, \tag{3.4}$$

where mole fraction, $\chi_y = \frac{n_y}{n_{\text{total}}}$.

The mole fraction (denoted by the Greek small letter Chi, χ) is a dimensionless number that expresses the composition of a mixture.

The mole fraction of a component, e.g., gas Y, in a mixture is defined as the ratio of the number of moles of Y (n_y) to the total number of moles of all components in the mixture (n_{total}).

Derivation of Eq. (3.4)

For a mixture of gases, the total pressure can be determined using

$$p_{\text{total}}V = n_{\text{total}}RT. \tag{3.5}$$

For component y in the mixture, its partial pressure can be solved using

$$p_yV = n_yRT. \tag{3.6}$$

If we divide Eq. (3.6) by Eq. (3.5), we have

$$\frac{p_y}{p_{\text{total}}} = \frac{n_y}{n_{\text{total}}},$$

which can be rearranged to give

$$p_y = \frac{n_y}{n_{\text{total}}} \times p_{\text{total}}.$$

Hence,

$$p_y = \chi_y \times p_{\text{total}}.$$

Example 3.7: At 101 kPa, a balloon was found to contain 0.2 moles of helium, 0.2 moles of oxygen and 0.5 moles of nitrogen. What is the partial pressure of oxygen?

Solution: Use the equation: $p_{O_2} = \chi_{O_2} \times p_{\text{total}}$, where mole fraction, $\chi_{O_2} = \frac{n_{O_2}}{n_{\text{total}}}$. p_{total} is 101 kPa. Find n_{total} by adding the amount of gases present:

$$\begin{aligned} n_{\text{total}} &= n_{\text{He}} + n_{\text{O}_2} + n_{\text{N}_2} \\ &= 0.2 + 0.2 + 0.5 \\ &= 0.9\,\text{mol}. \end{aligned}$$

Insert the data into the equation:

$$\begin{aligned} p_{\text{O}_2} &= \frac{n_{\text{O}_2}}{n_{\text{total}}} \times p_{\text{total}} \\ &= \frac{0.2}{0.9} \times 101 \\ &= 22.4\,\text{kPa} \end{aligned}$$

Example 3.8: An $800\,\text{dm}^3$ vessel contains 0.14 g of $N_2(g)$ and 0.24 g of $CH_4(g)$ at 20°C.
(i) What is the total pressure in the vessel?
(ii) Hence, calculate the partial pressure of each gas.

Solution: In this problem, we need to calculate the total pressure of the vessel using $pV = nRT$.
To calculate p_{T}, we need to find n_{T}:
Amount of $N_2 = \frac{0.14}{28.0} = 5 \times 10^{-3}\,\text{mol}$.
Amount of $CH_4 = \frac{0.24}{16.0} = 0.015\,\text{mol}$.
Hence,

$$n_{\text{T}} = n_{\text{N}_2} + n_{\text{CH}_4} = 0.02\,\text{mol}.$$

Thus,

$$p_{\text{T}} = \frac{n_{\text{T}}\text{RT}}{V} = \frac{0.02\times 8.31\times(273+20)}{800\times 10^{-3}} = 60.9\,\text{Pa}.$$

Also,

$$p_{\text{N}_2} = \frac{5\times 10^{-3}}{0.02} \times 60.9 = 15.2\,\text{Pa}.$$

and

$$p_{\text{CH}_4} = \frac{0.015}{0.02} \times 60.9 = 45.7\,\text{Pa}.$$

If we add up p_{N_2} and p_{CH_4}, we get p_{T}.

Example 3.9: $10\,\text{dm}^3$ of gas **V** (measured at 298 K and 300 kPa) and $8\,\text{dm}^3$ of gas **W** (measured at 298 K and 210 kPa) are mixed together in a 15 dm^3 vessel. Determine the total pressure in the vessel. Assume the gases do not react with each other and that temperature is held constant.

Approach: Assume the gases behave ideally (so as to apply the gas laws).

To determine the total pressure of the vessel, we will apply Dalton's Law:

$$p_{\text{total}} = p_{\mathbf{V}} + p_{\mathbf{W}}.$$

However, we need to first determine the partial pressure of the gases in the 15 dm^3 vessel. The following data are given to us:

- Before mixing — volume and pressure of each gas is given.
- After mixing — volume is given but not pressure (which is what we have to find out).

Since temperature is fixed with volume and pressure data given, we can use Boyle's Law to find the pressure of each gas after mixing, i.e., the partial pressure of the gas in the 15 dm^3 vessel.

Solution:

Step 1: Calculate the pressure of each gas in the 15 dm^3 vessel.
Let the partial pressures of **V** and **W**, after mixing, be $p_{\mathbf{V}}$ atm and $p_{\mathbf{W}}$ atm, respectively.
Applying Boyle's Law, for gas **V**,
$300 \times 10 = p_{\mathbf{V}} \times 15$
$p_{\mathbf{V}} = 200\,\text{kPa}.$
And for gas **W**,
$210 \times 8 = p_{\mathbf{W}} \times 15$
$p_{\mathbf{W}} = 112\,\text{kPa}.$

Step 2: Calculate total pressure.

$$\begin{aligned}\text{The total pressure in the 15 dm}^3\text{ vessel} &= p_{\mathbf{V}} + p_{\mathbf{W}}\\ &= 200 + 112\\ &= 312\,\text{kPa}.\end{aligned}$$

Example 3.10: A $6\,\text{dm}^3$ flask containing gas **A** at a pressure of $10\,\text{atm}$ is connected to a $14\,\text{dm}^3$ flask containing gas **B** at a pressure of $5\,\text{atm}$.

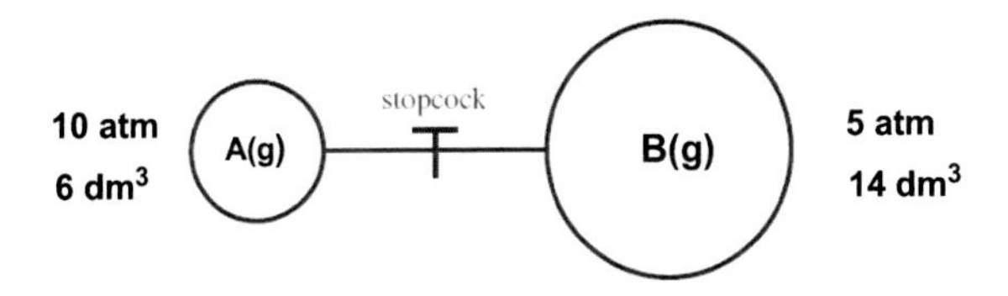

What is the final pressure in the flasks after they are connected at 25°C and the pressures are allowed to equilibrate?

Solution: The following assumptions hold for this question:

- The volume of connecting tube is negligible.
- The two gases do not react.
- There is no change in temperature on mixing.

After connection, the gases will equilibrate and the resultant volume is $20\,\text{dm}^3$. Let the new partial pressures of **A** and **B** be $p_{\mathbf{A}}$ atm and $p_{\mathbf{B}}$ atm, respectively. Using the formula, $p_1V_1 = p_2V_2$:

$$p_{\mathbf{A}} = (10 \times 6)/20 = 3\,\text{atm},$$
$$p_{\mathbf{B}} = (5 \times 14)/20 = 3.5\,\text{atm}.$$

According to Dalton's Law of Partial Pressure:
Final pressure, $p_{\text{total}} = p_{\mathbf{A}} + p_{\mathbf{B}} = 3 + 3.5 = 6.5\,\text{atm}$.

3.1.8 *Types of mathematical problems involving the gas laws*

When asked to determine any one of the variables, amount of gas, pressure, volume or temperature, refer to Fig. 3.3 for help in choosing the appropriate gas laws to use.

When a problem involves determining the M_r of the gas, its concentration or its density, use the Ideal Gas Equation and manipulate

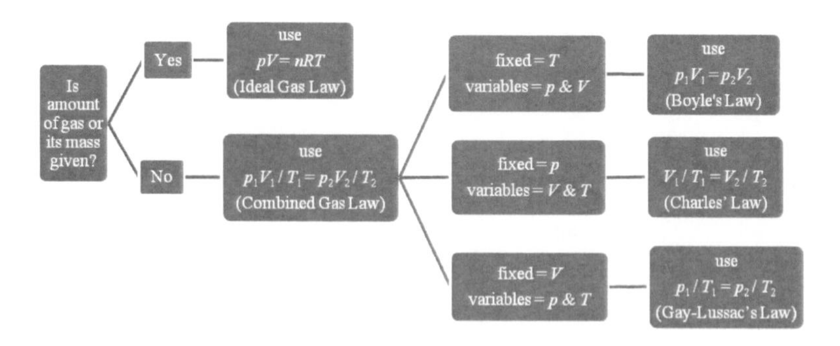

Fig. 3.3. Schema for selecting gas laws.

it so as to insert

$$n = \frac{\text{mass}}{\text{molar mass}}; \quad \text{concentration} = \frac{\text{amount in moles}}{\text{volume}};$$

$$\text{density} = \frac{\text{mass}}{\text{volume}}.$$

For questions that deal with a mixture of two or more gases, the common types of calculation require the determination of

- Total pressure using Dalton's Law of Partial Pressure.
- Partial pressure of a gas, p_y, by the following methods:
  - $p_y = \chi_y \times p_{\text{total}}$, where $\chi_y = \frac{n_y}{n_{\text{total}}}$,
  - Boyle's Law or Gay-Lussac's Law (when T or V is held constant, respectively).

3.2 Kinetic Theory of Gases

Q: When we apply any of the gas laws mentioned, we input the volume of the container as the volume occupied by the gas molecules to calculate the other properties, such as pressure, of the gas. Do we need to factor in the physical dimension (such as volume) of the gas molecules and subtract this value from the volume of the container to give the resulting volume of empty space that the molecules can occupy?

A: If we factor in the volume of the molecules, the mathematics involved will be made more complex but it will not improve the accuracy in applying the gas laws. Thus, for these laws, it is assumed that molecules have negligible volume. It is okay to make such an assumption since gas molecules are extremely small particles that are far apart from one another and they move freely anywhere inside a container.

Q: In Chap. 2 on Chemical Bonding, we have said that "like dissolves like". Thus, will gas molecules such as those of H_2O and N_2 tend not to mix with one another since H_2O molecules are polar and N_2 molecules are non-polar?

A: The very fact that these gases do mix suggests that intermolecular forces do not play a significant part in the behaviour of gases. It is plausible to assume that gas particles do not exert strong attractive forces on one another since they travel at very high speed and therefore hardly have the chance to interact with one another. From another perspective, the separation between gas particles is so large that it allows other types of gas particles to come in between.

If we consider the presence of weak attractive forces, the pressure either measured or calculated using the gas laws will be less than what it could actually be, since now, with molecules being attracted to one another, the impact of collision on the walls of the container will be less, leading to less force exerted on these walls.

The Kinetic Theory of Gases utilises these assumptions along with a few others so as to create a mathematical model that is used to study the behaviour of gases. What the theory seeks to do with all the assumptions in place is to give us a picture of how a gas will behave under perfect conditions in an ideal world. Accordingly, the gas that obeys all these assumptions is an ideal gas.

The assumptions applied to the Kinetic Theory of Gases are:

- The size of gaseous particles is negligible compared to the volume of the container it occupies. Thus, the particles are widely separated and can move anywhere in the container.
- The gas particles *do not exert attractive forces* on one another.

- The gas particles *move around randomly*, colliding occasionally with one another and the walls of the container.
- Collisions between the gas particles are perfectly *elastic*, i.e., there is no loss of kinetic energy upon collision.
- The average kinetic energy of particles in a gas is constant at constant temperature. The average kinetic energy is *proportional* to the absolute temperature.

The gas laws covered in this chapter assume that real gases behave like an ideal gas. An ideal gas obeys these laws totally under all types of conditions. On the other hand, real gases only obey these laws under conditions of *high temperature* and *low pressure.*

When precise measurements are made over other ranges of temperature and pressure, you will find that the gas laws do not hold up quite so well for real gases. Section 3.2.2 explains why this is so.

Nonetheless, even if an ideal gas is just a hypothetical gas that does not exist in real life, the concept of an ideal gas helps us to gain a better understanding of real gases by simplifying their complexities so that they can be more easily understood.

3.2.1 *Graphical plots representing ideal behaviour*

For a fixed amount of ideal gas, we can use the Ideal Gas Equation to generate graphical plots that show its ideal behaviour.

For instance, to generate a graph of pressure against volume (p versus V) at constant temperature, we first have to define the relationship between p and V. We can deduce this by inspecting the Ideal Gas Equation.

Rearranging $pV = nRT$, for p gives $p = \frac{nRT}{V}$. Since n, R and T are constants, we can say that $p \propto \frac{1}{V}$. Thus, p is inversely proportional to V and a plot of p versus V is shown in Fig. 3.4.

What is the graph for p versus $\frac{1}{V}$?

In mathematics, you would have come across the equation for a straight line:

$$y = mx + C, \tag{3.7}$$

where y and x are values on the axes, m stands for gradient and C stands for the intersection of the line with the y-axis.

From the relationship $p \propto \frac{1}{V}$, we can form the equation

$$p = \frac{k}{V}, \text{ where } k \text{ is a constant.} \tag{3.8}$$

Comparing Eqs. (3.7) and (3.8), a plot of p versus $\frac{1}{V}$ will be a straight line plot, passing through the origin as shown in Fig. 3.5.

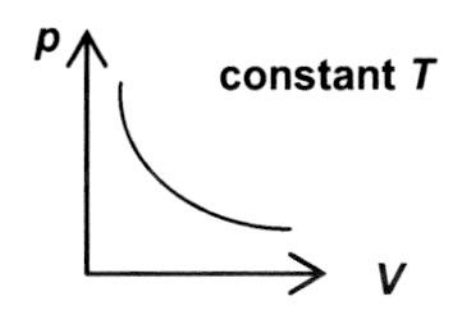

Fig. 3.4. A graph of p versus V.

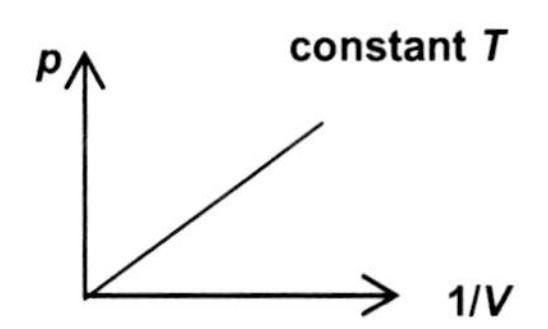

Fig. 3.5. A graph of p versus $1/V$.

Example 3.11: For a fixed amount of ideal gas, sketch the following graphs to depict its ideal behaviour:

(a) pV versus p at constant T,
(b) pV versus V at constant T,
(c) V versus T(K) at constant p,
(d) V versus T(°C) at constant p,
(e) ρ versus T(K) at constant p.

Solution:

(a) Applying $pV = nRT$:

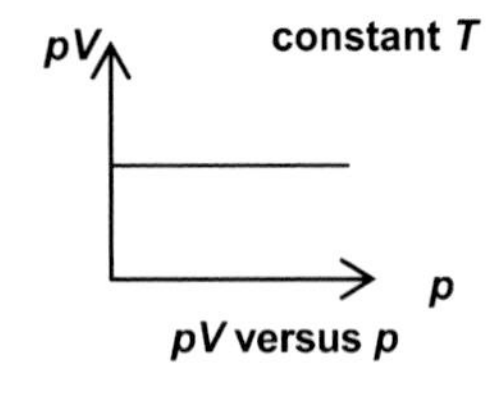

$$pV = \text{constant}.$$

Thus, for an ideal gas, pV is a constant at all values of p.

(b) Applying $pV = nRT$:

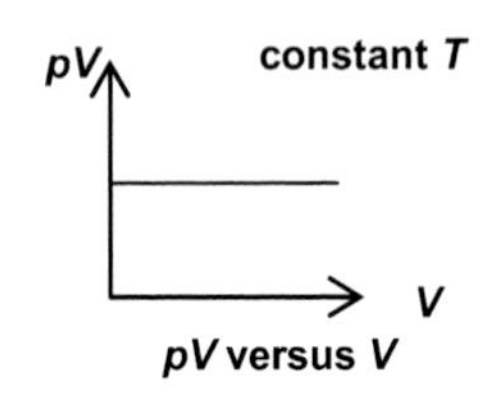

$$pV = \text{constant}.$$

Thus, for an ideal gas, pV is a constant at all values of V.

(c) Applying $pV = nRT$:

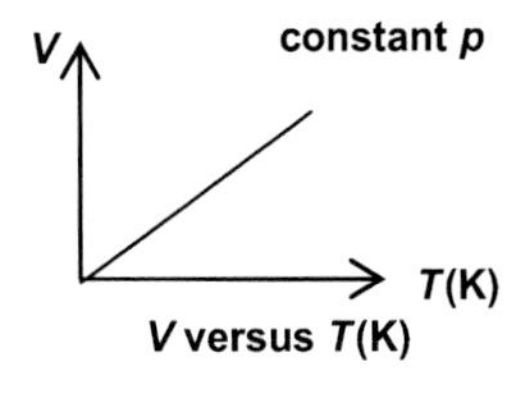

$$V = \frac{nRT}{p}.$$

Since n, R and p are constants, $V \propto T$. The plot of V versus T is a straight line passing through origin.

(d) From part (c), $V \propto T$. However, the unit for temperature is kelvin. Since $1\,\text{K} = 273.15°\text{C}$, we say that $V \propto (T + 273.15)$ when unit of T is degree Celsius. The plot of V versus $T(°\text{C})$ is a straight line with a y-axis intersection.

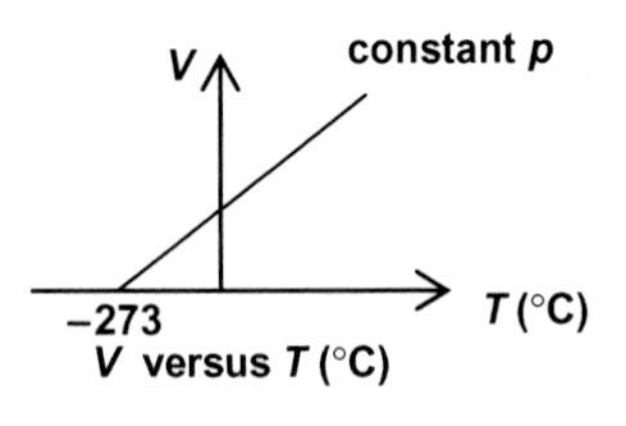

The temperature of $-273.15°\text{C}$ (i.e., 0 K) on the x-axis is known as "absolute zero" as any temperature below this point indicates a gas with negative volume.

(e) We can manipulate the Ideal Gas Equation to include the density term:

$$pV = nRT$$
$$p = \rho \times \frac{RT}{M}$$
$$\rho = \frac{pM}{RT}.$$

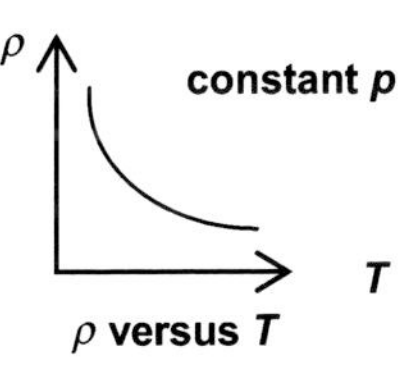

Since p, M and R are constants, $\rho \propto \frac{1}{T}$.

The plot of ρ versus T is a curve showing their reciprocal relationship.

Exercises: Which graph represents the behaviour of a fixed amount of an ideal gas?

Graph A

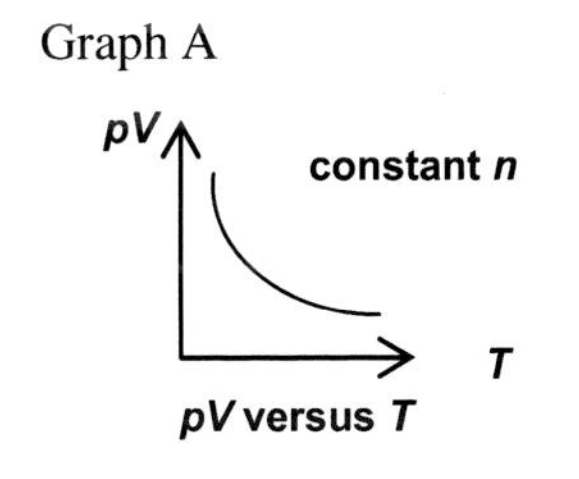

Graph B

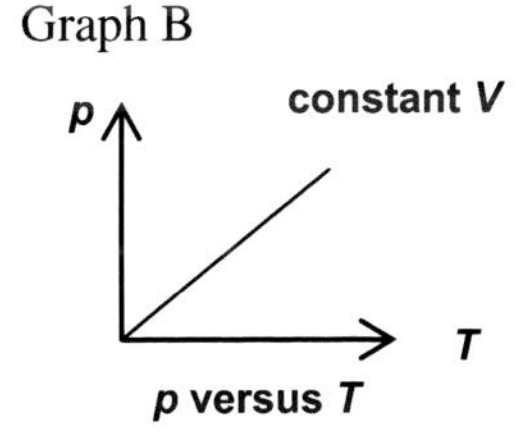

Graph C

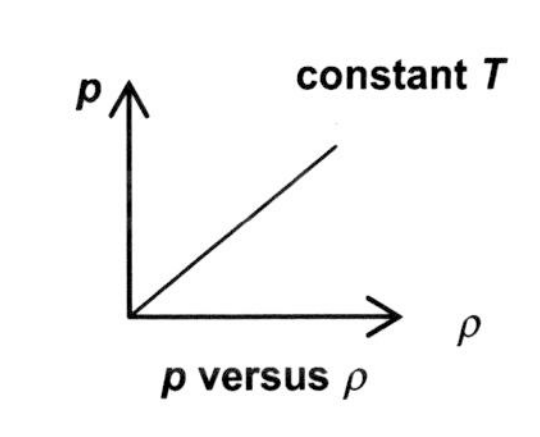

Graph D

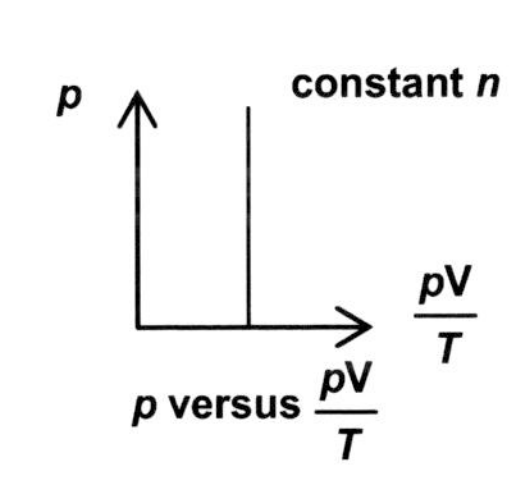

[Answer: Graphs B, C and D]

> Did you realise these plots illustrate the various gas laws?
> Boyle's Law: $p \propto \frac{1}{V}$
> Charles' Law: $V \propto T$
> Gay-Lussac's Law: $p \propto T$

3.2.2 *Deviation from ideal gas behaviour*

For a given amount of ideal gas, a plot of $\frac{pV}{RT}$ versus p will yield a horizontal straight line.

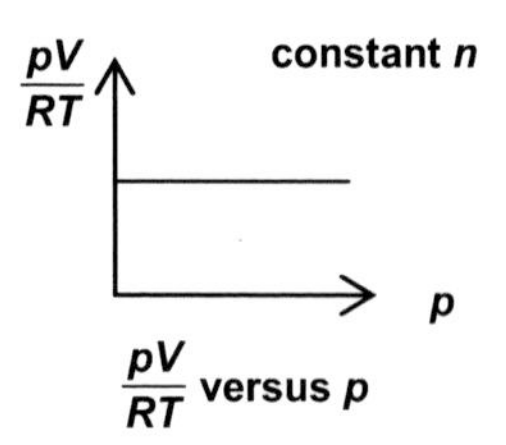

When experiments are performed on a real gas, taking nitrogen for instance, the plot of $\frac{pV}{RT}$ versus p does not yield a straight line. Instead, the graph shows a deviation from the ideal plot if nitrogen were to behave ideally (Fig. 3.6).

Figure 3.6 shows that nitrogen deviates greatly from ideal behaviour at conditions of

- high pressure, and
- low temperature.

These are precisely the kind of conditions under which we would expect the assumptions of an ideal gas to be invalid for a real gas.

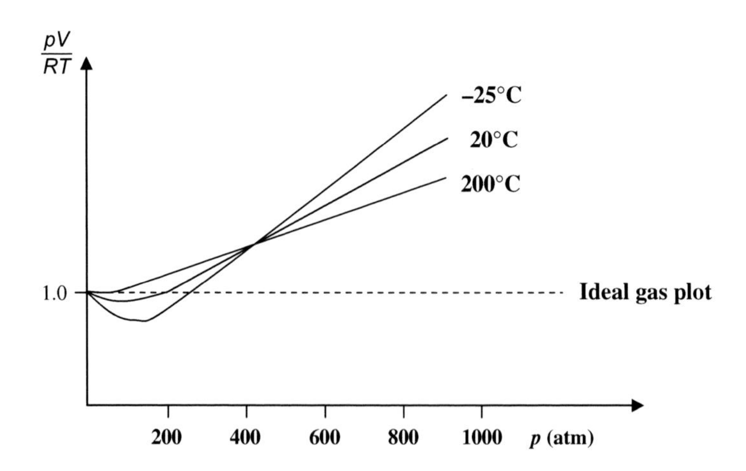

Fig. 3.6. Graph of $\frac{pV}{RT}$ versus p plotted at various temperatures for nitrogen.

Unlike ideal gases, the molecules of real gases actually experience attractive forces between one another, which allows real gases to be liquefied under high pressures. In addition, real gas molecules do occupy finite volumes. These features are neglected in the Ideal Gas Law, but they become increasingly significant when real gas molecules are close together. And it is at conditions of low temperature and high pressure that we find gas molecules very close together.

Q: So, it is actually the weak attractive force between the particles in the gas state that helps to "pull" the particles even closer together during compression?

A: You are right! A lot of students thought that it is the applied pressure ALONE that helps to "push" the particles together, this is incorrect. Without the attractive force between the particles in the gaseous state, the particles would not be "held on" together in the liquid state. This also explains why gas with extremely weak attractive force is difficult to be liquefied through the using of high pressure.

Q: Let me try explain this again: In the beginning, without any applied pressure, the gas does not liquefy because the attractive force between the particles is simply too weak to pull them together. But with applied pressure, the particles are now closer together, hence, the attractive force between the particles are now stronger. And this helps to hold the particles together in the liquid state. Am I right?

A: Bingo! To add on, once you removed the applied pressure, the liquid would change back to the gas state at the same temperature that you have liquefied it.

Q: So, how would you explain that decreasing temperature would cause the gas to condense to the liquid state?

A: When you decrease the temperature of a gas, you are decreasing the kinetic energy of the particles. With lower kinetic energy, the particles move slower. Hence, the attractive force between the slower moving particles would be stronger as they have "more time or chances" to attract each other. This would help to bring about liquefaction.

3.2.2.1 *Changes in pressure*

There are no intermolecular forces between ideal gas particles; thus, one cannot liquefy an ideal gas. Hence, when a pressure is applied, the volume of a container decreases proportionately according to $pV = nRT$.

If we apply high pressure to a vessel containing nitrogen, intermolecular forces pull the molecules closer together than they would otherwise be if it were an ideal gas (Fig. 3.7).

Thus, the $\frac{pV}{RT}$ (or pV) value for a real gas would be lower than for an ideal gas, accounting for the portion of the graph in Fig. 3.6 that dips below the ideal plot.

If pressure is continuously exerted on a real gas, there is a point at which the gas particles are so close together that any further increase in pressure leads to a volume decrease that is less than that in an ideal gas (Fig. 3.8).

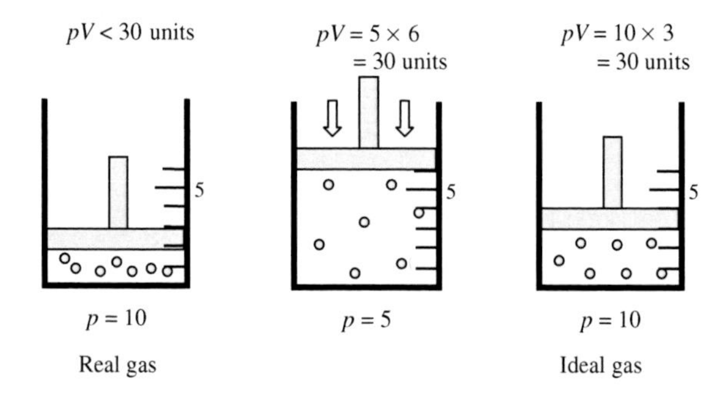

Fig. 3.7.

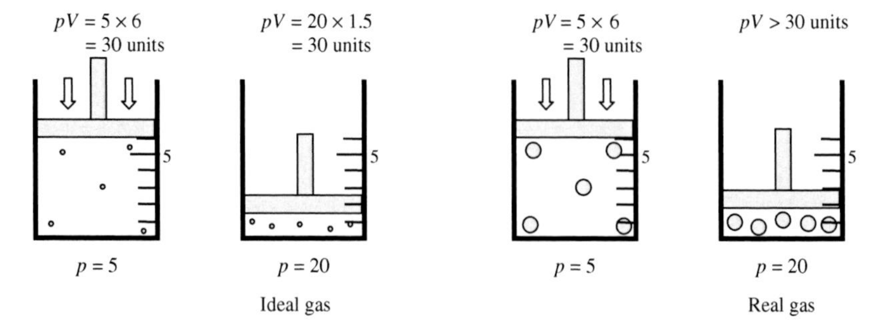

Fig. 3.8.

In this case, the $\frac{pV}{RT}$ (or pV) value for the real gas is higher than if it were an ideal gas, accounting for the portion of the graph in Fig. 3.6 that rises above the ideal plot. We say that the "size effect" is significant under high pressure conditions; in other words, the volume of the gas particles becomes significant.

3.2.2.2 *Changes in temperature*

When a gas is cooled to low temperatures, one expects the kinetic energy of the molecules to drop. Slower movement means that the gas particles are able to attract one another much more strongly. Now, when pressure is applied to this real gas at low temperature, there is a significant deviation from ideal behaviour, i.e., $pV < nRT$. The lower the temperature, the greater the deviation (Fig. 3.6).

If a real gas deviates greatly from ideal behaviour at high pressures and low temperatures, the reverse is true, that is, the real gas would approximate ideal behaviour at low pressures and high temperatures!

Example 3.12: Explain why real gases behave most like an ideal gas at low pressure and high temperature.

Solution: A real gas is expected to behave ideally at

- low pressure, and
- high temperature.

At low pressure the gas molecules are far apart, so the intermolecular forces of attraction between them are negligible. The volume occupied by the gas molecules can also be considered negligible as compared to the volume of the container.

At high temperature, the gas molecules have large enough kinetic energy to overcome the intermolecular forces of attraction between them and thus these can be considered negligible.

3.2.3 *Nature of gas*

Since real gases deviate from ideal behaviour because of the existence of attractive forces and the finite volumes that the molecules occupy,

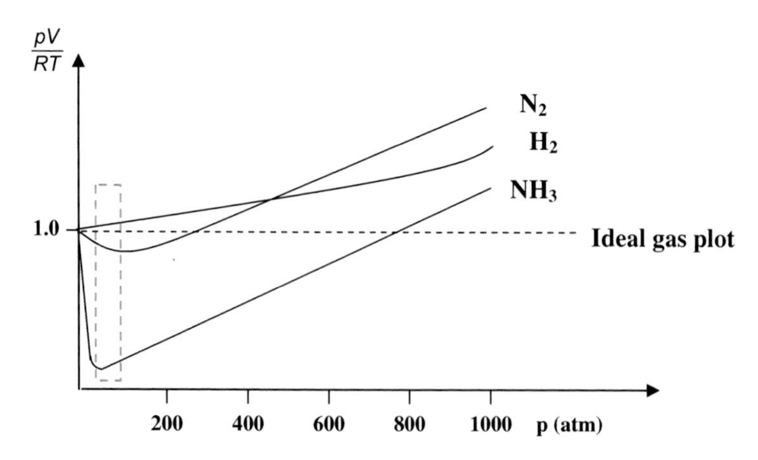

Fig. 3.9. Graphs of $\frac{pV}{RT}$ versus p plotted at 0°C for 1 mole of NH_3, H_2 and N_2.

we would expect greatest deviation for the real gas whose molecules (i) experience the greatest attractive forces, and (ii) are largest in size.

Figure 3.9 shows the plots of $\frac{pV}{RT}$ versus p for one mole each of H_2, N_2 and NH_3. We would expect these real gases to behave ideally at low pressures. Looking at the leftmost side of the x-axis (i.e., at low pressures) in Fig. 3.9, it is noted that deviation is greatest for NH_3, followed by N_2, and lastly H_2. This trend can be accounted for by two factors:

- type and magnitude of attractive forces, and
- size of the molecules.

3.2.3.1 *Type and magnitude of attractive forces*

Strong hydrogen bonding exists between the polar NH_3 molecules while weaker instantaneous dipole–induced dipole (id–id) interactions exist between the non-polar N_2 molecules, and likewise for the non-polar H_2 molecules. Thus, with stronger intermolecular forces of attraction between molecules of NH_3 than the other two gases, deviation from ideal behaviour is greatest for NH_3, i.e., $pV < nRT$ is most significant for NH_3 for a particular applied pressure.

3.2.3.2 *Size of molecules*

This effect is only prominent under very high pressure conditions because as the gas particles are being pushed closer together, the repulsive force between their electron clouds becomes more significant than the attractive force. Hence, this results in a volume decrease that is less than that for an ideal gas, i.e., $pV > nRT$. The deviation is actually attributed to the repulsion between the gaseous particles when they are close to one another.

In general, deviation is greatest for large, polar molecules with strong intermolecular forces of attraction and smallest for small, non-polar molecules. For example, in decreasing order of deviation: $Cl_2 > NH_3 > HCl > CO_2 > O_2 > N_2 > H_2 > He$.

Although Cl_2 is a non-polar molecule, its deviation from ideal behaviour is much greater than NH_3 because the id–id interactions between Cl_2 molecules are stronger than the hydrogen bonding between NH_3 molecules.

Example 3.13: List the gases He, Ne, O_2, H_2O, NH_3 and C_2H_6 in order of increasing deviation from ideal gas behaviour.

Solution: In order of increasing deviation: He, Ne, O_2, C_2H_6, NH_3, H_2O.

Explanation:

- H_2O deviates more from ideal behaviour than NH_3. The hydrogen bonding between gaseous H_2O molecules is more extensive than NH_3 (refer to Chap. 2).
- H_2O and NH_3 deviate more from ideal behaviour than the rest. Hydrogen bonding between molecules of H_2O and NH_3 is stronger than the id–id interactions between the non-polar molecules.
- Deviation increases in the order: $He < Ne < O_2 < C_2H_6$. The strength of the id–id interaction depends on the number of electrons the molecule has. The greater the number of electrons a molecule has, the more polarisable the electron cloud is, and hence the stronger the id–id interaction. Thus, with C_2H_6 having

the greatest number of electrons, the id–id interactions between its molecules are the strongest.

My Tutorial (Chapter 3)

1. (a) The critical point is a particular temperature and pressure condition whereby supercritical fluid exists. Supercritical fluid is a special substance which is neither solid, liquid nor gas, and carbon dioxide is one of the most widely used supercritical fluids. Calculate the volume of 1 mol of carbon dioxide at its critical point, assuming that it obeys the ideal gas equation.

$$(T_c = 304\,\text{K},\ p_c = 74 \times 10^5\,\text{N m}^{-2},\ R = 8.31\,\text{J K}^{-1}\,\text{mol}^{-1}.)$$

(b) The actual volume at the critical point of 1 mol of carbon dioxide is different from the value predicted by the ideal gas equation. Explain the given phenomenon.

(c) Van der Waals' equation $(p+a/V^2)(V-b) = RT$ can be used to account for the effect of gas imperfection.

(i) What physical meaning do the constants a and b have?

(ii) Explain how the constants would be different for gases such as hydrogen, nitrogen and ammonia.

(d) A sample of urine containing 0.120 g of urea, NH_2CONH_2, is treated with an excess of nitrous acid. The urea reacts according to the following equation:

$$NH_2CONH_2 + 2HNO_2 \longrightarrow CO_2 + 2N_2 + 3H_2O.$$

The gas produced is passed through aqueous potassium hydroxide and the final volume measured. What is this volume at room temperature and pressure?

2. (a) (i) List the main assumptions of the kinetic theory for ideal gases.

(ii) Suggest under what conditions real gases might deviate from ideality.

(b) Experimentally, it has been found that the rate at which a gas diffuses through a porous barrier at constant temperature and pressure is inversely proportional to the square root of its molar mass:

$$\text{Rate}\,\alpha(1/M_r)^{1/2}.$$

Carboxylic acids are known to dimerise in the gas phase. A sample of ethanoic acid vapour takes twice as long to diffuse through a porous barrier as the same amount of neon gas. Use this information to calculate:

(i) the apparent molar mass of the ethanoic acid vapour, and
(ii) the mole fraction of dimeric ethanoic acid in the vapour phase under these conditions.

(c) (i) Suggest how increasing the pressure and increasing the temperature will affect the rate of diffusion of neon.
(ii) Assuming that only a small amount of gas is allowed to diffuse, explain how increasing the pressure and temperature affects the rate of diffusion of the ethanoic acid vapour.

3. Gaseous xenon tetrafluoride, XeF_4, at a partial pressure of 2.0 kPa, and hydrogen, at a partial pressure of 10 kPa, are exploded in an enclosed container producing xenon and hydrogen fluoride.

(a) Write the balanced equation for the reaction between xenon tetrafluoride and hydrogen.
(b) Calculate, in terms of partial pressure, how much of hydrogen reacted with all the xenon tetrafluoride present.
(c) Hence, calculate the total pressure of the mixture of gases in the enclosed container, assuming temperature remains constant.
(d) Suggest reasons why fluorides of xenon have been synthesised whereas fluorides of the other noble gases, helium, neon, argon, are relatively uncommon.

4. A plot of ρ/p against pressure p at a given temperature for an unknown gas X is shown below.

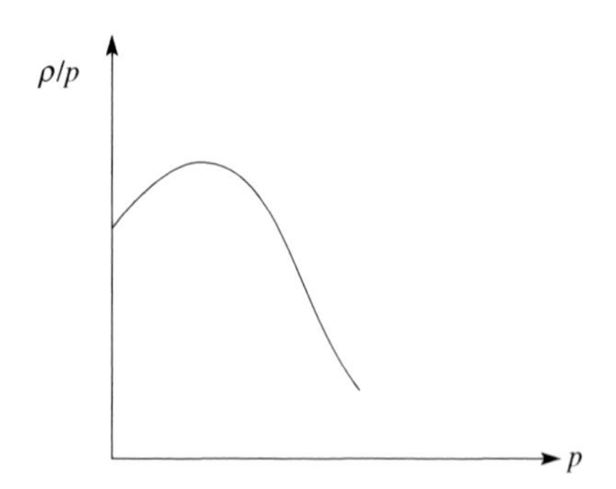

(a) Is the gas behaving ideally?
(b) Give an explanation for the graph obtained.
(c) Using the axes provided above, give a sketch of ρ/p against pressure p for gas X under a higher temperature.
(d) Using the axes provided above, give a sketch of ρ/p against pressure p for gas Y, which has a higher boiling point than X.

5. A sample of pure organic liquid W of mass 0.146 g is vapourised in a gas syringe at 127°C and occupies a volume of 100 cm^3 at a pressure of 101 kPa.

   (a) Calculate the relative molecular mass of W from the experimental data.
   (b) The accurate composition by mass of W is C: 52.2%; H: 13%; O: 34.8%. Determine the empirical formula of W.
   (c) What are the likely formula and relative molecular mass of W based on part (b)?
   (d) Suggest two possible reasons for the difference between the values of the relative molecular masses in parts (a) and (c).

CHAPTER 4

Chemical Thermodynamics

Q: What is "thermodynamics"?

A: Thermodynamics is the study of the interaction between matter and energy. It is concerned principally with **energy changes** and the **flow of energy** from one substance to another.

Q: Why must we know about thermodynamics?

A: It can be used to **predict the behaviours of chemical systems** — to determine whether or not a change is possible.

Energy is defined as the ability to do work. There are two fundamental types of energy that particles in matter possess:

- Kinetic energy (K.E.): the energy of motion arising from rotations, vibrations and translational movements of particles. The temperature of a system is a measure of the average kinetic energy of that system.
- Potential energy (P.E.): the energy that particles store within the electrostatic attractions or repulsions that they experience with one another.

According to the First Law of Thermodynamics, energy can neither be created nor destroyed; it can only be transferred. The ways to change K.E. and P.E. from one form to another are through heat or work. Energetics is thus the study of heat changes that occur during chemical reactions and phase changes.

Since many reactions are carried out under constant pressure, the energy content of a substance at **constant pressure** is termed its enthalpy ($\boldsymbol{H}$). However, it is not possible to measure the enthalpies of all particles in a system. Instead, a standard set of conditions is fixed wherein elements, taken to be in their most stable form at 298 K and 1 bar, have zero enthalpy. With this standard, we can compare enthalpies, in terms of enthalpy changes ($\Delta \boldsymbol{H}$) measured as the **flow of heat** in and/or out of a system. The units for $\Delta \boldsymbol{H}$ are J mol^{-1} or kJ mol^{-1}.

As the enthalpy of a substance is a measure of its energy content, it also **reflects its stability**.

Q: Why is the pressure kept constant when we define the energy content?

A: For chemists, since most of our reactions happen under conditions of constant atmospheric pressure, it is logical to define the energy content (H) by keeping the pressure constant. There are other energy contents of a system defined by keeping other conditions constant. For example, under conditions of constant volume, we define the energy content as Internal Energy, whereas under conditions of constant temperature and pressure, we term it the Gibbs Free Energy.

Q: What is meant by the "enthalpy of a substance reflects its stability"?

A: If ΔH is negative, it means that the energy content of the products is lower than that of the reactants; and the lower the energy content, the more stable the substance is, and *vice versa*.

Q: Why is a substance, with higher energy content, less stable?

A: Recall that the energy content of the particles in a substance is in the form of kinetic energy and potential energy. If the particles have higher kinetic energy, it means that they are moving faster and electrostatic attractive forces may not be able to hold the particles together. The particles may just simply "fly away" — just like in melting and boiling. Likewise, if there is a larger amount of potential energy stored, this means that the particles are further apart. Recall in Chap. 1 on Atomic Structure that we discussed that if two particles are an infinite distance apart, the electrostatic attractive force between them is negligible. So would you expect such a system to be stable?

4.1 Energy Changes in Chemical Reactions

A chemical reaction involves the rearrangement of particles (atoms, ions or molecules). During this process, old bonds are broken and new bonds are formed. The **breaking of bonds requires energy (an endothermic process)** and the **formation of bonds evolves energy (an exothermic process)**.

The difference between the quantity of heat needed to break the bonds in the reactants and that evolved during the formation of new bonds in the products, under the condition of constant pressure, gives rise to the heat change (or enthalpy change) of a reaction.

Q: Where does the energy required to break the bonds come from and where does it go?

A: Good question. The source of energy required for bond breaking is in the form of K.E. and this K.E. is transformed into the P.E. stored in the bond. What happens when the K.E. is large? The particles vibrate very fast, which means that it can increase the distance of separation. This resultant increase in bond length means that the P.E., which is being stored within the bond, has increased. The reverse is true when a bond forms: P.E. is released as K.E.. So remember that the energy required to break a bond is stored as P.E. in the bond.

Energy level diagrams help to illustrate the relative energies of the reactants and products, and hence, the enthalpy change of a reaction.

For an exothermic reaction, heat is lost from the system to the surroundings. The energy content of the reactants is higher than that of the products, as depicted by the negative value of ΔH. Since the products have a lower energy content, they are said to be more energetically stable compared to the reactants. Examples of exothermic reactions include the combustion of fuels and the neutralisation reaction between strong acids and alkalis.

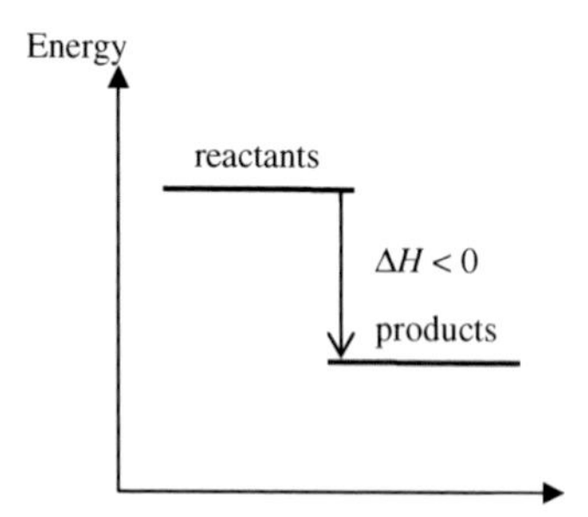

For an endothermic reaction, heat is absorbed by the system from the surroundings. The energy content of the reactants is lower than that of the products, as depicted by the positive value of ΔH. The products are less energetically stable compared to the reactants.

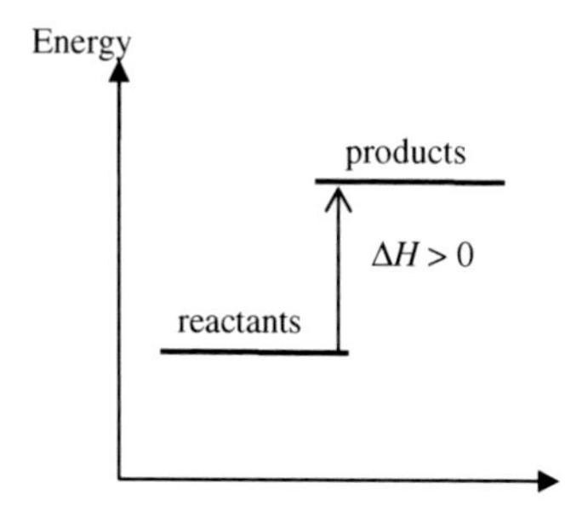

Examples of endothermic reactions include photosynthesis and thermal decomposition.

The magnitude and sign of ΔH provides us with important information. Other than informing us of the *stability of the product relative to the reactant*, the ΔH value also reflects the *strength of bonds in the product relative to the reactant.* A negative ΔH (exothermic) indicates that the bonds in the product are stronger than those in the reactant, and *vice versa.*

Q: If an exothermic reaction gives rise to products that are more stable than reactants, do all exothermic reactions possibly occur?

A: No. Many reactions are energetically feasible (exothermic) but they occur only very slowly due to the high activation energy involved — the energy barrier that needs to be surmounted before successful reactions can occur. Such reactions are said to be kinetically non-feasible. (More details are discussed in Chap. 5 on Reaction Kinetics.)

An example of a kinetically non-feasible reaction is the conversion of diamond to graphite:

$$\mathrm{C\,(diamond)} \longrightarrow \mathrm{C\,(graphite)}, \quad \Delta H = -1.83\,\mathrm{kJ\,mol^{-1}}.$$

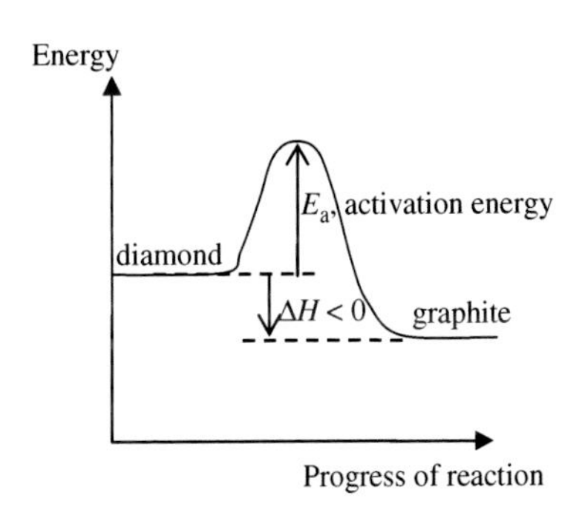

Diamond is energetically less stable than graphite, but it is kinetically stable — hence the expression, "Diamonds are forever".

On the other hand, petrol is energetically unstable, in terms of its reaction with oxygen in air, but it is kinetically stable until the change is triggered by a spark or a flame.

Thus, the enthalpy change of a reaction is an indication of the energetic (thermodynamic) stability, not the kinetic stability. It says nothing about the rate of a reaction.

Q: So if thermodynamics regards a reaction or change as possible, it may not be observed?

A: Yes, the reaction must occur fast enough to be observed and this would be where the study of kinetics comes in.

4.1.1 *Standard enthalpy changes*

This section covers common enthalpy terms and their definitions that you need to be well-versed in to aid your understanding of this chapter. For comparison of enthalpy changes, the standard conditions of

1 bar and 298 K are adhered to. The superscript ($^{\ominus}$) is used to denote standard conditions.

- **Standard enthalpy change of reaction ($\Delta H_{rxn}{}^{\ominus}$)**

This is the energy change when molar quantities of reactants, as specified by a chemical equation, react to form products under standard conditions. For example,

$$N_2(g) + 3H_2(g) \longrightarrow 2NH_3(g), \quad \Delta H_{rxn}{}^{\ominus} = -100\,kJ\,mol^{-1}. \quad (4.1)$$

The negative sign of ΔH indicates that heat is evolved from the reaction. More specifically, 100 kJ of heat is evolved when 1 mol of N_2 gas reacts with 3 mol of H_2 gas to form 2 mol of NH_3 gas. The units "kJ mol^{-1}" are used to refer to "per mole of stated equation of specific stoichiometry."

Example 4.1: Determine the value of $\Delta H_{rxn}{}^{\ominus}$ for the following reaction:

$$\tfrac{1}{2}N_2(g) + \tfrac{3}{2}H_2(g) \longrightarrow NH_3(g). \quad (4.2)$$

Solution: $\Delta H_{rxn}{}^{\ominus}$ for the reaction is –50 kJ mol^{-1}. This is because the molar quantities in Eq. (4.2) are actually half of those in Eq. (4.1). And the unit is still kJ mol^{-1}. Hence, it is important to state the appropriate equation for a particular ΔH value quoted.

- **Standard enthalpy change of formation ($\Delta H_f{}^{\ominus}$)**

This is the energy change when **one mole of a pure compound** in a specified state is formed from its **constituent elements** in their naturally occurring states under standard conditions. For example,

$$H_2(g) + \frac{1}{2}O_2(g) \longrightarrow H_2O(l),$$
$$C(s) + 2H_2(g) \longrightarrow CH_4(g).$$

Note that the value of $\Delta H_f{}^{\ominus}$ for elements in their naturally occurring states at 298 K and 1 bar is **zero**, i.e.,

$$\Delta H_f{}^{\ominus}[H_2(g)] = 0, \quad \Delta H_f{}^{\ominus}[C(s)] = 0.$$

Q: Can we call this enthalpy change of reaction?

A: Yes, you can, but then it would no longer be specific. If you term it *enthalpy change of formation of water*, you can directly formulate the reaction equation from its definition, which would then be helpful in the formulation of the energy cycle. This is why you need to know the definitions of each kind of enthalpy change.

- **Standard enthalpy change of combustion ($\Delta H_c^{\ominus}$)**

This is the **energy evolved** when **one mole of a substance** is **completely burned in oxygen** under standard conditions. It is an exothermic quantity. For example,

$$CH_4(g) + 2O_2(g) \longrightarrow CO_2(g) + 2H_2O(l), \quad \Delta H_c^{\ominus} = -866\,\mathrm{kJ\,mol^{-1}},$$

$$Mg(s) + \tfrac{1}{2}O_2(g) \longrightarrow MgO(s), \quad \Delta H_c^{\ominus} = -601\,\mathrm{kJ\,mol^{-1}}.$$

- **Standard enthalpy change of neutralisation ($\Delta H_{neut}^{\ominus}$)**

This is the energy change when an acid and alkali are neutralised to form **one mole of water** (in a **dilute** aqueous solution) under standard conditions.

Q: Why is the term "energy change" used here? Why not "evolve" or "absorb"?

A: $\Delta H_{neut}^{\ominus}$ can be either endothermic or exothermic, depending on the reaction system in consideration. This is because the acid or base involved in a neutralisation reaction can be either strong or weak. Strong acid or strong base refers to acid or base that have fully dissociated, whereas weak acid or base refers to them being partially dissociated. (More on this is covered in Chap. 7 on Ionic Equilibria.)

- For strong acid–strong base neutralisation, e.g., HCl(aq) reacting with NaOH(aq):
  - There is complete ionisation of the substances in water, forming ions.

- The neutralisation reaction can be given as:

$$H^+(aq) + OH^-(aq) \longrightarrow H_2O(l), \quad \Delta H_{neut}^{\ominus} = -57.0\,kJ\,mol^{-1}.$$

- For neutralisation involving a weak acid or a weak base, or both:
  - Weak acid and weak alkali undergo partial ionisation in water.
  - As ionisation is an endothermic process, energy is required for the complete ionisation of such weak acids and alkalis.
  - Thus, the value of $\Delta H_{neut}^{\ominus}$ in these cases is less exothermic than that for a strong acid — strong base neutralisation.

  For example,

$$CH_3COOH(aq) + NaOH(aq) \longrightarrow CH_3COO^-Na^+(aq) + H_2O(l),$$
$$\Delta H_{neut}^{\ominus} = -55.2\,kJ\,mol^{-1}.$$

- $\Delta H_{neut}^{\ominus}$ for a weak acid–weak base neutralisation is the least exothermic as energy has to be used to bring about complete ionisation of both substances. For example,

$$CH_3COOH(aq) + NH_3(aq) \longrightarrow CH_3COO^-NH_4^+(aq),$$
$$\Delta H_{neut}^{\ominus} = -50.4\,kJ\,mol^{-1}.$$

Q: Why do different acid molecules have different degrees of ionisation?

A: It depends on the strength of the bond that one needs to break. Breaking a H–Cl bond requires a different amount of energy to that needed to break a O–H bond in CH_3COOH. Knowing the chemical bonding topic well will help your understanding.

Q: Why is the term "dilute" emphasised?

A: In Chap. 7 on Ionic Equilibria, you will learn that the concentration of an acid or base affects its degree of dissociation. Thus, it is important to realise that all these chemistry topics are interrelated. So do not learn one topic and forget the rest! Integrate them and you will be able to see a more fuller picture.

- **Standard enthalpy change of atomisation ($\Delta H_{\text{at}}{}^{\ominus}$) of an element**

This is the **energy absorbed** to form **one mole of gaseous atoms** from the element in the defined physical state under standard conditions. This is an endothermic quantity since it involves bond breaking.

For example,

$$\tfrac{1}{2}\text{H}_2(\text{g}) \longrightarrow \text{H}(\text{g}), \quad \Delta H_{\text{at}}{}^{\ominus} = +218\,\text{kJ mol}^{-1},$$

$$\text{Mg}(\text{s}) \longrightarrow \text{Mg}(\text{g}), \quad \Delta H_{\text{at}}{}^{\ominus} = +148\,\text{kJ mol}^{-1}.$$

- **Standard enthalpy change of atomisation ($\Delta H_{\text{at}}{}^{\ominus}$) of a compound**

This is the **energy absorbed** when **one mole of a compound** in the defined physical state under standard conditions is converted into its **constituent gaseous atoms**. The higher the $\Delta H_{\text{at}}{}^{\ominus}$ value, the stronger the bond strength. For example,

$$\text{CH}_4(\text{g}) \longrightarrow \text{C}(\text{g}) + 4\text{H}(\text{g}),$$

$$\text{H}_2\text{O}(\text{l}) \longrightarrow 2\text{H}(\text{g}) + \text{O}(\text{g}).$$

One should always start off with the naturally occurring state of the compound at room temperature.

- **Bond energy** (BE) of an X–Y bond is the **average energy absorbed** when **one mole of X–Y bonds** are broken in the **gas phase** under standard conditions. This is an endothermic quantity since it involves bond breaking. The stronger the bond, the more endothermic the bond energy is. For example,

$$\text{Cl}_2(\text{g}) \longrightarrow 2\text{Cl}(\text{g}), \quad \text{BE}(\text{Cl–Cl}) = +244\,\text{kJ mol}^{-1}.$$

It is useful to bear in mind that some of the enthalpy change terms may overlap with one another. For example,

$$\tfrac{1}{2}\text{Cl}_2(\text{g}) \longrightarrow \text{Cl}(\text{g}), \quad \Delta H_{\text{at}}{}^{\ominus} = +122\,\text{kJ mol}^{-1},$$

$$\text{Cl}_2(\text{g}) \longrightarrow 2\text{Cl}(\text{g}), \quad 2 \times \Delta H_{\text{at}}{}^{\ominus} = +244\,\text{kJ mol}^{-1}.$$

Therefore,

$$\text{BE(Cl–Cl)} = 2\Delta H_{\text{at}}^{\ominus}[\text{Cl}_2(\text{g})].$$

In another example, the enthalpy term for the reaction $Mg(s) + \frac{1}{2}O_2(g) \longrightarrow MgO(s)$ may be represented by either $\Delta H_c^{\ominus}$ or $\Delta H_f^{\ominus}$. It is therefore important to be able to translate from the definition of an enthalpy term to an equation, and *vice versa*.

- **Bond dissociation energy** (BDE) of an X–Y bond is the **energy required** to **break one mole** of that **particular** X–Y bond in a **particular compound** in the **gas phase** under standard conditions.

For example, the BDE of C–Cl differs for each of the chlorine-containing molecules:

	CH_3Cl (H–C(H)(H)–Cl)	CCl_4 (Cl–C(Cl)(Cl)–Cl)	CH_3CH_2Cl (H_3C–C(H)(H)–Cl)
BDE(C–Cl)	$335\,\text{kJ mol}^{-1}$	$327\,\text{kJ mol}^{-1}$	$342\,\text{kJ mol}^{-1}$

The **strength of a bond is influenced by neighbouring atoms**. For example, if values for the C–Cl bond are obtained for many compounds, an average bond energy for the C–Cl bond can be obtained. Thus, bond energy is an *average value*, whereas bond dissociation energy is more specific to the compound under consideration.

Formulae in the equations must be accompanied by state symbols. As can be seen from Eqs. (4.3) and (4.4), there are differences in the $\Delta H^{\ominus}$ values for the two reactions. Less energy is released in the second reaction as a portion of it is used for the vapourisation of $H_2O(l)$ to $H_2O(g)$. Energy changes are also associated with phase changes and therefore we have to be cautious about the state of the substances.

$$H_2(g) + \tfrac{1}{2}O_2(g) \longrightarrow H_2O(l), \quad \Delta H^{\ominus} = -285.8\,\text{kJ mol}^{-1}, \tag{4.3}$$

$$H_2(g) + \tfrac{1}{2}O_2(g) \longrightarrow H_2O(g), \quad \Delta H^{\ominus} = -241.8\,\text{kJ mol}^{-1}. \tag{4.4}$$

4.1.2 *Calculating enthalpy changes from experimental data*

Let us suppose for instance that we want to determine the enthalpy change of neutralisation for the reaction between NaOH(aq) and H_2SO_4(aq):

$$2NaOH(aq) + H_2SO_4(aq) \longrightarrow Na_2SO_4(aq) + 2H_2O(l).$$

A typical school experimental set-up is shown below.

Procedure: A measured volume of NaOH(aq) of known concentration is placed in an insulated polystyrene cup. The initial temperature (T_i) of the solution is recorded before the start of the reaction. At a suitable time interval, a measured volume of H_2SO_4(aq) of known concentration is added to the cup. Continuous stirring is performed while the temperature of the resultant solution is recorded at regular time intervals.

Time, t (min)	0	1	2	3	4	5	6	7	8
Temperature (°C)									

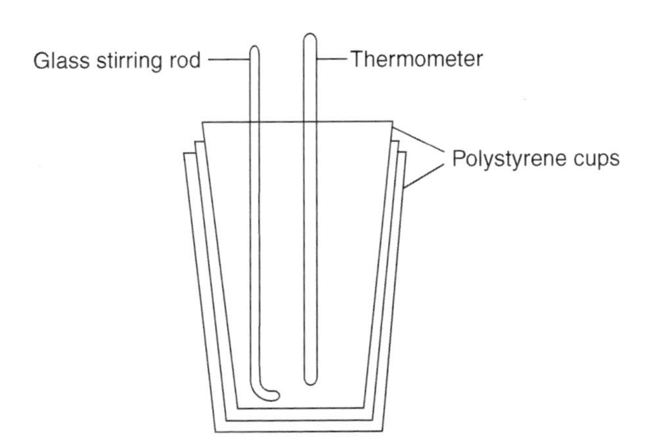

Treatment of results: With the data obtained, a temperature–time graph is plotted.

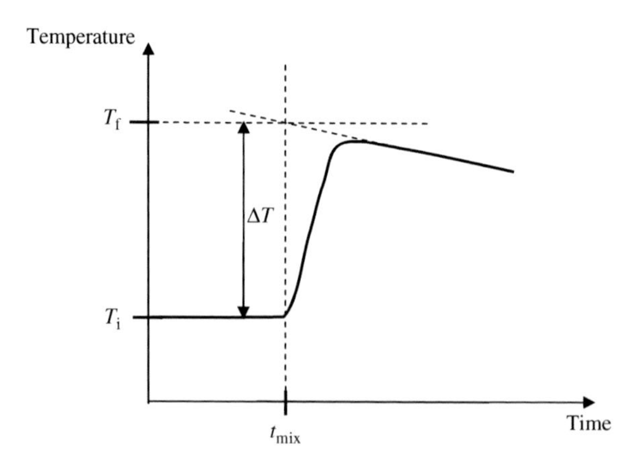

The maximum temperature change occurs at t_{mix} when the reactants are first mixed. A good estimate of the temperature (T_f) at t_{mix} is done by extrapolation. In so doing, it is assumed that heat loss to the surroundings has been accounted for. This is possible because the value of T_f corresponds to the heat that is being released at one go when all the reactants react at that specific moment in time.

Calculations: The heat change of the reaction (Q) is calculated using the formula:

$$\text{Heat absorbed/released} = mc\Delta T,$$

where m is mass of solution (g), c is the specific heat capacity of the solution (J g^{-1} K^{-1}), and $\Delta T = T_f - T_i$(K).

Since we usually measure volume for solutions rather than their mass, the formula can be rewritten to include a density term:

$$\text{Heat absorbed/released} = \boldsymbol{\rho} Vc\Delta T,$$

where $\boldsymbol{\rho}$ is the density of the solution (g cm^{-3}), and V is the volume of the solution (cm^3).

A simplification of the calculations includes the use of the following assumptions:

- The density of the solution is that of pure water, 1.00 g cm^{-3}.
- The specific heat capacity of the solution is that of water, 4.18 J g^{-1} K^{-1}.

Q: What is the specific heat capacity of a solution?

A: **Specific heat capacity** is the quantity of heat required to raise the temperature of a unit mass of the substance by 1 K. A similar term is the **heat capacity** of a mass of substance, which is the quantity of heat required to raise its temperature by 1 K.

Q: Why approximate both the density and specific heat capacity of the solution to that of pure water?

A: When the reaction happens, the majority of the heat energy that is released is transferred to the surrounding water molecules in the form of K.E. This is because the water molecules are present in huge amounts compared to the actual amount of reactants involved in the reaction. Thus, it is logical to approximate both the density and specific heat capacity of the solution to that of pure water.

After the heat change of reaction is obtained, ΔH_{rxn} is determined using the formula:

$$\Delta H_{\mathrm{rxn}} = +\frac{Heat\ absorbed}{Amount\ of\ limiting\ reagent};$$

$$\Delta H_{\mathrm{rxn}} = -\frac{Heat\ evolved}{Amount\ of\ limiting\ reagent}.$$

In summary, the set of calculations involve:

- determining the amount of limiting reagent
- calculating the amount of heat absorbed/evolved, and
- calculating the enthalpy change of the reaction.

Example 4.2: 40 cm^3 of 0.4 mol dm^{-3} NaOH(aq) is placed in an insulated polystyrene cup. The initial temperature recorded is 28.0°C. 40 cm^3 of 0.2 mol dm^{-3} H_2SO_4(aq) is then added. Upon mixing, the temperature rises to 30.7°C. Calculate the enthalpy change of neutralisation for this reaction.

Solution: Assume that:

- the density of solution is that of water, 1.00 g cm^{-3}, and

- the specific heat capacity of the solution is that of water, $4.18\,\text{J}\,\text{g}^{-1}\,\text{K}^{-1}$.

Total volume of resultant solution $= 40 + 40 = 80\,\text{cm}^3$.
Heat evolved $= mc\Delta T = \rho V c(T_\text{f} - T_\text{i}) = 1 \times 80 \times 4.18 \times (30.7 - 28.0) = 902.9\,\text{J}$.

In this example, the amount of limiting reagent is not calculated but rather the amount of H_2O is found. This is in accordance with the definition of ΔH_neut, which specifies the enthalpy change involved in the formation of one mole of water:

$$\Delta H_\text{neut} = -\frac{\textit{Heat evolved}}{\textit{Amount of water formed}}.$$

Q: Why must there be a negative sign in the above formula?
A: Since heat is evolved from the reaction, ΔHvalues for such exothermic reactions are negative values, an indication that the products have lower energy than the reactants (refer to Section 4.1).

$$2\text{NaOH(aq)} + \text{H}_2\text{SO}_4\text{(aq)} \longrightarrow \text{Na}_2\text{SO}_4\text{(aq)} + 2\text{H}_2\text{O(l)}.$$

Amount of H_2O formed = amount of NaOH reacted $= \frac{40}{1000} \times 0.4 = 0.016\,\text{mol}$.

Hence, $\Delta H_\text{neut} = -\frac{902.9}{0.016} = -56.4\,\text{kJ mol}^{-1}$.

Example 4.3: 2.00 g of powdered zinc is added to $50\,\text{cm}^3$ of $0.2\,\text{mol}\,\text{dm}^{-3}$ copper(II) nitrate solution in an insulated polystyrene cup. The maximum temperature rise recorded is 10.6°C. Calculate the enthalpy change for the reaction:

$$\text{Zn(s)} + \text{Cu}^{2+}\text{(aq)} \longrightarrow \text{Cu(s)} + \text{Zn}^{2+}\text{(aq)}.$$

Approach:

- Determine the amount of limiting reagent.
- Calculate the amount of heat absorbed/evolved.
- Calculate the enthalpy change of the reaction.

Solution: Amount of $Cu^{2+}(aq) = \frac{50}{1000} \times 0.2 = 0.01$ mol.
Molar mass of Zn $= 65.4\ \text{g mol}^{-1}$.
Amount of Zn(s) $= \frac{2}{65.4} = 0.0306$ mol.
Hence, the limiting reagent is $Cu^{2+}(aq)$.
Heat evolved $= 50 \times 4.18 \times 10.6 = 2215.4$ J.
Hence, $\Delta H_{\text{neut}} = \frac{-2215.4}{0.01} = -222\ \text{kJ mol}^{-1}$.

Q: Why was the mass of Zn not taken into account when calculating the heat change of the reaction?

A: The reaction between Zn and Cu^{2+} is exothermic. Heat produced from the reaction has to go somewhere. It is absorbed by the surrounding water molecules in the form of K.E., as it is present in large quantity. This translates to the temperature rise we are measuring.

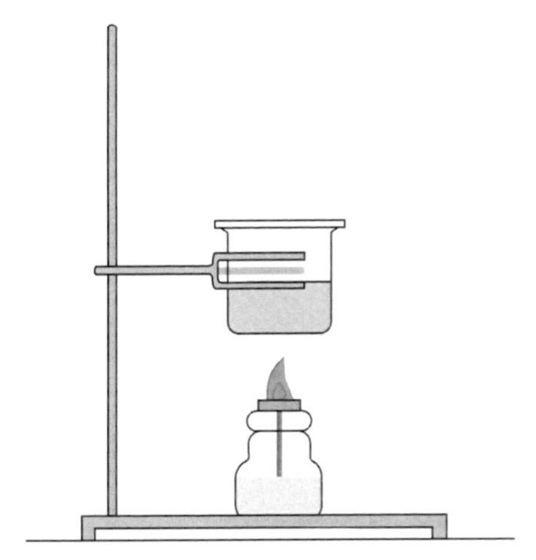

Another common school experiment involves heating water using the heat released from the combustion of an alcohol. Since the setup is not insulated, not all the heat released from the burning process is absorbed by the beaker of water.

Example 4.4: 5 g of hexanol is burnt to heat a beaker containing x cm^3 of H_2O. The temperature rise of the water is 62°C. Assume that 40% of the heat evolved from the combustion of hexanol is absorbed by the water. Determine the value of x, given that the enthalpy change of combustion of hexanol is $-3984\ \text{kJ mol}^{-1}$.

Approach:

(i) Calculate the amount of hexanol used.
(ii) Based on the given ΔH_c of hexanol, find the actual total amount of heat evolved from combustion.
(iii) Calculate, based on the conversion efficiency, the amount of heat absorbed by x cm^3 water.
(iv) Apply $Q = mc\Delta T$ to solve for m and hence x.

Solution: Molar mass of hexanol, $C_6H_{14}O = 102\,\text{g}\,\text{mol}^{-1}$.
Amount of hexanol $= \frac{5}{102} = 0.049\,\text{mol}$.
Heat evolved from the combustion of hexanol $= 3984 \times 0.049 = 1.953 \times 10^5\,\text{J}$.
Heat absorbed by water $= \frac{40}{100} \times 1.953 \times 10^5 = 7.812 \times 10^4\,\text{J}$.
Heat absorbed by water $= mc\Delta T = m \times 4.18 \times 62 = 7.812 \times 10^4\,\text{J}$.
Hence, $m = 301.4\,\text{g}$.
Since the density of water is $1\,\text{g cm}^{-3}$, the value of x is $301.4\,\text{cm}^3$.

4.1.3 *Calculation of enthalpy changes using Hess' Law*

Not all types of enthalpy changes of reactions can be determined directly from experimental data. Nonetheless, we can determine these enthalpy changes indirectly from other enthalpy changes using Hess' Law, which is derived from the Law of Conservation of Energy.

Hess' Law states that the enthalpy change of a reaction depends only on the initial and final states of the system and is independent of the reaction pathway taken.

Given the reaction: $A + B \longrightarrow C + D$, the following energy cycle illustrates Hess' Law:

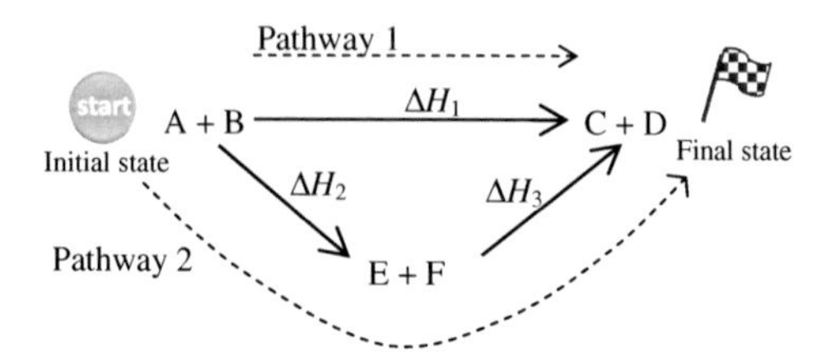

There are two ways to go from "Initial state" to "Final state": either via Pathway 1 (clockwise direction) or Pathway 2 (anticlockwise direction). Regardless of which pathway is chosen, the enthalpy change of the reaction (ΔH_{rxn}) is a fixed value.

By Hess' Law, $\Delta H_{\text{rxn}} = \Delta H_1 = \Delta H_2 + \Delta H_3$.

If we are interested in finding the enthalpy change for reaction 3: $\text{E} + \text{F} \longrightarrow \text{C} + \text{D}$, we can simply use the above energy cycle and apply Hess' Law:

By Hess' Law, ΔH_{rxn} for reaction 3 $= \Delta H_3 = -\Delta H_2 + \Delta H_1$.

As you notice, in computing ΔH_3, there is a negative sign attached to ΔH_2. This is in alignment with Pathway 2 in the energy cycle below; each ΔH term is attached to an arrow that represents the progress of the reaction from reactants to products. If a pathway runs in the opposite direction to these arrows, this indicates the backward reaction, which has a reverse ΔH term.

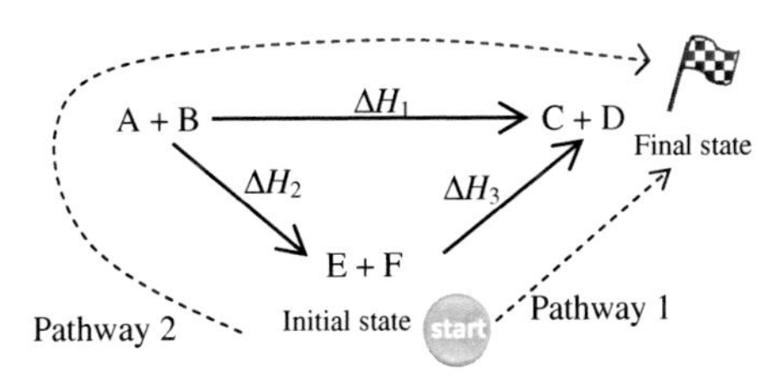

Q: Can I calculate ΔH_{rxn} for the reaction $\text{C} + \text{D} \longrightarrow \text{A} + \text{B}$ by just using "C+D" as the initial state and "A+B" as the final state?

A: Yes! It does not matter where you start off in an energy cycle. Just remember that when a reaction direction is reversed, the sign on the ΔH term also has to be reversed.

The calculated $\Delta H_{\text{rxn}} = -\Delta H_1 = -\Delta H_3 - \Delta H_2$.

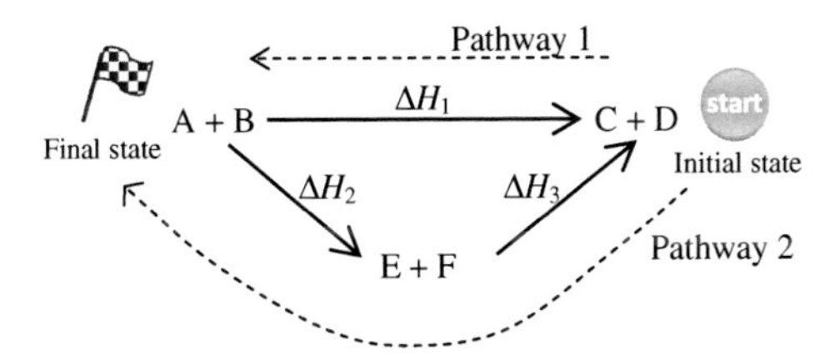

Now that we have seen the versatility in Hess' Law, the next thing to learn is the construction of energy cycles.

4.1.4 *Constructing energy cycles*

Example 4.5: In this section, we will work on finding the standard enthalpy change of combustion of propane, $\Delta H_c^\ominus[C_3H_8(g)]$, given the following data:

$$\Delta H_f^\ominus[C_3H_8(g)] = -104\,\text{kJ mol}^{-1},$$
$$\Delta H_f^\ominus[CO_2(g)] = -394\,\text{kJ mol}^{-1},$$
$$\Delta H_f^\ominus[H_2O(l)] = -286\,\text{kJ mol}^{-1}.$$

Approach:

(i) In order to construct the energy cycle, we need to first write out the balanced equations that illustrate the enthalpy terms given. Remember to input all state symbols.

Enthalpy Term	**Corresponding Balanced Chemical Equation**
$\Delta H_c^\ominus[C_3H_8(g)]$	$C_3H_8(g) + 5O_2(g) \longrightarrow 3CO_2(g) + 4H_2O(l)$
$\Delta H_f^\ominus[C_3H_8(g)]$	$3C(s) + 4H_2(g) \longrightarrow C_3H_8(g)$
$\Delta H_f^\ominus[CO_2(g)]$	$C(s) + O_2(g) \longrightarrow CO_2(g)$
$\Delta H_f^\ominus[H_2O(l)]$	$H_2(g) + \frac{1}{2}O_2(g) \longrightarrow H_2O(l)$

(ii) (a) Place, in the centre, the equation linked to the enthalpy term that is to be determined.

(b) Next, try to connect the remaining equations together by locating **common parts** to the main equation. Write down the corresponding ΔH terms on the arrows.

Q: I managed to come up with the energy cycle below. How do I link the remaining two equations that contain O_2 which is missing in my diagram?

$C_3H_8(g) + 5O_2(g) \xrightarrow{\Delta H_c^\ominus[C_3H_8(g)]} 3CO_2(g) + 4H_2O(l)$

$\Delta H_f^\ominus[C_3H_8(g)]$

$3C(s) + 4H_2(g)$

A: Just add the required number of moles of O_2 as shown below:

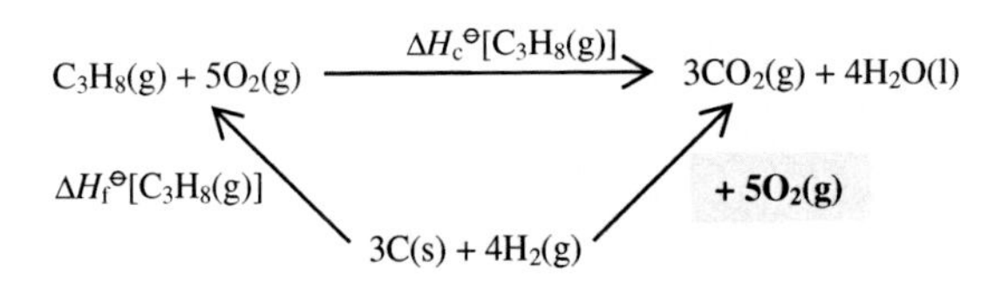

The final connected components in the energy cycle is actually the sum of Eqs. (4.1) and (4.2):

$$C(s) + O_2(g) \longrightarrow CO_2(g) \qquad (4.1) \times \mathbf{3}$$

$$H_2(g) + \tfrac{1}{2}O_2(g) \longrightarrow H_2O(l) \qquad (4.2) \times \mathbf{4}$$

Overall equation: $3C(s) + 4H_2(g) + 5O_2(g) \longrightarrow 3CO_2(g) + 4H_2O(l)$

Up to this point, the energy cycle should look like this:

$C_3H_8(g) + 5O_2(g) \xrightarrow{\Delta H_c^\ominus[C_3H_8(g)]} 3CO_2(g) + 4H_2O(l)$

$\Delta H_f^\ominus[C_3H_8(g)]$ $\quad + 5O_2(g)$ $\quad \Delta H_f^\ominus[CO_2(g)] + \Delta H_f^\ominus[H_2O(l)]$

$3C(s) + 4H_2(g)$

(iii) Balance the equations with respect to one another in the energy cycle. If the mole ratio of a reaction needs to be multiplied, the enthalpy value has to be multiplied by the same amount. This will ensure that the equations correctly represent the enthalpy terms involved.

$C_3H_8(g) + 5O_2(g) \xrightarrow{\Delta H_c^\ominus[C_3H_8(g)]} 3CO_2(g) + 4H_2O(l)$

$\Delta H_f^\ominus[C_3H_8(g)]$ $\quad + 5O_2(g)$ $\quad \mathbf{3\times}\Delta H_f^\ominus[CO_2(g)] + \mathbf{4\times}\Delta H_f^\ominus[H_2O(l)]$

$3C(s) + 4H_2(g)$

(iv) Determine the two pathways from "Initial state" to "Final state" and apply Hess' Law in your calculations.

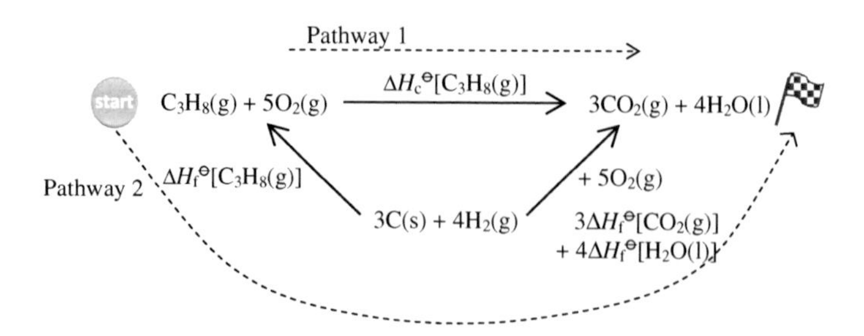

By Hess' Law,

$$\begin{aligned}\Delta H_c^\ominus[C_3H_8(g)] &= -\Delta H_f^\ominus[C_3H_8(g)] + 3\Delta H_f^\ominus[CO_2(g)] + 4\Delta H_f^\ominus[H_2O(l)]\\ &= -(-104) + 3(-394) + 4(-286)\\ &= -2222\,\text{kJ mol}^{-1}.\end{aligned}$$

Enthalpy terms are always accompanied by a sign (+ or −). In this case, the negative value of $\Delta H_c^\ominus[C_3H_8(g)]$ serves as a check on the calculations since the combustion reaction is an exothermic reaction.

Example 4.6: Determine the standard enthalpy change of formation of propane, $\Delta H_f^\ominus[C_3H_8(g)]$, given the following data:

$$\begin{aligned}\text{BE(C–C)} &= +350\,\text{kJ mol}^{-1},\\ \text{BE(C–H)} &= +412\,\text{kJ mol}^{-1},\\ \text{BE(H–H)} &= +436\,\text{kJ mol}^{-1},\\ \Delta H_{at}^\ominus[C(s)] &= +715\,\text{kJ mol}^{-1}.\end{aligned}$$

Approach: Referring to the definitions for both bond energies (BE) and enthalpy change of atomisation (ΔH_{at}), we are dealing basically with gaseous atoms as the end products. Thus, for every species in the main equation $3C(s) + 4H_2(g) \longrightarrow C_3H_8(g)$, we just have to convert them to gaseous atoms.

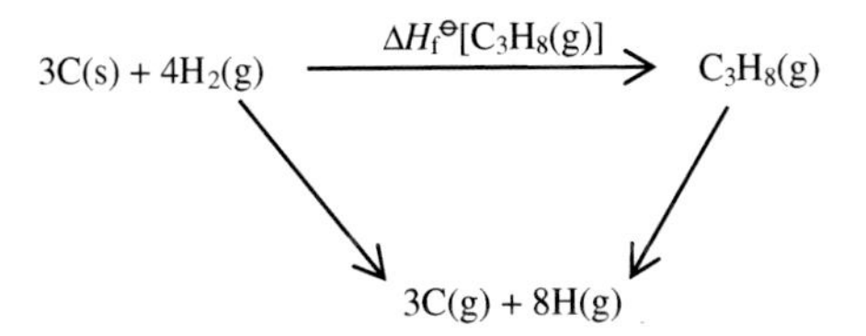

Make sure to balance the equations with respect to one another in the energy cycle. Write down the corresponding enthalpy terms on the arrows. To get an accurate count for the type and number of bonds to cleave, it is recommended that you draw the full constitutional/structural formula of the molecule. For instance, to convert the $C_3H_8(g)$ molecule into gaseous atoms, we can start by drawing its constitutional/structural formula:

```
    H   H   H
    |   |   |
H—C—C—C—H
    |   |   |
    H   H   H
```

Based on the constitutional/structural formula, there are 2 C–C bonds and 8 C–H bonds to cleave. Thus, the energy needed for atomisation of propane is 2×BE(C–C) and 8×BE(C–H). You should end up with the following energy cycle:

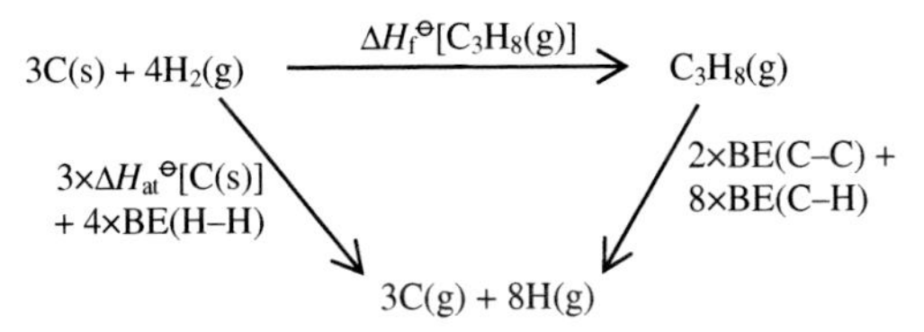

Next, determine the two pathways from "Initial state" to "Final state" and apply Hess' Law in your calculations.

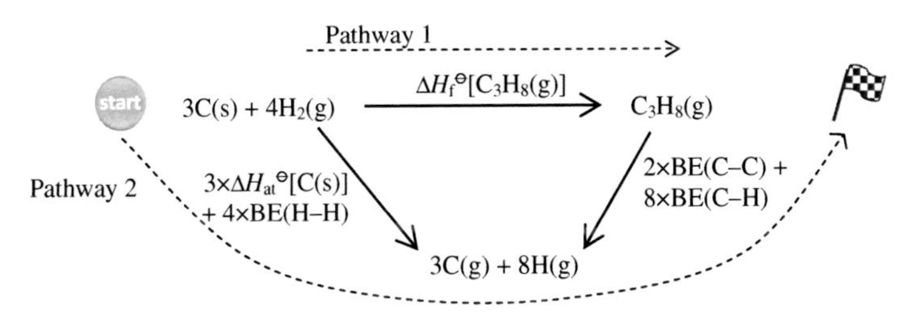

By Hess' Law,

$$\begin{aligned}\Delta H_f^{\ominus}[C_3H_8(g)] &= 3 \times \Delta H_{at}^{\ominus}[C(s)] + 4 \times BE(H\text{–}H) - 2 \times BE(C\text{–}C) \\ &\quad - 8 \times BE(C\text{–}H) \\ &= 3(715) + 4(436) - 2(350) - 8(412) \\ &= -107\,kJ\,mol^{-1}.\end{aligned}$$

The following are useful formulae that are derived from the energy cycles illustrated:

- $\Delta H_{rxn} = \Sigma n \Delta H_f(\text{product}) - \Sigma n \Delta H_f(\text{reactant})$,
- ΔH_{rxn} = energy absorbed in breaking bonds in reactants + energy released in forming bonds in products
  $= \Sigma n\text{BE(reactants)} - \Sigma n\text{BE(products)}$,

where n is the stoichiometric coefficient; Σ is the summation of all terms.

4.1.5 *Born–Haber cycle and Hess' law*

The Born–Haber cycle is an **energy level** diagram that has similar usage to an energy cycle. The main difference is that for a Born–Haber cycle, only two types of vertical arrows can be used:

↑ represents an endothermic reaction,
↓ represents an exothermic reaction.

The Born–Haber cycle is useful in determining lattice energy (L.E.) of ionic compounds from experimentally determined data since L.E. values cannot be obtained directly from experiments.

Q: What is the difference between an energy level diagram and an energy cycle?

A: An energy level diagram is a specific type of energy cycle. There are specific rules to obey when constructing an energy level diagram. If you are asked to construct a complex energy level diagram, it is advisable to construct an energy cycle first and then convert it to the energy level diagram.

The following are some other important enthalpy terms you need to be acquainted with that are commonly found in a Born–Haber cycle:

- **First ionisation energy (1st I.E.)**

This is the **energy required** to **remove one mole of electrons** from **one mole of gaseous atoms** in the ground state to form one mole of gaseous **singly charged cations.** It is an endothermic quantity. For example,

$$Mg(g) \longrightarrow Mg^{+}(g) + e^{-}, \qquad \Delta H = \text{1st I.E.} = +736\,\text{kJ mol}^{-1}.$$

Q: Is the ionisation energy always endothermic?
A: Yes. You need energy to break the attractive force the nucleus exerts on the electron. First I.E. data are very informative because different electrons in different orbitals have different I.E. (refer to Chap. 1).

The nth ionisation energy indicates the energy required to remove the nth electron after the removal of the preceding electrons:

$$X^{(n-1)+}(g) \longrightarrow X^{n+}(g) + e^{-}, \quad \Delta H = n\text{th I.E.}$$

- **First electron affinity (1st E.A.)**

This is the energy change when **one mole of electrons** is **added** to **one mole of gaseous atoms** to form one mole of gaseous **singly charged anions**. For example,

$$O(g) + e^{-} \longrightarrow O^{-}(g), \quad \text{1st E.A.} = -140\ \text{kJ mol}^{-1},$$
$$O^{-}(g) + e^{-} \longrightarrow O^{2-}(g), \quad \text{2nd E.A.} = +798\ \text{kJ mol}^{-1}.$$

Q: Why is the term "energy change" used in the definition of 1st E.A.?
Do you mean to say that 1st E.A. can be either endothermic or exothermic?

A: Yes. 1st E.A. is usually exothermic for many elements because a net attractive force results after the extra electron "sits" in the atom. But for some elements, there may be a net repulsive force instead. Hence, work has to be done to "force" this extra electron to reside in the atom (i.e., the process is endothermic). This also explains why the 2nd E.A. for oxygen is endothermic — an electron is added to a negative ion and the repulsive forces between like charges are greater than the attractive force experienced with the nucleus.

- **Lattice energy (L.E.)**

This is the **energy evolved** when **one mole** of a **solid ionic compound** is formed from its **constituent gaseous ions** under standard conditions.

$$\text{Lattice energy} \propto \frac{q_+ \times q_-}{r_+ + r_-},$$

where q_+ and q_- are the cationic and anionic charges, respectively, and $(r_+ + r_-)$ is the inter-ionic distance.

As the inter-ionic radius **increases,** lattice energy becomes **less exothermic.** As charge **increases**, lattice energy becomes **more exothermic** (refer to Chap. 2 on Chemical Bonding).

4.1.6 *Constructing Born–Haber cycles*

Example 4.7: In this example, we will work on finding the lattice energy of MgO(s) by constructing a Born–Haber cycle using the following data:

- ΔH_f[MgO(s)] = –602 kJ mol^{-1},
- ΔH_{at} [Mg(s)] = +148 kJ mol^{-1},
- ΔH_{at} [O_2(g)] = +249 kJ mol^{-1},
- 1st I.E. of magnesium = +736 kJ mol^{-1},
- 2nd I.E. of magnesium = +1450 kJ mol^{-1},
- 1st E.A. of oxygen = −140 kJ mol^{-1},
- 2nd E.A. of oxygen = +798 kJ mol^{-1}.

Approach:

(i) First, draw the baseline that consists of the most stable species, i.e., the ionic compound, MgO(s).

(ii) Draw the two energy levels that, when connected to the baseline, give the equations that represent ΔH_f[MgO(s)] and its L.E.. Since both terms are exothermic quantities, the equations are built downwards to the baseline.

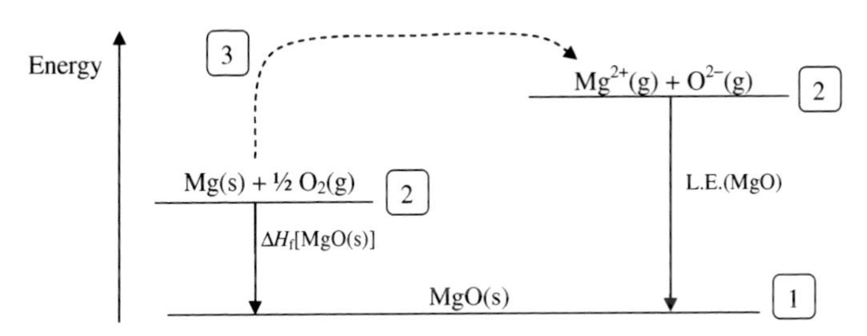

(iii) In a clockwise direction, attempt to convert the reactants Mg(s) and O_2(g) to the respective gaseous ions, Mg^{2+}(g) and O^{2-}(g). Work with each species in turn.

Let us first deal with the conversion of Mg(s) to Mg^{2+}(g), which requires the following steps:

$$\begin{array}{ll} \mathbf{Mg(s)} \longrightarrow \mathrm{Mg(g)}, & \Delta H_{\mathrm{at}}[\mathrm{Mg(s)}], \\ \mathrm{Mg(g)} \longrightarrow \mathrm{Mg^{+}(g)} + e^{-}, & \text{1st I.E. (Mg)}, \\ \mathrm{Mg^{+}(g)} \longrightarrow \mathbf{Mg^{2+}(g)} + e^{-}, & \text{2nd I.E. (Mg)}. \end{array}$$

At this point, the Born–Haber cycle should look like this:

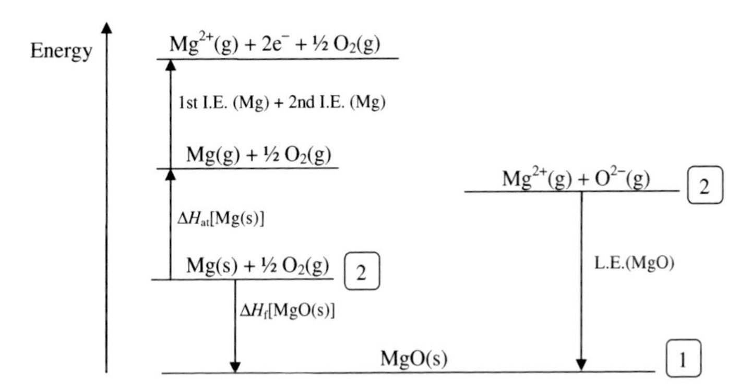

The atomisation and ionisation processes are endothermic and thus the equations are built upwards. Just as for an energy cycle, equations in the segments of the Born–Haber cycle have to be balanced — both in terms of mass and charge. Input relevant enthalpy terms along the way.

(iv) Similar to Step (iii), we now deal with the conversion of $O_2(g)$ to $O^{2-}(g)$, which requires the following steps:
$\mathbf{\frac{1}{2}O_2(g)} \longrightarrow O(g)$, $\Delta H_{at}[O_2(g)]$,
$O(g) + e^- \longrightarrow O^-(g)$, 1st E.A.(O),
$O^-(g) + e^- \longrightarrow \mathbf{O^{2-}(g)}$, 2nd E.A.(O).

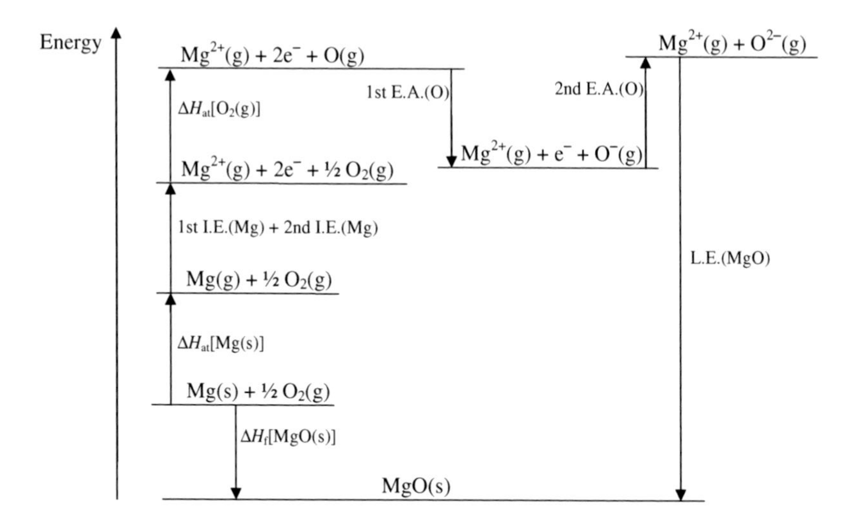

Behold! You have yourself a completed Born–Haber cycle.

(v) Determine the two pathways from "Initial state" to "Final state" and apply Hess' Law in your calculations. Remember that it does not matter where you start off in an energy cycle. Just do not forget that when the direction of a reaction is reversed, the sign on the ΔH term has to be reversed.

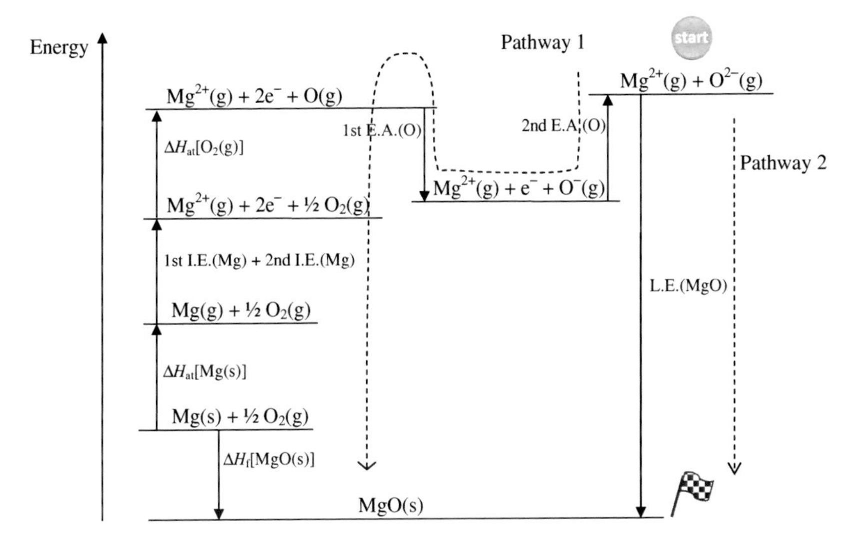

By Hess' Law,

$$\begin{aligned} \text{L.E.(MgO)} &= -\text{2nd E.A.(O)} - \text{1st E.A.(O)} - \Delta H_{\text{at}}[\text{O}_2(\text{g})] \\ &\quad - \{\text{1st I.E.(Mg)} + \text{2nd I.E.(Mg)}\} \\ &\quad - \Delta H_{\text{at}}[\text{Mg(s)}] + \Delta H_{\text{f}}[\text{MgO(s)}] \\ &= -(+798) - (-140) - (+249) - (+736 + 1450) \\ &\quad - (+148) + (-602) \\ &= -3843\,\text{kJ mol}^{-1}. \end{aligned}$$

Q: Why is the L.E. determined from the Born–Haber cycle called the experimental L.E.?

A: This is because most of the enthalpy values used in constructing the Born–Haber cycle are obtained experimentally.

Q: Why is the experimentally determined L.E. of MgO(s) using the Born–Haber cycle different (more exothermic) than the theoretical value of –3795 kJ mol^{-1}?

A: The more exothermic experimental L.E. suggests that the bonding in the ionic compound is stronger than that predicted by theoretical calculation. In theoretical calculation, the ions are assumed to be spherical. But in reality, due to the polarisation of anion's electron cloud by the cation, there is some degree of covalent character in the ionic bond. This covalent character might strengthen the ionic bond, which in this case, it actually does!

Exercise: Construct the Born–Haber cycle to determine the lattice energy of $FeCl_2(s)$. Input relevant enthalpy terms on the arrows.

Solution:

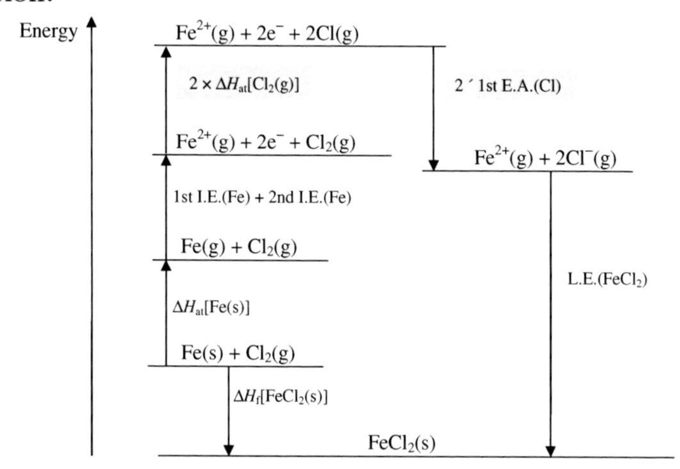

If you are to look at the two shaded enthalpy terms, you will find that the coefficient of two is needed for each. This is aligned with the definition associated with each term.

For instance, by definition, ΔH_{at} of an element specifies the enthalpy change in forming one mole of gaseous atoms:

$$\tfrac{1}{2}\text{Cl}_2(\text{g}) \longrightarrow \text{Cl}(\text{g}), \quad \Delta H_{\text{rxn}} = \Delta H_{\text{at}}[\text{Cl}_2(\text{g})],$$

$$\text{Cl}_2(\text{g}) \longrightarrow 2\text{Cl}(\text{g}), \quad \Delta H_{\text{rxn}} = 2 \times \Delta H_{\text{at}}[\text{Cl}_2(\text{g})].$$

Thus, be sure to remember the definitions for the various enthalpy terms and the corresponding equations. Recognise also that sometimes an equation can be represented by interchangeable enthalpy terms. For instance,

$$\text{Cl}_2(\text{g}) \longrightarrow 2\text{Cl}(\text{g}), \quad \Delta H_{\text{rxn}} = 2 \times \Delta H_{\text{at}}[\text{Cl}_2(\text{g})] = \text{BE(Cl–Cl)}.$$

4.1.7 *Energetics involving aqueous ionic compounds*

Recall in Chap. 2, that it was mentioned that ionic compounds have a wide range of solubilities in water. In this section, we will address why this is so by considering the process of dissolving an ionic compound as occurring in two stages:

- **First stage:** involves the separation of ions from the crystal lattice. In order to do this, energy is needed to overcome the attractive forces holding the ions in the lattice. This amount of energy required is numerically equivalent to the magnitude of lattice energy (–L.E.).
- **Second stage:** involves the hydration of the ions, i.e., the formation of ion–dipole interactions between the ions and water molecules (the term "solvation" is used for solvents other than water). Energy is released in bond formation and it is termed the hydration energy. The stronger the ion–dipole interactions, the more energy will be evolved. The magnitude of the hydration energy therefore depends on the charge density of the ion.

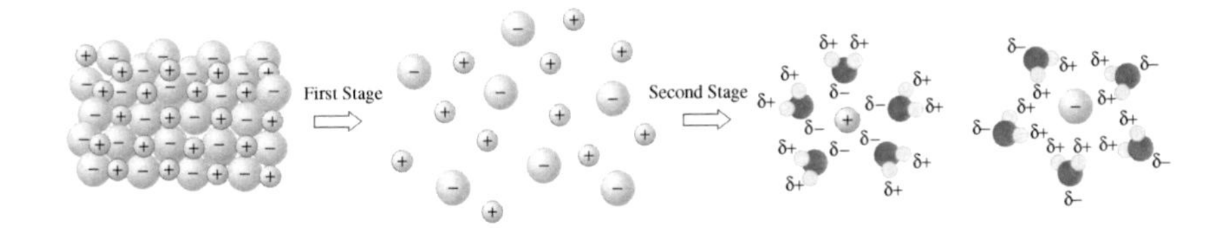

- **Standard enthalpy change of hydration ($\Delta H_{\mathrm{hyd}}^{\ominus}$):**

This is the **energy evolved** when **one mole of gaseous ions** is hydrated under standard conditions.

A less exothermic ΔH_{hyd} decreases solubility but a less exothermic L.E. enhances solubility.

As you can see, these two terms work against each other in determining the solubility of an ionic compound.

It is now time to introduce a new energy term associated with the analysis of the solubility of ionic compounds — the enthalpy change of solution.

- **Standard enthalpy change of solution ($\Delta H_{soln}^{\ominus}$):**

This is the energy change when **one mole of solute** is completely dissolved in enough solvent so that no further heat change takes place on adding more solvent (**infinite** dilution) under standard conditions.

The relationship among these terms is highlighted in the formula below and depicted clearly in the energy cycle:

$$\Delta H_{soln} = \Delta H_{hyd}[\text{cation}] + \Delta H_{hyd}[\text{anion}] - \text{L.E.}$$

$MX(s) \xrightarrow{\Delta H_{soln}} M^{+}(aq) + X^{-}(aq)$

$MX(s) \xrightarrow{-\text{L.E.}} M^{+}(g) + X^{-}(g) \xrightarrow{\Delta H_{hyd}[\text{cation}] + \Delta H_{hyd}[\text{anion}]} M^{+}(aq) + X^{-}(aq)$

As a good indicator, if an ionic compound is to dissolve in water, the energy released during the hydration process must be sufficient to offset the energy needed to overcome the strong ionic bonds and hence break up the crystal lattice.

In general, a less exothermic ΔH_{soln} signifies a less soluble salt.

- When ΔH_{hyd} is more exothermic than L.E., ΔH_{soln} will be exothermic, indicating the salt is relatively soluble.
- When ΔH_{hyd} is less exothermic than L.E., ΔH_{soln} will be endothermic, indicating the salt is relatively insoluble.

Take heed though that ΔH_{soln} only provides an indication of a compound's solubility in relation to another. A positive ΔH_{soln} need not necessarily mean that the salt is insoluble in water, and *vice versa.*

Example 4.8: Calculate the ΔH_{soln} of sodium chloride given that its lattice energy is $-776\,\text{kJ mol}^{-1}$ and the enthalpy change of hydration for Na^{+} and Cl^{-} are $-390\,\text{kJ mol}^{-1}$ and $-381\,\text{kJ mol}^{-1}$, respectively.

Solution:

$$\begin{aligned}\Delta H_{soln} &= \Delta H_{hyd}[Na^{+}(g)] + \Delta H_{hyd}[Cl^{-}(g)] - \text{L.E.}(NaCl)\\ &= (-390) + (-381) - (-776)\\ &= +5\ \text{kJ mol}^{-1}.\end{aligned}$$

We know for sure that NaCl(s) dissolves in water and it is undeniable that this process is spontaneous — yet the calculated ΔH_{soln} is a positive value!

Q: Is the mixing of two liquids similar to the dissolving of a solid in a solvent? Do they have "similar" energy processes involved during the mixing process as that of the dissolution process?

A: When you are mixing two liquids, essentially, you are removing one liquid molecule (lets call this solute) from their "friends" (you need to break bonds here, so you need energy) and trying to put this molecule in-between some other different molecules (lets call this solvent). But in order to put the solute molecule in, you again need to break bonds so as to create a "hole" to accommodate this incoming solute molecule, which means you need energy again. So, where does all these energies that are needed come from? Now, when this solute molecule is "sitting" amongst the solvent molecules, it would be able to interact with the solvent molecules. Thus, bond forms and energy is released. If this energy that is released is sufficient enough to overcome both the bonds that are needed to break within the solute and solvent, the mixing process would likely take place. Hence, you see the similarities between mixing two liquids and dissolving a solid.

Q: So, basically if the interaction between the solute and solvent is of similar strength to that within the solute and solvent, the energy that is released would be sufficient to compensate for that needed to overcome the bonds within the solute and solvent?

A: You are right! This is the basis behind the cliche "like dissolves like". It is important for organic chemistry when we mix two organic compounds together. Whether they mix well or not depends on the nature of the two compounds. For details, you can refer to *Understanding Advanced Organic and Analytical Chemistry* by K. S. Chan and J. Tan.

Q: Why did you say "if this energy that is released is sufficient enough to overcome the bonds that are needed to break both within the solute and solvent, the mixing process would likely take place"? Do you mean to say that the consideration of energy change is not sufficient to decide whether the mixing would take place?

A: Indeed! Other than enthalpy consideration, there is also the entropy factor that we need to take into account in any processes (refer to section 4.2).

In general, an exothermic ΔH term does not guarantee the spontaneity of a reaction. A reaction with an endothermic ΔH can also be spontaneous. As a matter of fact, the spontaneity of a reaction is determined by not one but two thermodynamic terms: (i) enthalpy and (ii) entropy (which we will cover in the next section).

Exercise: Construct an energy level diagram that illustrates the relationship among the terms, ΔH_{soln}, ΔH_{hyd} and L.E., for sodium chloride.

Solution:

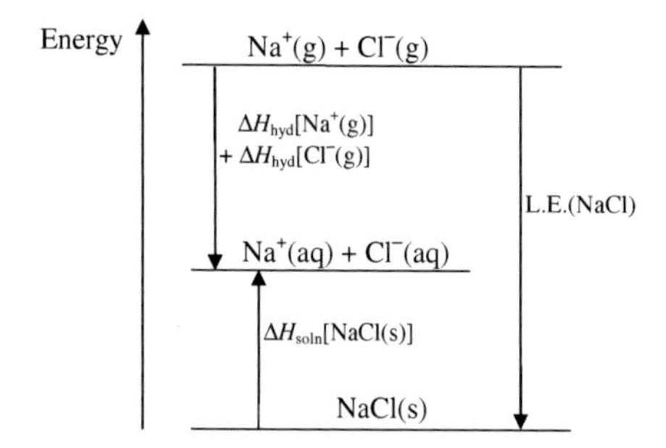

4.2 Entropy

The main objective of studying chemical thermodynamics is to be able to predict whether or not a reaction is spontaneous.

A spontaneous reaction (or change) is one that occurs **without any outside assistance**.

Examples of spontaneous processes include:

- A waterfall cascades downhill, but never uphill, spontaneously.
- Water freezes spontaneously at temperature below 0°C; ice melts spontaneously above 0°C (at 1 bar).

- A lump of sugar dissolves spontaneously in a cup of coffee, but dissolved sugar does not spontaneously reappear in its original form.
- Expansion of a gas in an evacuated bulb is a spontaneous process. The reverse — the gathering of all the molecules into one bulb — is not.
- Heat flows from a hotter object to a colder one, but the reverse never happens spontaneously.

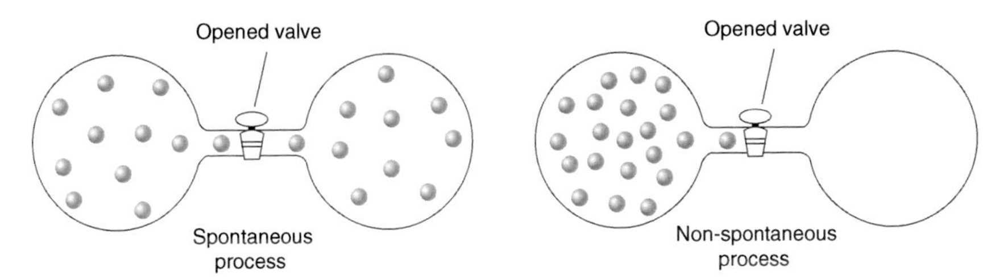

A spontaneous change is ***irreversible***, as it cannot be brought back to its original state again unless some outside influence is brought to bear.

In chemical reactions, we find that a larger number of exothermic reactions are spontaneous. There is a *tendency* to decrease the energy of a system in spontaneous processes. One example is the combustion of methane:

$$CH_4(g) + 2O_2(g) \longrightarrow CO_2(g) + 2H_2O(l), \quad \Delta H^\circ = -890.4 \text{ kJ mol}^{-1}.$$

But the assumption that spontaneous processes always decrease a system's energy fails in a number of cases:

- Consider a solid-to-liquid phase transition such as the melting of ice:
$$H_2O(s) \longrightarrow H_2O(l), \quad \Delta H^\circ = +6.01 \text{ kJ mol}^{-1}.$$
Our experience tells us that ice melts spontaneously above 0°C even though the process is endothermic.
- Consider the cooling that results when ammonium nitrate dissolves in water:
$$NH_4NO_3(s) \longrightarrow NH_4{}^+(aq) + NO_3{}^-(aq), \quad \Delta H^\circ = +25 \text{ kJ mol}^{-1}.$$
The dissolution of the ammonium salt is spontaneous, yet it is also endothermic.

- The decomposition of HgO is an endothermic reaction that is non-spontaneous at room temperature, but it becomes spontaneous when the temperature is raised:

$$2\,\text{HgO(s)} \longrightarrow 2\,\text{Hg(l)} + \text{O}_2\text{(g)}, \quad \Delta H^\ominus = +90.7\ \text{kJ mol}^{-1}.$$

Hence, **exothermicity *favours* the spontaneity of a reaction but *does not* guarantee it**. It is possible for endothermic reactions to be spontaneous; it is also possible for exothermic reactions to be non-spontaneous.

Consideration of enthalpy changes alone is not enough to predict the spontaneity or non-spontaneity of a reaction. It becomes necessary to look for another thermodynamic quantity (in addition to enthalpy) to help predict the direction of chemical reactions. This quantity turns out to be **entropy**.

Q: Water becomes solid ice at 0°C. This only happens in the freezer where work is being done, so how can you say that there is no external help? And as such, how is this transformation spontaneous without external help?

A: There is a fallacy here. The freezer creates the condition of 0°C, temperature at which water becomes solid ice spontaneously. The water cannot "resist" becoming ice and no one is forcing it or helping it to change and the change is IRREVERSIBLE in that setting. So it seems like the freezer is intervening but in reality it is not. It simply creates the conditions and the rest is left to water alone.

4.2.1 *What is entropy?*

Entropy (S) describes the degree of randomness or disorder in a system.

The entropy of a system increases with temperature. For a crystal lattice at 0 K, its entropy is zero since the particles are stationary and arranged in a perfect order.

When heat is applied to the crystal, a higher degree of disorder is brought about through two ways:

- The particles' kinetic energy increases and they move around more randomly, leading to greater disorder and larger entropy.
- With a greater volume of space, together with greater freedom of movement, there are many more varied options for the arrangement of the particles — all of which are not in a perfect order.

Entropy is closely linked to statistical probability. In general, an event that brings about disorder is more highly likely to occur (spontaneously) than one that brings about order.

Consider, for example, the melting of ice:

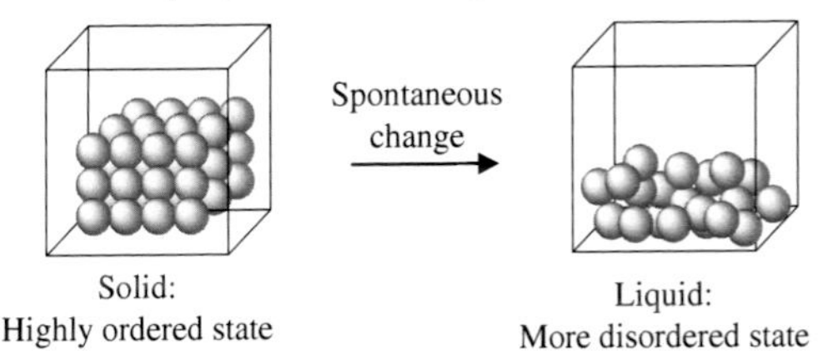

Just like enthalpy, entropy (S) is a **state function** and therefore, entropy change (ΔS) from a certain initial state to a final state is expressed as:

$$\Delta S = S_{\mathrm{f}} - S_{\mathrm{i}},$$

where S has the units of J mol^{-1} K^{-1}.

$\Delta S > 0$, indicates that there is an increase in the randomness or disorder of a system. A reaction is said to be spontaneous.

4.2.2 *Factors affecting entropy of a chemical system*

4.2.2.1 *Effect of temperature*

Particles can vibrate, rotate and translate (move about) as well, and all these movements involve energy. Energy comes in quanta for each of these changes. When a particle absorbs a particular amount of energy, this energy can be distributed into various energy states.

An increase in temperature will result in an increase in entropy of a system. In general, an **increase in temperature means more packets of energy are *available* to spread out within the system**.

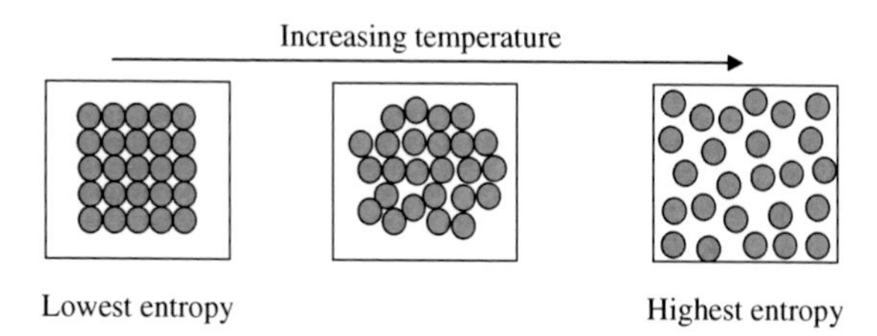

4.2.2.2 *Effect of phase change*

For the same number of moles of particles, the degree of order decreases from solid to liquid to gas. In other words, the magnitude of entropy of these phases increases in the order: $S_{\text{solid}} < S_{\text{liquid}} \ll S_{\text{gas}}$. In the gaseous phase, the particles have the greatest amount of kinetic energy; they can move more randomly compared to the liquid phase. In addition, a gaseous particle occupies greater spatial volume, which also increases its degree of disorder.

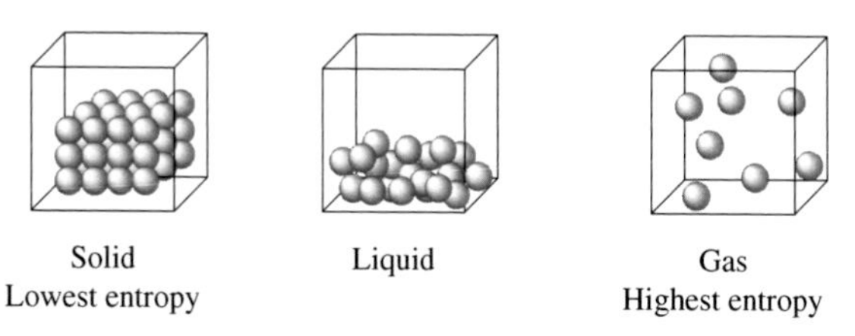

Q: What is "phase"? Is it the same as the physical state, that is, solid, liquid and gas?

A: Phase is not equivalent to physical state! A phase is defined as a state of matter that is homogeneous in terms of chemical composition and also physical state. Take for instance, a layer of oil floating above water. There are two different phases clearly being demarcated by what we call a phase boundary, but there is only one physical state, which is the liquid state.

4.2.2.3 *Effect of the number of particles*

The entropy of a system increases if the products of a reaction contain more **gaseous** molecules than the reactants.

An example is the combustion of propane:

$$C_3H_8(g) + 5O_2(g) \longrightarrow 3CO_2(g) + 4H_2O(g).$$

Lower entropy Higher entropy

4.2.2.4 *Effect of mixing*

The entropy of a system increases when gases mix.

When mixed, each gas will expand to occupy the entire container. The expansion of a gas results in an increase in its entropy since the molecules will now have a greater volume of space to move about, leading to a greater degree of disorder.

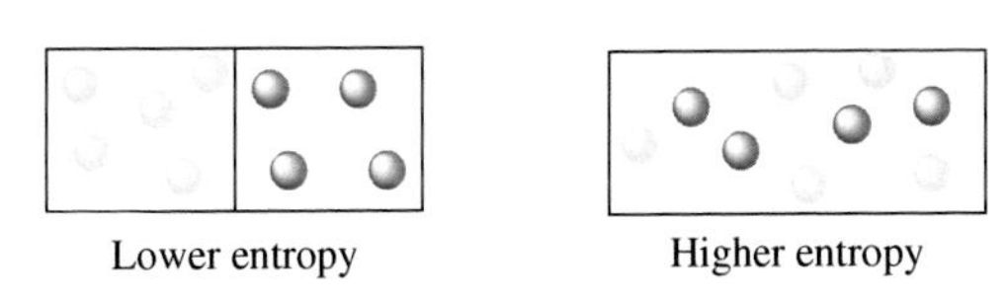

For liquids that are able to mix with each other (i.e., miscible liquids with similar polarities), there is also an increase in entropy of the system. Upon mixing, the molecules have a greater volume of space to move about, leading to a greater degree of disorder, and thus an increase in entropy.

The same can be said when a solute dissolves in a solvent.

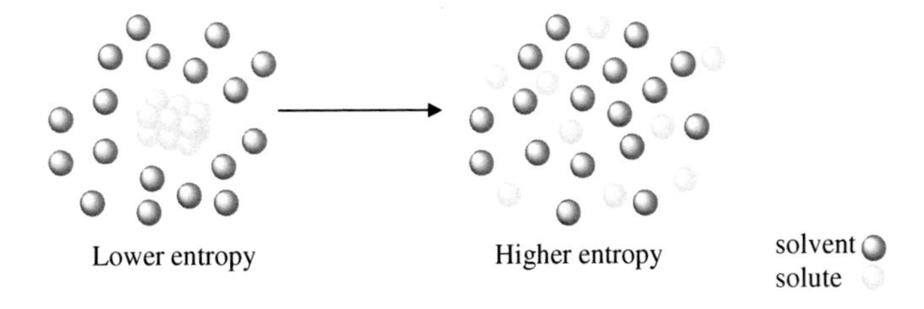

Example 4.9: For each of the following processes or reactions, predict whether the ΔS of the system is positive or negative at a temperature of 298 K and pressure of 1 bar, unless otherwise stated:

(a) $H_2O(l)$ at 298 K $\longrightarrow$ $H_2O(l)$ at 308 K.
(b) $H_2O(s)$ at 273 K $\longrightarrow$ $H_2O(l)$ at 273 K.
(c) $F_2(g) \longrightarrow 2F(g)$.
(d) $Li^+(g) + Cl^-(g) \longrightarrow LiCl(s)$.
(e) 1 mol of $H_2(g)$ is pumped into a sealed vessel of volume 24 dm^3 containing 1 mol of $O_2(g)$ at room temperature.

Solution:

(a) $\Delta S > 0$.
Substances have higher entropy at higher temperature as the larger amount of K.E. allows the particles to occupy a greater number of different energy states. In addition, the particles also occupy greater spatial volume. These increase the degree of disorder.
(b) $\Delta S > 0$.
Liquids have higher entropy than solids. Particles in a liquid are more randomly arranged than in a solid [the explanation is similar to (a)].
(c) $\Delta S > 0$.
Two moles of gaseous particles have a higher entropy than one mole of gaseous particles as there are more ways of arranging the particles in the container. In addition, there are more particles present where energy can be distributed.
(d) $\Delta S < 0$.
The gaseous ions, which are more disordered, crystallise to form the more ordered solid LiCl, resulting in a decrease in entropy of the system.
(e) $\Delta S > 0$.
As $H_2(g)$ expands to occupy the entire containing vessel, mixing of the two gases occurs, and there is an increase in the disorder, and hence the entropy of the system. The two gases mix completely and spontaneously.

4.2.3 *Predicting the spontaneity of a reaction*

We have studied two factors, enthalpy (H) and entropy (S), and tried to use them to predict whether or not a physical or chemical event will be spontaneous. At times, these two factors work together.

For example, when gasoline burns in oxygen, $\Delta H < 0$ and $\Delta S > 0$. Since both of these favour a spontaneous change, the two factors complement each other.

In other situations, the effects of enthalpy and entropy are in opposition, for example, in the melting of an ice-cream or the boiling of a kettle of water. The endothermic nature of these changes tends to make them non-spontaneous, but the increase in the randomness of molecules tends to make them spontaneous.

In addition, when hydrogen gas combusts with oxygen to form water in a rocket engine, the enthalpy and entropy changes are also in opposition. In this case, the exothermic nature of the reaction is sufficient to overcome the negative value of ΔS and cause the reaction to be spontaneous.

Hence, when enthalpy and entropy antagonise each other, their relative importance in determining spontaneity is far from obvious. Moreover, the situation is made worse as temperature becomes a third factor that can influence the direction in which a change is spontaneous. Specialists in thermodynamics defined a quantity called the **Gibbs free energy, *G***, in honour of Josiah Willard Gibbs (1839–1903):

$$G = H - TS.$$

The Gibbs free energy term enables us to predict the spontaneity of a reaction that is governed by two thermodynamic terms (i) enthalpy and (ii) entropy.

The relationship is expressed in the following equation for the standard *Gibbs free energy change of reaction*:

$\Delta G^{\ominus} = \Delta H^{\ominus} - T\Delta S^{\ominus}$, where $\Delta G^{\ominus}$ and $\Delta H^{\ominus}$ are in kJ mol^{-1},
T is in K, and
$\Delta S^{\ominus}$ is in J mol^{-1}K^{-1}.

At **constant temperature and pressure**, a reaction is considered **spontaneous** if there is a **decrease in the free energy of the system.**

- $\Delta G^{\ominus} < 0$: Forward reaction is thermodynamically spontaneous and is thus feasible.

- $\Delta G^{\ominus} > 0$: Forward reaction is thermodynamically non-spontaneous and is thus not feasible.
  Backward reaction is thermodynamically spontaneous.
- $\Delta G^{\ominus} = 0$: Both forward and backward reactions are at equilibrium.

4.2.4 *Relationship between $\Delta G^{\ominus}, \Delta H^{\ominus}, \Delta S^{\ominus}$ and temperature*

It is not always the case that $\Delta H^{\ominus}$ and $\Delta S^{\ominus}$ both contribute to the spontaneity of a reaction. Sometimes, they oppose each other, as in the melting of ice; $\Delta H^{\ominus}$ is positive and $\Delta S^{\ominus}$ is positive. In these cases, temperature is the determining factor for the spontaneity of a reaction.

The table below shows the conditions under which a reaction may be spontaneous:

Enthalpy	**Entropy**	**Is Reaction Spontaneous?**
$\Delta H^{\ominus} < 0$	$\Delta S^{\ominus} > 0$	Always ($\Delta G^{\ominus} < 0$)
$\Delta H^{\ominus} < 0$	$\Delta S^{\ominus} < 0$	Only at low temperature when $\Delta H^{\ominus} > T\Delta S^{\ominus}$
$\Delta H^{\ominus} > 0$	$\Delta S^{\ominus} > 0$	Only at high temperature when $\Delta H^{\ominus} < T\Delta S^{\ominus}$
$\Delta H^{\ominus} > 0$	$\Delta S^{\ominus} < 0$	Never ($\Delta G^{\ominus} > 0$)

Take note that using $\Delta G^{\ominus}$ to predict the thermodynamic spontaneity is highly dependent on the conditions, i.e., temperature and pressure, in which the $\Delta G^{\ominus}$ is being defined. Meaning? $\Delta G^{\ominus}$ predicts the thermodynamic spontaneity under standard conditions of 298 K and 1 bar. Thus, if $\Delta G^{\ominus}$ at 298 K and 1 bar predicts the reaction as thermodynamically non-spontaneous, this same reaction may be spontaneous at another set of temperature and pressure conditions.

Example 4.10: For each of the following reactions, predict the temperature conditions, if any, under which each reaction will be spontaneous:

(a) $A(s) \longrightarrow B(s) + C(g)$, $\Delta H^\ominus = +350\,kJ\,mol^{-1}$.
(b) $2A(g) + B(g) \longrightarrow 2C(g)$, $\Delta H^\ominus = -200\,kJ\,mol^{-1}$.
(c) $A(l) \longrightarrow B(l) + C(g)$, $\Delta H^\ominus = -300\,kJ\,mol^{-1}$.
(d) $2A(s) + 2B(g) \longrightarrow C(g)$, $\Delta H^\ominus = +150\,kJ\,mol^{-1}$.

Solution:

(a) $A(s) \longrightarrow B(s) + C(g)$, $\Delta H^\ominus = +350\,kJ\,mol^{-1}$; $\Delta S^\ominus > 0$.
The above endothermic reaction is thermodynamically spontaneous at high T, when $T > \Delta H^\ominus/\Delta S^\ominus$.
(b) $2A(g) + B(g) \longrightarrow 2C(g)$, $\Delta H^\ominus = -200\,kJ\,mol^{-1}$; $\Delta S^\ominus < 0$.
The above exothermic reaction is thermodynamically spontaneous at low T, when $T < \Delta H^\ominus/\Delta S^\ominus$.
(c) $A(l) \longrightarrow B(l) + C(g)$, $\Delta H^\ominus = -300\,kJ\,mol^{-1}$; $\Delta S^\ominus > 0$.
The above reaction is thermodynamically spontaneous at all T. ($\Delta G^\ominus < 0$ for all temperatures.)
(d) $2A(s) + 2B(g) \longrightarrow C(g)$, $\Delta H^\ominus = +150\,kJ\,mol^{-1}$; $\Delta S^\ominus < 0$.
The above reaction is thermodynamically non-spontaneous at all T. ($\Delta G^\ominus > 0$ for all temperatures.)

Example 4.11: Calculate the change in standard free energy and determine whether the reaction below will take place at (a) 50°C, and (b) 550°C:

$$Fe_2O_3(s) + 3H_2(g) \longrightarrow 2Fe(s) + 3H_2O(g).$$

	Fe_2O_3	H_2	Fe	H_2O
$\Delta H_f^\ominus$ ($kJ\,mol^{-1}$)	−822	0	0	−242
$\Delta S^\ominus$ ($kJ\,mol^{-1}\,K^{-1}$)	0.090	0.131	0.027	0.189

Solution:
$\Delta S^\ominus$ for the reaction $= 2(0.027) + 3(0.189) - 0.090 - 3(0.131) = +0.138\,kJ\,mol^{-1}\,K^{-1}$.
$\Delta H^\ominus$ for the reaction $= [0 + 3(-242)] - (-822 + 0) = +96\,kJ\,mol^{-1}$.

(a) At 50°C, using $\Delta G^\ominus = \Delta H^\ominus - T\Delta S^\ominus$:
$\Delta G^\ominus = +96 - (323)(0.138) = +51.4\,kJ\,mol^{-1}$.

Since $\Delta G^{\ominus}$ is **positive**, the reaction is **not** thermodynamically spontaneous at 323 K and will therefore not occur.

(b) At 550°C, using $\Delta G^{\ominus} = \Delta H^{\ominus} - T\Delta S^{\ominus}$:
$\Delta G^{\ominus} = +96 - (823)(0.138) = -17.6\,\text{kJ mol}^{-1}$.
Since $\Delta G^{\ominus}$ is **negative**, the reaction is thermodynamically spontaneous at 823 K and will therefore occur.

Note: It is assumed that $\Delta H^{\ominus}$ and $\Delta S^{\ominus}$ do not change with temperature.

Example 4.12:

Substance	$MgCO_3$(s)	MgO(s)	CO_2(g)
$\Delta G_f^{\ominus}$ (kJ mol^{-1} K^{-1})	−1012	−569	−394

(a) Using the standard $\Delta G_f^{\ominus}$ given above, calculate the free energy change $\Delta G^{\ominus}{}_{rxn}$ for the reaction:

$$MgCO_3(s) \longrightarrow MgO(s) + CO_2(g).$$

Determine whether the above reaction takes place spontaneously.

(b) Given that $\Delta H^{\ominus}{}_{rxn}$ in the above reaction is $+101\,\text{kJ mol}^{-1}$, and $\Delta S^{\ominus}{}_{rxn}$ for the reaction is $+0.159\,\text{kJ mol}^{-1}\,\text{K}^{-1}$, estimate the temperature at which the reaction will take place spontaneously.

Solution:

(a) Using the formula $\Delta G^{\ominus}{}_{rxn} = \Sigma m\Delta G_f^{\ominus}(\text{products}) - \Sigma n\Delta G_f^{\ominus}(\text{reactants})$:

$$\Delta G^{\ominus}{}_{rxn} = (-569 - 394) - (-1012) = +49\ \text{kJ mol}^{-1}.$$

The reaction will **not** take place spontaneously because $\mathbf{\Delta G^{\ominus}{}_{rxn} > 0}$.

(b) Use $\Delta G^{\ominus}{}_{\text{rxn}} = \Delta H^{\ominus}{}_{\text{rxn}} - T\Delta S^{\ominus}{}_{\text{rxn}}$. For the reaction to be thermodynamically spontaneous,

$$\Delta G^{\ominus}{}_{\text{rxn}} = 101 - T(0.159) < 0$$
$$101 < T(0.159)$$
$$T > \frac{101}{0.159} = \mathbf{635\ K}.$$

A temperature of above 635 K or 362°C is needed for the reaction to take place. (It is assumed that $\Delta H^{\ominus}{}_{\text{rxn}}$ and $\Delta S^{\ominus}{}_{\text{rxn}}$ do not change with temperature.)

My Tutorial (Chapter 4)

1. (a) During physical training, people may suffer from twisted ankles. In such cases ice should be applied to shrink the blood vessels around the sprain in order to minimize any internal bleeding. But storing ice for such usage is neither convenient nor economical. So, trainers often use cold packs consisting of a divided plastic bag containing ammonium nitrate and water. (Given that $\Delta H^{\ominus}{}_{\text{soln}}(NH_4NO_3) = +26.0\ \text{kJ mol}^{-1}$.)

   (i) Determine the heat change when 30.0 g of ammonium nitrate dissolves in water.

   (ii) What is the final temperature if 30.0 g of ammonium nitrate is added to 200 g of water at 298 K? Assume that the heat capacity of ammonium nitrate solution is $4.0\ \text{J g}^{-1}\ \text{K}^{-1}$.

   (b) Ammonium nitrate is widely used as an explosive because it can decompose according to the following equation:

$$2NH_4NO_3(s) \longrightarrow 2N_2(g) + 4H_2O(g) + O_2(g).$$

   (i) With reference to the Data Booklet, determine the enthalpy change for this decomposition. (Assume that the lattice energy of NH_4NO_3 is $-653\ \text{kJ mol}^{-1}$.)

   (ii) What features of its decomposition make ammonium nitrate explosive?

2. (a) With reference to the Data Booklet, calculate the standard enthalpy change of formation of calcium(I) chloride. [Assume $CaCl(s) \longrightarrow Ca^+(g) + Cl^-(g)$.]
   (b) The standard enthalpy change of formation of calcium(II) chloride is $-795\,kJ\,mol^{-1}$. Calculate the enthalpy change for the reduction of calcium(II) chloride to calcium(I) chloride by calcium metal.
   (c) Experimentally, making calcium(I) chloride by simply reducing calcium(II) chloride with calcium metal has not been fruitful. Theoretically, is it possible to make calcium(I) chloride by this method? Explain.

3. With dwindling reserves of fossil fuels, scientists have suggested using methanol (CH_3OH) as a possible alternative fuel for motor cars. The advantages of using methanol as a fuel are that it burns cleanly, giving out fewer pollutants than gasoline, and is less of a fire hazard in an accident.

   (a) Using your Data Booklet, calculate the enthalpy change when one litre of methanol is burnt in excess oxygen. (Assume that the density of methanol is $0.79\,g\,cm^{-3}$ and ΔH_{vap} (methanol) is $+38.3\,kJ\,mol^{-1}$.)
   (b) The heat change on running one litre of petrol is approximately $-33{,}000\,kJ$. Comment on the significance of your answer in part (a) to the design and operation of cars which run on methanol.
   (c) Explain why methanol is less fire hazardous?

4. (a) Define the term *standard enthalpy change of formation.*
   (b) Cyclohexene reacts with hydrogen to form cyclohexane C_6H_{12} as follows:

$$C_6H_{10} + H_2 \longrightarrow C_6H_{12}.$$

   Calculate ΔH for this reaction, given that the enthalpy change of formation of cyclohexene and cyclohexane is $-36\,kJ\,mol^{-1}$ and $-156\,kJ\,mol^{-1}$, respectively.

(c) Benzene undergoes a similar reaction with hydrogen to form cyclohexane:

$$C_6H_6 + 3H_2 \longrightarrow C_6H_{12}.$$

Assuming that benzene contains three C=C bonds of the type found in ethene, predict the value of ΔH for this reaction.

(d) The actual value of ΔH for the reaction in part (c) is $-210\,\text{kJ mol}^{-1}$. What can you deduce from this about the stability of the benzene ring? Use an energy level diagram to illustrate your answer.

5. (a) State Hess' Law.

(b) (i) Using the following data:

$S(s) + O_2(g) \longrightarrow SO_2(g)$, $\Delta H^{\ominus} = -297\text{ kJ mol}^{-1}$,
$C(s) + O_2(g) \longrightarrow CO_2(g)$, $\Delta H^{\ominus} = -393\text{ kJ mol}^{-1}$,
$CS_2(l){+}3O_2(g) \longrightarrow CO_2(g){+}2SO_2(g)$, $\Delta H^{\ominus} = -\ 297\text{ kJ mol}^{-1}$,

calculate the enthalpy change for the reaction between carbon and sulfur to form carbon disulfide:

$$C(s) + 2S(s) \longrightarrow CS_2(l).$$

(ii) Is the enthalpy change you have calculated equal to the standard enthalpy change of formation of carbon disulfide? Explain your answer.

(iii) Metal sulfide ores are usually roasted in air to form the oxide before reduction to the metal with carbon. Explain this practice with reference to your answer in part (b)(i).

(c) In a typical experiment using acid and alkali of equal volumes, the following results are obtained:

Concentration of sulfuric(VI) acid = $1.00\,\text{mol dm}^{-3}$,
Concentration of sodium hydroxide = $2.00\,\text{mol dm}^{-3}$,
Volume of sulfuric(VI) acid used = $25.0\,\text{cm}^3$,
Initial temperature of sulfuric(VI) acid = $21.0°\text{C}$,
Final temperature of sodium hydroxide = $23.0°\text{C}$,
Highest temperature reached after mixing = $35.6°\text{C}$,
Specific heat capacity of water = $4.2\,\text{J g}^{-1}\,\text{K}^{-1}$.

(i) Calculate the molar enthalpy change of neutralization of sulfuric(VI) acid with sodium hydroxide from these results. Indicate the assumptions you have made.

(ii) Calculate the percentage error in your result assuming that temperatures were accurate to $\pm 0.1^\circ C$.

(iii) Account for the fact that a similar experiment using hydrochloric acid gave an identical value for the molar enthalpy change of neutralisation, while an experiment with methanoic acid gave a value of $-55.0\,kJ\,mol^{-1}$.

(d) Describe how you would carry out an experiment to determine the molar enthalpy change of sulfuric(VI) acid with sodium hydroxide. Your account should mention the measurements which you would make and the reasons behind your choice of apparatus.

CHAPTER 5

Reaction Kinetics

Chemical kinetics is the study of reaction rates, the factors which affect them and the mechanisms by which chemical reactions occur. Why should we know about reaction kinetics?

The importance of kinetic studies:

- Kinetics studies **provide information on how *quickly* and economically a product can be made**. Such information is essential when considering the economics of a manufacturing process in the chemical industry. The industrial chemist can select a set of reaction conditions which would optimise yield of the desired product in the shortest time feasible.
- Kinetics studies also **provide information on how *fast* a chemical product will "work"**. This enables the manufacturer to adjust the properties of the product (e.g., the setting time for a glue product) or to provide recommended instructions for the safe use of the product (e.g., finding the time that must elapse between spraying a pesticide and eating a vegetable crop).
- Kinetics studies **also provide information on the reaction mechanism**, that is, the way in which a reaction occurs — either in one step or in a sequence of steps. Knowing the reaction mechanism helps the industrial chemist to modify a product, thereby increasing its effectiveness. For example, by slightly altering the molecular structure of a drug, its effects may be made more rapid and longer lasting.

5.1 Qualitative Analysis of Reaction Rates

The Collision Theory is derived from the Kinetic Theory of Gases and it is used to explain how reactions occur and the different rates at which they occur.

A chemical reaction involves the rearrangement of particles (atoms, ions or molecules). In this rearrangement process, old bonds break and new bonds form.

According to the Collision Theory, a number of conditions must be satisfied before a reaction can occur:

(i) the reactant particles must **collide** with one another,
(ii) with a **minimum amount of energy** (known as activation energy, E_a), and
(iii) in the **correct orientation**.

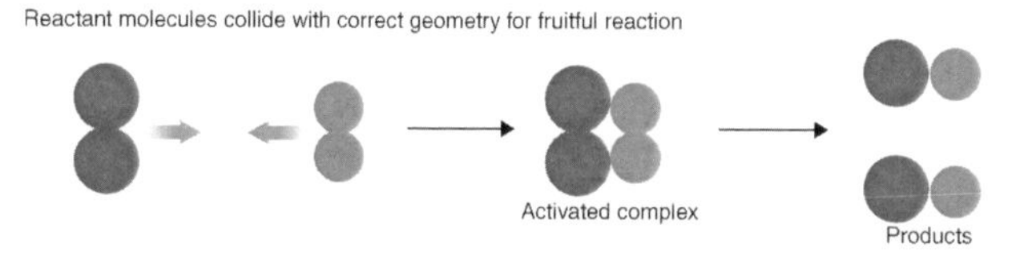

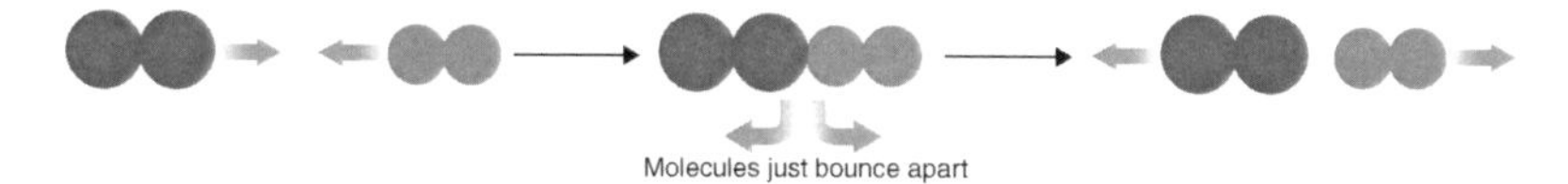

The **activation energy** E_a is the *minimum amount of energy* that the reactant particles must acquire before they can collide *to result* in a reaction. It is needed to break existing bonds in the reactant species or to overcome inter-electronic repulsion before forming new bonds in the products. E_a can be viewed as the "push" needed to set a reaction in motion.

The energy profile diagrams shown in Fig. 5.1 show the energy changes for an exothermic and an endothermic reaction. In both diagrams, E_a is represented as the energy barrier that needs to be overcome. The source of activation energy comes from the kinetic energy of the particles.

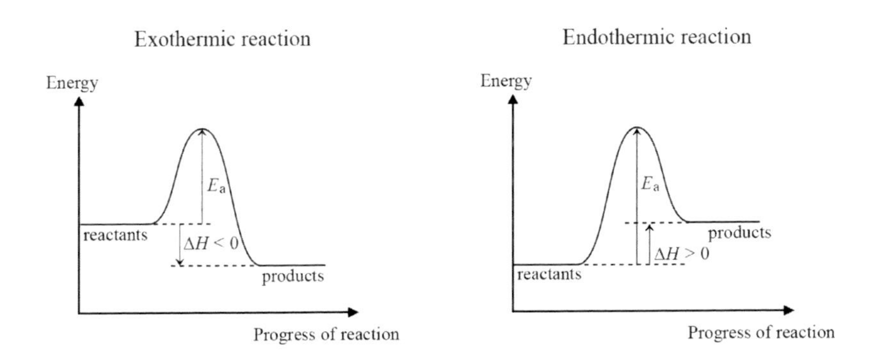

Fig. 5.1. Energy profile diagrams for an exothermic reaction (left) and endothermic reaction (right).

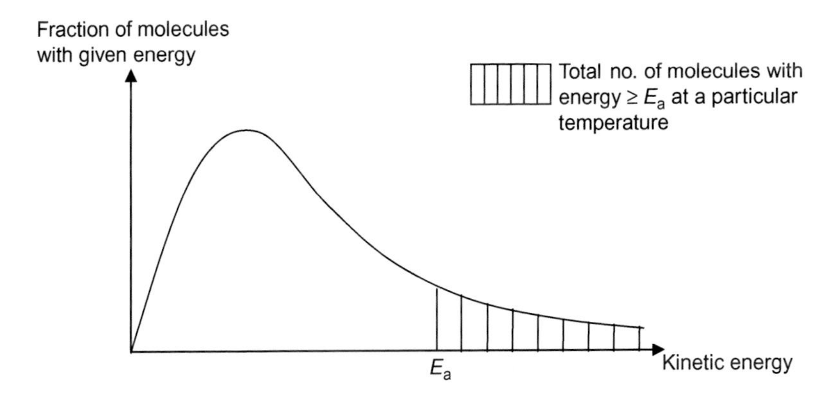

Fig. 5.2. Maxwell–Boltzmann energy distribution curve.

Q: Why is E_a labelled as the minimum energy needed? Shouldn't all the reactant particles have the same energy?

A: The particles in a system are constantly moving around, vibrating or rotating in space. Energy is constantly being transferred from one particle to another when they collide. As a result, not all particles possess the same specific amount of energy at a specific time.

The Maxwell–Boltzmann energy distribution curve (see Fig. 5.2) is a useful representation of the amount of kinetic energy possessed by a particular fraction of particles. Although at a specific point of time, the kinetic energy of a particle is constantly changing (due to collision), if the temperature of the system is maintained at a constant level, then the system of particles has a fixed distribution profile of energy states. This distribution profile of energy states is dependent on the temperature.

The area under the profile indicates the total number of particles in the system. The highest peak in the profile gives the most probable kinetic energy that the system would have at a particular temperature. If the temperature of the system is increased, then the whole distribution profile shifts to the right, indicating an increase in the kinetic energy of the system.

5.1.1 *Factors affecting reaction rates*

There are various factors that can influence the rate of a chemical reaction. Some of these factors are:

(i) temperature,
(ii) concentration of reactants,
(iii) physical states of reactants,
(iv) catalysts.

5.1.1.1 *Effect of temperature*

The average kinetic energy of a system is directly proportional to the temperature of that system. As temperature rises, the **average kinetic energy** of the reacting particles **increases**. This leads to an increase in the frequency of collisions. In addition, the increase in temperature also leads to a **significant increase in the number of reactant particles having energy greater than or equal to the activation energy** (E_a). Consequently, the **frequency of effective collisions increases** and so reaction rate increases. Subsequently, we can see that an increase in reaction rate brought about by an increase in temperature can be mathematically translated as an **increase in the rate constant value**.

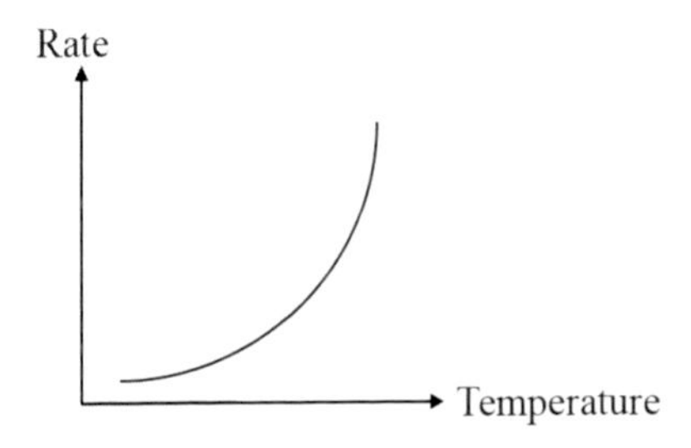

The evidence of this in the Maxwell–Boltzmann distribution curve corresponds to a lower maximum of the profile at the higher temperature, with the profile displaced towards the region of higher kinetic energy such that the total shaded area, representing the number of molecules with energy greater than or equal to E_a, is larger.

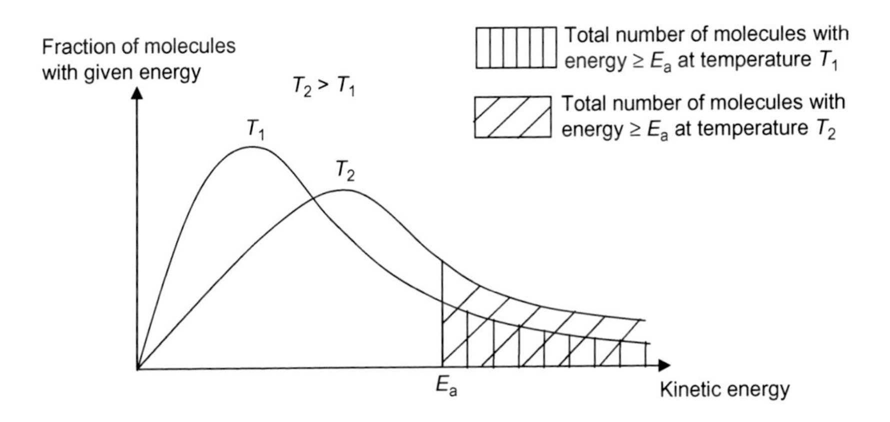

In general, ***increasing the temperature increases the rate of a reaction***. For many reactions, the rate is approximately doubled for every 10 K rise in temperature.

5.1.1.2 *Effect of concentration of reactants*

Picture yourself shopping at the mall on a crowded weekend. The chances of bumping into someone during the peak shopping period are so high that we tend to "react" to the annoyances.

Similarly, when the concentration of a reactant increases, the frequency of collisions increases.

When the reactant particles, especially those with kinetic energy greater than or equal to E_a, bump into one another more often, the chances of a successful reaction occurring will be much higher, leading to an increase in the frequency of collisions and hence the reaction rate.

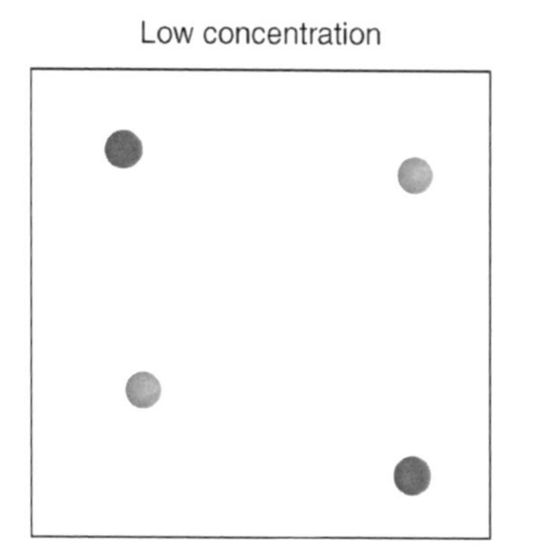

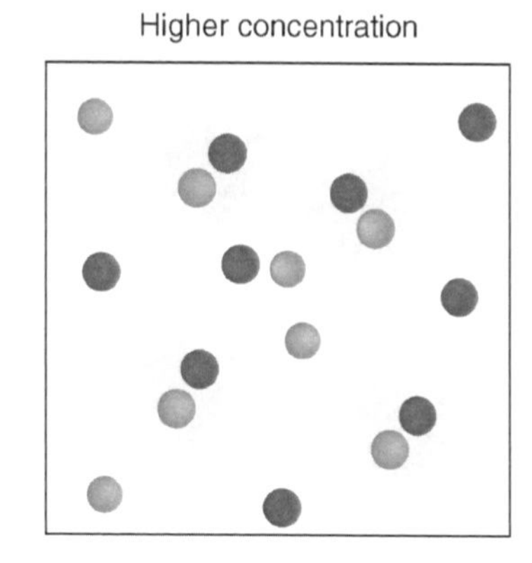

When it comes to dealing with gases, we often use pressure as a quantifiable measure. The effect of pressure changes on reaction rate can be simply accounted for in a similar manner as concentration changes.

Recall the ideal gas law can be expressed in terms of pressure as:

$$p = \left(\frac{n}{V}\right) RT,$$

where (n/V) is a concentration term expressed in $\mathrm{mol\,m^{-3}}$.

Here, we see that the partial pressure of a gas is directly proportional to its concentration (n/V).

Thus, an increase in the pressure on a gaseous system simply means an increase in the effective concentration of the reactants, and with a higher frequency of effective collisions, the reaction rate increases.

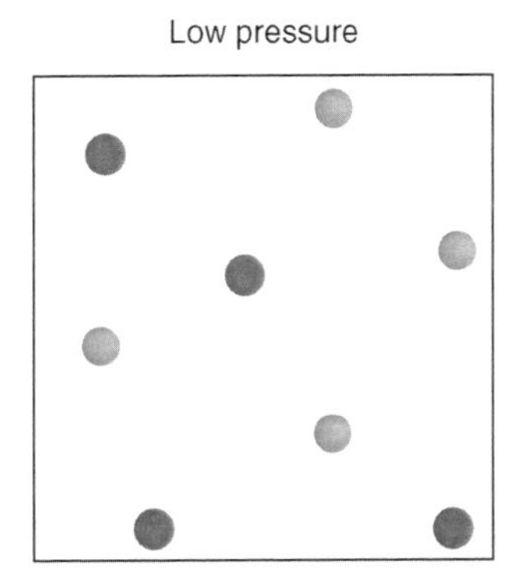

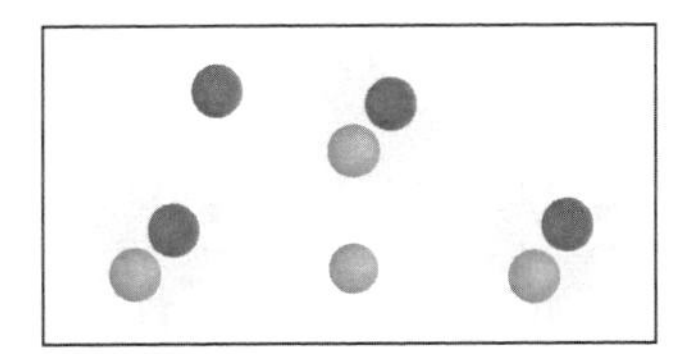

Pressure changes have a negligible effect on the volume of solids and liquids and consequently their concentrations, since they are not as compressible as gases.

5.1.1.3 *Effect of physical state of reactants*

Have you ever observed that it is much faster to dissolve a sachet of sugar grains compared to a sugar cube? A substance reacts more quickly if it is presented in smaller pieces.

Water molecules will have higher chances of interacting with more sugar molecules in a shorter time when the same quantity of sugar is presented in granular form. For a sugar cube to completely dissolve, the water molecules can only interact with the exposed sugar molecules at the surface, and when only these are dislodged from the lattice, can the inner sugar molecules then interact with the water molecules (just like "peeling of an onion"). As you can imagine, it will take a longer time for the sugar cube to dissolve in this way.

Given a fixed mass of a substance, there is greater net surface area available for reactions to take place when the size of the substances are smaller.

The following serves to illustrate this:

Surface area of sugar cube

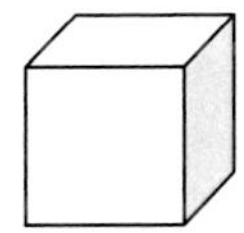

Surface area of cube, now broken into smaller chunks

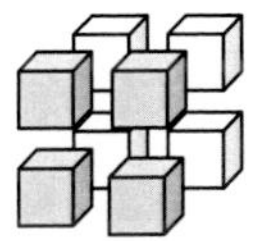

Reactions involving liquids or gases occur much faster than those involving solids. Dissolving a substance is a means of providing much greater surface area for reactions to take place as we are dealing with the smallest particles available — ions or molecules.

For instance,

(i) $NaCl(s) + AgNO_3(s) \rightarrow$ *no reaction.*
(ii) $NaCl(aq) + AgNO_3(aq) \rightarrow NaNO_3(aq) + AgCl(s)$ (*immediate reaction*)

The ratio of surface area to mass is greater in small particles than in large particles. This implies that the area over which the solid can come into contact with liquid or gaseous reactants is greater.

5.1.1.4 *Effect of catalysts*

A catalyst is a substance that increases the rate of a chemical reaction without actually being consumed in the reaction itself. A substance that causes a decrease in rate is known as an inhibitor.

A catalyst does its job by:

- **orientating** reactant particles so that they achieve the **correct collision geometry**;
- **locally increasing concentrations** of the reactant particles (for heterogeneous catalysts);
- **weakening the intra-molecular bonds** of the reactant molecules;
- **facilitating ease of transfer of electrons** in an oxidation-reduction reaction (homogeneous catalysts).

A catalysed reaction proceeds via an **alternative reaction mechanism** compared to an uncatalysed reaction.

A catalyst is said to provide an alternative pathway of lower E_a as shown in the energy profile diagram (Fig. 5.3).

The following Maxwell–Boltzmann distribution curve (Fig. 5.4) shows the distribution of kinetic energies of reacting particles and the activation energies of both the catalysed and uncatalysed reactions.

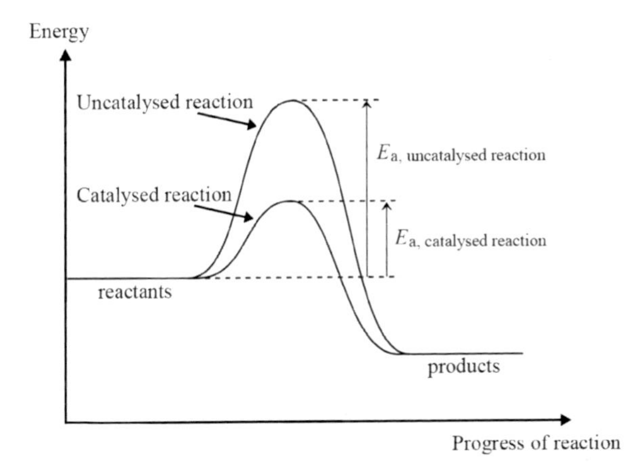

Fig. 5.3. Energy profile diagrams for a catalysed and uncatalysed reaction.

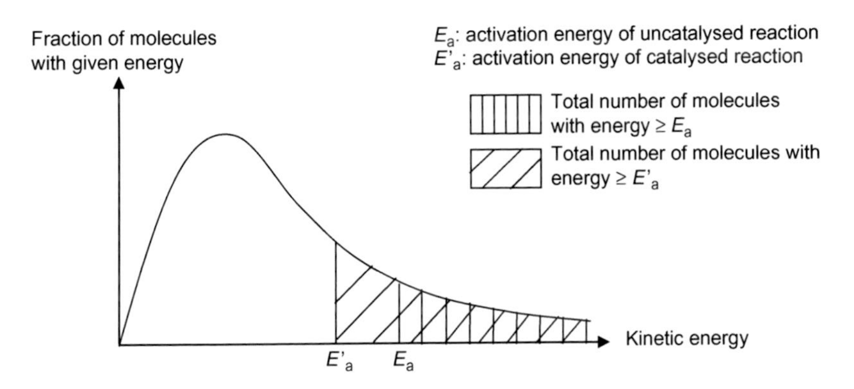

Fig. 5.4. Maxwell–Boltzmann energy distribution curve showing the effect of a catalyst on the reaction rate.

Notice the greater proportion of reacting particles having sufficient kinetic energy greater than activation energy for the catalysed reaction as indicated by the larger shaded area in the diagram.

Thus, having a **greater number of reactant particles with kinetic energy greater than or equal to the activation energy (E_a) results in an increase in the frequency of effective collisions**. This leads to an **increase in the reaction rate**. Subsequently, we will see that an increase in reaction rate brought about by a lowered E_a due to the presence of a catalyst can be mathematically translated as an **increase in the rate constant value**.

Further discussion on the type and role of catalysts will be covered later in this chapter.

5.2 Quantitative Analysis of Reaction Rates

In an earlier section, we discussed qualitatively the factors affecting reaction rate. We know that an increase in concentration will lead to an increase in reaction rate but sometimes we need to, and want to, put numbers on things. How fast and how much is really enough?

It will be good to know how much of a concentration change is needed to bring about a significant difference in reaction time and how fast the reaction can arrive at completion — a few hours or even days?

In this section, we will cover the quantitative measurements of reaction rates and the treatment of the experimental data obtained.

5.2.1 *Rate of reaction*

Rate of reaction is defined as the **change in concentration of a reactant or product with time**. It is analogous to speed, which is the change in distant with time!

Mathematically, it can be expressed as:

$$\text{Rate of reaction} = \frac{d[\text{product}]}{dt} \quad \text{or} \quad \text{Rate of reaction} = -\frac{d[\text{reactant}]}{dt}.$$

The negative sign is to account for the decreasing concentration of reactant over time.

The rate of reaction has the units of $\mathbf{mol\,dm^{-3}\ time^{-1}}$, where time is usually in seconds or minutes.

When experimental values of concentration and time are plotted graphically, we may obtain the following graphs shown in Fig. 5.5.

As a reaction progresses, the amount of product formed, and hence its concentration, increases over time. This is indicated by the

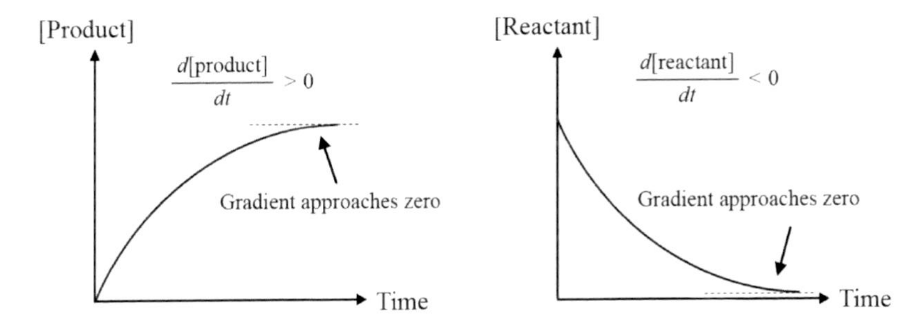

Fig. 5.5. Graphs of [product] against time (left) and [reactant] against time (right).

upward sloping curve (see Fig. 5.5, left). Meanwhile, more reactants are used up to form the products and we get a downward sloping curve when plotting [reactant] over time (see Fig. 5.5, right).

The rate of reaction at any specific time t can be determined mathematically by equating it to be:

- the gradient of the tangent at time $= t$ on a [product]–time graph, or
- the negative of the gradient of the tangent at time $= t$ on a [reactant]–time graph (since concentration of reactant decreases over time).

A steeper gradient indicates a faster rate of reaction. A gradient of zero value indicates that the rate of reaction is zero, i.e., the reaction has stopped.

Recall from the earlier section that an increase in concentration of a reactant leads to a faster rate of reaction. Looking at both graphs, you will notice that the slopes are steepest at the start of the reaction when the concentrations of the reactants are at their maximum. As the reaction progresses, the gradients of the curves become gentler until they reaches zero, indicating that the reaction has stopped.

For most reactions, rate decreases with time, since reactants are consumed as the reaction proceeds. From this, we can derive the following relationship between rate and time:

$$\textbf{rate} \propto \frac{\mathbf{1}}{\textbf{time}}.$$

Q: Do you mean to say that there are reactions where the rate actually increases with a decrease in the concentration of reactant?

A: Yes, there are such reactions. One example is autocatalysis wherein a **product that is formed in the reaction serves as its catalyst.** More details are covered in Section 5.5.3.

When we have a reaction of the following form: $X + Y \rightarrow Z$, we can measure the rate of reaction by following (i) the rate of consumption of X, (ii) the rate of consumption of Y, or (iii) the rate of formation of Z.

However, not all reactions have the reactants reacting in stoichiometric ratio of 1:1. Let us consider the following reaction: $X + 3Y \rightarrow 2Z$.

You will notice that for every mole of X being used up, the amount of Y will be used up three times as fast, and Z is formed twice as fast as X is consumed.

Which of these substances' concentration changes should we thus equate to the rate of the overall reaction? It may be any one of these three substances, but we must first set a reference:

$$\begin{aligned}\text{Rate of consumption of X} &= 1/3 \times \text{Rate of consumption of Y}\\ &= 1/2 \times \text{Rate of formation of Z.}\end{aligned}$$

We now have *rate of reaction* defined as:

$$\text{Rate of reaction} = -\frac{d[\mathrm{X}]}{dt} = -\frac{1}{3}\frac{d[\mathrm{Y}]}{dt} = \frac{1}{2}\frac{d[\mathrm{Z}]}{dt}.$$

Thus, in general, for a given reaction: $w\mathrm{W} + x\mathrm{X} \rightarrow y\mathrm{Y} + z\mathrm{Z}$,

The rate of reaction can be equated to any of these substances as follows:

$$\text{Rate of reaction} = -\frac{1}{w}\frac{d[\mathrm{W}]}{dt} = -\frac{1}{x}\frac{d[\mathrm{X}]}{dt} = \frac{1}{y}\frac{d[\mathrm{Y}]}{dt} = \frac{1}{z}\frac{d[\mathrm{Z}]}{dt}.$$

Q: So, are you saying that when we define the so-called rate of reaction, it may not necessary be numerically equivalent to the quantitative change in the concentration of reactant or product with time?

A: You are absolutely right! The rate of reaction is a concept pertaining to a particular reaction stoichiometric equation and this value

MAY NOT necessarily be NUMERICALLY EQUIVALENT to the quantitative change in the concentration of reactant with time (rate of consumption) or that of product with time (rate of formation)! Nevertheless, even though there may be a numerical difference between the rate of reaction and, for instance, the rate of consumption, take note that the difference would just be a constant factor.

In the study of kinetics, we are concerned with the initial rate of reaction of the reactants. The initial rate of reaction is defined as the change in concentration of a reactant or a product at time $t = 0$. We make use of the initial rate because at other times the rate would be affected by the concentration of the reactants and we might not know its value.

It is at this time of instantaneous mixing that the reaction rate is the fastest since at the very start of the reaction:

- the concentration of the reactants is at a maximum, and
- only an infinitely small amount of (i) the reactant has been used up, and (ii) the product has been formed.

$$\text{Recall that rate of reaction} = -\frac{d[\text{reactant}]}{dt} = \frac{d[\text{product}]}{dt}.$$

From a concentration–time graph, the initial rate of reaction is obtained by calculating the gradient of the tangent drawn to the curve at $t = 0$.

Q: Why can we not just calculate the average rate of reaction? Calculating the average rate is easier than drawing a good tangent.

A: The average rate of reaction is defined as the change in concentration of a reactant or product over a specified time interval, which can be short or covers the entire duration of a reaction. Hence, it becomes very subjective to define the range to calculate the average rate. And do not forget that if the rate of the reaction is affected significantly by the concentration of the reactant, then it becomes difficult to pinpoint what value of concentration the rate depends on. As such, an average rate of reaction does not necessary inform us of the maximum reaction rate attainable.

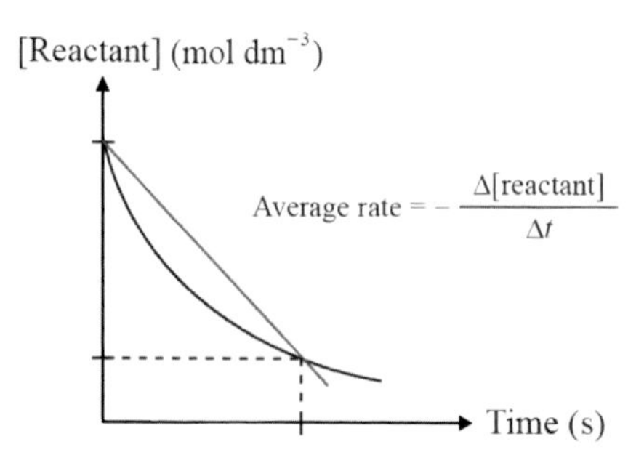

However, the average rate can still be useful if we measure it over a very short time interval and a good time would be at the point of initial mixing of the reactants. This is the very same idea as the measuring of the gradient of the tangent to the curve at $t = 0$! Recall that gradient of tangent $= -\Delta[\text{reactant}]/\Delta t$. Thus, you see that initial rate is actually a specific average rate and determining it is more objective.

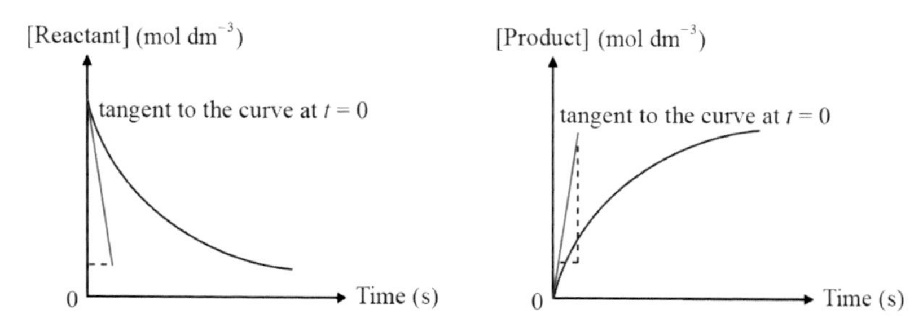

We now understand how it is possible to extract information on reaction rates from the experimental data. But wouldn't it be cumbersome to always have to perform a reaction just to get quantified figures and then repeat the experiment again but with adjustments done to the concentration of reactants to improve the reaction time?

That is why there is something good about scientific experiments. The reproducibility and accuracy of such controlled activities allow us to devise mathematical relationships that we can use for future predictions or in the design of similar experiments. When it comes to kinetics, we have the mathematical expression known as the rate equation.

5.2.2 *The rate equation (Rate law)*

A rate equation informs us of the quantified relationship between the rate of reaction and the concentration of the reactants.

It is expressed as:

$$\text{Rate} = k[\text{A}]^x[\text{B}]^y,$$

where:

- rate has the units of mol dm^{-3} time^{-1},
- x and y are known as the order of reaction, and
- k is the rate constant for a given reaction. It is temperature-dependent and its units depend on the overall order of reaction.

The order of reaction with respect to a given reactant is the power to which the reactant's concentration is raised in the rate equation. The sum of the orders of the reaction (i.e., $x + y$) gives the overall order of reaction. These are experimentally determined quantities; they are not equivalent to the stoichiometric coefficients of the reactants in the balanced chemical equation. It is not necessary that all reactants appear in the rate equation.

The Arrhenius Equation

The Arrhenius equation gives the quantitative basis of the relationship between the activation energy, temperature and the rate at which a reaction proceeds:

$$k = A\exp(-E_\text{a}/RT),$$

where

k = rate constant,

A = Arrhenius constant,

E_a = activation energy,

(*Continued*)

(Continued)

R = molar gas constant,
T = temperature in Kelvin.

For a particular given reaction, both A and E_a are constants. k is a temperature-dependent constant for a given reaction.

This equation is useful in quantitative calculations for finding E_a and the others. From the Arrhenius equation, it can be observed that

- an increase in temperature T leads to a larger k and therefore the reaction rate is increased, and
- using a catalyst leads to a smaller E_a and hence a larger k.

5.2.2.1 *Common types of order of reaction*

With reference to an arbitrary reaction A+B → C, the rate equation takes the general form of: Rate = $k[A]^m[B]^n$, where k is the rate constant.

Common values of m and n are 0, 1 and 2 but are not limited to these only.

Combinations of these values give rise to different types of rate equation as follows:

- When both $m = 0$ and $n = 0$.

The order of reaction with respect to each reactant is **zero**. This means that reaction rate is independent of [A] and [B], i.e., changing [A], [B] or both, does not affect the reaction rate.

This translates to a downward sloping straight line on the [reactant]–time graph with a constant gradient and hence a constant reaction rate.

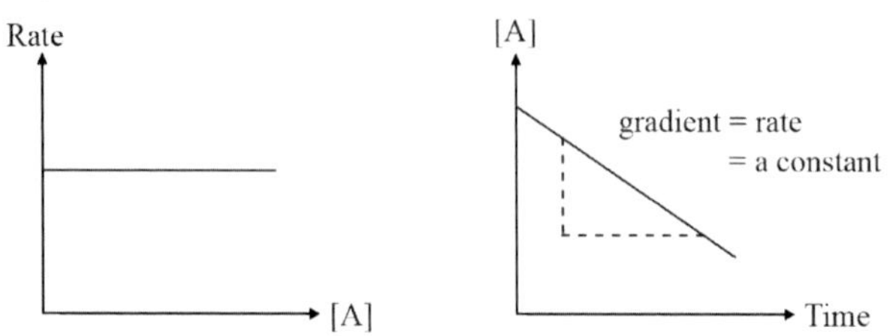

The reaction is known as an overall **zero-order reaction**.

The rate equation is expressed as: Rate = k, and the units of k are $\mathbf{mol\,dm^{-3}\,s^{-1}}$.

Did you know?

For a zero-order kinetics, we have:

$$\text{Rate} = -\frac{d[\text{A}]}{dt} = k.$$

Let C_o = [A] at time zero and C = [A] at any time t. Then,

$$-\frac{dC}{dt} = k \quad \text{and} \quad -dC = kdt.$$

Integrating, we get

$$-\int_{C_\text{o}}^{C} dC = k\int_0^t dt$$
$$-[C]_{C_\text{o}}^{C} = k[t]_0^t.$$

Hence,

$$C = C_\text{o} - kt.$$

At $t = t_{1/2}$, $C = C_\text{o}/2$, and therefore $t_{1/2} = C_\text{o}/2k$.

- When $m = 1$ and $n = 0$.

The order of reaction with respect to reactant A is **one**. This means that the reaction rate is directly proportional to [A], i.e., rate $\propto$ [A]. When [A] is doubled, the reaction rate is doubled.

The order of reaction with respect to reactant B is zero. This means that the reaction rate is independent of [B].

In all, the reaction is said to be an overall **first-order reaction**. The rate equation is expressed as: Rate = k[A] and the unit of k is $\mathbf{s^{-1}}$.

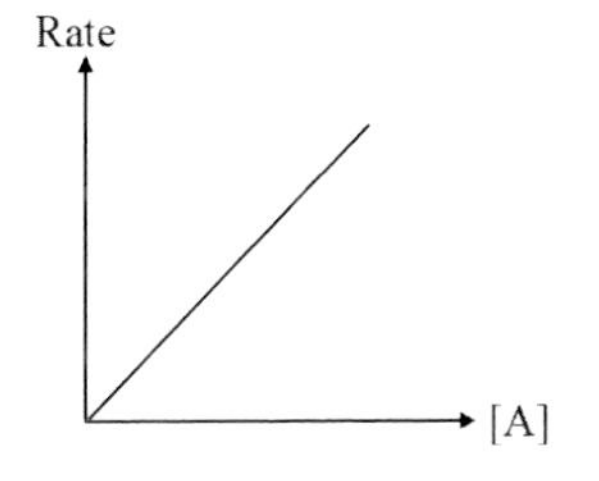

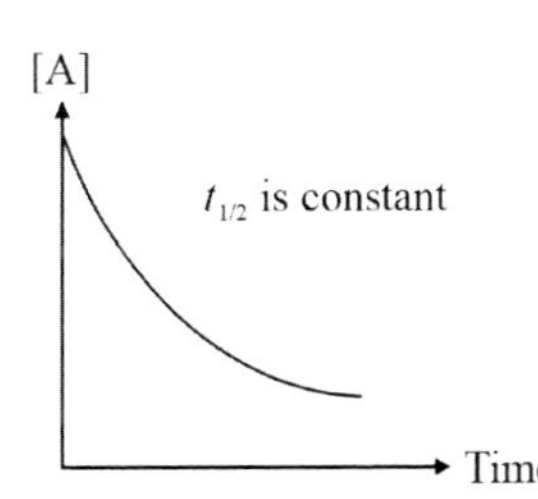

Q: What is $t_{1/2}$?
A: The half-life of a reactant ($t_{1/2}$) is the time taken for its concentration to decrease to half of its original value.

The notion of half-life is one useful concept for describing reaction rate. Why? It is sometimes meaningless to ask how long it takes for a reaction to be completed because when the concentration of a reactant becomes so small, theoretically, the reaction time will tend to infinity. The concept of half-life was originally coined for use in radioactive decay, but it has been applied to other fields such as the study of kinetics.

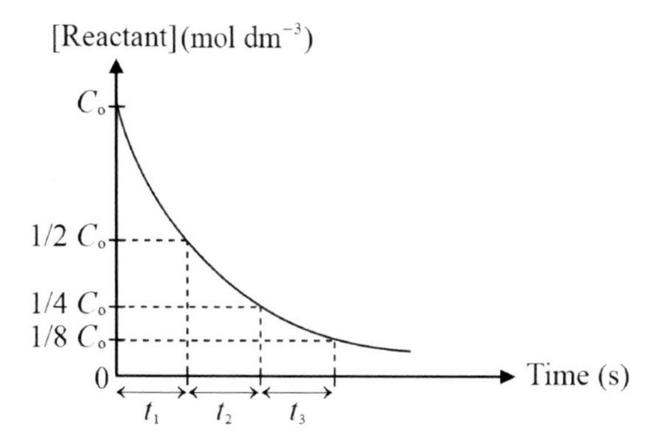

A first-order reaction has a constant half-life. This means that the half-life is independent of initial concentration of reactant and it is related to the rate constant by the equation: $t_{1/2} = \frac{\ln 2}{k}$.

Did you know?

For a first-order kinetics, we have:

$$\text{Rate} = -\frac{d[\text{A}]}{dt} = k[\text{A}]$$

Let $C_o = [\text{A}]$ at time zero and $C = [\text{A}]$ at any time t. Then,

$$-\frac{dC}{dt} = kC \quad \text{and} \quad -\frac{dC}{C} = kdt.$$

(*Continued*)

(*Continued*)

Integrating, we get

$$-\int_{C_o}^{C} \frac{dC}{C} = k \int_0^t dt$$

$$-\ln[C]_{C_o}^{C} = k[t]_0^t.$$

Hence,

$$-\ln C + \ln C_o = kt.$$

Rearranging gives

$$\ln C = -kt + \ln C_o \quad \text{or} \quad \ln \frac{C_o}{C} = kt.$$

At $t = t_{1/2}$, $C = C_o/2$, and therefore $\ln 2 = kt_{1/2}$.

Thus, if we find that the half-life is constant from the [reactant]-time plot, then the reaction is first order with respect to that reactant.

- When $m = 2$ and $n = 0$.

The order of reaction with respect to A is **two**. Rate $\propto$ $[A]^2$. When [A] is doubled, the rate is quadrupled (increased by four times). The order of reaction with respect to B is zero. The overall order of reaction $= m + n = 2 + 0 = 2$.

The reaction is said to be an overall **second-order reaction**. The rate equation is expressed as: Rate $= k[A]^2$ and the units of k are $\mathbf{mol^{-1}\ dm^3\ s^{-1}}$.

Did you know?

For a second-order kinetics, we have:

$$\text{Rate} = -\frac{d[A]}{dt} = k[A]^2.$$

Let $C_o = [A]$ at time zero and $C = [A]$ at any time t. Then,

$$-\frac{dC}{dt} = kC^2 \quad \text{and} \quad -\frac{dC}{C^2} = kdt.$$

Integrating, we get

$$-\int_{C_o}^{C} \frac{1}{C^2} dC = k \int_0^t dt$$

$$\left[\frac{1}{C}\right]_{C_o}^{C} = k[t]_0^t.$$

Hence,

$$\frac{1}{C} = \frac{1}{C_o} + kt.$$

At $t = t_{1/2}$, $C = C_o/2$, and therefore $t_{1/2} = 1/(kC_o)$.

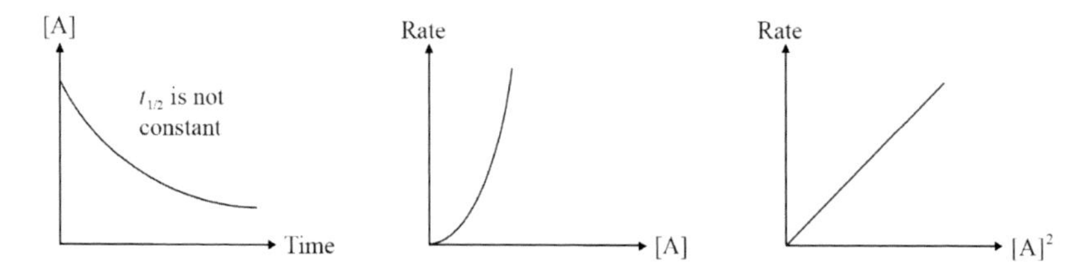

An overall second-order reaction is also observed when both $m = 1$ and $n = 1$.

In this case, the order of reaction with respect to each reactant is **one**, i.e., rate $\propto$ [reactant].

The rate equation is expressed as: Rate $= k$[A][B] and the units of k are $\mathbf{mol^{-1}\ dm^3\ s^{-1}}$.

Q: For the overall second-order reaction whereby rate $= k$[A][B], since rate $\propto$ [A], will the graph of [A] against time indicate a constant half-life?

A: It depends! If [B] used is relatively high enough such that [B] $\gg$ [A], then the relative changes in [B] are so much smaller that [B] can essentially be regarded as a constant throughout the reaction.

Q: For this same experiment, if we map out the [B] with respect to time, would it give us a constant half-life?

A: No! This is because the [A] is not in excess relative to [B] here. Thus, the changing [A] would affect rate. So in order for the reaction to be pseudo-1st order with respect to [B], the [A] must be in excess relative to [B].

The rate equation can then be simplified as: Rate $\approx k'[\mathrm{A}]$ where $k' = k[\mathrm{B}] = \text{constant}$.

This reaction is regarded as a **pseudo-first-order** reaction with respect to A.

In this case, the [A]–time plot will indicate a constant half-life.

Q: For a second order reaction whereby rate $= \mathrm{k}[\mathrm{A}]^2[\mathrm{B}]^0$, will the graph of [B] against time be a liner straight line?

A: Yes, provided the [A] is high enough at the beginning. If [A] is not high enough, then the [B] against time plot will follow the profile of the [A] against time plot. Similarly, for rate $= k[A][B]^0$, the [B] against time plot will follow that of [A] if [A] is not in excess relative to [B].

5.2.3 *Experimental methods used to determine order of reaction*

Since rate is affected by concentration, then in order to determine the order of reaction with respect to a particular reactant, one needs to experimentally monitor the concentration of the reactant of interest with respect to time. Experiments performed to find the order of reaction involve either

- discontinuous measurement, or
- continuous measurement.

Essentially, we want to determine how concentration changes affect the reaction rate. Firstly, we need to relate these concentration changes to a particular physical property whose changes can be measured during the progress of the reaction.

Examples of physical properties of substances that can be linked to concentration changes are visible changes such as formation of product precipitate, colour changes, volume of gas evolved, pressure of gas and electrical conductivity.

5.2.3.1 *Experiment involving discontinuous measurement*

This method entails performing separate sets of the same experiment by using different starting concentrations of the reactants.

The time taken for a prominent visible change to occur is noted for each experiment. Such experiments are known as "clock reactions."

Example of a clock reaction: Reaction of sodium thiosulfate with HCl(aq)

$$Na_2S_2O_3(aq) + 2HCl(aq) \longrightarrow \underset{\text{pale yellow}}{S(s)} + SO_2(g) + 2NaCl(aq) + H_2O(l).$$

The rate of reaction between sodium thiosulfate and dilute HCl can be conveniently studied by measuring the time taken for a definite, small amount of sulfur to be formed.

How do we quantify this amount? A good gauge is observing a sufficient amount of yellow sulfur masking an image on a piece of white paper placed underneath the reaction beaker (see Fig. 5.6).

The basic idea is that different concentrations of the reactants result in different rates of formation of the sulfur product and thus the times taken for its formation will differ accordingly.

For each experiment, a measured volume of one reagent is introduced into a beaker, which has a paper with an "X" placed underneath it. As soon as the other reagent is added to the solution in the beaker, a stopwatch is started to note the time taken for the "X" to be masked by the sulfur formed when viewed through the mouth of the beaker. This time is then tabulated.

To determine the order of reaction with respect to one reactant, a minimum of two sets of experiments are needed in which its concentration is varied while that of the other reactants are held constant. Since there are two reactants whose order of reaction are to be determined, a typical procedure will entail a set of three experiments to

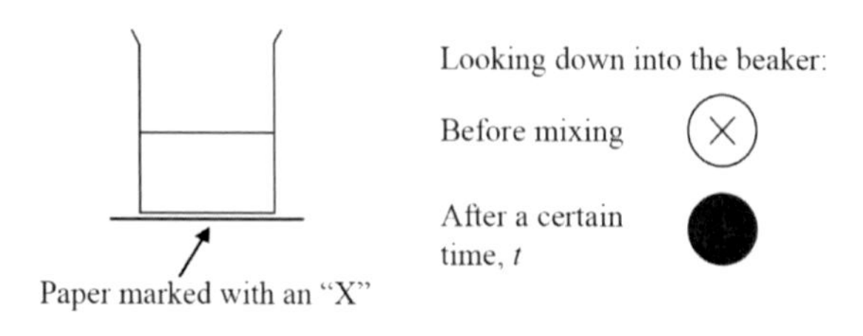

Fig. 5.6.

be performed using varying volumes of reagents (of known concentration).

Experiment	Volume of $Na_2S_2O_3$ (cm^3)	Volume of HCl (cm^3)	Volume of H_2O (cm^3)	Time Δt Taken for "X" to be Masked (s)
1	10	10	10	
2	20	10	0	
3	10	20	0	

Q: Why do we need to add water, which is not a reactant?

A: Note that the total volume of the two reactants is not the same in all three experiments. It is $20\,cm^3$ for Experiment 1 and $30\,cm^3$ for Experiments 2 and 3. An appropriate volume of water is added to the first mixture to ensure that the total volumes of all three mixtures are the same. In this way, the **initial concentration of each reactant in the reaction mixture is then directly proportional to its volume used**. If this were not done, we would never get two different experimental sets whereby the concentration of one reactant changes while the rest are being kept constant. All initial concentrations would be different! Try figuring this out yourself.

Let us say we have the results tabulated as follows:

Experiment	Volume of $Na_2S_2O_3$ (cm^3)	Volume of HCl (cm^3)	Volume of H_2O (cm^3)	Time Δt Taken for "X" to be Masked (s)
1	10	10	10	40
2	20	10	0	20
3	10	20	0	20

The order of reaction with respect to each reactant is calculated using what we call the **initial rates method**.

The initial rates method for deducing order of reaction

Recall that average rate is a good approximation of initial rate provided it is measured at a very short time interval and at the point of initial mixing of the reactants, i.e., measuring the gradient of the tangent to the curve at $t = 0$.

If the time elapsed is too long, then the concentration of each of the reactants changes to a value that may be very different from its original value. And remember that different concentrations lead to different rates of reaction. Thus, the measured rate cannot be linked to the original concentrations of the reactants.

But in this case, the initial rate of the reaction is approximated by the average rate of the reaction since the **approximation is a good one due to the sufficiently small time interval**.

Q: How do we determine if the time interval is small enough?

A: This is a very good question. The time interval we use cannot be too short as this will cause large errors of uncertainty in the time measurement as well as in obtaining observable data. Neither can it be too long because by then the concentrations of the reactants may have changed so significantly that the average rate does not tie in with the value of concentration that we have in mind.

In practice, the average rate of **reacting a fixed concentration of a reactant or forming a fixed concentration of product** is actually measured.

In fact, since we are measuring the average time taken for the formation of a fixed amount of sulfur product,

$$\text{Initial rate of reaction} = \frac{d[\text{product}]}{dt}$$

can be re-expressed as:

$$\text{Rate} = \frac{\Delta[\text{product}]}{\Delta t}.$$

The fact that the amount of sulfur formed is considered fixed translates to a fixed concentration of sulfur formed (i.e., $\Delta[\text{product}] =$ constant):

$$\text{Rate} = \frac{\text{Fixed amount of S(s) formed}}{\text{Time taken } \Delta t}.$$

Hence, the derivation of the following relationship:

$$\text{Initial rate} \propto \frac{1}{\Delta t}.$$

We can now compute a column for the initial rates in the experiments:

Experiment	**Volume of $Na_2S_2O_3$ (cm^3)**	**Volume of HCl (cm^3)**	**Volume of H_2O (cm^3)**	**Time Δt Taken for "X" to be Masked (s)**	**Initial Rate (s^{-1})**
1	10	10	10	40	0.025
2	20	10	0	20	0.050
3	10	20	0	20	0.050

With the total volume of the mixture kept constant: Initial [reactant] $\propto V_{\text{reactant}}$ used.

First, we assume the rate equation takes the general form: Rate $= k[S_2O_3{}^{2-}]^m[H^+]^n$. Next, we have to determine the values of m and n, in turn, using the data from the table.

How do we go about doing this? Firstly, we need to recognise that the order of reaction with respect to a given reactant tells us exactly how the rate of reaction is dependent on its concentration. To determine this exact dependency, we have to keep all other factors constant, i.e., the concentrations of the other reactants have to be held constant.

- Therefore, to find the value of m, we compare the results from Experiments 1 and 2.

  In these experiments, only the initial $[S_2O_3{}^{2-}]$ differs while $[H^+]$ is held constant.

  It can be seen that when $[S_2O_3{}^{2-}]$ is doubled (since volume of $S_2O_3{}^{2-}$ is doubled), the rate of reaction also doubles.

  Hence, rate $\propto [S_2O_3{}^{2-}]$ and the order of reaction with respect to $S_2O_3{}^{2-}$ is one, and therefore $m = 1$.
- To find the value of n, we compare the results from Experiments 1 and 3.

In these experiments, only the initial $[H^+]$ differs while $[S_2O_3^{2-}]$ is held constant.

It can be seen that when $[H^+]$ is doubled (since volume of H^+ is doubled), the rate of reaction also doubles.

Hence, rate $\propto [H^+]$ and the order of reaction with respect to H^+ is one, and therefore $n = 1$.

- In determining m and n, the following information can also be obtained:
  - Rate $\propto [S_2O_3^{2-}][H^+]$.
  - Rate equation: Rate $= k[S_2O_3^{2-}][H^+]$.
  - Overall order of reaction $= 2$.

  We can go on to calculate the rate constant k by using the rate equation as we will see in the following example.

Example 5.1: The reaction kinetics of the following reaction was studied by monitoring the rate of formation of I_2:

$$H_2O_2(aq) + 2H^+(aq) + 2I^-(aq) \rightarrow 2H_2O(l) + I_2(aq).$$

Using the results shown in the table,

(i) find the order of reaction with respect to each reactant using the initial rates method, and

(ii) hence, state the rate equation and use it to calculate the rate constant, k.

Experiment	**Initial $[H_2O_2]$ ($mol\,dm^{-3}$)**	**Initial $[I^-]$ ($mol\,dm^{-3}$)**	**Initial $[H^+]$ ($mol\,dm^{-3}$)**	**Initial Rate of Formation of I_2 ($mol\,dm^{-3}\,min^{-1}$)**
I	0.010	0.015	0.20	2.0×10^{-4}
II	0.015	0.020	0.10	4.0×10^{-4}
III	0.030	0.015	0.20	6.0×10^{-4}
IV	0.010	0.030	0.20	4.0×10^{-4}

Approach:

(i) To determine the order of reaction with respect to each reactant:

- pick two sets of data wherein concentration of one reactant changes and others are held constant.

(ii) To calculate the numerical value of k:

- pick any one set of experimental data,
- insert values for [reagents] and reaction rate into the rate equation to solve for k, and
- do not forget the units for k.

Solution for (i):

- To determine the order of reaction with respect to H_2O_2, compare Experiments I and III: when $[H_2O_2]$ is tripled, the reaction rate also triples.
  Hence, rate $\propto [H_2O_2]$.
  The order of reaction with respect to H_2O_2 is **1**.
- To determine the order of reaction with respect to I^-, compare Experiments I and IV: when $[I^-]$ is doubled, the reaction rate also doubles.
  Hence, rate $\propto [I^-]$.
  The order of reaction with respect to I^- is **1**.
- To determine the order of reaction with respect to H^+, you will find that there are no two sets of experiments that we can inspect to determine how rate is affected solely by changes in its concentration.
  We need to employ a mathematical approach which simply compares two sets of rate equations.
  Comparing Experiments I and II, we have

$$\frac{\text{Rate}_{\text{II}}}{\text{Rate}_{\text{I}}} = \frac{4.0 \times 10^{-4}}{2.0 \times 10^{-4}} = \frac{k[H_2O_2]_{\text{II}}^x [I^-]_{\text{II}}^y [H^+]_{\text{II}}^z}{k[H_2O_2]_{\text{I}}^x [I^-]_{\text{I}}^y [H^+]_{\text{I}}^z} \tag{5.1}$$

where x and y are found to be equal to one and this simplifies Eq. (5.1), giving:

$$\frac{\text{Rate}_{\text{II}}}{\text{Rate}_{\text{I}}} = \frac{4.0 \times 10^{-4}}{2.0 \times 10^{-4}} = \frac{(0.015)(0.020)(0.10)^z}{(0.010)(0.015)(0.20)^z}$$

$$\left(\frac{0.10}{0.20}\right)^z = 1$$

$$\therefore z = 0.$$

Hence, the order of reaction with respect to H^+ is zero.

Solution for (ii): Rate equation:

$$\text{Rate} = k[H_2O_2][I^-].$$

Using data from Experiment I:

$$\begin{aligned} k &= \text{Rate} \div [H_2O_2][I^-] \\ &= 2.0 \times 10^{-4}/(0.010)(0.015) \\ &= 1.33\,\text{mol}^{-1}\,\text{dm}^3\,\text{min}^{-1}. \end{aligned}$$

As seen from the above example, not all the reactants appeared in the rate equation. Furthermore, the orders of reaction do not follow that of the stoichiometric coefficents in the chemical equation.

Exercise: The initial rate data of the following reaction are shown in the table:

$$2A + 3B + C \longrightarrow 3D + E.$$

Experiment	Initial [A] (mol dm^{-3})	Initial [B] (mol dm^{-3})	Initial [C] (mol dm^{-3})	Initial Rate (mol dm^{-3} s^{-1})
1	0.10	0.02	0.04	0.010
2	0.10	0.03	0.04	0.015
3	0.20	0.02	0.08	0.080
4	0.20	0.02	0.16	0.320

(i) Determine the order of reaction with respect to each reactant.
(ii) Hence, state the rate equation and use it to calculate the rate constant, k.

Solution:

(i) Order of reaction with respect to A is 1. Order of reaction with respect to B is 1. Order of reaction with respect to C is 2.
(ii) Rate $= k[A][B][C]^2$; $k = 3125\,\text{mol}^{-3}\,\text{dm}^9\,\text{s}^{-1}$.

5.2.3.2 *Experiment involving continuous measurement*

This method entails performing one set of experiments and monitoring concentration changes over a period of time. There are different ways to monitor the changes. One of them involves the sampling and titration method.

Sampling and titration method

For the sampling and titration method, samples of the reaction mixture are drawn at regular time intervals, and a subsequent test is performed to determine the concentration of reactants at sampling time.

Q: Won't the reaction continue in the samples drawn?
A: Yes, it will. That is why, as soon as a sample at time t is drawn, the reaction is made to stop — a process known as quenching the reaction.

Quenching can be achieved by adding:

- a large amount of cold water to the sample so that reaction rate will be so slow as to be insignificant, or
- a "quenching reagent" that actually removes one of the reactants (by reacting with it) so that the primary reaction is stopped.

Q: How does "large amount of cold water" slow down the reaction?
A: Firstly, introducing cold water decreases the rate because when temperature decreases, rate of reaction decreases. Secondly, increasing volume increases distance of separation between the reactant particles, this thus decreases the frequency of collision.

The aim of such experiments is to obtain, at regular time intervals, the concentration of remaining unreacted reactant or the concentration of product formed so that a concentration–time graph can be plotted and used for the analysis of reaction rate.

We can choose to follow the concentration of any one of the substances in the chemical equation. Our choice will be based on which substances can be easily, yet accurately, analysed.

Example: Determine the order of reaction using a [reactant]–time graph

The following equation represents the base-catalysed decomposition of hydrogen peroxide:

$$2H_2O_2(aq) \xrightarrow{OH^-} 2H_2O(l) + O_2(g).$$

A measured volume of $H_2O_2(aq)$ of known concentration is placed in a conical flask. As soon as NaOH(aq) is added to the reaction vessel, a stopwatch is started. Samples of the reaction mixture are drawn, using a pipette, at regular time intervals. The sample is quenched by adding $H_2SO_4(aq)$ so that the composition at time of sampling can be analysed. $H_2SO_4(aq)$ stops the reaction by removing the base catalyst in an acid–base neutralisation reaction.

To determine the concentration of unreacted $H_2O_2(aq)$ at each time of sampling, a titration is performed on the quenched sample using a standard solution of potassium manganate(VII).

The $H_2O_2(aq)$ in the quenched sample will react with $MnO_4^-(aq)$ as follows:

$$\underset{\text{purple}}{2MnO_4^-(aq)} + 5H_2O_2(aq) + 6H^+(aq) \longrightarrow \underset{\text{pale pink}}{2Mn^{2+}(aq)} + 5O_2(g) + 8H_2O(l).$$

The end point of titration is reached when one drop of the purple $MnO_4^-(aq)$ results in a permanent tinge of pink.

The volume of MnO_4^- used is proportional to the $[H_2O_2]$ remaining. We can then plot a graph of volume of MnO_4^- against time (see Fig. 5.7) which will give us similar information to plotting a graph of $[H_2O_2]$ against time.

Once the graph is plotted, the shape of the curve gives us crucial information about the order of reaction with respect to $H_2O_2(aq)$ (since this is the reactant whose changes in concentration we are following).

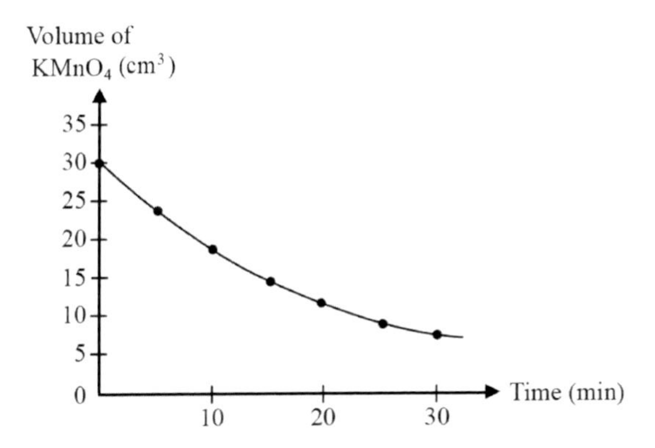

Fig. 5.7. Graph of volume of $KMnO_4$ against time.

We then apply the following **half-life method** to deduce the order of reaction.

Q: How do you prove that $[H_2O_2]$ is directly proportional to volume of MnO_4^-?

A: No. of moles of $H_2O_2 \propto$ no. of moles of MnO_4^-.
No. of moles of $MnO_4^- = [MnO_4^-]$ x volume of MnO_4^-.
Since $[MnO_4^-]$ is constant, therefore no. of moles of $MnO_4^- \propto$ volume of MnO_4^-.
$[H_2O_2] =$ No. of moles of H_2O_2/Volume of H_2O_2.
Since volume of H_2O_2 is constant, $[H_2O_2] \propto$ no. of moles of H_2O_2.
Therefore, $[H_2O_2]$ is directly proportional to volume of MnO_4^-.

The half-life method for deducing order of reaction

The half-life of a reaction ($t_{1/2}$) is defined as the time taken for the concentration of a reactant to be reduced by half, i.e., from C_o to $C_o/2$.

Based on the type of plot obtained from a [reactant]–time graph, the order of reaction with respect to that reactant can be deduced:

- If a downward sloping straight line graph is obtained, then it is a zero-order reaction with respect to H_2O_2(aq).
- If a curve is obtained, determine if the half-life is a constant.

- If half-life is constant, then it is a first-order reaction with respect to $H_2O_2(aq)$.
  <u>Half-life of a first-order reaction:</u> $t_{1/2}$ is a constant (i.e., $t_1 = t_2 = t_3$).
  It is independent of initial concentration of reactant:

$$t_{1/2} = (\ln 2)/k.$$

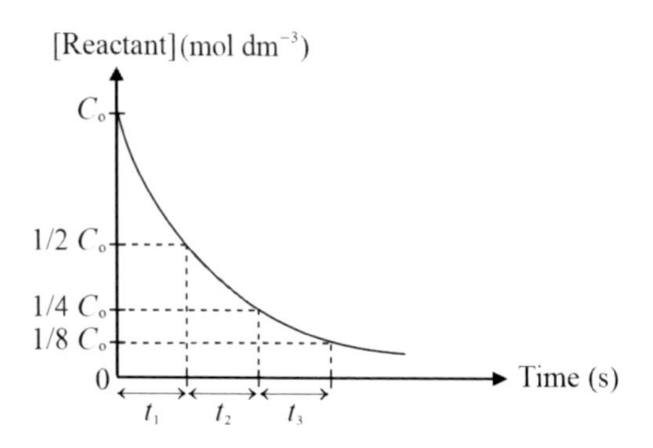

- If half-life is not constant, then it is neither a first- nor zero-order reaction with respect to $H_2O_2(aq)$.
  It may be a second-order reaction if we get an upward sloping straight line that passed through the origin when the graph of rate against $[A]^2$ is plotted.

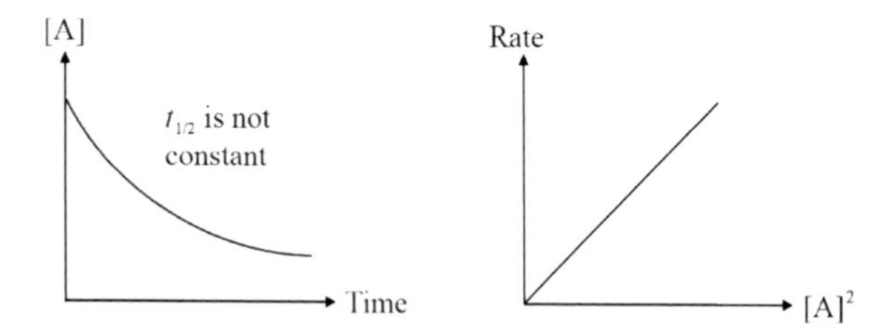

Q: How do we determine if half-life is a constant?
A: Basically, you just need to note two Δt values:

- the time taken for [reactant] to decrease from C_o to $1/2C_o = t_1$, and
- the time taken for [reactant] to decrease from $1/2C_o$ to $1/4C_o = t_2$.

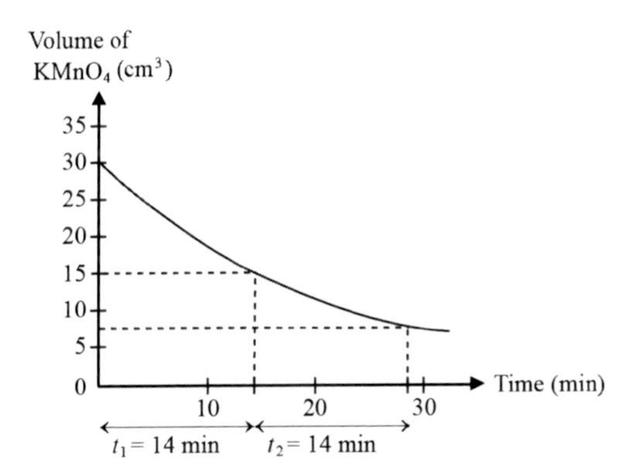

Fig. 5.8. Graph of volume of $KMnO_4$ against time.

If $t_1 = t_2$, then $t_{1/2}$ is considered a constant.
This, in turn, means that the order of reaction with respect to the reactant is 1.

With reference to the graph plotted in Fig. 5.8, since $t_1 = t_2 =$ 14 min, the **half-life of the reaction is constant** and reaction is therefore **first order with respect to H_2O_2(aq)**.

Half-life is related to rate constant and this relationship is useful. As seen in this example, the rate constant for a first-order reaction can be calculated if the half-life value is known. For instance, $t_{1/2} = \frac{\ln 2}{k} \Rightarrow k = \frac{\ln 2}{14} = 0.0495\,\text{min}^{-1}$.

Q: Is it necessary to always use the very initial value of concentration when we determine $t_{1/2}$ for a first-order reaction?

A: It is theoretically not necessary but experimentally necessary. Why? For a first-order reaction, no matter what value of concentration you used as the starting point, $t_{1/2}$ would still be the same theoretically. But unfortunately in real life, this may not be so. Why? Let us assume that the rate equation is Rate $= k$[A][B]. If we want to show that the reaction is first-order with respect to A, we need to monitor the changes in [A] with respect to time (but note that this is only possible if there is a physical attribute for reactant A to be traceable). In addition, we have to ensure that [B] is always a constant. This can be achieved by using a large excess of reactant B. And the condition whereby there is a large excess of [B], which is required to show that the reaction is first-order with respect to A, is probably true only at the very beginning of the experiment. At a later stage of the experiment, [B] may not be sufficiently large to not affect the rate of the reaction and this translates into inaccurate measurements that would not be useful and conclusive enough to find the order of reaction with respect to A.

Q: Can we also monitor the concentration of product formed during the course of reaction and use it to determine order of reaction?

A: Yes, you can. The [product]–time plot is just an inversion of the [reactant]–time plot and it reflects the effect of concentration of reactant on reaction rate.

Example: Determine the order of reaction using a [product]–time graph

Let us look at the hypothetical example of the decomposition of compound X:

$$\mathrm{X(aq)} \longrightarrow \mathrm{2Y(g) + Z(g)}.$$

The kinetics for this reaction is studied by monitoring the concentration changes of the products formed by simply measuring the volume of gas evolved at regular time intervals:

Time, t (s)	**0**	**100**	**200**	**300**	**400**	**500**	**600**	**700**	**800**	∞
Volume, V_t (cm^3)	0	17.9	30.0	38.2	45.0	50.0	52.5	54.4	56.3	60.0

Plotting a graph of volume of gas collected against time corresponds to plotting a [product]–time graph since $V_t \propto$ [product]. We can then deduce the order of reaction using the **half-life method**.

Q: What is the significance of $t = \infty$?

A: This is time taken for the reaction to reach completion. With that said, V_∞ = total volume of gas evolved at the end of the reaction, and $V_\infty \propto$ total amount of reactant reacted, i.e., the initial amount of reactant.

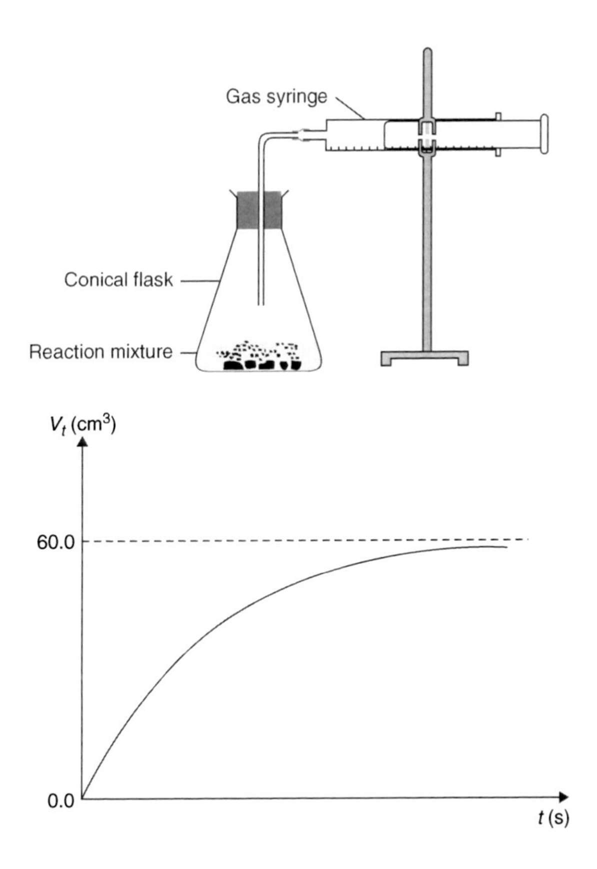

Fig. 5.9. Graph of V_t against time.

Since V_t = volume of gas evolved at time t, i.e., $V_t \propto$ amount of reactant reacted at time t, $(V_\infty - V_t) \propto$ amount of reactant that remained unreacted at time t.

Since the amount of reactant is proportional to its concentration, $(V_\infty - V_t) \propto$ [reactant] and thus plotting the graph of $(V_\infty - V_t)$ against time is similar to plotting the graph of [reactant] against time (see Fig. 5.10).

Time, t (s)	**0**	**100**	**200**	**300**	**400**	**500**	**600**	**700**	**800**	∞
Volume, V_t (cm^3)	0	17.9	30.0	38.2	45.0	50.0	52.5	54.4	56.3	60.0
$(V_\infty - V_t)$ (cm^3)	60.0	42.1	30.0	21.8	15.0	10.0	7.5	5.6	3.7	0.0

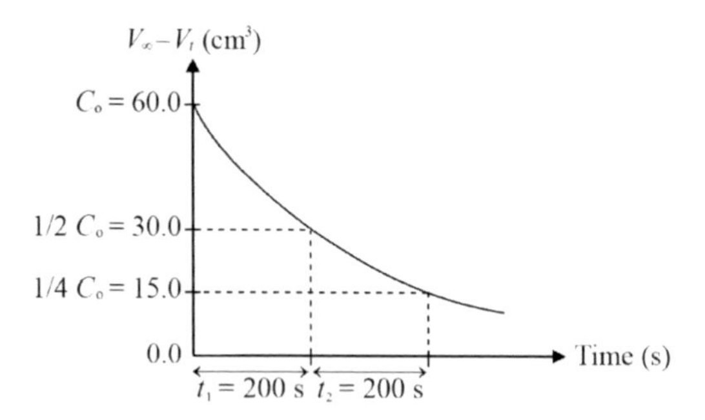

Fig. 5.10. Graph of $(V_\infty - V_t)$ against time.

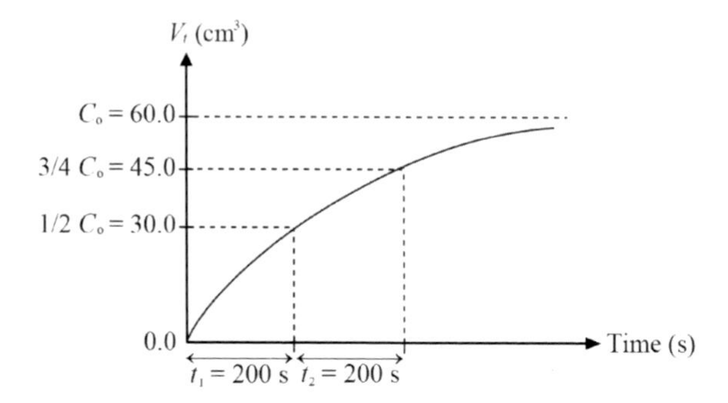

Fig. 5.11. Graph of V_t against time.

Now, compare Figs. 5.10 and 5.11. Do you see that the [product]–time plot is an inversion of the [reactant]–time plot?

Based on the [reactant]–time graph:	**Based on the [product]–time graph:**
Two sets of half-life can be obtained as the time taken for:	This same time period corresponds to:
(i) initial [reactant], C_o to decrease to $\frac{1}{2}C_o$	(i) increase in [product] from 0 to $\frac{1}{2}C_o$
(ii) $\frac{1}{2}C_o$ to decrease to $\frac{1}{4}C_o$.	(ii) increase in [product] from $\frac{1}{2}C_o$ to $\frac{3}{4}C_o$.

Hence, with reference to the graph plotted in Fig. 5.11, since $t_1 = t_2 = 200\,\text{s}$, the **half-life of the reaction is constant** and reaction is therefore **first order with respect to X(aq)**.

Example: Determine the order of reaction with respect to two reactants

Consider the case of two reactants such that $A + B \longrightarrow C$.

The rate equation takes the general form: Rate $= k[A]^m[B]^n$.

We can follow the concentration changes of each reactant in turn. However, this is not often practised. Instead, we first decide which reactant's concentration changes we can monitor easily; say, A.

Next, we perform the same type of experiment twice, using:

- the same initial concentrations of A, but
- different initial concentrations of B.

Initial [B] must be much greater than initial [A] so that, at any time during the reaction, [B] hardly changes and is thus considered essentially constant relative to the changes in [A]. We can then determine how rate is affected by [A] in this so-called pseudo-mth order reaction.

When [B] is a constant, the rate equation

$$\text{Rate} = k[A]^m[B]^n$$

can be re-expressed as

$$\text{Rate} = k'[A]^m,$$

where $k' = k[B]^n$.

From the two experiments, we can plot two graphs of [A] against time.

- To determine the order of reaction with respect to A, we use the half-life method (refer to Sec. 5.2.3.2).
- To determine the order of reaction with respect to B, we use the initial rates method (refer to Sec. 5.2.3.1).

Example: Acid-catalysed hydrolysis of methyl ethanoate in water

$$CH_3COOCH_3 + H_2O \xrightarrow{H^+} CH_3COOH + CH_3OH.$$

The rate equation takes the form of: Rate $= k[CH_3COOCH_3]^x[H^+]^y$.

Two sets of experiments are carried out with different initial $[H^+]$. The following results are obtained:

	Initial $[H^+]$ = 0.200 mol dm^{-3}	**Initial $[H^+]$ = 0.400 mol dm^{-3}**
Time (min)	**$[CH_3COOCH_3]$ (mol dm^{-3})**	
0	0.100	0.100
20	0.076	0.063
40	0.061	0.040
60	0.050	0.025
80	0.040	0.016
100	0.031	0.010
120	0.025	0.006
140	0.021	0.003

(i) First, plot the two graphs of $[CH_3COOCH_3]$ against time (see Fig. 5.12).

(ii) **Determine the order of reaction with respect to the ester by the half-life method.**

We just need to use one of the graphs since both will give the same information about half-life.

Using Curve 2 with $[H^+]$ = 0.400 mol dm^{-3}, t_1 = 30 min; t_2 = 30 min.

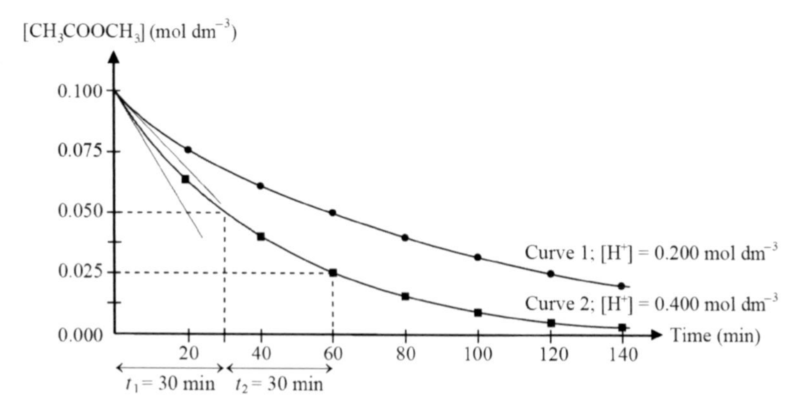

Fig. 5.12. Graph of $[CH_3COOCH_3]$ against time.

Since $t_1 = t_2 = 30\,\text{min}$, the half-life of the reaction is constant and the reaction is therefore first order with respect to CH_3COOCH_3.
Half-life is 60 min for Curve 1 with $[H^+] = 0.200\,\text{mol}\,\text{dm}^{-3}$.

(iii) **Determine the order of reaction with respect to H^+ using the initial rates method.**

In this method, we compare the changes in initial $[H^+]$ with that of the initial rate of reaction to establish the relationship between the two quantities. (Note that the comparison is made valid as at $t = 0$, [ester] is the same for both experiments.)

How do we obtain values of the initial rate of reaction from experimental data?

- For each graph, draw a tangent to the curve at time $t = 0$.
- Measure the gradient of this tangent. This corresponds to the initial rate.

Gradient of the tangent to Curve 1 at the start of the reaction

$$= \frac{0.100 - 0.082}{0 - 10} = -1.8 \times 10^{-3}\,\text{mol}\,\text{dm}^{-3}\,\text{min}^{-1}.$$

Gradient of the tangent to Curve 2 at the start of the reaction

$$= \frac{0.100 - 0.082}{0 - 5} = -3.6 \times 10^{-3}\,\text{mol}\,\text{dm}^{-3}\,\text{min}^{-1}.$$

When initial $[H^+]$ is doubled, the initial reaction rate also doubles.
Thus, the reaction is first order with respect to H^+.
The rate equation is therefore: **Rate = $\boldsymbol{k}$[CH_3COOCH_3][H^+].**

How can we determine the value of k?
There are two approaches in determining k.

- **First approach:**
  Substitute the known values into the rate equation:

$$\text{Rate} = k[CH_3COOCH_3][H^+].$$

You can make use of either one of the experiments.

For instance, take the data of Experiment 2:

$$\text{Initial rate} = 3.6 \times 10^{-3}\,\text{mol dm}^{-3}\,\text{min}^{-1},$$

$$\text{Initial } [\text{CH}_3\text{COOCH}_3] = 0.100\,\text{mol dm}^{-3},$$

$$\text{Initial } [\text{HCl}] = 0.400\,\text{mol dm}^{-3}.$$

Thus,

$$\text{Rate} = k[\text{CH}_3\text{COOCH}_3][\text{H}^+]$$

$$3.6 \times 10^{-3} = k \times 0.100 \times 0.400$$

$$k = 0.09\,\text{mol}^{-1}\,\text{dm}^3\,\text{min}^{-1}.$$

You will obtain the same value for k if you use the data from Experiment 1.

- **Second approach:**
  Use the expression $t_{1/2} = (\ln 2)/k$.

  Q: Wait a minute! This is only applicable to first-order reactions. In this example, the overall order of reaction is two. Furthermore, the initial $[\text{H}^+]$ is not very much greater than that of the ester. So how can we use this expression?

  A: You have learnt well! The reason why we can use $t_{1/2} = (\ln 2)/k$ is because H^+ is a catalyst here. Being a catalyst, its concentration is essentially unchanged throughout the whole reaction. Thus,

  $$\text{Rate} = k'[\text{CH}_3\text{COOCH}_3],$$

  where $k' = k[\text{H}^+]$.

  We have here a pseudo-first-order reaction!

Taking the data from any experiment, say, Experiment 2, we can substitute $t_{1/2} = 30\,\text{min}$ into the equation to get:

$$t_{1/2} = \frac{\ln 2}{k'} = \frac{\ln 2}{k[\text{H}^+]} = 30$$

and hence

$$k = \frac{\ln 2}{0.4 \times 30} = 0.058\,\text{mol}^{-1}\,\text{dm}^3\,\text{min}^{-1}.$$

The value of k calculated from both approaches differs because of different sources of inaccuracy arising from the plotting of tangents and so forth.

The half-life expression shows the dependence of $t_{1/2}$ on $[H^+]$, i.e., when $[H^+]$ doubles, $t_{1/2}$ is halved. This explains why $t_{1/2}$ for Experiment 1 is twice that of Experiment 2.

Q: Why is it that the order of reaction with respect to H_2O is not being determined?

A: As water is used as the solvent in this reaction (i.e., it is present in large excess), its concentration remains essentially constant throughout the reaction. The total amount of water reacted is negligible compared to the total amount of water present. It is not possible to change its concentration and hence its order cannot be determined.

5.3 Rate Equation and Reaction Mechanism

Thermochemical data help us to understand the feasibility of a reaction. Kinetics informs us how fast the reaction can proceed and factors that affect its rate. The reaction mechanism gives us insight as to how reactants combine to form products.

There are important terms to be acquainted with when dealing with the reaction mechanism:

- A reaction may not necessarily proceed in one step but rather in a series of steps known as elementary steps. An **elementary step** is the simplest step which cannot be further broken down into simpler steps.

  For example, for the reaction $H_2O_2(aq)+2H^+(aq)+2I^-(aq) \longrightarrow 2H_2O(l)+I_2(aq)$, it is highly unlikely that the five reacting particles will collide at the same time. The reaction must have occurred in several elementary steps that involve the formation of **intermediates**, which subsequently react in later steps to generate the final products:

$$\begin{array}{ll} \text{Step 1: } H_2O_2 + I^- \longrightarrow H_2O + OI^- & \text{(slow)} \\ \text{Step 2: } H^+ + OI^- \longrightarrow HOI & \text{(fast)} \\ \text{Step 3: } HOI + H^+ + I^- \longrightarrow H_2O + I_2 & \text{(fast)} \end{array}$$

For this example, OI^- and HOI are the intermediates or reaction intermediates.

- The **molecularity** of an elementary step is the number of reactant species that react in that step. For instance, both Steps 1 and 2 are bimolecular. Step 3 is termolecular, which involves three reacting species. A unimolecular step consists of only one reactant species.
- The **collection of elementary steps is called the reaction mechanism**, which gives a complete description of how reactants are converted into products.
- When all the elementary steps of a proposed reaction mechanism are summed up, you get the overall chemical equation.

$$
\begin{array}{ll}
\text{Step 1: } H_2O_2 + I^- \longrightarrow H_2O + OI^- & \text{(slow r.d.s.)} \\
\text{Step 2: } H^+ + OI^- \longrightarrow HOI & \text{(fast)} \\
\text{Step 3: } HOI + H^+ + I^- \longrightarrow H_2O + I_2 & \text{(fast)} \\
\hline
\text{Overall equation: } H_2O_2 + 2H^+ + 2I^- \longrightarrow 2H_2O + I_2 &
\end{array}
$$

- In any particular reaction mechanism, there is bound to be one particular step that is the slowest step. This slowest step is known as the **rate-determining step** (r.d.s.). It is usually indicated by the word "slow".
- The **overall rate of the reaction is equal to (or controlled by) the rate of the rate-determining step**, i.e., the rate of formation of the final products, H_2O and I_2, depends on how quickly Step 1 proceeds and not how quickly Step 3 proceeds.

  Consequently, **the rate equation for the overall reaction** is obtained from the **rate-determining step (slow step)** and not from the balanced chemical equation, i.e., rate equation: Rate = $k[H_2O_2][I^-]$.
- **The order of reaction with respect to each reactant in the rate-determining step is equal to the coefficient of the respective reactant in the rate-determining step** (**and preceding steps leading to it**).

  In this case, H_2O_2 and I^- react in the mole ratio of 1:1 in the rate-determining step. This corresponds to Rate $\propto [H_2O_2]$ and Rate $\propto [I^-]$.

The reaction mechanism for a given reaction can be proposed based on the experimentally determined rate equation.

Example 5.5:

$$H_2O_2(aq) + 2H^+(aq) + 2I^-(aq) \longrightarrow 2H_2O(l) + I_2(aq);$$
$$\text{Rate} = k[H_2O_2][I^-].$$

Is the proposed mechanism below plausible for the given reaction?

$$\begin{array}{ll} \text{Step 1: } H_2O_2 + I^- \longrightarrow HOI + OH^- & \text{(slow r.d.s.)} \\ \text{Step 2: } HOI + I^- \longrightarrow I_2 + OH^- & \text{(fast)} \\ \text{Step 3: } 2OH^- + 2H^+ \longrightarrow 2H_2O & \text{(fast)} \end{array}$$

Solution: Yes, this is a plausible mechanism. The stoichiometry of the reactants in the rate-determining step coincides with the order of reaction with respect to each species in the rate equation. In addition, when the three steps are summed up, the overall chemical equation is obtained.

Example 5.6:

$$BrO_3{}^-(aq) + 5Br^-(aq) + 6H^+(aq) \longrightarrow 3Br_2(aq) + 3H_2O(l);$$
$$\text{Rate} = k[BrO_3{}^-][Br^-][H^+]^2.$$

Is the proposed mechanism below plausible for the given reaction?

$$\begin{array}{ll} \text{Step 1: } Br^- + H^+ \longrightarrow HBr & \text{(fast)} \\ \text{Step 2: } BrO_3{}^- + H^+ \longrightarrow HBrO_3 & \text{(fast)} \\ \text{Step 3: } HBrO_3 + HBr \longrightarrow HBrO_2 + HBrO & \text{(slow)} \\ \text{Step 4: } HBrO_2 + HBr \longrightarrow 2HBrO & \text{(fast)} \\ \text{Step 5: } HBrO + HBr \longrightarrow Br_2 + H_2O & \text{(fast)(Occurs three times)} \end{array}$$

Solution: Yes, this is a plausible mechanism. The stoichiometry of the reactants in the rate-determining step, and the preceding steps leading to it, coincides with the order of reaction with respect to each species in the rate equation. In addition, when the five steps are summed up, the overall chemical equation is obtained.

Q: Why must we include the species in the preceding steps before the rate-determining step?

A: The overall rate depends on [HBr] and [$HBrO_3$]. In turn, the formation of HBr depends on [H^+] and [Br^-], while the formation of $HBrO_3$ depends on [H^+] and [BrO_3^-]. Therefore, Rate $= k[BrO_3^-][Br^-][H^+]^2$.

Hence, the ***rate equation involves the reactants (not intermediates) in the overall equation* that take part in the elementary steps preceding and including the rate-determining step.**

5.4 Transition State Theory

The Collision Theory was formulated to understand how reactants collide to form products. Although it was mentioned that a minimum activation energy is required to get the reaction going, the model does not account for what happens to the initial amount of energy once the reaction has started. All we know is that energy is being conserved according to the First Law of Thermodynamics. In what form the energy is being conserved, we do not know.

The Transition State Theory was then formulated to account for what specifically happens to the initial kinetic energy that the reactant particles possess, during the progress along the reaction pathway (also known as the reaction coordinate). It focuses on the species existing at the point where the reactants are just about to change into products. This particular point is known as the **transition state** for the reaction and it has the highest energy content over the duration of the reaction. The species that is present at this transition state is known as the **activated complex**.

When two reacting particles approach each other in a collision, there is inter-electronic repulsion between their interacting electron clouds. Such a repulsive force decreases the speed of the two approaching particles. Thus, if the approaching particles did not have a sufficient amount of energy to do work against the repulsive force, the particles would simply "touch and go". But if there were a sufficient amount of energy, then inter-penetration of the electron clouds would occur. At this point, there could be a rearrangement

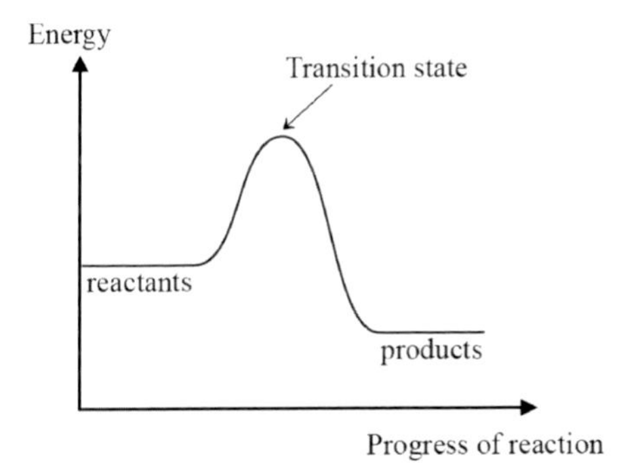

Fig. 5.13. Energy profile diagram for a single-step reaction.

of valence electrons, with the partial breaking of old bonds and the partial formation of new bonds. This would result in the formation of an activated complex, which is the arrangement of atomic nuclei and bonding electrons at the point of maximum potential energy. Part of the kinetic energy that the reacting particles have initially would be converted to electrostatic potential energy and stored in the bonds. At this point, the electrostatic potential energy would be at its maximum, as depicted in Fig. 5.13 for a single-step reaction:

$$\underset{\text{reactants}}{\text{A–B} + \text{X}} \rightleftharpoons \underset{\text{activated complex}}{[\text{A---B---X}]^{\ddagger}} \longrightarrow \underset{\text{products}}{\text{A+B–X}}\,.$$

For a reaction that consists of more than one mechanistic step, there would be more than one transition state, as shown in Fig. 5.14.

For such a reaction mechanism, other than the activated complexes that are being generated, there is additional species known as a **reactive intermediate**. Unlike a reactive intermediate, an activated complex cannot be isolated. Once formed, the activated complex (at the transition state) can transform into the products or "fall back" to regenerate the reactants.

Q: So according to the Collision Theory, reactants must possess a minimum energy known as the activation energy and collide with the correct geometry. This energy is in the form of kinetic energy. The MAIN DIFFERENCE between the Transition State Theory

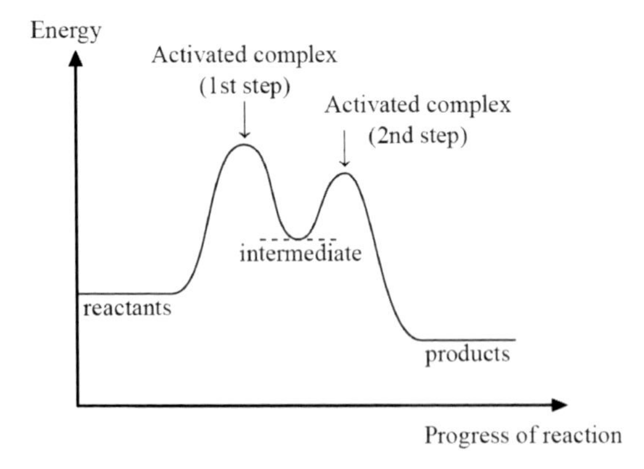

Fig. 5.14. Energy profile diagram for a two-step reaction.

and the Collision Theory is that the former "tells" us more about what the activation energy is used for, what happens to it during the progress of a reaction, whether there is any energy "left over," and lastly, the transformation process (structural changes, bond breakage, bond formation, and so on) of reactants to products. Is that the gist of it?

A: Yes! That's very good. We will quote some examples below to show the usefulness of the Transition State Theory especially in understanding organic reactions.

Q: From the energy profile diagram in Fig. 5.14, how would we know which of the peaks belongs to the rate-determining step in the reaction mechanism?

A: For a reaction mechanism with more than one step, the peak for the rate-determining step is the one in which the activated complex has the highest potential energy. For Fig. 5.14, the first peak corresponds to the rate-determining step.

Example: In the hydrolysis of bromomethane with NaOH, it is found that the reaction is first-order with respect to OH^-. But for the hydrolysis of 2-bromomethylpropane with NaOH, the reaction is found to be zero-order with respect to OH^-. Further mechanistic studies result in the proposal of the following two mechanisms:

Reaction: $CH_3Br + OH^- \longrightarrow CH_3OH + Br^-$.

Mechanism:

$$HO^- + H_3C{-}Br \longrightarrow [HO\cdots CH_3\cdots Br]^- \longrightarrow HO{-}CH_3 + Br^-$$

Transition state

Reaction:

$$CH_3C(CH_3)(Br)CH_3 + OH^- \longrightarrow CH_3C(CH_3)(OH)CH_3 + Br^-.$$

Mechanism:

Step 1: $$CH_3{-}C(CH_3)_2{-}Br \xrightarrow{\text{slow}} CH_3{-}C^+(CH_3)_2 + Br^-$$

Step 2: $$CH_3{-}C^+(CH_3)_2 + OH^- \xrightarrow{\text{fast}} CH_3{-}C(CH_3)_2{-}OH$$

5.5 Catalysis

5.5.1 *Homogeneous catalysis*

The catalyst and the reactants are in the *same phase*, either liquid or gaseous, in a reaction involving *homogeneous catalysis*.

In a homogeneous catalytic reaction, the catalyst **takes part in the reaction** by being **converted into an intermediate** ion or compound, which is subsequently consumed to form the products. The catalyst is able to perform its role because of its ability to exist in more than one oxidation state. By the time the reaction is completed, the catalyst will have been regenerated.

Example: Reaction between peroxodisulfate ions and iodide ions catalysed by Fe^{2+}

$$S_2O_8{}^{2-}(aq) + 2I^-(aq) \longrightarrow 2SO_4^{2-}(aq) + I_2(aq).$$

The uncatalysed reaction involves a direct collision between two similarly charged particles. The electrostatic repulsion between the two negatively charged ions partly causes the reaction to have a high activation energy.

When the reaction is catalysed by Fe^{2+}, it proceeds via an alternative pathway that consists of the following steps:

$$\text{Step 1:}\quad \underset{\text{catalyst}}{2Fe^{2+}(aq)} + S_2O_8{}^{2-}(aq) \longrightarrow 2SO_4{}^{2-}(aq) + \underset{\text{intermediate}}{2Fe^{3+}(aq)}.$$

$$\text{Step 2:}\quad 2I^-(aq) + \underset{\text{intermediate}}{2Fe^{3+}(aq)} \longrightarrow I_2(aq) + \underset{\text{catalyst}}{2Fe^{2+}(aq)}.$$

Transition metal ions are particularly effective at acting as homogeneous catalysts because they can *exist in different oxidation states and they can change their oxidation states easily*, facilitating the formation and breakdown of intermediate compounds between the ions and the reactants (refer to Chap. 11).

Q: So, if originally the non-catalysed reaction is a single-step reaction with a single peak in the energy profile diagram, for the catalysed reaction, there would be two peaks because it has become a two-step reaction?

A: Yes, absolutely spot on.

Example: An environmental concern: The catalytic oxidation of atmospheric sulfur dioxide by atmospheric oxides of nitrogen

Man-made sulfur dioxide gas comes from the combustion of fossil fuel. The atmospheric sulfur dioxide can be oxidised to sulfur trioxide but the reaction is very slow in the absence of a catalyst. If you recall in the contact process of making sulfuric acid, a catalyst known as vanadium pentoxide (V_2O_5) is needed. But in the atmosphere, another equally obnoxious pollutant, NO_2, can accelerate the conversion of SO_2 to another secondary pollutant, SO_3, as follows:

$$\begin{aligned}
\text{Step 1:}\quad & SO_2(g) + NO_2(g) \longrightarrow SO_3(g) + NO(g).\\
\text{Step 2:}\quad & NO(g) + 1/2O_2(g) \longrightarrow NO_2(g).\\
\text{Overall:}\quad & SO_2(g) + 1/2O_2(g) \longrightarrow SO_3(g).
\end{aligned}$$

The large amount of SO_3 formed in the atmosphere then reacts with water to form sulfuric acid:

$$SO_2(g) \xrightarrow{\text{oxidation}} SO_3(g) \xrightarrow{H_2O(l)} H_2SO_4(aq).$$

This results in acid rain and amplifies the problem of atmospheric SO_2 as a pollutant.

5.5.2 *Heterogeneous catalysis*

The catalyst and the reactants are in different phases in a reaction involving heterogeneous catalysis. The catalyst is usually in the solid phase and it provides active sites at which the reaction can take place. The reactants are usually liquids or gases. For example, the Haber process:

$$N_2(g) + 3H_2(g) \rightleftharpoons 2NH_3(g), \quad \text{Fe}/250\,\text{atm}/450^\circ\text{C}.$$

Three likely steps can be put forward for any heterogeneous catalysis:

Step 1: Adsorption of the reactant particles onto the active sites on the surface of the catalyst. This is facilitated by the formation of weak bonds between the reactant particles and the catalyst. The adsorption increases the surface concentration of the reactant.

Step 2: Reaction at the surface.

- The activation energy of the catalysed process is lower because the intra-molecular bonds within the reactant particles are weakened by the adsorption effects, thereby reducing the energy required to disrupt them.
- The reactant particles are brought into close contact and are properly orientated for reaction.

Step 3: Desorption of the reactants or products from the surface.

Example: Catalytic hydrogenation of alkenes

Consider $CH_2{=}CH_2(g) + H_2(g) \xrightarrow{Ni(s)} CH_3{-}CH_3(g)$.

1. Adsorption of reactants

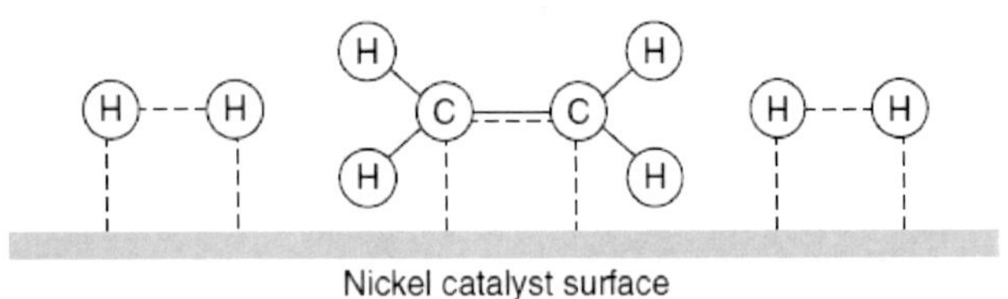

2. Chemical reaction

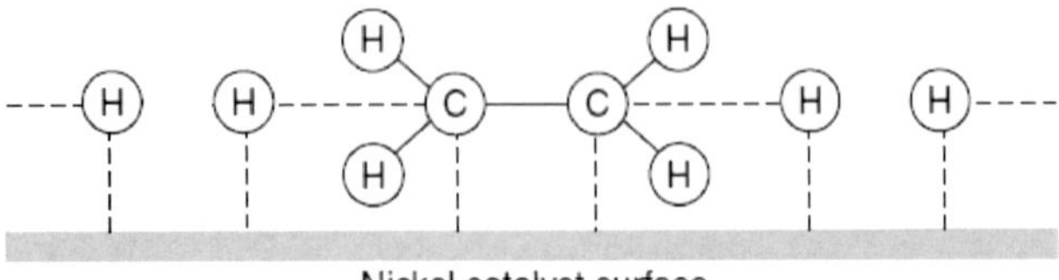

3. Desorption of products

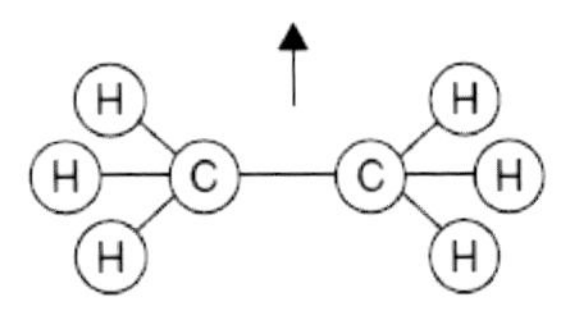

Example: Catalytic removal of oxides of nitrogen in the exhaust gases from car engines

Pollutants emitted from the combustion of fuel in car engines make up a substantial amount of man-made pollutants. Thus, to minimise emissions, cars are fitted with catalytic converters to remove three main pollutants (CO, NO_x and noncombustible hydrocarbons) from exhaust gases. These pollutants are converted into less harmful products, such as CO_2, N_2 and H_2O, through a "three-way" catalytic converter.

The catalytic converter consists of a ceramic honeycomb structure coated with the precious metals platinum (Pt), palladium (Pd) and rhodium (Rh), which acts as catalysts. A honeycomb structure is used so as to maximise the surface area on which heterogeneously catalysed reactions take place, as the noble metals used are very expensive.

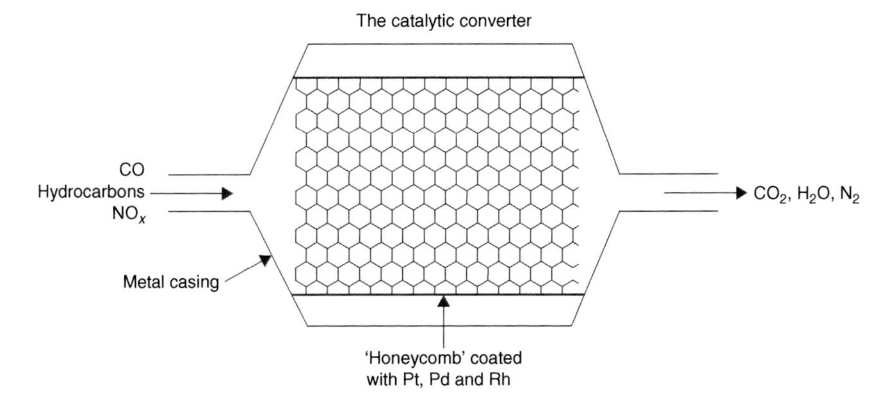

The chemical reactions that occur in the catalytic converter are as follows:

- As the gases enter, the oxides of nitrogen (NO and NO_2) are reduced to N_2 by the excess CO present, with rhodium acting as the catalyst:

$$NO + CO \xrightarrow{Rh} 1/2N_2 + CO_2.$$

- The CO is also oxidised to CO_2 with platinum and palladium as the catalysts:

$$1/2O_2 + CO \xrightarrow{Pt/Pd} CO_2.$$

- Noncombustible hydrocarbons are oxidised to CO_2 and H_2O with platinum and palladium as the catalysts.

The catalytic activity of catalytic converters is destroyed by the presence of lead. This is because lead is preferentially adsorbed on the active sites, making it unavailable for the reactants to be adsorbed. Hence, cars fitted with catalytic converters must run on unleaded petrol.

5.5.3 *Autocatalysis*

An autocatalytic reaction involves the formation of a product that acts as a catalyst for the reaction. The reaction is said to have undergone autocatalysis.

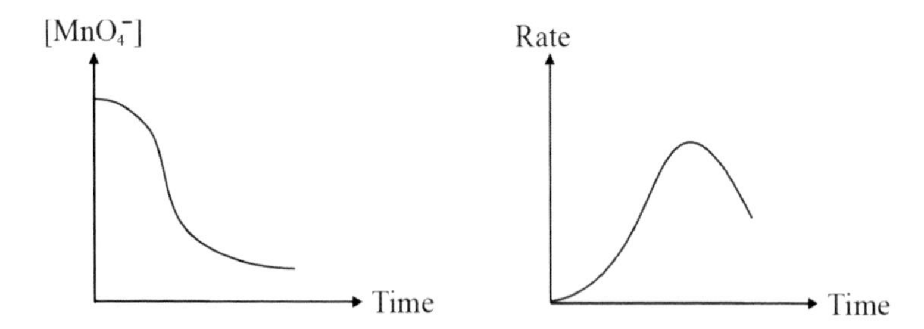

Fig. 5.15. Graph of $[MnO_4^-]$ against time (left) and graph of rate against time (right) for an autocatalytic reaction.

The oxidation of ethanedioate ions by manganate(VII) ions is one such autocatalytic reaction. The reaction is accelerated by the Mn^{2+} ions produced during the reaction:

$$2MnO_4^-(aq) + 5C_2O_4^{2-}(aq) + 16H^+(aq) \longrightarrow 2Mn^{2+}(aq) + 10CO_2(g) + 8H_2O(l).$$

Initially, the reaction is slow since it is not catalysed. As the Mn^{2+} ions are produced, they increase the rate of the reaction by acting as an autocatalyst (shown by the increasing steepness of the gradient of the graph in Fig. 5.15). Towards the end of the reaction, the concentration of the reactants has fallen to a low level and so the rate of the reaction decreases, even though there is an adequate supply of catalyst.

5.5.4 *Enzymes (biological catalysts)*

Enzymes, which consist of proteins or polypeptides, are an important class of biological catalyst, speeding up chemical reactions in living systems. Without them, most biochemical reactions would be too slow to sustain life. Examples of enzymes include amylase (for hydrolysing starch) and lipase (for hydrolysing fats and lipids).

- Characteristics of enzymes:

  (i) Nature and size: Enzymes are globular proteins that fold in a particular conformation, creating a three-dimensional active site. They have relative molecular masses of around 10^5 to 10^7.

(ii) Efficiency: Enzymes are required in very small amounts — they are very effective catalysts. This is because enzyme molecules are regenerated during their catalytic activity. A typical enzyme molecule may be regenerated a million times in one minute.

(iii) Specificity: Due to the special conformation of the three-dimensional active site where only certain molecules can be adsorbed, enzymes are very specific to a particular reaction or type of reaction.

(iv) Temperature: High temperatures of above 50°C result in a change in the three-dimensional active site; this denatures the enzyme, rendering it inactive. Thus, enzymes operate most effectively at body temperature of about 37°C.

(v) Sensitivity to pH: pH change results in a change in the three-dimensional active site; this denatures the enzyme, rendering it inactive. Different enzymes have differing optimum pH levels.

- Enzymatic action: Like other non-biological catalysts, enzymes catalyse reactions by providing an alternative reaction pathway with a lower activation energy. To bring about this, the enzyme forms a complex with the substrate or substrates (reactants) of the reaction. Thus, a simple picture of the enzyme's action is:

$$\begin{aligned}\text{substrate} + \text{enzyme} &\longrightarrow \text{enzyme/substrate complex}\\ &\longrightarrow \text{enzyme} + \text{products.}\end{aligned}$$

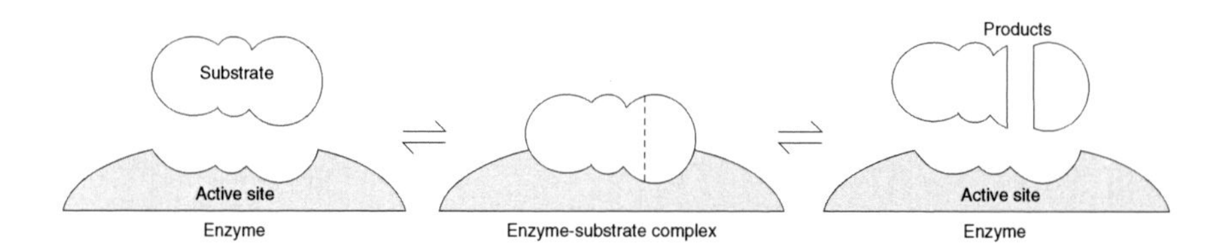

A theory which accounts for their specificity is the lock and key model. According to this model, the substrate and the enzyme molecule have complementary shapes such that they fit together like a lock and its key. This is how it works:

> First, the substrate (reactant) fits the active site on the enzyme (catalyst) in the same way a key fits a lock. An enzyme-substrate complex is formed and this weakens the intramolecular bonding within the substrate, lowering the activation energy. Then, the enzyme–substrate complex breaks apart as the products formed are no longer of the right shape to remain in the active site, leaving the enzyme free to catalyse further reactions.

- The relationship between the enzyme, substrate and rate of reaction: In an enzyme catalysed reaction, a fixed amount of enzyme has a limited number of active sites in each enzyme.

  When [enzyme] $\ll$ [substrate], the rate of reaction is directly proportional to [enzyme], i.e., reaction is first order with respect to the enzyme.

  At low [substrate], the active sites are not fully filled and the rate is proportional to [substrate]. (In most cases, the reaction is first order with respect to the substrate.)

  At higher [substrate], the active sites are saturated and any increase in substrate concentration will not have any effect on the reaction rate. Under such conditions, we term it Saturation Kinetics.

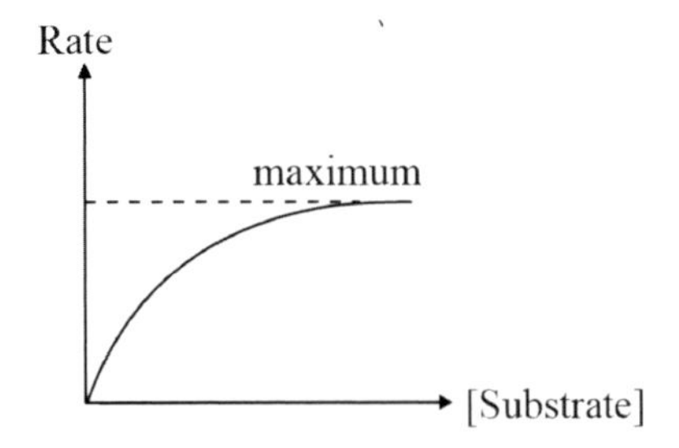

My Tutorial (Chapter 5)

1. Acid rain contains dissolved sulfuric(VI) acid that has been formed by the atmospheric oxidation of sulfur dioxide by other pollutants produced from combustion. The mechanisms involved, together with their rate constants (k), are shown below. (All rate

constants have units of $mol^{-1}\,dm^3\,min^{-1}$.)

$$HO\cdot + SO_2 \longrightarrow HOSO_2\cdot, \quad k = 7.2 \times 10^{10}, \quad (1)$$

$$HOSO_2\cdot + O_2 \longrightarrow HO_2\cdot + SO_3, \quad k = 1.44 \times 10^{10}, \quad (2)$$

$$SO_3 + H_2O \longrightarrow H_2SO_4, \quad k = 2.16 \times 10^{8}, \quad (3)$$

$$HO_2\cdot + NO\cdot \longrightarrow HO\cdot + NO_2\cdot, \quad k = 3.0 \times 10^{11}. \quad (4)$$

(a) Suggest why oxidation of SO_2 to H_2SO_4 occurs more rapidly in air which contains other gaseous pollutants produced from combustion.

(b) Explain the meaning of the term *rate constant.*

(c) Laboratory experiments have been done to investigate the reaction between SO_2 and $HO\cdot$ [Eq. (1)] under conditions similar to those present in polluted air. The data in the table below show how the concentration of $HO\cdot$ changes with time, for an experiment in which the sulfur dioxide concentration is $5 \times 10^{-5}\,mol\,dm^{-3}$.

Concentration of HO· (10^{-8} mol dm^{-3})	**2.0**	**1.5**	**1.1**	**0.8**	**0.6**	**0.4**
Time (t) (10^{-8} min)	0	8.33	16.67	25.00	33.33	41.67

(i) Plot these data on a piece of graph paper.

(ii) Using the half-life method, show that the reaction is first order with respect to $HO\cdot$.

(iii) Explain what is meant by the statement: "Reaction is first order with respect to $HO\cdot$".

2. The reaction between bromate(V) and bromide ions in acid solution to form bromine follows the equation:

$$BrO_3^-(aq) + 5Br^-(aq) + 6H^+(aq) \rightarrow 3Br_2(aq) + 3H_2O(l).$$

The rate of this reaction can be studied using the initial rates method. A small fixed amount of phenol and an acid–base indicator are added to colour the solution. The phenol undergoes a rapid initial reaction with the bromine, resulting in a clear solution. When this reaction is completed, the next portion of bromine bleaches the indicator. Finally, a white precipitate is slowly formed and the solution eventually turns orange as excess bromine is produced. The time, t taken for the solution to turn colourless can be used as a measure of the rate of reaction. A study of the kinetics of the above reaction reveals that it is second order with respect to H^+ and first order with respect to $BrO_3{}^-$ and Br^-.

(a) What is the overall order of the reaction?
(b) Using the Collision Theory, explain why this reaction is unlikely to occur in a single step.
(c) By what factor will the rate of reaction be reduced if an equal volume of water is added to a sample of the reacting mixture?
(d) Calculate the rate of bromine formation if the initial concentrations of bromate(V), bromide and H^+ are 0.04, 0.2 and 0.1 mol dm^{-3}, respectively, and the rate constant is $2 \times 10^{-2}\,\text{dm}^9\,\text{mol}^{-3}\,\text{min}^{-1}$.
(e) Suggest why the orders with respect to Br^- and H^+ are not equal to the coefficients in the stoichiometric equation for the reaction.
(f) Suggest why the time t is inversely proportional to the rate constant k, given the conditions in part (d).
(g) The activation energy, E_a, for the reaction is related to the time, t, by the equation (where T is temperature in Kelvin):

$$\ln t = \text{constant} + E_a/RT.$$

Use the data below to calculate a value for E_a.

T (°C)	26	49
t (s)	148	24

3. The orders of reaction for the alkaline hydrolysis of two bromoalkanes are given below:

Reactants	Order with Respect to the Bromoalkane	Order with Respect to the OH^-
(I) $CH_3CH_2CH_2Br$ and OH^-	1	1
(II) $(CH_3)_3CBr$ and OH^-	1	0

(a) In the context of reaction (I), write down the rate equation and explain what is meant by the terms *rate of reaction*, *rate constant* and *overall order*.

(b) Explain the effect on the rate of each reaction if only the concentration of hydroxide ion is doubled.

(c) Briefly explain how you would study the rate of one of these reactions in the laboratory.

(d) In one such experiment on reaction (II) it is found that the time taken for the concentration of the $(CH_3)_3CBr$ to fall to half of its starting concentration is 45 min. Determine the rate constant of the reaction and how long it would take for the concentration of the $(CH_3)_3CBr$ to fall to 1/8 of its starting concentration. Explain your answers.

4. (a) The first step in a possible mechanism for the reaction $2NO(g) + O_2(g) \rightarrow 2NO_2(g)$ is as follows:

$$2NO \rightleftharpoons N_2O_2.$$

(i) Draw dot-and-cross diagrams of NO and N_2O_2. What feature of the electronic structure of NO suggests that this is a likely first step in this reaction?

(ii) Explain why the enthalpy change for this step is $-163\,\text{kJ mol}^{-1}$, given that the average bond energy for the N–N bond in compounds of nitrogen is $+163\,\text{kJ mol}^{-1}$.

(iii) Explain why this step does not control the rate of the reaction. Assuming there is one further step in the reaction, write an equation for this.

(b) (i) Sketch the Maxwell–Boltzmann distribution of molecular energies at a temperature T_1 K. On the same axes, sketch the curve which shows the distribution of molecular energies at a higher temperature T_2 K.

(ii) Use these graphs to explain how the rate of a gas phase reaction changes with increasing temperature.

(c) For a gaseous reaction, state and explain what effect the addition of a catalyst would have on:

(i) the energy distribution of the gas molecules,
(ii) the activation energy for the reaction,
(iii) the rate of reaction,
(iv) the overall order of reaction for both homogeneous and heterogeneous catalysis, and
(v) the individual order of reaction for both homogeneous and heterogeneous catalysis.

(d) Thioethanamide reacts with sodium hydroxide as follows:

$$CH_3CSNH_2 + 2OH^- \longrightarrow CH_3CO_2{}^- + HS^- + NH_3.$$

The
reaction is first order with respect to both thioethanamide and hydroxide ions.

(i) Write the rate equation for this reaction.
(ii) Given that the reaction occurs in two stages and the rate determining step is

$$CH_3CSNH_2 + OH^- \longrightarrow CH_3CONH_2 + HS^-,$$

write an equation for the second step in the reaction.

CHAPTER 6

Chemical Equilibria

It is wrong to think that all chemical reactions proceed to completion and all reactions are irreversible. In fact, there are many chemical reactions that are reversible. For such reversible reactions, not all the reactants are converted into products. We would simply have a mixture of unreacted reactants and products co-existing.

What then can be used to characterise a reversible reaction? How do we know if one system is more reversible than another? What are the factors that affect a reversible reaction? Can we modify a reversible reaction to optimise the formation of the desired products that manufacturers want? These are the questions that we will seek to answer in this chapter.

6.1 Reversible Reactions

What we usually consider as irreversible reactions, i.e., reactions that proceed in one direction (as commonly represented by a single-headed arrow, $\rightarrow$) are in actual fact reversible reactions under specific conditions. All chemical reactions are reversible to a certain extent. Products formed in these reactions combine and react to re-form the original reactants. Both forward and backward reactions continue for as long as there are reactants combining to give products.

For example, given the reaction A+B → C+D (forward reaction), the products C and D can react to re-form A and B: C + D → A + B (backward reaction).

As we can now see, A can be a reactant and also a product. For clarity's sake, the term "reactants" is conventionally used for the species on the left-hand side of the equation and "products" are those species on the right-hand side.

Equations of reversible reactions are represented with a double-headed arrow ($\rightleftharpoons$):

$$\mathrm{A + B \rightleftharpoons C + D}.$$

The extent of reversibility depends on the magnitude of the activation energy (E_a) of the backward reaction. For an exothermic reversible reaction, if E_a of the backward reaction is too large, the backward reaction is essentially non-occurring compared to the forward reaction (see Fig. 6.1). Hence, one can assume the forward reaction proceeds almost to completion. In contrast, for an endothermic reversible reaction, higher E_a for the forward reaction makes it less likely for it to proceed as compared to the backward reaction.

The energy profile diagram for a reversible reaction is shown in Fig. 6.2. As can be seen from the diagram, E_a of the backward reaction is comparable to that of the forward reaction. Or in short, ΔH of reaction is about zero!

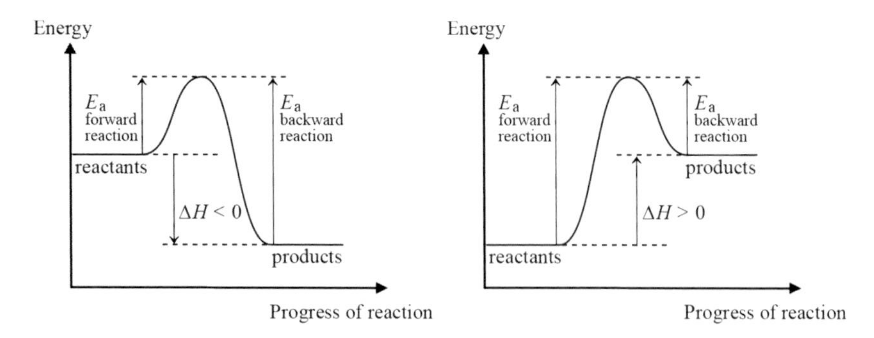

Fig. 6.1. Energy profile diagrams for an exothermic reversible reaction (left) and an endothermic reversible reaction (right).

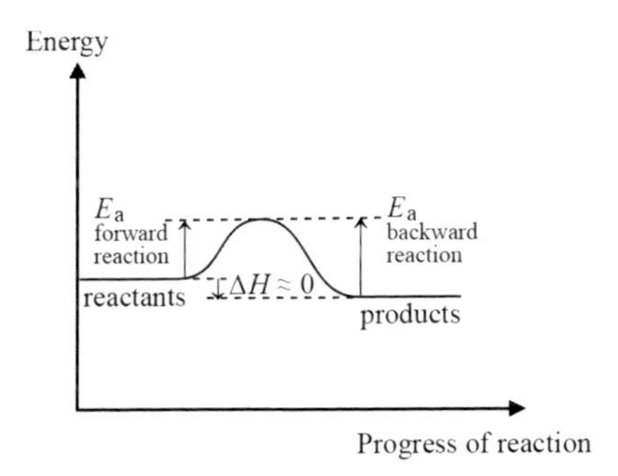

Fig. 6.2. Energy profile diagram for a reversible reaction.

6.2 Equilibrium Systems

Consider the following reaction: $A + B \rightleftharpoons C + D$.

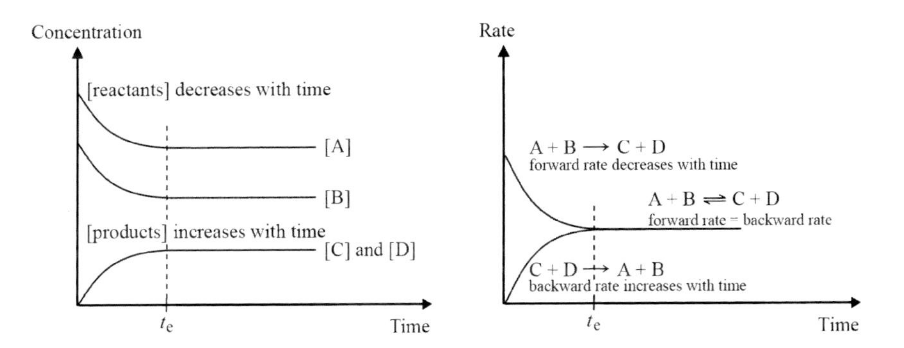

At the start of the reaction, with just A and B present, only the forward reaction will occur. The rate of the forward reaction (determined by the gradient of the tangent drawn to the concentration versus time plot) is at its peak since [reactants] is at its highest while the rate of the backward reaction is zero.

As the reaction progresses, the rate of the forward reaction (R_f) decreases as [A] and [B] decrease, since these are used to form the products C and D.

At the same time, as soon as C and D are formed, the backward reaction proceeds, albeit at a slow rate initially since [products] is low. This helps to regenerate A and B. As [C] and [D] increase over time, the rate of the backward reaction (R_b) also increases.

There will come a point in time (t_e) when both the forward and backward reactions occur at the same rate, i.e., forward rate (R_f) = backward rate (R_b). When this happens, the concentration of every reactant and product remains constant, and the system is in a state of balance, known as equilibrium.

More specifically, the system is said to be in **dynamic equilibrium**. As the rates of these reactions are equal, it may seem that the reactions have completely stopped but this is not true.

Although the concentrations of the substances remain unchanged (as indicated by the term "equilibrium"), there is still activity going on: both forward and backward reactions are continually occurring (as indicated by the term "dynamic") but since they proceed at the same rate, each species is formed as fast as it is consumed, resulting in a constant concentration term.

Q: For a reaction in which the forward reaction is exothermic and is at equilibrium, will there be a continuous production of heat energy?

A: No! If the forward reaction is exothermic, the backward reaction will be endothermic. A system at equilibrium means that the forward rate equals to the backward rate. There will not be any net absorption or production of heat energy. Hence, temperature will be at a constant value.

6.3 Equilibrium Constants K_c and K_p

Every reversible reaction will attain equilibrium at different times; some take minutes while others take weeks. The composition of a system at equilibrium also differs. How can we then determine if equilibrium has been achieved?

Consider the following reaction:

$$a\text{A} + b\text{B} \rightleftharpoons c\text{C} + d\text{D}.$$

The mass action expression or reaction quotient Q_c is expressed as the ratio of the concentration of the products to the reactants, each raised to the power corresponding to their stoichiometric coefficient in the balanced chemical equation:

$$Q_\mathrm{c} = \frac{[\mathrm{C}]^c[\mathrm{D}]^d}{[\mathrm{A}]^a[\mathrm{B}]^b}.$$

Q_c embraces the concentration values of the reactants and products at any point of time. Its value therefore continuously changes with the progress of the reaction until equilibrium is reached.

At equilibrium, Q_c becomes a constant value, which is more appropriately known as the equilibrium constant, K_c:

$$K_\mathrm{c} = \frac{[\mathrm{C}]^c[\mathrm{D}]^d}{[\mathrm{A}]^a[\mathrm{B}]^b} = \text{constant at a fixed temperature},$$

where [A], [B], [C] and [D] are concentrations of species at equilibrium.

The unit of K_c varies depending on the terms in its expression but since units of concentration are in terms of mol dm^{-3}, the unit of K_c can be computed as "$\mathrm{mol}^{(c+d)-(a+b)}\ \mathrm{dm}^{-3[(c+d)-(a+b)]}$".

The equilibrium constant K_c can also be conceptualised from the following mathematical derivation. Consider the **elementary** reaction:

$$a\mathrm{A} + b\mathrm{B} \rightleftharpoons c\mathrm{C} + d\mathrm{D}.$$

For the forward reaction: $a\mathrm{A} + b\mathrm{B} \rightarrow c\mathrm{C} + d\mathrm{D}$, $\mathrm{Rate_f} = k_\mathrm{f}[\mathrm{A}]^a[\mathrm{B}]^b$.

For the backward reaction: $c\mathrm{C} + d\mathrm{D} \rightarrow a\mathrm{A} + b\mathrm{B}$, $\mathrm{Rate_b} = k_\mathrm{b}[\mathrm{C}]^c[\mathrm{D}]^d$.

When equilibrium is achieved, $\mathrm{Rate_f} = \mathrm{Rate_b} \Rightarrow k_\mathrm{f}[\mathrm{A}]^a[\mathrm{B}]^b = k_\mathrm{b}[\mathrm{C}]^c[\mathrm{D}]^d$.

Rearranging, we get $k_\mathrm{f}/k_\mathrm{b} = [\mathrm{C}]^c[\mathrm{D}]^d/[\mathrm{A}]^a[\mathrm{B}]^b$ = constant (which is defined as K_c), i.e.,

$$K_\mathrm{c} = \frac{k_\mathrm{f}}{k_\mathrm{b}} = \frac{[\mathrm{C}]^c[\mathrm{D}]^d}{[\mathrm{A}]^a[\mathrm{B}]^b}.$$

The advantage of this definition is that it allows us to perceive the equilibrium constant K_c as the ratio of the rate constants.

Importantly, the Equilibrium Law expresses K_c as a relationship between the concentrations of products and reactants in a system at equilibrium, and it **provides us with a quantifying means to determine the position of the equilibrium.**

The magnitude of the equilibrium constant informs us of the relative proportion of products to reactants, providing us with information on the **extent of reaction** (but not information on the reaction rate).

- When $K_c > 1$, there is a higher proportion of products to reactants. The formation of products is favoured, i.e., the position of equilibrium lies more to the right.
- When $K_c < 1$, there is a high proportion of reactants as not many of these are converted to products, i.e., the position of equilibrium lies towards the left.
- A K_c value close to 1 indicates that the concentrations of both reactants and products are almost the same.

Q: The $\Delta G^{\ominus}$ of a reaction indicates the thermodynamic spontaneity of a reaction. How is it linked to the equilibrium constant, K_c?

A: Good question! The more negative the $\Delta G^{\ominus}$, the more spontaneous is the reaction likely to be. The relationship between $\Delta G^{\ominus}$ and K_c is through the equation, $\Delta G^{\ominus} = -\text{RT} \ln K_c$. If the K_c is a value that is greater than 1, then $\Delta G^{\ominus}$ is a negative value. This would mean that the reaction is thermodynamically spontaneous and the position of the equilibrium lies more to the right hand side of the reaction equation. Vice versa.

Reaction	**K_c Value**
$NH_3(aq) + H_2O(l) \rightleftharpoons NH_4^+(aq) + OH^-(aq)$	1.8×10^{-5} mol dm^{-3} (at 25°C)
$N_2(g) + 3H_2(g) \rightleftharpoons 2NH_3(g)$	1.7×10^{2} mol^{-2} dm^{6} (at 227°C)
$N_2(g) + 3H_2(g) \rightleftharpoons 2NH_3(g)$	4.1×10^{8} mol^{-2} dm^{6} (at 25°C)

At 25°C, the formation of $NH_3(g)$ is thermodynamically spontaneous, whereas the ionisation of ammonia does not occur as readily. If you notice, there is a temperature value tied to the K_c value as stipulated. As the rate constants are temperature-dependent [assuming the Arrhenius rate constant, $k = A\exp(-E_a/RT)$], so too is K_c that is derived from them.

Altering temperature actually creates a stress or disturbance in an equilibrium system and thus brings changes to its composition. We will cover this aspect in Section 6.4 on "Le Chatelier's Principle," which helps us to predict the effects of changing conditions on equilibrium systems.

The magnitude of Q_c, relative to that of K_c, indicates where the position of equilibrium lies:

- When $Q_c = K_c$, the system is at equilibrium, i.e., $R_f = R_b$; there is no change in the position of equilibrium.
- If $Q_c < K_c$, then $R_f > R_b$, i.e., the position of equilibrium shifts to the right.
- If $Q_c > K_c$, then $R_f < R_b$, i.e., the position of equilibrium shifts to the left.

Q: Why when $Q_c < K_c$, does the position of equilibrium shift towards the product side?

A: From a kinetics perspective, $Q_c < K_c$ implies that the concentrations of the reactant particles are greater than the product particles as compared to if the system is at equilibrium. Thus, we expect a higher effective collision frequency for the forward reaction than that for the backward reaction. As a result, more products form and the position of equilibrium shifts towards the product side.

Q: But as more products form, wouldn't the rate of the backward reaction increase and deplete the amount of products present, which in turn cause the backward rate to decrease?

A: There is a fallacy here. As more products form, the rate of the backward reaction does increase. But this increase does not deplete the concentration of the products because the rate of the forward reaction is still greater than the rate of the backward reaction at this time (refer to the rate versus time plot in Section 6.2). In fact, the concentration of the products continues to increase and fuel an increase in the rate of the backward reaction. All these changes pertaining to the concentrations and rate will stop once a state of dynamic equilibrium is established.

Example 6.1:

$$CO(g) + 2H_2(g) \rightleftharpoons CH_3OH(g).$$

At a particular temperature, a system contains 0.5 mol of CO(g), 0.8 mol of H_2(g) and 0.9 mol of CH_3OH(g) in a 5 dm^3 vessel. Is this system at equilibrium? If not, in which direction does the reaction proceed? The given value of K_c is 54 at this particular temperature.

Solution: As the system may not be at equilibrium, we calculate Q_c and compare it against K_c:

$$Q_c = \frac{[CH_3OH]}{[CO][H_2]^2} = \frac{\frac{0.9}{5}}{\frac{0.5}{5} \times \left(\frac{0.8}{5}\right)^2} = 70.3\,\text{mol}^{-2}\,\text{dm}^6.$$

Since $Q_c > K_c$, the backward reaction proceeds at a higher rate than the forward reaction. The position of equilibrium shifts to the left.

Back to the reversible reaction: $a\text{A} + b\text{B} \rightleftharpoons c\text{C} + d\text{D}$.

Equilibrium of the above system comprising the species A, B, C and D can be attained via either the forward or the backward reaction. If you start off with A and B reacting only, without any C and D present, equilibrium can be achieved. Similarly, an equilibrium can still be achieved if you start off with only C and D, without any A and B present.

It must be noted that the expression for K_c and thus its value are dependent on the balanced chemical equation written. Whenever K_c is quoted, it must be accompanied by the balanced chemical equation.

Q: If we start off with 1 mole of A and 1 mole of B, would the equilibrium concentrations of A, B, C and D be the same as if we have started off with 1 mole of C and 1 mole of D?

A: It may not be! The equilibrium concentrations of the species when at equilibrium, really depends on the value of K_c.

Example 6.2: At 500 K, an equilibrium mixture is found to contain 0.053 mol dm^{-3} of N_2(g), 1.00 mol dm^{-3} of H_2(g) and 3.00 mol dm^{-3} of NH_3(g).

Based on the data given, calculate K_c for the following reactions:

(i) $N_2(g) + 3H_2(g) \rightleftharpoons 2\,NH_3(g)$,
(ii) $2N_2(g) + 6H_2(g) \rightleftharpoons 4NH_3(g)$,
(iii) $2NH_3(g) \rightleftharpoons N_2(g) + 3H_2(g)$.

Solution:

(i) $K_{c,1} = \frac{[NH_3]^2}{[N_2][H_2]^3} = \frac{3^2}{0.053\times 1} = 1.7 \times 10^2\,\text{mol}^{-2}\,\text{dm}^6$,
(ii) $K_{c,2} = \frac{[NH_3]^4}{[N_2]^2[H_2]^6} = \frac{3^4}{(0.053)^2\times 1} = 2.9 \times 10^4\,\text{mol}^{-4}\,\text{dm}^{12}$,
(iii) $K_{c,3} = \frac{[N_2][H_2]^3}{[NH_3]^2} = \frac{0.053\times 1}{3^2} = 5.9 \times 10^{-3}\,\text{mol}^2\,\text{dm}^{-6}$.

- When the equation is written in the reverse order, the K_c value is inverted, i.e., $K_{c,3} = \frac{1}{K_{c,1}}$
- When the coefficients in the equation are multiplied by a factor of n, the K_c value is raised to the power of n, i.e., $K_{c,2} = {K_{c,1}}^2$.

For gases, we usually express their concentrations in terms of partial pressures since these are much easier to measure.

The equilibrium constant, expressed in terms of the partial pressures of the substances, is known as K_p (subscript "p" stands for partial pressure). Consider the following equation:

$$N_2(g) + 3H_2(g) \rightleftharpoons 2NH_3(g).$$

Instead of expressing the equilibrium constant as $K_c = \frac{[NH_3]^2}{[N_2][H_2]^3}$, we can express it as:

$$K_p = \frac{{p_{NH_3}}^2}{p_{N_2}{p_{H_2}}^3}$$

where p_X is partial pressure of gas X at equilibrium.

The unit of partial pressure can be in terms of either Pa or bar, and thus the unit of K_p varies depending on the terms in its expression. For instance, the unit for $K_p = \frac{{p_{NH_3}}^2}{p_{N_2}{p_{H_2}}^3}$ is Pa^{-2} or bar^{-2}.

6.3.1 *Writing K_c or K_p for heterogeneous equilibria*

An equilibrium system in which all the species are in the same phase is in **homogeneous equilibrium**, e.g., $N_2(g) + 3H_2(g) \rightleftharpoons 2NH_3(g)$.

Heterogeneous equilibrium refers to an equilibrium system that contains species in different phases.

In writing the K_c or K_p expression for heterogeneous equilibria, the concentrations and partial pressures of pure solids and liquids (but not aqueous solutions) are excluded.

For instance, given the equilibrium $Ti(s) + 2Cl_2(g) \rightleftharpoons TiCl_4(l)$, we have the following expressions:

$$K_c = \frac{1}{[Cl_2(g)]^2} \quad \text{and} \quad K_p = \frac{1}{{p_{Cl_2}}^2}.$$

Q: Why are the terms involving solids and liquids not included in the equilibrium constant expression for heterogeneous equilibrium?

A: For a given temperature, the saturated vapour pressures of solids (and that of liquids) are constant. In addition, even though their actual amounts may change, both solids and liquids have constant concentrations, as explained in the following.

Recall that concentration = amount (in mol)/volume. Since amount (in mol) is calculated as mass/molar mass, concentration = mass/(molar mass × volume) = density/molar mass. For pure solids and liquids, both density and molar mass are constant values.

Q: What is saturated vapour pressure?

A: In a closed vessel containing a liquid or solid, the particles at the surface tend to evaporate and the gas particles formed tend to condense back into the liquid or solid phase. In time, an equilibrium is established between the gas particles and their condensed form. The pressure exerted by the gas particles at this equilibrium is known as the saturated vapour pressure. As saturated vapour pressure is dependent only on temperature, it is a constant at a fixed temperature.

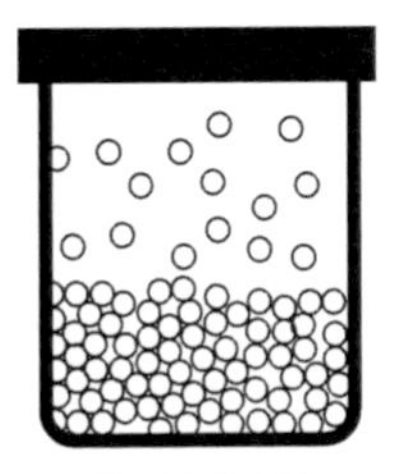

Pure solvent

Exercise 6.1:

(i) Write the K_p expression and state its units for $CO_2(g) + H_2(g) \rightleftharpoons CO(g) + H_2O(l)$.
(ii) Write the K_c expression and state its units for $Ag^+(aq) + Cl^-(aq) \rightleftharpoons AgCl(s)$.

Solution:

(i) $K_p = \frac{p_{CO}}{p_{CO_2} p_{H_2}}$; unit of K_p is Pa^{-1} or bar^{-1}.
(ii) $K_c = \frac{1}{[Ag^+][Cl^-]}$; units of K_c are $mol^{-2}\,dm^6$.

6.3.2 *Calculations involving K_c*

Calculations that seek to solve for K_c or K_p values require the following information:

- the balanced chemical equation,
- the K_c or K_p expression, and
- equilibrium concentrations of all species.

Example 6.3: Calculating K_c given concentration data

1.5 mol of $CH_3OH(l)$ and 1.5 mol of $CH_3COOH(l)$ are allowed to react in a $0.5\,dm^3$ vessel to form the ester $CH_3COOCH_3(l)$:

$$CH_3OH(l) + CH_3COOH(l) \rightleftharpoons CH_3COOCH_3(l) + H_2O(l).$$

The equilibrium mixture is found to contain 0.3 mol of $CH_3COOCH_3(l)$ and 0.3 mol of $H_2O(l)$ at 298 K. Calculate K_c for the reaction.

Approach:

- Write the K_c expression: $K_c = \frac{[CH_3COOCH_3(l)][H_2O(l)]}{[CH_3OH(l)][CH_3COOH(l)]}$.
- Do we have values for the equilibrium concentrations of species? No, but we can find these using the following steps:

Step 1: Construct an "I.C.E." table that shows the Initial concentration, Change in concentration and Equilibrium concentration of the species concerned.
Step 2: Fill in all the known values into the table.

	$CH_3OH(l)$	+ $CH_3COOH(l)$	⇌ $CH_3COOCH_3(l)$	+ $H_2O(l)$
Initial conc. (mol dm^{-3})	1.5/0.5 = 3.0	1.5/0.5 = 3.0	0	0
Change in conc. (mol dm^{-3})				
Equilibrium conc. (mol dm^{-3})			0.3/0.5 = 0.6	0.3/0.5 = 0.6

Since equilibrium [CH_3COOCH_3] is 0.6 mol dm^{-3}, the change in [CH_3COOCH_3] is +0.6 mol dm^{-3} (the "+" sign indicates a gain).

Based on the stoichiometric ratios in the balanced equation, 0.3 mol of ester is produced from 0.3 mol each of CH_3OH and CH_3COOH.

Thus, the change in concentration of these reactants is −0.6 mol dm^{-3} (the "−" sign indicates a loss).

Hence, equilibrium [CH_3OH] and [CH_3COOH] = 3.0 − 0.6 = 2.4 mol dm^{-3}.

	$CH_3OH(l)$	+ $CH_3COOH(l)$	⇌ $CH_3COOCH_3(l)$	+ $H_2O(l)$
Initial conc. (mol dm^{-3})	3.0	3.0	0	0
Change in conc. (mol dm^{-3})	−0.6	−0.6	+0.6	+0.6
Equilibrium conc. (mol dm^{-3})	2.4	2.4	0.6	0.6

Step 3: Calculate K_c and determine its units:

$$K_c = \frac{[CH_3COOCH_3(l)][H_2O(l)]}{[CH_3OH(l)][CH_3COOH(l)]} = \frac{0.6 \times 0.6}{2.4 \times 2.4} = 0.0625.$$

K_c in this instance has no units.

Q: Must we start off with stoichiometric ratio of the reactants in order for the system to achieve equilibrium?

A: Not necessary! You can start off with non-stoichiometric amounts of CH_3OH and CH_3COOH, lets say with 2 moles of CH_3OH and 1 mole of CH_3COOH. The system will achieve equilibrium. Importantly, though you started off with non-stiochiometric ratio of the reactants, when the reaction proceeds, the reactants will still react in accordance to the stoichiometric ratio as given by the balanced equation. The main thing to take note is that the ratio of the reactants at equilibrium will not be in stoichiometric ratio for a system in which the start-off ratio of the reactants is non-stoichiometric.

Q: So, if we start off with stoichiometric ratio of reactants, when the reaction occurs according to the stoichiometric ratio of the reaction equation, at equilibrium, we would have stoichiometric ratio of the reactants left unreacted?

A: Yes, you are right!

Example 6.4: Calculating concentration data given a K_c value

2 mol of $CH_3OH(l)$ and 2 mol of $CH_3COOH(l)$ are allowed to react to form the ester $CH_3COOCH_3(l)$:

$$CH_3OH(l) + CH_3COOH(l) \rightleftharpoons CH_3COOCH_3(l) + H_2O(l).$$

Given that K_c is 0.0625 at 298 K,

(i) determine the equilibrium amount of $CH_3COOCH_3(l)$, and
(ii) hence determine the percentage yield of ester.

Approach:

- Since the value of K_c is given, we start off by writing the K_c expression, i.e., $K_c = \frac{[CH_3COOCH_3(l)][H_2O(l)]}{[CH_3OH(l)][CH_3COOH(l)]} = 0.0625$.
- Although concentration of the species and the volume of the vessel are not stated, we can still construct an I.C.E. table in terms of the amount of species and assign "x mol" as the amount of ester produced at equilibrium.

Solution (i): Let the equilibrium amount of ester be x mol.

	$CH_3OH(l)$ +	$CH_3COOH(l)$ $\rightleftharpoons$	$CH_3COOCH_3(l)$ +	$H_2O(l)$
Initial amount (mol)	2	2	0	0
Change in amount (mol)	$-x$	$-x$	$+x$	$+x$
Equilibrium amount (mol)	$2-x$	$2-x$	x	x

Since volume is not given, assume the volume of mixture to be V dm^3 so that concentration can be expressed in terms of V:

$$K_c = \frac{[CH_3COOCH_3(l)][H_2O(l)]}{[CH_3OH(l)][CH_3COOH(l)]}$$

$$= \frac{\left(\frac{x}{V}\right)\left(\frac{x}{V}\right)}{\left(\frac{2-x}{V}\right)\left(\frac{2-x}{V}\right)} = 0.0625.$$

For the quadratic equation:

$$ax^2 + bx + c = 0;$$

$$x = \frac{-b \pm \sqrt{b^2 - 4ac}}{2a}.$$

Solving the quadratic equation, we get

$$x = 0.4 \text{ or } -0.7.$$

Rejecting the negative value, the equilibrium amount of ester is 0.4 mol.

Solution (ii): The balanced chemical equation gives the theoretical yield, which is, ideally, the maximum amount of product obtainable if the reaction proceeds to completion.

Since the reaction does not go to completion, the actual yield attained is less than the theoretical yield.

Percentage yield is thus the ratio, expressed as a percentage, of the equilibrium amount of product obtained to the theoretical yield.

Percentage yield

$$= \frac{\text{actual yield}}{\text{theoretical yield}} \times 100\%.$$

Percentage yield $= \frac{0.4}{2.00} \times 100\% = 20\%$.

6.3.3 *Calculations involving K_p*

As seen in earlier examples involving expressions for K_c, there are cases when the numerical value for concentration is not given up front. However, we can arrive at a value for concentration through manipulation of data (if the amount and volume are known) or express it in terms of an unknown V (when data is given only in terms of moles).

It is the same when we work with questions involving K_p.

To find the partial pressure of gas W, p_W, we can use the following:

- If the number of moles of each of the species and total pressure p_T, of the system are given, we can find the partial pressure of W using:

$$p_W = \chi_W p_T = \frac{n_W}{n_T} \times p_T,$$

  where χ_W is the mole fraction of W.

- If the amount of gas (n_W), temperature (T) and volume (V) of the reaction vessel are all known, we can apply the Ideal Gas equation, assuming that W is an ideal gas:

$$p_W = (n_W RT)/V.$$

Example 6.5:

$$2A(g) + B(g) \rightleftharpoons 2C(g).$$

Calculate K_p for the reaction given that the equilibrium mixture contains 0.2 mol of A(g), 0.5 mol of B(g) and 0.3 mol of C(g) and the pressure of the reaction vessel is 4 bar.

Solution:

$$p_A = \frac{0.2}{0.2 + 0.5 + 0.3} \times 4 = 0.8\,\text{bar},$$

$$p_B = \frac{0.5}{0.2 + 0.5 + 0.3} \times 4 = 2.0\,\text{bar},$$

$$p_C = \frac{0.3}{0.2 + 0.5 + 0.3} \times 4 = 1.2\,\text{bar},$$

$$K_p = \frac{{p_C}^2}{{p_A}^2 p_B} = 1.13\,\text{bar}^{-1}.$$

There are gases that break down to form smaller gaseous components in a process called dissociation. The degree of dissociation tells us the extent of the dissociation. It can be expressed as a percentage or fraction of gas that has dissociated.

Example 6.6: Calculations involving degree of dissociation of a gas

Dinitrogen tetraoxide dissociates into its monomer on heating:

$$N_2O_4(g) \rightleftharpoons 2NO_2(g).$$

In a particular reaction, it is found that only 40% of $N_2O_4(g)$ has dissociated and the total pressure of the equilibrium mixture is 2 bar. Calculate the value of K_p.

Method 1

Approach: 40% of $N_2O_4(g)$ dissociates into $NO_2(g)$. This means that 60% of the initial amount of $N_2O_4(g)$ remains in its original form, i.e., undissociated. Thus, we need to define the initial amount of $N_2O_4(g)$, which was not given.

Solution: Let w be the initial amount of $N_2O_4(g)$. Since 1 mol of $N_2O_4(g)$ dissociates to give 2 mol of $NO_2(g)$, $0.4w$ mol of $N_2O_4(g)$ dissociates to give $2 \times 0.4w = 0.8\,w$ mol of $NO_2(g)$.

	$N_2O_4(g)$	$\rightleftharpoons$	$2NO_2(g)$
Initial amount (mol)	w		0
Change in amount (mol)	$-0.4w$		$+0.8w$
Equilibrium amount (mol)	$0.6w$		$0.8w$

Total amount of gases n_T at equilibrium $= 0.6w + 0.8w = 1.4w$. Total pressure p_T at equilibrium $= 2$ bar.

$$p_{N_2O_4} = \frac{n_{N_2O_4}}{n_T} \times p_T = \frac{0.6w}{1.4w} \times 2 = 0.857\,\text{bar}$$

$$p_{NO_2} = \frac{n_{NO_2}}{n_T} \times p_T = \frac{0.8w}{1.4w} \times 2 = 1.143\,\text{bar}.$$

Hence,

$$K_p = \frac{{p_{NO_2}}^2}{p_{N_2O_4}} = \frac{(1.143)^2}{0.857} = 1.52\,\text{bar}.$$

Method 2

Approach: Instead of defining the initial amount of $N_2O_4(g)$, we can choose to work with defining the initial pressure of $N_2O_4(g)$.

Solution: Let y be the initial partial pressure of $N_2O_4(g)$. Note that the initial pressure is not 2 bar.

	$N_2O_4(g)$	$\rightleftharpoons$	$2NO_2(g)$
Initial partial pressure (bar)	y		0
Change in partial pressure (bar)	$-0.4y$		$+0.8y$
Equilibrium partial pressure (bar)	$0.6y$		$0.8y$

Total pressure, p_T at equilibrium = 2 bar.

$$\begin{aligned} p_T = p_{N_2O_4} + p_{NO_2} &= 2, \\ 0.6y + 0.8y &= 2, \\ y &= 1.429\,\text{bar}. \end{aligned}$$

Hence,

$$K_p = \frac{{p_{NO_2}}^2}{p_{N_2O_4}} = \frac{(0.8y)^2}{0.6y} = \frac{0.64 \times 1.429}{0.6} = 1.52\,\text{bar}.$$

Exercise:

(i) At a particular temperature, 2.5 mol of $CO(g)$ and 2.5 mol of $H_2(g)$ are allowed to react in a 2 dm^3 vessel:

$$CO(g) + 3H_2(g) \rightleftharpoons CH_4(g) + H_2O(g).$$

The equilibrium mixture was found to contain 0.4 mol of $CH_4(g)$. Calculate K_c for the reaction.

(ii) 4.0 mol of $NOCl(g)$ is allowed to decompose. It is found that only 28% of $NOCl(g)$ has dissociated and the total pressure of the equilibrium mixture is 2 bar. Calculate the value of K_p.

$$2NOCl(g) \rightleftharpoons 2NO(g) + Cl_2(g).$$

[**Answer:** (i) $0.139\,\text{mol}^{-2}\,\text{dm}^6$, and (ii) 3.71×10^{-2} bar.]

6.4 Le Chatelier's Principle

Le Chatelier's Principle is a helpful rule for predicting the effects of changes applied to a system that is already at equilibrium.

It states that, "When a system in equilibrium is subjected to a change, the system responds in such a way as to counteract the imposed change and reestablish the equilibrium state."

The change can be brought about by changes in *concentration, pressure, volume, presence of catalyst, or temperature.*

The principle helps us to predict in which direction the equilibrium position will shift (favouring either the forward or backward reaction) in response to the change.

When a change in conditions is introduced to an equilibrium system, the system will no longer be in equilibrium since the change will affect the rates of both forward and backward reactions to different extents. The system will then readjust itself to attain a new equilibrium where the concentrations of reactants and products become constant again. These concentration values will, however, be **different from the previous equilibrium state**.

However, the values of K_c and K_p remain unchanged as these are only affected by changes in temperature.

Q: Why are the equilibrium constants only affected by a temperature change?

A: Remember that K_c can be perceived as the ratio of rate constants? Now, rate constant can be perceived as $k = A\exp(-E_a/RT)$. So do you see now that temperature affects the rate constant and thus the equilibrium constant?

6.4.1 *Effect of concentration changes*

Let us consider the reaction $PCl_3(g) + Cl_2(g) \rightleftharpoons PCl_5(g)$.

6.4.1.1 *Effect of adding a reactant (at constant volume)*

When extra $Cl_2(g)$ is added at constant volume to the equilibrium mixture, its concentration increases.

According to Le Chatelier's Principle, the system will attempt to remove some of the extra $Cl_2(g)$ by favouring the forward reaction. The equilibrium position will shift to the right. More $PCl_3(g)$ and $Cl_2(g)$ will react to form $PCl_5(g)$ until a new equilibrium is established.

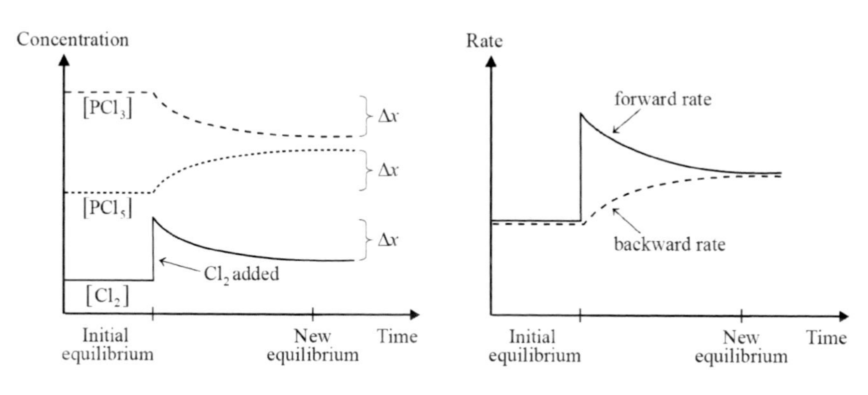

This new equilibrium mixture has a different composition compared to the initial equilibrium mixture before the change is introduced, but K_c remains unchanged. At the new position of equilibrium, $[Cl_2]$ increases (as not all the $Cl_2(g)$ added has been completely removed) while $[PCl_3]$ decreases (as it was consumed in order to remove the additional Cl_2 that has been added) and $[PCl_5]$ increases.

Q: How do we use kinetics to explain the way the system responds to the change?

A: When extra $Cl_2(g)$ is added while keeping the volume constant, the forward rate rapidly increases because the effective collisional frequency increases as there are now more Cl_2 molecules in the system. The increased amount of PCl_5 formed soon results in an increase in the backward rate. As time passes, the forward rate and backward rate become equal again, but at a higher value than the previous equilibrium state.

Q: But shouldn't the new forward rate be lower than the old forward rate since $[PCl_3]$ has decreased because it was being consumed to remove the Cl_2?

A: Based on the decrease in $[PCl_3]$, the forward rate should be lower. But do not forget that at the new equilibrium position, $[Cl_2]$ is higher than at the old equilibrium position. So the higher $[Cl_2]$ and lower $[PCl_3]$ together ensures a new higher forward rate. Now, if you are still not convinced, the $[PCl_5]$ at the new equilibrium position is higher than at the old equilibrium position. This leads to a new higher backward rate. Since both the forward rate and backward rate at the new equilibrium position are the same, then a new higher backward rate also means a new higher forward rate!

6.4.1.2 *Effect of removing a reactant (at constant volume)*

With reference to the same reaction, $PCl_3(g) + Cl_2(g) \rightleftharpoons PCl_5(g)$, when $PCl_5(g)$ is removed at constant volume from the equilibrium mixture, its concentration decreases.

According to Le Chatelier's Principle, the system will attempt to replace the removed $PCl_5(g)$ by favouring the forward reaction. The equilibrium position will shift to the right. More $PCl_3(g)$ and $Cl_2(g)$ will react to form $PCl_5(g)$ until a new equilibrium is established.

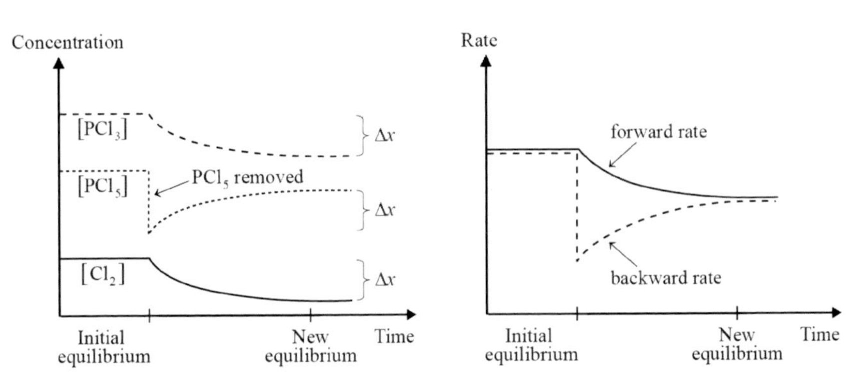

Not all the $PCl_5(g)$ removed has been completely replenished. In all, $[PCl_5]$ changes by the same extent as both $[PCl_3]$ and $[Cl_2]$, but K_c remains unchanged.

Q: How do we use kinetics to explain the way the system responds to the change?

A: When some $PCl_5(g)$ is removed, the backward rate decreases rapidly because the effective collisional frequency decreases as there are now fewer PCl_5 molecules in the system. The forward rate also decreases as there are smaller amounts of PCl_3 and Cl_2 formed. The backward rate slowly picks up as more PCl_5 forms. As time passes, the forward rate and backward rate become equal again, but at a lower value than the previous equilibrium state.

Removal of a species can also be achieved by introducing a new substance that will react with it (see Example 6.7 below).

Example 6.7: What will be observed when $OH^-(aq)$ is added to the following system at equilibrium?

$$\underset{\text{yellow}}{2CrO_4^{2-}(aq)} + 2H^+(aq) \rightleftharpoons \underset{\text{orange}}{Cr_2O_7^{2-}(aq)} + H_2O(l).$$

Solution: The added $OH^-(aq)$ will react with $H^+(aq)$ and cause $[H^+]$ to decrease. According to Le Chatelier's Principle, the system will attempt to increase $[H^+]$ by favouring the backward reaction. The equilibrium position will shift to the left. The solution will thus change colour from orange to yellow.

6.4.2 *Effect of pressure changes*

Changes in pressure only affect reactions that involve gases. A pressure change can be introduced through the following ways:

(a) addition/removal of a gaseous component of the equilibrium mixture under conditions of constant volume,
(b) expansion or compression of the reaction vessel,
(c) addition of an inert gas at constant volume, and
(d) addition of an inert gas at constant pressure.

6.4.2.1 *Addition/removal of a gaseous component of the equilibrium mixture*

Addition or removal of a gaseous component in the equilibrium mixture affects the partial pressure of that particular component. The effect of partial pressure changes is similar to that of concentration changes since $p \propto n/V$ (assuming ideal behaviour).

Let us consider the same reaction: $PCl_3(g) + Cl_2(g) \rightleftharpoons PCl_5(g)$.

When extra $Cl_2(g)$ is **added** at constant volume to the equilibrium mixture, its partial pressure increases since n increases.

According to Le Chatelier's Principle, the system will attempt to **remove some of the extra** $Cl_2(g)$ by favouring the forward reaction. The equilibrium position will shift to the right. More $PCl_3(g)$ and $Cl_2(g)$ will react to form $PCl_5(g)$ until a new equilibrium is established.

Does this sound familiar? Refer to Sec. 6.4.1 for more details.

6.4.2.2 *Expansion or compression of reaction vessel*

Expansion or compression of the reaction vessel affects the total pressure (p_T) of the system:

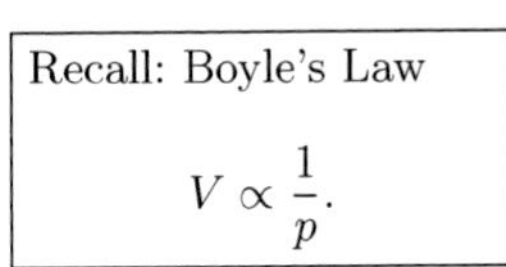

- Expansion (an increase in volume) causes p_T of system to decrease, which means that the partial pressure of each of the component gases also decreases.
- Compression (a decrease in volume) causes p_T of system to increase, which means that the partial pressure of each of the component gases also increases.

The effect of changes in p_T of a gaseous system depends on the relative number of gaseous molecules on the right and left sides of the balanced chemical equation. There are two different scenarios as shown in the examples below:

- **Changing the total pressure of a system in which the number of gas molecules on each side of the chemical equation is different:**

$$PCl_3(g) + Cl_2(g) \rightleftharpoons PCl_5(g).$$

According to Le Chatelier's Principle, when the **total pressure** of the system is **increased**, concentrations of all species (both reactants and products) also increase. The system will attempt to **decrease the overall pressure, by favouring the reaction that decreases the overall number of gaseous molecules**, i.e., the forward reaction is favoured.

Recall:
$pV = nRT$, $p \propto n$.

The equilibrium position will shift to the right. More $PCl_3(g)$ and $Cl_2(g)$ will react to form $PCl_5(g)$ until a new equilibrium is established.

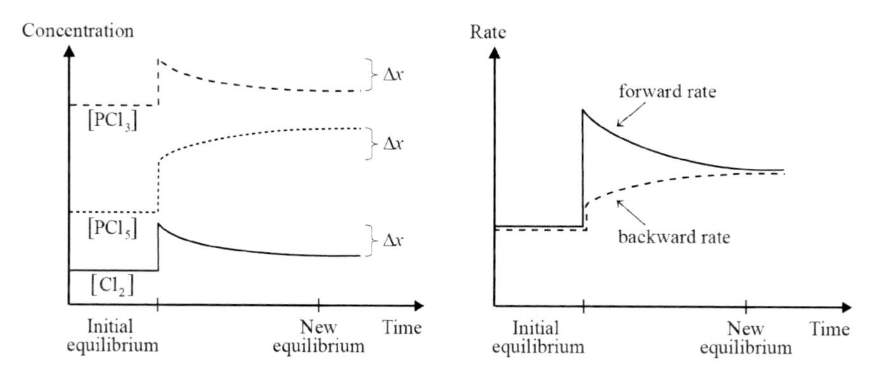

Note: Δx = change in [substance]

Since $p \propto n/V$ (assuming ideal behaviour), when p_T of the system is increased, concentrations of all species (both reactants and products) also increase. In terms of the number of moles, the new equilibrium mixture has a higher percentage of $PCl_5(g)$ but lower percentages of $PCl_3(g)$ and $Cl_2(g)$, as compared to the old equilibrium position, but the equilibrium constant remain unchanged.

The reverse can be said when total pressure is decreased.

Q: How do we use kinetics to explain the way the system responds to the change?

A: When the total pressure increases due to a decrease in the volume of the container, the volume of the system shrinks. The gaseous particles now occupy a smaller volume. As a result, the effective collisional frequencies of both the forward and backward reactions increase. These lead to increases in the rates of both the forward and backward reactions. BUT the rate of the reaction that involves a greater number of particles colliding is increased by a greater extent. So, in this case, since the forward reaction involves the collision of two particles whereas the backward reaction only one, the percentage increase of the forward rate is higher than the backward rate. As time passes, the forward rate and backward rate become equal again, but at a higher value than the previous equilibrium state.

- **Changing the total pressure of a system in which the number of gas molecules on each side of the chemical equation is the same:**

$$H_2(g) + I_2(g) \rightleftharpoons 2HI(g).$$

According to Le Chatelier's Principle, when the total pressure of the system is increased, concentrations of all species (both reactants and products) also increase. The system will attempt to decrease the overall pressure by favouring the reaction that decreases the overall number of gaseous molecules.

However, since both forward and backward reactions produce the same number of gaseous molecules, none of these reactions is favoured over the other.

The equilibrium position will not shift as the system remains at equilibrium. In terms of the number of moles, the equilibrium composition is unchanged and so is the equilibrium constant, BUT both the forward and backward rates do increase!

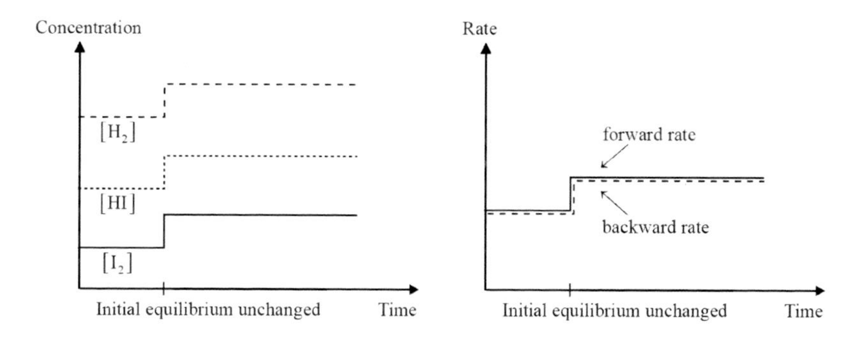

Q: How do we use kinetics to explain the way the system responds to the change?

A: When the total pressure increases, the volume of the system shrinks. The gaseous particles now occupy a smaller volume. As a result, the effective collisional frequencies of both the forward and backward reactions increase. These lead to increases in the rates of both the forward and backward reactions. Since both reactions involve the collision of the same number of particles, both rates increase by the same extent. As time passes, the forward rate and backward rate remain equal, but at a higher value than the previous equilibrium state.

6.4.2.3 *Addition of an inert gas at constant volume*

Consider the reaction: $A(g) + B(g) \rightleftharpoons C(g)$.

When an inert gas is added at constant volume to the equilibrium mixture, the total pressure of the system increases:

$$p_T = p_A + p_B + p_C + p_{\text{inert gas}}.$$

The increase in p_T is due to an increase in n_T since $n_T = n_A + n_B + n_C + n_{\text{inert gas}}$.

However, there are no changes in the partial pressures of the reacting gases since for each gas, concentration is unchanged (i.e., both n and V are constant).

Thus, the equilibrium position will not shift as the system remains at equilibrium.

> Recall:
>
> $$p_A = \frac{n_A}{n_T} \times p_T,$$
>
> where n_A is the amount of A(g) and n_T is the total amount of gas molecules.

The equilibrium composition is unchanged and so is the equilibrium constant.

Q: When inert gas particles are added into the system, wouldn't the inert gas particles take up the space in between the reactant particles and inhibit them from reacting?

A: Take note that in the gaseous state, the separation between two gas particles is huge. Moreover, all the gas particles are constantly in rapid motion. Thus, **introducing inert gas particles at constant volume does not affect the position of equilibrium no matter what the chemical equation for the reaction is!**

Q: If two reactant particles are about to collide and react, when an inert gas particle comes in between, wouldn't this decrease the rate of reaction?

A: Well, why not you think in another perspective in which an inert gas particle can collide with the reactant particles and caused them to "meet" with each other. In such a scenario, wouldn't this increase the rate of reaction? Now, when inert gas is introduced into a system, the collision of the inert gas particles with the reactants in helping them to react is statistically nullified by the collisions that "destroy" the reaction. So, overall there is no collisional effect.

6.4.2.4 *Addition of an inert gas at constant pressure*

When an inert gas is added at constant pressure to the equilibrium mixture, the volume of the vessel increases (i.e., expansion takes place) to keep total pressure constant.

The expansion leads to a decrease in the concentrations of all gases and also their partial pressures (since the same number of molecules now occupy a bigger volume). Whether there is a change in the position of equilibrium depends on the following:

- **If the numbers of gas molecules are not the same for both sides of the equation:**

$$PCl_3(g) + Cl_2(g) \rightleftharpoons PCl_5(g).$$

According to Le Chatelier's Principle, the system will attempt to counteract the change by favouring the reaction that increases the overall number of gaseous molecules, i.e., the backward reaction is favoured.

The equilibrium position will shift to the left. More $PCl_5(g)$ will dissociate to form $PCl_3(g)$ and $Cl_2(g)$ until a new equilibrium is established.

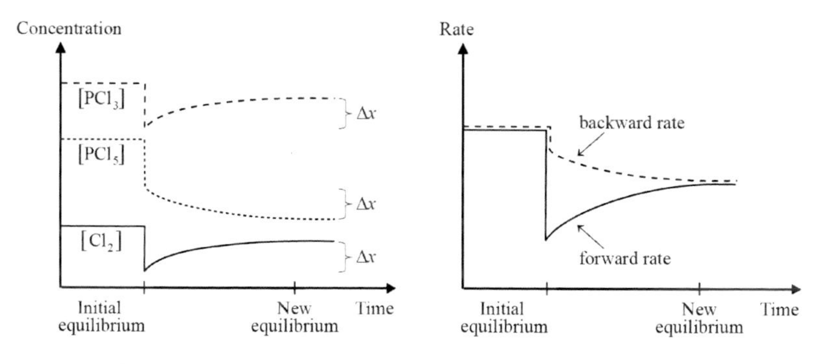

The concentrations of all species (both reactants and products) decrease. In terms of the number of moles, the new equilibrium mixture has a higher percentage of $PCl_3(g)$ and $Cl_2(g)$ but a lower percentage of $PCl_5(g)$ as compared to the old equilibrium position. The equilibrium constant remains unchanged.

Q: How do we use kinetics to explain the way the system responds to the change?

A: When the volume increases due to the addition of an inert gas at constant pressure, the concentration of each reacting gas decreases. As a result, the effective collisional frequencies of both the forward and backward reactions decrease. These lead to decreases in the rates of both the forward and backward reactions. BUT the rate of the reaction that involves a greater number of particles colliding is decreased by a greater extent. So, in this case, since the forward reaction involves the collision of two particles whereas the backward reaction only one, the percentage decrease of the forward rate is higher than the backward rate. As time passes, the forward rate and backward rate become equal again, but at a lower value than the previous equilibrium state.

- **If the numbers of gas molecules are the same for both sides of the equation:**

$$H_2(g) + I_2(g) \rightleftharpoons 2HI(g).$$

The addition of inert gas at constant pressure results in an increase in the volume of the system. The concentrations of the reacting gases decrease. According to Le Chatelier's Principle, the system will attempt to counteract the change by favouring the reaction that increases the overall number of gaseous molecules.

However, since both forward and backward reactions produce the same number of gaseous molecules, none of these reactions is favoured over the other. Thus, the equilibrium position will not shift as the system remains at equilibrium. In terms of the number of moles, the equilibrium composition is unchanged and so is the equilibrium constant.

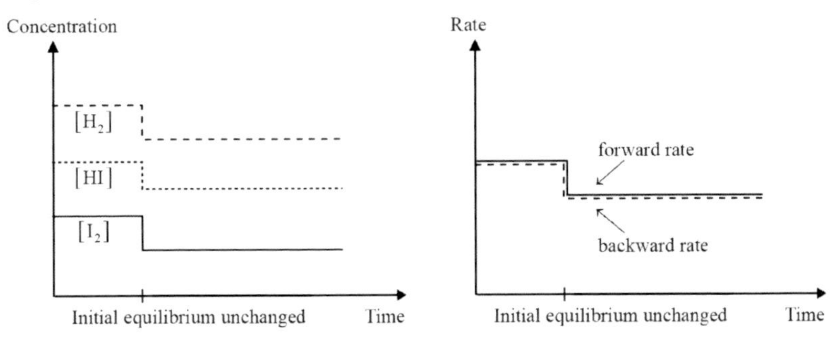

Q: How do we use kinetics to explain the way the system responds to the change?

A: When inert gas is added at constant pressure, the volume of the system expands. The gaseous particles now occupy a bigger volume. As a result, the effective collisional frequencies of both the forward and backward reactions decrease. These lead to decreases in the rates of both the forward and backward reactions. Since both reactions involve the collision of the same number of particles, both rates decrease by the same extent. As time passes, the forward rate and backward rate remain equal, but at a lower value than the previous equilibrium state.

6.4.3 *Effect of temperature changes*

Increasing the temperature of an equilibrium system will result in the following:

- a change in the equilibrium position, and
- **a change in the value of the equilibrium constant.**

Consider the hypothetical reaction below whereby the forward reaction is exothermic:

$$a\mathrm{A} + b\mathrm{B} \rightleftharpoons c\mathrm{C} + d\mathrm{D}, \quad \Delta H < 0.$$

- **Effect of an increase in temperature**
  According to Le Chatelier's Principle, when there is an **increase in temperature**, the system will attempt to **decrease** the **overall temperature** by **favouring** the **reaction** that **absorbs heat**, i.e., the **endothermic** reaction.

  Since the backward reaction is endothermic, the equilibrium position will shift to the left. Greater amounts of C and D will react to form A and B until a new equilibrium is established.

  This new equilibrium mixture will contain more of the reactants A and B, and less of the products C and D.

  As a result, the value of the equilibrium constant becomes smaller (recall: $K_\mathrm{c} = \frac{[\mathrm{C}]^c[\mathrm{D}]^d}{[\mathrm{A}]^a[\mathrm{B}]^b}$).

  Q: How do we use kinetics to explain the way the system responds to the change?

  A: When the temperature of the system is increased, the rates of both the forward and backward reactions increase as the rate constant is temperature dependent ($k = A\exp(-E_\mathrm{a}/RT)$). **The percentage increase in rate is greater for the reaction with the greater activation energy** (refer to Chap. 5 on Reaction Kinetics), i.e., the backward reaction for the reaction here. So as time passes, the position of equilibrium shifts to the left. This explains how the system "gets rid" of the added heat.

Q: I am still not convinced! Can you please prove it to me?
A: Look at the following calculations which assume the value of A to be one:

E_a (kJ mol^{-1})	at T = 300 K	at T = 400 K	Percentage Change in Rate Constant
500 (backward)	0.818	0.860	5.13
200 (forward)	0.923	0.942	2.06

As can be seen from the above, an increase in temperature leads to an increase in the rate constant for both the forward and backward reactions. BUT there is a greater percentage increase for the reaction that has a higher activation energy (E_a).

- **Effect of a decrease in temperature**
According to Le Chatelier's Principle, when there is a **decrease in temperature**, the system will attempt to **increase** the **overall temperature** by **favouring** the **reaction** that **releases heat**, i.e., the **exothermic** reaction.
Since the forward reaction is exothermic, the equilibrium position will shift to the right. Greater amounts of A and B will react to form C and D until a new equilibrium is established.
This new equilibrium mixture will contain less of the reactants A and B, and more of the products C and D.
As a result, the value of the equilibrium constant becomes larger.

Similar arguments can be applied to the case where the forward reaction is endothermic:

$$a\mathrm{A} + b\mathrm{B} \rightleftharpoons c\mathrm{C} + d\mathrm{D}, \quad \Delta H > 0.$$

Just remember that,

- an increase in temperature favours the endothermic reaction, and
- a decrease in temperature favours the exothermic reaction.

6.4.4 *Effect of temperature changes on the value of the equilibrium constant*

K_c can be expressed as a ratio of the rate constants of a reversible reaction as follows:

$$K_c = k_f/k_b.$$

Since the rate constant is dependent on temperature, $k = A\exp(-E_a/RT)$, so too is the equilibrium constant. How is it actually affected? We know that reaction rate increases when temperature increases. This means that equilibrium can be attained at a faster rate if temperature is raised.

We also know that when temperature is raised, the endothermic reaction is favoured, i.e., there is a greater proportional increase for its rate constant as compared to the exothermic reverse reaction.

The reverse is true when temperatures are lowered. Both the reaction rates decrease and equilibrium is attained at a slower rate. The exothermic reaction is favoured, i.e., there is a smaller proportional decrease for its rate constant as compared to the endothermic forward reaction.

All in all, given that $K_c = k_f/k_b$, and that a larger rate constant k indicates a faster rate:

- For an exothermic forward reaction $a\text{A} + b\text{B} \rightleftharpoons c\text{C} + d\text{D}$, $\Delta H < 0$,
  - an increase in temperature will result in K_c becoming smaller since the rate of the backward reaction has increased by a greater proportion than the rate of the forward reaction, i.e., $\Delta k_f < \Delta k_b$;
  - a decrease in temperature will result in K_c becoming larger since the rate of the backward reaction has decreased by a greater proportion than the rate of the forward reaction, i.e., $|\Delta k_f| < |\Delta k_b|$.
- For an endothermic forward reaction $a\text{A} + b\text{B} \rightleftharpoons c\text{C} + d\text{D}$, $\Delta H > 0$,
  - an increase in temperature will result in K_c becoming larger since the rate of the forward reaction has increased by a greater proportion than the rate of the backward reaction, i.e., $\Delta k_f > \Delta k_b$;

- a decrease in temperature will result in K_c becoming smaller since the rate of the forward reaction has decreased by a greater proportion than the rate of the backward reaction, i.e., $|\Delta k_f| > |\Delta k_b|$.

Example 6.8:

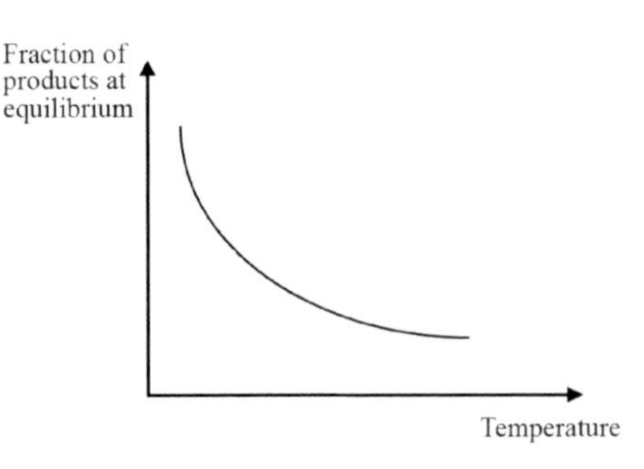

For which system is the profile of the graph correct?

(a) $4NH_3(g) + 5O_2(g) \rightleftharpoons 4NO(g) + 6H_2O(g)$, $\Delta H = -910\,kJ\,mol^{-1}$.
(b) $2NOBr(g) \rightleftharpoons 2NO(g) + Br_2(g)$, $\Delta H = +30\,kJ\,mol^{-1}$.

Solution: The profile of the graph fits system (a). Based on the graph, the proportion of products at equilibrium decreases as temperature increases. This implies that the backward reaction is favoured at higher temperatures.

According to Le Chatelier's Principle, when there is an increase in temperature, the system will attempt to decrease the overall temperature by favouring the endothermic reaction.

Therefore, the backward reaction is endothermic and thus the forward reaction is exothermic as shown here:

$$4NH_3(g) + 5O_2(g) \rightleftharpoons 4NO(g) + 6H_2O(g), \quad \Delta H = -910\,kJ\,mol^{-1}.$$

6.4.5 *Effect of catalyst*

A catalyst only aids a reaction in attaining equilibrium in a shorter time. It does not cause any changes to the position of the equilibrium and hence the value of the equilibrium constant.

This is because a catalyst lowers the activation energies of both the forward and backward reactions to the same extent. This, in turn, leads to the rates of both the forward and backward reactions being increased to the same extent.

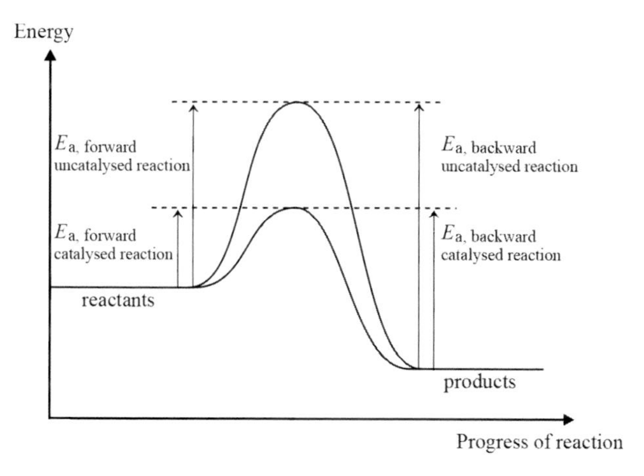

Example 6.9:

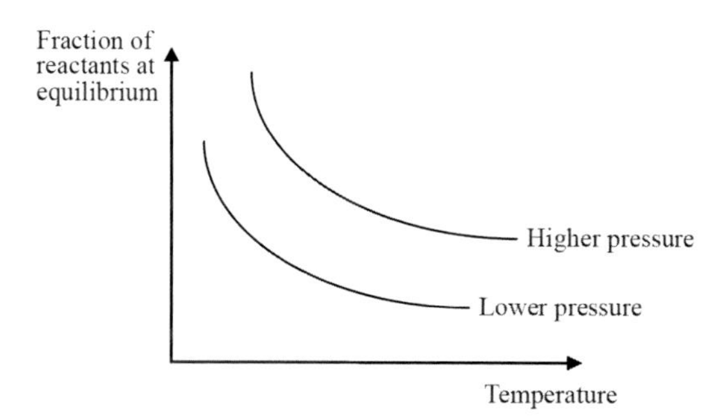

Which of the following system(s) will give the graphs shown?

(a) $4NH_3(g) + 5O_2(g) \rightleftharpoons 4NO(g) + 6H_2O(g)$, $\Delta H = -910\,\text{kJ mol}^{-1}$.
(b) $2NOBr(g) \rightleftharpoons 2NO(g) + Br_2(g)$, $\Delta H = +30\,\text{kJ mol}^{-1}$.
(c) $N_2(g) + 3H_2(g) \rightleftharpoons 2NH_3(g)$, $\Delta H = -92\,\text{kJ mol}^{-1}$.
(d) $H_2(g) + I_2(g) \rightleftharpoons 2HI(g)$, $\Delta H = +53\,\text{kJ mol}^{-1}$.
(e) $N_2O_4(g) \rightleftharpoons 2NO_2(g)$, $\Delta H = +57\,\text{kJ mol}^{-1}$.

Solution: According to Le Chatelier's Principle, when there is an increase in temperature, the system will attempt to decrease the overall temperature by favouring the endothermic reaction.

Based on the graph, the proportion of reactants at equilibrium decreases as temperature increases. This implies that the **forward reaction is endothermic.**

According to Le Chatelier's Principle, when there is an increase in pressure, the system will attempt to decrease the overall pressure by favouring the reaction that decreases the overall number of gaseous molecules.

Since the graph indicates that at a given temperature, the proportion of reactants at equilibrium increases when pressure is increased, the backward reaction must be favoured at higher pressures. This implies that the **number of molecules on the left-hand side of the equation is lower than that on the right-hand side.**

Based on the two features discussed, the systems to which the graphs correspond to are:

(b) $2NOBr(g) \rightleftharpoons 2NO(g) + Br_2(g), \quad \Delta H = +30\,kJ\,mol^{-1}$, and
(e) $N_2O_4(g) \rightleftharpoons 2NO_2(g), \quad \Delta H = +57\,kJ\,mol^{-1}$.

Exercise: Predict how the position of equilibrium will be affected when the following changes are made to the equilibrium system:

$$CO(g) + 3H_2(g) \rightleftharpoons CH_4(g) + H_2O(g), \quad \Delta H = -206\,kJ\,mol^{-1}.$$

(i) Temperature is increased.
(ii) Total pressure is increased.
(iii) Some CO is removed.
(iv) A catalyst is added.
(v) An inert gas is added at constant volume.

Solution:

(i) It shifts to the left.
(ii) It shifts to the right.
(iii) It shifts to the left.
(iv) It does not shift.
(v) It does not shift.

Exercise: For which of these systems is the formation of products favoured when

(i) temperature is decreased;

(ii) pressure is decreased;
(iii) both temperature and pressure are increased?

(a) $2SO_2(g) + O_2(g) \rightleftharpoons 2SO_3(g)$, $\Delta H = -198\,\text{kJ mol}^{-1}$.
(b) $N_2(g) + O_2(g) \rightleftharpoons 2NO(g)$, $\Delta H = +181\,\text{kJ mol}^{-1}$.
(c) $2CO_2(g) \rightleftharpoons 2CO(g) + O_2(g)$, $\Delta H = +566\,\text{kJ mol}^{-1}$.
(d) $N_2(g) + 2O_2(g) \rightleftharpoons 2NO_2(g)$, $\Delta H = +66\,\text{kJ mol}^{-1}$.
(e) $2O_3(g) \rightleftharpoons 3O_2(g)$, $\Delta H = -427\,\text{kJ mol}^{-1}$.

Solution:

(i) (a) and (e).
(ii) (c) and (e).
(iii) d only.

The discussion on Le Chatelier's Principle and the changes that affect an equilibrium system help us to understand why certain conditions are employed in trying to achieve production yield, which is especially important in the manufacturing industry amidst other considerations such as cost, time and environmental concerns.

One classic textbook example is the manufacturing conditions for the production of $NH_3(g)$, which will be discussed in the following section.

6.5 The Haber Process

The aim of the manufacturing industry is to produce the maximum yield of products in the shortest, most efficient time possible and at minimum cost.

$NH_3(g)$ is an important industrial product which has various uses such as in the manufacturing of fertilizers, household cleaners and in producing nitrogen-containing derivatives like nitric acid, which are useful for bomb making.

$$N_2(g) + 3H_2(g) \rightleftharpoons 2NH_3(g), \quad \Delta H^{\ominus}{}_{(298\,\text{K})} = -92.4\,\text{kJ mol}^{-1}.$$

To maximise the yield of $NH_3(g)$ and minimise production cost, the following conditions are applied to the production process:

- *Temperature: 450°C*
  Reason for choice: Lowering the temperature of the reaction vessel will favour the forward reaction since it is exothermic, and hence a higher yield of $NH_3(g)$ can be obtained.
  On the other hand, the rate of production can be too slow at low temperatures. Therefore, it has been found that at 450°C, a substantial yield and production rate can be achieved. Higher temperatures than this will result in lower yield and high operating costs. The latter comes from the investment in expensive reactors that can withstand the high temperature and the continuous requirement for large amounts of energy needed to maintain the high temperature.
- *Pressure: about 200 atm*
  Reason for choice: Increasing the pressure of the reaction vessel will favour the forward reaction since it will decrease the overall number of gaseous molecules and hence the pressure. This will increase the yield of $NH_3(g)$. However, operating at very high pressures will increase production cost, so a moderate pressure of 200 atm is used.

- *The use of catalyst: finely divided iron catalyst with aluminium oxide as promoter*
  Reason for use: Although the use of catalyst does not affect the yield of $NH_3(g)$, it will enable production to be faster.

My Tutorial (Chapter 6)

1. Ethanol, which is an important motor fuel nowadays, can be manufactured by direct catalytic hydration of ethene with steam, using a phosphoric acid as a catalyst. Assume that the reaction of ethene (C_2H_4) with steam to give ethanol (C_2H_5OH, which is gaseous at the temperature of the reaction), is at equilibrium.

   (a) Write the equation for the reaction (with state symbols).
   (b) Write the expression for the *equilibrium constant* K_c for the reaction in terms of the concentrations of reactants and products.

(c) Use the information below to calculate the equilibrium concentration of ethanol vapour under these conditions:

$$\begin{aligned} \text{Temperature} &= 570\,\text{K}, \\ \text{Pressure} &= 60 \times 10^5\,\text{Pa}, \\ K_c &= 24\,\text{dm}^3\,\text{mol}^{-1} \text{ at } 570\,\text{K}, \\ \text{equilibrium } [\text{H}_2\text{O(g)}] &= 0.050\,\text{mol dm}^{-3}, \\ \text{equilibrium } [\text{C}_2\text{H}_4\text{(g)}] &= 0.45\,\text{mol dm}^{-3}. \end{aligned}$$

(d) The enthalpy change for the forward reaction is $\Delta H = -46\,\text{kJ mol}^{-1}$. Why is a temperature of 570 K used for the reaction, rather than

(i) a higher temperature?
(ii) a lower temperature?

(e) Why is a pressure of 60×10^5 Pa used for the reaction, rather than

(i) a higher pressure?
(ii) a lower pressure?

2. The Haber process is an important industrial process for the synthesis of ammonia, an important precursor for making industrial fertilizers. Iron is a common catalyst used in this process. The process does not go to completion on its own and would reach the following dynamic equilibrium state:

$$N_2(g) + 3H_2(g) \rightleftharpoons 2NH_3(g).$$

(a) Define the terms *partial pressure* of a gas in a mixture of gases and *dynamic equilibrium*.

(b) In an equilibrium mixture consisting of nitrogen, hydrogen and ammonia at 800 K, the partial pressures of the three gases are 20.4, 57.2 and 11.7 atm, respectively.

(i) Calculate the total pressure in the system.
(ii) What is the mass of ammonia present under these conditions in a vessel of volume $200\,\text{m}^3$? (Take $R = 8.20 \times 10^{-5}\,\text{m}^3\,\text{atm mol}^{-1}\,\text{K}^{-1}$.)
(iii) Write an expression for K_p for the formation of ammonia.
(iv) Calculate the value of K_p under these conditions.

(v) What would the effect on the equilibrium yield of ammonia be under the conditions in part (b)(ii) if a better catalyst were used?

(vi) Predict and explain the effect of an increase in temperature on the value of K_p.

(vii) Predict and explain the effect of an increase in temperature on the rate of the forward reaction.

(viii) Give two reasons why a new catalyst might be preferred to the existing one even though it costs more.

(c) The gases are passed through a conversion chamber containing granulated iron as a catalyst. Describe and explain the effect of the iron on:

(i) the rate of the production of ammonia, and
(ii) the amount of ammonia in the equilibrium mixture.

(d) The equilibrium mixture formed is then passed into a refrigeration plant. Explain why this is done and what happens after this stage?

3. At high temperatures, phosphorus pentachloride is a gas that dissociates as follows:

$$PCl_5(g) \rightleftharpoons PCl_3(g) + Cl_2(g).$$

(a) Write an expression for the equilibrium constant K_p for this equilibrium.

(b) At a given temperature, the degree of dissociation of an original sample of $PCl_5(g)$ is 0.52 and the system reaches equilibrium. If the total equilibrium pressure is 2 atm, calculate the values of the equilibrium partial pressures of PCl_5 and PCl_3.

(c) Hence calculate the value of K_p and give its units.

4. This question concerns the reaction

$$H_2(g) + I_2(g) \rightleftharpoons 2HI(g),$$

which is slow even at high temperature.

(a) (i) Using your Data Booklet, calculate ΔH for the reaction between hydrogen and iodine.

(ii) Sketch an energy level diagram for this reaction.

(b) Sketch the energy profile diagram and indicate on your sketch:

(i) ΔH for the reaction,

(ii) the activation energy for the forward reaction ($E_{a(f)}$), and

(iii) the activation energy for the reverse reaction ($E_{a(b)}$).

(c) For an analogous reaction involving the formation of hydrogen chloride

$$H_2(g) + Cl_2(g) \longrightarrow 2HCl(g),$$

explain how you would expect the activation energy of the forward reaction to be compared with that shown for the formation of HI.

(d) The reaction for the formation of hydrogen iodide does not go to completion but reaches an equilibrium state.

(i) Write an expression for the equilibrium constant, K_c, for this reaction.

(ii) A mixture of 2.9 mol of H_2 and 2.9 mol of I_2 is prepared and allowed to reach equilibrium in a closed vessel of 250 cm^3 capacity at 700°C. The resulting equilibrium mixture is found to contain 4.5 mol of HI. Calculate the value of K_c at this temperature.

(iii) Explain why the formation of hydrogen chloride goes to completion as compared to the formation of hydrogen iodide.

(e) In an experiment to establish the equilibrium concentration in (d)(ii), the reaction is allowed to reach equilibrium at 723 K and then quenched by addition of a known large volume of water. The concentration of iodine in this solution was then determined by titration with standard sodium thiosulfate solution.

(i) Explain the purpose of quenching.

(ii) Write an equation for the reaction between sodium thiosulfate and iodine.

(iii) What indicator would you use? Give the colour change when the end point is reached.

(iv) In this titration and in titrations involving potassium manganate(VII), a colour change occurs during the reaction. Why is an indicator usually added in iodine/thiosulfate titrations but not in titrations that involve potassium manganate(VII)?

(f) The rate expression for the forward reaction between hydrogen and iodine is

$$\text{Rate} = k[H_2][I_2].$$

(i) What is the order of the reaction with respect to iodine?

(ii) When 0.20 mol each of H_2 and I_2 are mixed at 600°C in a vessel of 500 cm^3 capacity, the initial rate of formation of HI is found to be 2.3×10^{-5} mol dm^{-3} s^{-1}. Calculate a value for k at 600°C, stating the units.

CHAPTER 7

Ionic Equilibria

There are three important characteristics that an acid may display:

- Acid reacts with metal to give off hydrogen gas.
- Acid reacts with carbonate/hydrogencarbonate to give off carbon dioxide gas.
- Acid reacts with a base to give salt and water.

Not all acids display all the above characteristics. Some acids may just display one out of three. For example, water reacts with sodium metal to give off hydrogen gas but it does not react with carbonates and bases.

Hence, based on the above characteristics, there are not one but three common definitions for acids and bases:

- the Arrhenius Theory of acids and bases,
- the Brønsted–Lowry Theory of acids and bases, and
- the Lewis Theory of acids and bases.

In the late 1800s, the Arrhenius Theory of acids and bases was conceptualised and it defined an acid as a substance that *releases* H^+ ions when *dissolved in water* (e.g., HCl) and a base as a substance that *releases* OH^- ions in the *presence of water* (e.g., NaOH). However, this definition of acids and bases is confined to those acid–base reactions that occur in aqueous solutions.

There are a few limitations to the Arrhenius Theory. For one, there are acid–base reactions that do not occur in water but in other mediums. For instance, the reaction between gaseous hydrogen chloride and ammonia can be considered an acid–base reaction as it produces a salt. However, NH_3 is not considered a base if we apply Arrhenius Theory since it does not contain the $-OH$ group, but it does behave as a base.

In the early 1900s, an alternative definition for acids and bases was offered under the Brønsted–Lowry Theory. Based on the theory, **an acid is a proton donor and a base is a proton acceptor.** Under this theory, the basic properties of substances such as NH_3 can be accounted for: NH_3 is a base that accepts a proton from HCl to form NH_4^+ in the acid–base reaction.

A further generalisation of the acid–base definition involves the perspective of electron flow to account for how bases actually accept protons. According to the Lewis Theory of acids and bases, NH_3 accepts a proton by actually donating a pair of electrons to it and as a result, a dative covalent bond is formed.

$$H^+ + :NH_3 \longrightarrow [H \leftarrow NH_3]^+$$

A Lewis base is an electron-pair donor and a Lewis acid is a substance that accepts a pair of electrons from a base (electron-pair acceptor). A dative covalent bond results from the sharing of the pair of electrons.

Based on this theory, BF_3 is a Lewis acid which accepts a pair of electrons from the Lewis base NH_3, and a dative covalent bond forms between them:

$$\underset{\text{Lewis acid}}{BF_3} + \underset{\text{Lewis base}}{:NH_3} \longrightarrow H_3N \longrightarrow BF_3.$$

In general, the Lewis Theory of acids and bases is the broadest of the three theories since it includes all the possible acids and bases ascribed by the other two theories. And most importantly, it provides a mechanism for showing how a base acts and inform us that an acid cannot be an acid in the absence of a base! In addition, take note that there are many organic reactions that can be perceived as acid-base in nature based on Lewis Theory (refer to *Understanding Advanced Organic and Analytical Chemistry* by K.S. Chan and J. Tan).

For this chapter, we will just focus our attention on the Brønsted–Lowry Theory of acids and bases.

7.1 The Brønsted–Lowry Theory of Acids and Bases

Based on the theory,

- an acid is a proton donor;
- a base is a proton acceptor; and
- an acid–base reaction involves the acid transferring a proton to the base.

Examples:

$$\underset{\text{acid}}{\mathrm{HCl(g)}} + \underset{\text{base}}{\mathrm{NH_3(g)}} \longrightarrow \mathrm{NH_4Cl(s)},$$

$$\underset{\text{acid}}{\mathrm{CH_3COOH(aq)}} + \underset{\text{base}}{\mathrm{OH^-(aq)}} \longrightarrow \mathrm{CH_3COO^-(aq)} + \mathrm{H_2O(l)}.$$

To be considered a proton donor, a substance must have a hydrogen atom that can be lost. To be considered a proton acceptor, a substance must have a lone pair of electrons to form a dative covalent bond with the proton.

A substance can be either a base or an acid depending on what they react with, i.e. the role of an acid or base is a relative one.

H_2O functions as the base in the presence of HCl but it behaves as an acid when in the presence of NH_3. Substances like water, which are capable of either accepting or donating a proton, are known as amphiprotic substances. Thus,

$$\underset{\substack{\text{acid}\\\text{proton donor}}}{HCl(aq)} + \underset{\substack{\text{base}\\\text{proton acceptor}}}{H_2O(l)} \longrightarrow Cl^-(aq) + H_3O^+(aq),$$

$$\underset{\substack{\text{acid}\\\text{proton donor}}}{H_2O(l)} + \underset{\substack{\text{base}\\\text{proton acceptor}}}{NH_3(aq)} \rightleftharpoons OH^-(aq) + NH_4^+(aq).$$

Q: Can we simply write the equation as $HCl(aq) \rightarrow Cl^-(aq) + H^+(aq)$ to indicate that HCl is acidic?

A: For the purpose of simplicity, yes. In reality, hydrogen ions do not exist in solution. An H^+ ion has such a high charge density that it actually bonds to at least one water molecule when in aqueous solution, i.e., H^+ binds with H_2O to form H_3O^+ (a hydronium ion).

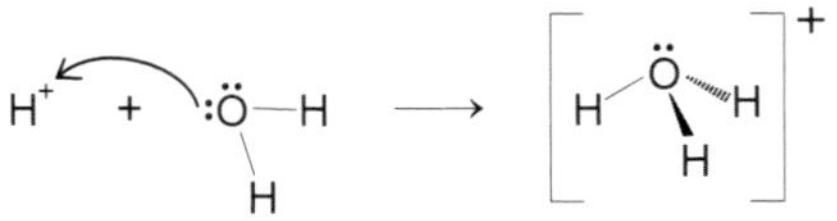

It is common to find the symbol "$H^+(aq)$" used for simplicity's sake in many texts, including this one. It is fine to use it but we must bear in mind that when we write "$H^+(aq)$", we are actually referring to "$H_3O^+(aq)$".

For the dissociation of HCl molecules in water, it is more meaningful to write the equation as:

$$HCl(aq) + H_2O(l) \longrightarrow Cl^-(aq) + H_3O^+(aq).$$

This is because the H–Cl bond does not automatically break up when an HCl molecule "plunges" into the water. The "loss" of a proton from an HCl molecule is mediated by the lone pair of electrons from a water molecule. One can actually visualise the lone pair of electrons "extracting" the proton.

From this perspective, it is important to note that a Brønsted–Lowry acid, by itself, cannot function as an acid unless a base is present, and *vice versa*.

7.2 Conjugate Acid–Base Pairs

Consider the reversible reaction below:

$$\underset{\text{acid}}{H_2O(l)} + \underset{\text{base}}{NH_3(aq)} \rightleftharpoons \underset{\text{conjugate acid}}{NH_4^+(aq)} + \underset{\text{conjugate base}}{OH^-(aq)} .$$

- The acid H_2O donates a proton, leaving behind OH^-. OH^- is known as the conjugate base of the acid H_2O. In the backward reaction, OH^- acts as a base, accepting a proton from NH_4^+ to form H_2O.
- The base NH_3 accepts the proton, forming NH_4^+. NH_4^+ is known as the conjugate acid of the base NH_3. In the backward reaction, NH_4^+ acts as an acid, donating a proton to OH^- and forming NH_3.
- There are two conjugate acid–base pairs in one acid–base reaction:
    - H_2O and OH^- form one pair;
    - NH_3 and NH_4^+ form another pair; and
    - The members of a conjugate acid–base pair differ from each other in terms of one proton.

7.3 The pH Scale

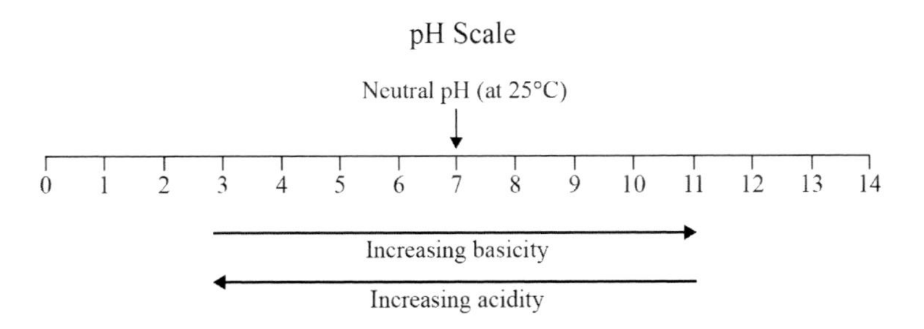

The pH scale was developed to measure the acidity of solutions based on the concentration of hydrogen ions in the solution. Since most concentrations of H^+ ions range between 10^0 and 10^{-14} mol dm^{-3}, the pH scale is expressed as a range from 0 to 14.

The pH of a solution is defined as the negative $\log_{10}$ of the hydrogen ion concentration in mol dm^{-3}:

$$\mathbf{pH = -lg[H^+(aq)].}$$

- pH has no units.
- The greater $[H^+]$, the smaller the pH value.

By knowing $[H^+]$, we can calculate the pH of an acid.

Q: What if we are asked to calculate the pH of NaOH(aq)? NaOH(aq) is a base and it does not produce H^+ ions. How is it possible to assign a pH value to NaOH(aq) when we do not have a value for $[H^+]$?

A: It turns out that for all aqueous solutions, be they acidic, neutral or basic, there are both H^+ and OH^- ions present. The concentrations of these differ depending on the type of solution:

- for neutral solutions, $[H^+] = [OH^-]$;
- for acidic solutions, $[H^+] > [OH^-]$; and
- for basic solutions, $[H^+] < [OH^-]$.

How does the above conclusion come about? It has got to do with the dissociation of water, which is discussed in the following section.

7.4 The Dissociation Constant of Water, K_w

Water is amphiprotic. By itself, water actually dissociates, although to a small extent as follows:

$$H_2O(l) + H_2O(l) \rightleftharpoons H_3O^+(aq) + OH^-(aq), \quad \Delta H > 0. \quad (7.1)$$

For simplicity, Eq. (7.1) can be written as:

$$H_2O(l) \rightleftharpoons H^+(aq) + OH^-(aq).$$

The above reaction is known as the **autoionisation of water**.

The K_c expression for this system at equilibrium is:

$$K_c = \frac{[H_3O^+(aq)][OH^-(aq)]}{[H_2O(l)][H_2O(l)]}.$$

Since pure water is involved, we can regard $[H_2O(l)]$ as constant and incorporate it into the equilibrium constant, which is better known as the **ionic product of water ($\boldsymbol{K_w}$)**:

$$K_w = [H_3O^+][OH^-] \text{ or simply } K_w = [H^+][OH^-].$$

The units for K_w is mol^2 dm^{-6}. **At 25°C, $\boldsymbol{K_w = [H_3O^+][OH^-] = 1.0 \times 10^{-14}}$ mol^{2}dm^{-6} for all aqueous solutions.**

Just as for K_c, K_w is temperature dependent. Its value becomes larger as temperature increases.

Q: Why does K_w become larger as temperature increases?
A: The dissociation of H_2O is endothermic. At higher temperatures, the endothermic reaction is favoured and as increasingly more H_2O molecules dissociate, $[H_3O^+]$ and $[OH^-]$ will also be greater, leading to a larger K_w.

This also implies that the pH of pure water decreases with increasing temperature.

Q: How do we calculate the pH of pure water at 25°C?
A: Given that pure water is neutral, we can equate $[H_3O^+] = [OH^-]$:

$$K_w = [H_3O^+][OH^-] = 1.0 \times 10^{-14}$$
$$[H_3O^+] = [OH^-] = 1.0 \times 10^{-7}\,\text{mol dm}^{-3}.$$

Inserting values into the pH expression,

$$\begin{aligned} \text{pH} &= -\lg[H_3O^+] \\ &= -\lg(1.0 \times 10^{-7}) \\ &= 7. \end{aligned}$$

Pure water has a pH of 7 **only** at 25°C. At higher temperatures, the pH of pure water is less than 7.

Q: Why is the pH of pure water temperature dependent?

A: Recall that the dissociation of water is endothermic. More H_2O molecules will dissociate at higher temperatures, leading to greater $[H_3O^+]$ and $[OH^-]$. Since $pH = -lg[H_3O^+]$, the larger the $[H_3O^+]$, the smaller the pH value.

Q: Does that mean that pure water becomes more acidic (since $[H_3O^+]$ has increased) at higher temperature?

A: No! You cannot say that because the pH range at higher temperature is no longer equal to 14. It is actually less than 14, which means that the defined neutral pH at a higher temperature is going to be less than 7. Thus, although the pH of pure water is lower than that compared against the pH scale from 0 to 14 at 25°C, you cannot say that the water is more acidic. Also, do not forget that as $[H_3O^+]$ increases with temperature, $[OH^-]$ also increases. $[H_3O^+]$ is always equal to $[OH^-]$ for pure water at any temperature!

As K_w is a constant at a given temperature, the product of $[H_3O^+]$ and $[OH^-]$ must be equal to the value of K_w for that temperature. For instance, if $[H_3O^+]$ becomes larger, $[OH^-]$ will become smaller but their product is always equal to 1.0×10^{-14} mol^2 dm^{-6} at 298 K.

Q: With considerations at the molecular level, why when $[H_3O^+]$ becomes larger, will $[OH^-]$ become smaller?

A: Imagine that you have pure water at 25°C. Due to the autoionisation of water, $[H_3O^+] = [OH^-]$ and the system is at equilibrium. What happens when you add more H_3O^+? The chances of an OH^- ion encountering an H_3O^+ ion increases. This causes the rate of the backward reaction [Eq. (7.1)] to increase. As time passes, the position of equilibrium shifts to the left, resulting in lower $[OH^-]$.

Thus, at 25°C (298 K),

- for neutral solutions, $[H_3O^+] = [OH^-]$ at $pH = 7$;
- for acidic solutions, $[H_3O^+] > [OH^-]$ at $pH < 7$; and
- for basic solutions, $[H_3O^+] < [OH^-]$ at $pH > 7$.

However, an acid need not necessarily have an acidic pH. According to Brønsted–Lowry's definition, water is an acid but its pH is neutral.

Although water does not react with zinc to give hydrogen gas, it does react with a reactive metal such as sodium. It does show characteristic features of an acid.

Example 7.1: Calculating pH of a strong acid.

(i) Calculate the pH of a $0.05\,\text{mol dm}^{-3}$ HCl solution at 298 K.
(ii) What is the concentration of OH^- ions in this solution?

Solution (i): Since HCl is a strong acid that completely dissociates in water, we can equate $[H_3O^+] = [HCl] = 0.05\,\text{mol dm}^{-3}$:

$$\begin{aligned} \text{pH} &= -\lg[H_3O^+] \\ &= -\lg 0.05 \\ &= 1.3. \end{aligned}$$

If the pH of a solution is given, $[H^+]$ can be found by using the formula:

$$\text{pH} = -\lg[H^+]$$

$$[H^+] = 10^{-\text{pH}}$$

Is this answer reasonable? Yes, we would expect the solution of the strong acid HCl to have a low pH, much smaller than 7, at 298 K.

Solution (ii):

Method 1

Since HCl is a strong acid that completely dissociates in water, $[H_3O^+] = [HCl] = 0.05\ \text{mol dm}^{-3}$.

We can substitute this value into the K_w expression:

$$K_w = [H_3O^+][OH^-] = 1.0 \times 10^{-14}\,\text{mol}^2\,\text{dm}^{-6} \text{ (at 298 K)}$$
$$1.0 \times 10^{-14} = 0.05 \times [OH^-]$$
$$[OH^-] = 2.0 \times 10^{-13}\,\text{mol dm}^{-3}.$$

Method 2

pH is defined as the negative $\log_{10}$ of the H^+ concentration in mol dm^{-3}. We can also apply similar definitions to other p-notations such as pOH, pK_w, pK_a and pK_b.

In this case, we can have pOH defined as the negative $\log_{10}$ of the hydroxide ion concentration in mol dm^{-3}, i.e., $\mathbf{pOH = -lg[OH^-]}$.

Since $K_w = [H_3O^+][OH^-] = 1.0 \times 10^{-14}$ $mol^2\,dm^{-6}$ (at 298 K), and taking logarithms on both sides of the K_w expression, we can derive that

$$\mathbf{p}K_\mathbf{w} = \mathbf{pH} + \mathbf{pOH} = \mathbf{14}\ (\mathbf{at\ 298\,K}).$$

Thus, in answering part (ii), we can make use of the pH value calculated in part (i) to determine $[OH^-]$, i.e.,

$$\text{pH} + \text{pOH} = 14$$

$$\text{pOH} = 14 - 1.3 = 12.7.$$

Since $\text{pOH} = -\lg[OH^-]$,

$$\begin{aligned}[OH^-] &= 10^{-\text{pOH}}\\ &= 10^{-12.7}\\ &= 2.0 \times 10^{-13}\,\text{mol}\,\text{dm}^{-3}.\end{aligned}$$

Q: In Example 7.1, $[H_3O^+(aq)]$ is taken to be from the dissociation of HCl. Why isn't the $[H_3O^+]$ that arises from the autoionisation of water taken into consideration?

A: As the given HCl solution is quite concentrated, we can regard the contribution of H_3O^+ ions from the autoionisation of water to be negligible and thus it can be ignored in the calculation of pH of the solution. The $[H_3O^+]$ arising from the dissociation of acid is $0.05\,\text{mol}\,\text{dm}^{-3}$, which is much more significant than the $[H_3O^+]$ arising from the autoionisation of water, which is $\sim 1.0 \times 10^{-7}\,\text{mol}\,\text{dm}^{-3}$. In addition, the presence of the H_3O^+ ions from the dissociation of HCl further suppresses the autoionisation of water. We will discuss more of this in Sec. 7.14.3 on "The Common Ion Effect."

Example 7.2: Calculating pH of a strong base.

Calculate the pH of a $0.05\,\text{mol}\,\text{dm}^{-3}$ NaOH solution at 298 K.

Solution: As the NaOH solution is quite concentrated, the $[OH^-]$ arising from the autoionisation of water is negligible.

Since NaOH is a strong base that completely ionises in water, $[OH^-] = [NaOH] = 0.05\,\text{mol}\,\text{dm}^{-3}$. We can substitute this value into the K_w expression:

$$K_w = [H_3O^+][OH^-] = 1.0 \times 10^{-14}\,\text{mol}^2\,\text{dm}^{-6}$$

$$1.0 \times 10^{-14} = [H_3O^+] \times 0.05$$

$$[H_3O^+] = 2.0 \times 10^{-13}\,\text{mol}\,\text{dm}^{-3}.$$

$$\begin{aligned} \text{pH} &= -\lg[H_3O^+] \\ &= -\lg(2.0 \times 10^{-13}) \\ &= 12.7. \end{aligned}$$

Is this answer reasonable? Yes, we would expect the solution of the strong base NaOH to have a high pH, much greater than 7, at 298 K.

Now, we have seen that it is quite easy to calculate the pH of basic solutions. The next problem we need to solve is the calculation of pH of weak acidic solutions and that of weak basic solutions too.

Q: Can't we just do the same, i.e., equate $[H_3O^+] = [\text{weak acid}]$ and likewise equate $[OH^-] = [\text{weak base}]$?

A: Unfortunately, we can't. The following sections will explain the difference between weak and strong acids and bases.

7.5 Strength of Acids

The strength of an acid depends on the extent of its dissociation in aqueous solution.

- A **strong** acid **completely dissociates** in aqueous solution. 1 mol of HCl will provide 1 mol of H_3O^+ and 1 mol of Cl^- upon dissociation in water, i.e., $[HCl] = [H_3O^+]$:

$$HCl(aq) + H_2O(l) \longrightarrow H_3O^+(aq) + Cl^-(aq). \quad (7.2)$$

Equation (7.2) can also be written as:

$$HCl(aq) \longrightarrow H^+(aq) + Cl^-(aq)$$

- A **weak** acid undergoes **partial dissociation** in aqueous solution. 1 mol of CH_3COOH will provide less than 1 mol of H_3O^+ and CH_3COO^- each upon partial dissociation in water, i.e., initial $[CH_3COOH] > [H_3O^+]$ formed from dissociation

$$CH_3COOH(aq) + H_2O(l) \rightleftharpoons CH_3COO^-(aq) + H_3O^+(aq). \quad (7.3)$$

Equation (7.3) can also be written as:

$$CH_3COOH(aq) \rightleftharpoons CH_3COO^-(aq) + H^+(aq).$$

Although a weak acid does not fully dissociate in water, the amount of base required to completely neutralise it is the same as that needed to neutralise a strong acid of the same concentration:

$$HCl(aq) + NaOH(aq) \longrightarrow NaCl(aq) + H_2O(l),$$
$$\Delta H_{neut,1}^{\ominus} = -57.0\,kJ\,mol^{-1},$$
$$CH_3COOH(aq) + NaOH(aq) \longrightarrow CH_3COO^-Na^+(aq) + H_2O(l),$$
$$\Delta H_{neut,2}^{\ominus} = -55.2\,kJ\,mol^{-1}.$$

Compared to $\Delta H_{neut,1}^{\ominus}$, $\Delta H_{neut,2}^{\ominus}$ is less exothermic because some of the energy has to be expended to dissociate the undissociated weak acid molecules.

The extent of an acid's dissociation in aqueous solution, and hence the acid's strength, can be quantified using the concept of acid dissociation constant (K_a).

7.6 Acid Dissociation Constant K_a and pK_a

The K_a concept is not new as it is similar to the equilibrium constant you learned in Chap. 6. Consider the partial dissociation of a weak acid CH_3COOH:

$$CH_3COOH(aq) + H_2O(l) \rightleftharpoons CH_3COO^-(aq) + H_3O^+(aq).$$

The K_c expression for this system at equilibrium is:

$$K_c = \frac{[H_3O^+(aq)][CH_3COO^-(aq)]}{[CH_3COOH(aq)][H_2O(l)]}.$$

Basicity is defined as the number of moles of OH^- required to react with 1 mole of acid

A monobasic acid (e.g., HCl and CH_3COOH) donates only one H^+ ion per molecule to a base.

A dibasic acid, such as H_2SO_4 and ethanedioic acid, $(COOH)_2$, can donate 2 such H^+ ions.

Can you think of a tribasic acid?

$[H_2O]$ is essentially constant and it is incorporated into the equilibrium constant which is now known as the acid dissociation constant (K_a):

$$\text{Acid dissociation constant } K_a = \frac{[CH_3COO^-(aq)][H_3O^+(aq)]}{[CH_3COOH(aq)]},$$

where $K_a = K_c[H_2O]$.

- The units for K_a are mol dm^{-3}. Just like K_c, K_a is temperature dependent.
- A stronger acid dissociates to a greater degree and hence its K_a value is larger.
- Although the magnitude of K_a serves to measure the strength of acids, it is more convenient to use the corresponding pK_a value for comparison.

The relationship between K_a and pK_a is given as follows:

$$\mathbf{p}\boldsymbol{K_a} = -\mathbf{lg}\,\boldsymbol{K_a}.$$

The smaller the pK_a value, the stronger the acid.

Q: Isn't the strength of an acid determined by its concentration?

A: The concentration of an acid does not give any indication of its strength. For instance, 0.001 mol dm^{-3} of HCl implies that its concentration is much diluted. Nevertheless, it is still a strong acid which undergoes complete dissociation in aqueous solution. You cannot find any undissociated HCl molecule in water. 0.500 mol dm^{-3} of CH_3COOH may have a higher concentration than 0.001 mol dm^{-3} of HCl, but it is a weaker acid than the latter since it only partially dissociates in water.

Q: Does concentration affect the degree of dissociation?

A: Yes. If you add a weak acid into lots of water, all of its molecules may be considered dissociated because the dissociated species have very little chance to meet each other and recombine.

Now, if you increase the amount of weak acid molecules in the same amount of water, more dissociate but at the same time, chances that the dissociated species meet and recombine also increase.

Imagine if you keep on adding more weak acid: the degree of dissociation is going to decrease. A pure weak acid does not dissociate because of the lack of a proton acceptor. Thus, a higher concentration of weak acid does not imply a greater degree of dissociation and a lower pH value.

Q: Can degree of dissociation be used to compare acid strength?

A: Yes, provided that you are comparing two weak acids of the same concentrations. Only then would it be fair to conclude that the stronger acid is the one that dissociates more and gives a higher $[H_3O^+]$.

Q: Can we use pH as an indication of acid strength?

A: Here again, you must compare weak acids of the same concentrations. The explanation is the same as that above.

In calculations involving a weak acid, there are normally three terms involved:

- pH (derived from $[H^+]$),
- K_a, and
- initial [acid].

Two of these will be given while the third has to be calculated.

The three general types of questions center around the K_a expression:

$$HA(aq) \rightleftharpoons A^-(aq) + H^+(aq),$$

$$K_a = \frac{[A^-][H^+]}{[HA]}. \longleftarrow$$

$[H^+]$ is related to pH. Ignoring H^+ from autoionisation of water, we have $[A^-] = [H^+]$.

- Given initial [HA] and K_a, calculate pH (see Example 7.3).
- Given initial [HA] and pH, calculate K_a (see Example 7.4).
- Given K_a and pH, calculate initial [HA] (see Example 7.5).

Example 7.3: Calculating pH of a weak acid (given initial [acid] and K_a)

Calculate the pH of a 0.10 mol dm^{-3} ethanoic acid solution at 298 K (given that K_a of ethanoic acid is 1.8×10^{-5} mol dm^{-3} at 298 K).

Solution:

Step 1: Determine equilibrium concentrations of the species using an "I.C.E." table and the following assumptions:

- Let equilibrium $[H^+] = y$ mol dm^{-3}.
- Ignore the H^+ contribution from the autoionisation of water, $[CH_3COO^-] = [H^+] = y$ mol dm^{-3}.

	$CH_3COOH(aq)$	$\rightleftharpoons$	$CH_3COO^-(aq)$	+	$H^+(aq)$
Initial conc. (mol dm^{-3})	0.10		—		—
Change in conc. (mol dm^{-3})	$-y$		$+y$		$+y$
Equilibrium conc. (mol dm^{-3})	$(0.10 - y)$		y		y

Step 2: Solve for y using the K_a expression:

$$K_a = \frac{[CH_3COO^-][H^+]}{[CH_3COOH]},$$

$$1.8 \times 10^{-5} = \frac{y^2}{(0.10 - y)}.$$

Since CH_3COOH is a weak acid with a small K_a, we can assume $y \ll 0.10$ such that $(0.10 - y) \approx 0.10$. Therefore,

$$1.8 \times 10^{-5} = \frac{y^2}{0.10}$$

$$y = 1.34 \times 10^{-3}.$$

Equilibrium $[H^+] = 1.34 \times 10^{-3}\,\text{mol dm}^{-3}$.

Step 3: Calculate pH:

$$\begin{aligned} pH &= -\lg[H^+] \\ &= -\lg(1.34 \times 10^{-3}) \\ &= 2.87. \end{aligned}$$

Q: Initially, why is $[H^+]$ taken to be zero in the I.C.E. table?

A: In reality, before any CH_3COOH dissociates, $[H^+] = 10^{-7}\,\text{mol dm}^{-3}$ because of the autoionisation of water. But we ignore it!

Q: Why is the autoionisation of water ignored in the above calculation? Isn't water a weak acid too?

A: Both water and ethanoic acid are weak acids. But ethanoic acid is still a stronger acid than water, and dissociates to a greater extent. The chances of these H^+ ions produced from the dissociation meeting an OH^- ion from the autoionisation of water are high. This results in a smaller extent of dissociation of the water molecules. From another perspective, we may say that the dissociation of water has been suppressed by the presence of the stronger weak acid.

Example 7.4: Calculating K_a of a weak acid (given initial [acid] and pH)

(i) Calculate K_a of a 0.01 mol dm^{-3} solution of a weak acid HA at 298 K (given that the pH of HA is 3.7 at 298 K).

(ii) Hence, determine if the weak acid HA is a stronger acid than ethanoic acid from Example 7.3.

Solution (i):

Step 1: Calculate equilibrium $[H^+]$ from the pH value given:

$$\begin{aligned} pH &= -\lg[H^+] \\ [H^+] &= 10^{-pH} \\ &= 10^{-3.7} \\ &= 2 \times 10^{-4}\,\text{mol}\,\text{dm}^{-3}. \end{aligned}$$

Step 2: Determine equilibrium [HA] and hence solve for K_a:

$$HA(aq) \rightleftharpoons A^-(aq) + H^+(aq),$$

$$\begin{aligned} \text{Equilibrium [HA]} &= \text{initial [HA]} - \text{equilibrium } [H^+] \\ &= 0.01 - 2 \times 10^{-4} \\ &= 9.8 \times 10^{-3}\,\text{mol dm}^{-3}. \end{aligned}$$

Ignoring the H^+ contribution from the autoionisation of water,

$$[A^-] = [H^+] = 2 \times 10^{-4}\,\text{mol}\,\text{dm}^{-3}.$$

Thus,

$$K_a = \frac{[A^-][H^+]}{[HA]} = \frac{(2 \times 10^{-4})^2}{9.8 \times 10^{-3}} = 4.08 \times 10^{-6}\,\text{mol}\,\text{dm}^{-3}.$$

Solution (ii): HA is a weaker acid than ethanoic acid since it has a smaller K_a value, which indicates that its extent of dissociation is less than that of ethanoic acid.

Example 7.5: Calculating initial concentration of a weak acid (given K_a and pH)

A weak acid HA, whose pH is 5.3, has a K_a of 4.2×10^{-6} mol dm^{-3} at 298 K. Calculate the initial concentration of HA at 298 K.

Solution:

Step 1: Calculate equilibrium $[H^+]$ from the pH value given:

$$\begin{aligned} \text{pH} &= -\lg[\text{H}^+] \\ [\text{H}^+] &= 10^{-\text{pH}} \\ &= 10^{-5.3} \\ &= 5.012 \times 10^{-6}\,\text{mol}\,\text{dm}^{-3}. \end{aligned}$$

Step 2: Solve for initial [HA] using the K_a expression. Let initial $[\text{HA}] = y\,\text{mol}\,\text{dm}^{-3}$:

$$\text{HA(aq)} \rightleftharpoons \text{A}^-\text{(aq)} + \text{H}^+\text{(aq)},$$

$$\begin{aligned} \text{Equilibrium [HA]} &= \text{initial [HA]} - \text{equilibrium } [\text{H}^+] \\ &= y - 5.012 \times 10^{-6}\,\text{mol}\,\text{dm}^{-3}. \end{aligned}$$

Ignoring the H^+ contribution from the autoionisation of water,

$$[\text{A}^-] = [\text{H}^+] = 5.012 \times 10^{-6}\,\text{mol}\,\text{dm}^{-3}.$$

Thus,

$$\begin{aligned} K_a = \frac{[\text{A}^-][\text{H}^+]}{[\text{HA}]} &= \frac{(5.012 \times 10^{-6})^2}{(y - 5.012 \times 10^{-6})} \\ &= 4.2 \times 10^{-6}\,\text{mol}\,\text{dm}^{-3}. \end{aligned}$$

Since HA is a weak acid, assume equilibrium $[\text{H}^+] \ll y$ such that $(y - 5.012 \times 10^{-6}) \approx y$. Thus,

$$K_a = \frac{(5.012 \times 10^{-6})^2}{y} = 4.2 \times 10^{-6}$$

$$y = 5.98 \times 10^{-6}.$$

Hence, the initial concentration of HA is $5.98 \times 10^{-6}\,\text{mol}\,\text{dm}^{-3}$.

7.7 Strength of Bases

The strength of a base depends on the extent of its ionisation in aqueous solution.

- A **strong** base **completely ionises** in aqueous solution:

$$NaOH(aq) \longrightarrow Na^+(aq) + OH^-(aq).$$

Examples of strong bases: NaOH, KOH and $Ba(OH)_2$.

1 mol of monoacidic base (e.g., NaOH) requires 1 mol of H^+ for complete neutralisation.

1 mol of diacidic base [e.g., $Ba(OH)_2$] requires 2 mol of H^+ for complete neutralisation.

- A **weak** base undergoes **partial ionisation** in aqueous solution:

$$NH_3(aq) + H_2O(l) \rightleftharpoons NH_4{}^+(aq) + OH^-(aq),$$

$$CO_3{}^{2-}(aq) + H_2O(l) \rightleftharpoons HCO_3{}^-(aq) + OH^-(aq).$$

Examples of weak bases: NH_3, Na_2CO_3 and amines such as ethylamine ($CH_3CH_2NH_2$).

Although a weak base does not fully ionise in water, the amount of acid required to completely neutralise it is the same as that needed to neutralise a strong base of the same concentration.

Q: Since ammonia is a weak base that partially ionizes in water to give ions. Does the formation of the ions increase the solubility of the ammonia molecules?

A: Certainly! The high solubility of ammonia is not just due to the hydrogen bonding formed with the water molecules but also because of the higher solubility of the NH_4^+ and OH^- ions formed from the hydrolysis of NH_3 molecules in water. The formation of the ions from the partial ionization does increase the solubility of the ammonia molecules as the ions can form stronger ion-dipole interaction with water.

Q: Does this also apply to the weak acid that ionizes in water too?

A: Yes! Weak acid such as ethanoic acid (CH_3COOH) is very soluble in water not just because of the strong hydrogen bonds formed between the ethanoic acid and water molecules. It is also partly because of the higher solubility of the ions that are formed from the partial dissociation of the ethanoic acid molecules.

Q: Wait a minute, the formation of ions from the dissociation/ionization would increase solubility. So, would solubility affects degree of dissociation/ionization, i.e. the acidic or basic strength?

A: Certainly! If a compound such as ethanoic acid or ammonia does not have "solubility issue" in the first place, then the dissociation/ionization would further enhance its solubility. But if the compound has a "problem" dissolving in the first place, like $Mg(OH)_2$ or benzoic acid (C_6H_5COOH), then the low solubility would affect its basic or acidic strength.

7.8 Base Dissociation Constant K_b and pK_b

The extent of a base's ionisation in aqueous solution and hence the base's strength can be quantified using the concept of the base dissociation constant K_b.

Consider the partial ionisation of the weak base NH_3:

$$NH_3(aq) + H_2O(l) \rightleftharpoons OH^-(aq) + NH_4{}^+(aq).$$

The K_c expression for this system at equilibrium is:

$$K_c = \frac{[NH_4{}^+(aq)][OH^-(aq)]}{[NH_3(aq)][H_2O(l)]}.$$

$[H_2O]$ is essentially constant and it is incorporated into the equilibrium constant which is now known as the base dissociation constant (K_b), since we are dealing with bases:

$$\text{Base dissociation constant } K_b = \frac{[NH_4{}^+(aq)][OH^-(aq)]}{[NH_3(aq)]},$$

where $K_b = K_c[H_2O]$.

- The units for K_b are mol dm^{-3}. K_b is temperature dependent.
- The larger the K_b value, the stronger is the base.
- The relationship between K_b and pK_b is given as follows:

$$\mathbf{p}\boldsymbol{K}_\mathbf{b} = -\mathbf{lg}\,\boldsymbol{K}_\mathbf{b}.$$

The smaller the pK_b value, the stronger the base.

Example 7.6: Calculating pH of a weak base

Calculate the pH of a 0.05 mol dm^{-3} sodium ethanoate (CH_3COONa) solution at 298 K (given that the K_b of ethanoate ion is 5.56×10^{-10} mol dm^{-3} at 298 K).

Approach:

(i) Firstly, calculate equilibrium $[OH^-]$.
(ii) Next, calculate $[H^+]$ by using the relationship: $K_w = [H^+][OH^-] = 1.0 \times 10^{-14}$ mol^2 dm^{-6} at 298 K.
(iii) Lastly, calculate pH by using the formula: pH $= -\lg[H^+]$.

Solution: Let equilibrium $[OH^-] = y$ mol dm^{-3}.

Ignoring the OH^- contribution from the autoionisation of water, $[OH^-] = [CH_3COOH] = y$ mol dm^{-3}.

	$CH_3COO^-(aq)$	+ $H_2O(l)$ ⇌	$CH_3COOH(aq)$	+ $OH^-(aq)$
Initial conc. (mol dm^{-3})	0.05		—	—
Change in conc. (mol dm^{-3})	$-y$		$+y$	$+y$
Equilibrium conc. (mol dm^{-3})	$(0.05 - y)$		y	y

$$K_b = \frac{[CH_3COOH][OH^-]}{[CH_3COO^-]},$$

$$5.56 \times 10^{-10} = \frac{y^2}{(0.05 - y)}.$$

Since CH_3COO^- is a weak base, assume $y \ll 0.05$ such that $(0.05 - y) \approx 0.05$. Hence,

$$5.56 \times 10^{-10} = \frac{y^2}{0.05}$$

$$y = 5.27 \times 10^{-6}.$$

Equilibrium $[OH^-] = 5.27 \times 10^{-6}\,\text{mol dm}^{-3}$. At 298 K, $K_w = [H^+][OH^-] = 1.0 \times 10^{-14}\,\text{mol}^2\,\text{dm}^{-6}$:

$$[H^+] \times (5.27 \times 10^{-6}) = 1.0 \times 10^{-14}$$

$$\begin{aligned}[H^+] &= 1.897 \times 10^{-9}\,\text{mol dm}^{-3},\\ \text{pH} &= -\lg[H^+]\\ &= -\lg(1.897 \times 10^{-9})\\ &= 8.72.\end{aligned}$$

Alternatively, when $[OH^-]$ has been determined, proceed to calculate pOH and insert its value into the relationship "pH + pOH = 14 at 298 K" to find the pH value. You should arrive at the same answer.

Example 7.7: Calculating the degree of ionisation of a weak base

Referring to Example 7.6, calculate the degree of ionisation of CH_3COO^- in a 0.05 mol dm^{-3} sodium ethanoate solution, given that its pH is 8.72 at 298 K.

Solution: The degree of ionisation of a base can be defined as the ratio of the amount of base ionised to the initial amount of base, i.e.,

$$\begin{aligned}&\text{Degree of ionisation of } CH_3COO^-\\ &= \frac{\text{Amount of } CH_3COO^- \text{ ionised in } 1\,\text{dm}^3 \text{ of solution}}{\text{Initial amount of } CH_3COO^- \text{ in } 1\,\text{dm}^3 \text{ of solution}}\\ &= \frac{[OH^-]\text{formed at equilibrium}}{0.05}\\ &= \frac{5.27 \times 10^{-6}}{0.05}\\ &= 1.05 \times 10^{-4}.\end{aligned}$$

Is this answer reasonable? Yes, as the weak base undergoes partial ionisation, the degree of ionisation is expected to be a value between 0 and 1. For strong bases that completely ionise in solution, the degree of ionisation is 1.

Similar calculations can be done to determine the degree of dissociation of acids.

7.9 Complementary Strengths of a Conjugate Acid–Base Pair

Consider the conjugate acid–base pair of ethanoic acid and ethanoate ion:

$$CH_3COOH(aq) + H_2O(l) \rightleftharpoons CH_3COO^-(aq) + H_3O^+(aq).$$

The K_a of ethanoic acid is given as:

$$K_a = \frac{[CH_3COO^-][H_3O^+]}{[CH_3COOH]}, \tag{7.4}$$

and based on the ionisation of ethanoate ion,

$$CH_3COO^-(aq) + H_2O(l) \rightleftharpoons CH_3COOH(aq) + OH^-(aq),$$

the K_b of ethanoate ion is:

$$K_b = \frac{[CH_3COOH][OH^-]}{[CH_3COO^-]}. \tag{7.5}$$

When we multiply Eqs. (7.4) and (7.5), we will arrive at the simplified expression:

$$K_a \times K_b = [H_3O^+][OH^-],$$

which gives us the relationship:

$$\boldsymbol{K_a \times K_b = K_w} \text{ (recall: } K_w = [H_3O^+][OH^-]).$$

Take note that the relationship only holds true if the K_a and K_b values belong to a conjugate acid–base pair. Take for instance, the autoionisation of water:

$$H_2O(l) + H_2O(l) \rightleftharpoons OH^-(aq) + H_3O^+(aq).$$

The K_a of water is given as:

$$K_a = \frac{[H_3O^+][OH^-]}{[H_2O]}.$$

The K_b of water is given as:

$$K_b = \frac{[H_3O^+][OH^-]}{[H_2O]}.$$

If you now take $K_a \times K_b = \frac{[H_3O^+]^2[OH^-]^2}{[H_2O]^2}$, you find that it is not equal to $K_w = [H_3O^+][OH^-]$! Why?

This is because if H_2O acted as an acid, its conjugate base would be OH^-, which means that the hydrolysis of OH^- would be as follows:

$$OH^-(aq) + H_2O(l) \rightleftharpoons H_2O(l) + OH^-(aq).$$

Correspondingly,

$$K_b = \frac{[H_2O][OH^-]}{[OH^-]} = [H_2O], \quad \text{and not } K_b = \frac{[H_3O^+][OH^-]}{[H_2O]}.$$

On the other hand, if H_2O acted as a base, then its conjugate acid would be H_3O^+, which means that the dissociation of H_3O^+ would be as follows:

$$H_3O^+(aq) + H_2O(l) \rightleftharpoons H_2O(l) + H_3O^+(aq).$$

Correspondingly,

$$K_a = \frac{[H_3O^+][H_2O]}{[H_3O^+]} = [H_2O], \quad \text{and not } K_a = \frac{[H_3O^+][OH^-]}{[H_2O]}.$$

To continue with our previous discussion, since $K_a \times K_b = K_w$, then $\mathbf{p}\boldsymbol{K}_\mathbf{w} = \mathbf{p}\boldsymbol{K}_\mathbf{a} + \mathbf{p}\boldsymbol{K}_\mathbf{b}$.

Given that at 25°C, $K_w = K_a \times K_b = 1.0 \times 10^{-14}\ \text{mol}^2\,\text{dm}^{-6}$, then

$$pK_a + pK_b = pK_w = 14.$$

Just as how K_w, a fixed value at a given temperature, reflects the $[H^+]$ and $[OH^-]$ in all aqueous solutions, so too does it specify the relative values of K_a and K_b of a conjugate acid–base pair. If K_a is large, then K_b must be small, and *vice versa.*

Essentially, this means that:

- If an acid is strong, its conjugate base is weak. Example: HCl is a strong acid. Cl^- is a weak conjugate base.
- If an acid is weak, its conjugate base is strong. Example: CH_3COOH is a weak acid. CH_3COO^- is a strong conjugate base.

Q: If CH_3COO^- is a strong conjugate base, does this mean that it is a strong base?

A: A strong conjugate base is not equivalent to stating that it is a strong base. Likewise, a strong conjugate acid does not mean it is a strong acid.
When the term "strong conjugate base" is used, the word "strong" is used in association with the strength of its conjugate acid. The weaker the weak acid, the stronger is its conjugate base.
Usually, the conjugate base of a weak acid is a weak base; and the conjugate acid of a weak base is a weak acid.

Q: Essentially, how do we know whether two species make a conjugate acid-base pair?

A: A conjugate acid differs from its conjugate base simply by a difference of a H^+ eg. HCl and Cl^-.

Exercise: Calculate the K_b of NH_3 at 25°C given that the K_a of $NH_4{}^+$ is $5.7 \times 10^{-10}\,\text{mol}\,\text{dm}^{-3}$ at the same temperature. (Answer: $1.75 \times 10^{-5}\,\text{mol}\,\text{dm}^{-3}$.)

7.10 Hydration and Hydrolysis

When an acid and a base react, a salt and water is formed. For example,

$$\text{HCl(aq)} + \text{NaOH(aq)} \longrightarrow \underset{\text{salt}}{\text{NaCl(aq)}} + \text{H}_2\text{O(l)}.$$

When NaCl(s) dissolves in water, the ions become surrounded by water molecules. We say that hydration occurs when the ions are attracted to surrounding water molecules through ion–dipole interactions.

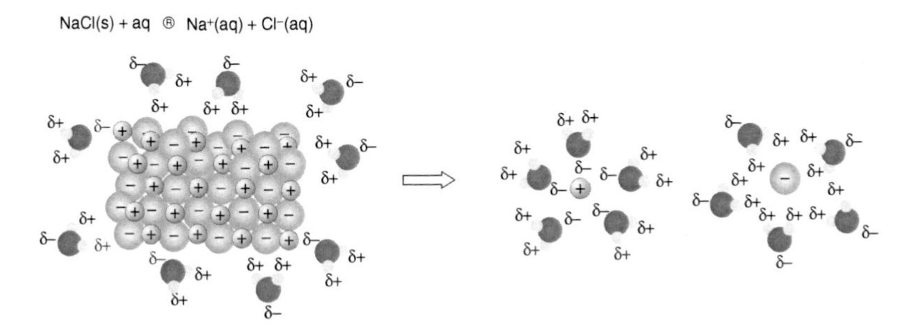

We know that NaCl is neutral and its solution at 25°C has a pH of 7.

However, not all salts are neutral. There are certain salts, which when dissolved in water, react with water to produce either acidic or alkaline solutions. This phenomenon is termed salt hydrolysis.

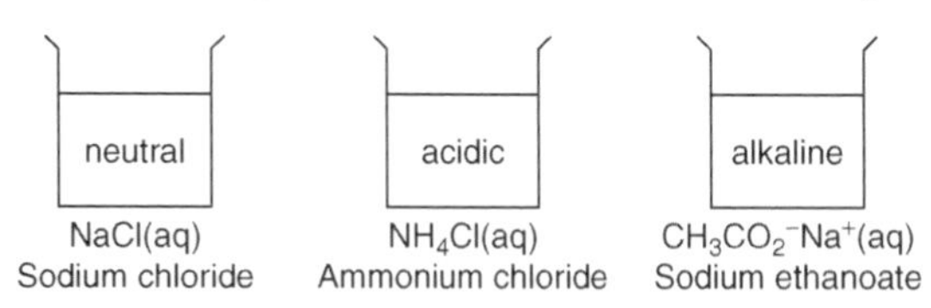

Salt hydrolysis is a reversible reaction and it occurs for salts that consist of any or all of the following species:

- an anion that is a strong conjugate base of a weak acid,
- a cation that is a strong conjugate acid of a weak base,
- a cation that has high charge density such as Al^{3+}, Cr^{3+} and Fe^{3+}.

Q: How can we tell if the anion or cation is a strong conjugate base or acid, respectively?

A: If we look at the chemical formula of the salt, for example NaCl, the anion Cl^- actually comes from the acid HCl, and the cation Na^+ has a low charge density.

To determine if the anion Cl^- is a strong or weak conjugate base, we just have to determine the strength of the acid it originates from. We know that HCl is a strong acid, and recalling the complementary strength of a conjugate acid-base pair, its conjugate base Cl^- must be a weak conjugate base. As such, Cl^- does not undergo hydrolysis.

As for the cation Na^+, its charge density is too low to break up water molecules. Hence, it does not undergo hydrolysis.

Q: What is charge density?

A: Charge density can be defined as the amount of charge per unit surface area (or volume or length). In this case, it is specified as the amount of charge per unit length. The greater the charge and smaller the size of the cation, the higher its charge density, and hence the stronger its polarising power. Therefore, a unipositive charge cation is unlikely to break up a water molecule unless it has a very small size.

7.10.1 *Classifying type of salt based on strengths of acid and base that form it*

(i) **A strong acid reacts with a strong base to give a neutral salt.**
Example:

$$HCl(aq) + KOH(aq) \longrightarrow KCl(aq) + H_2O(l).$$

Both K^+ and Cl^- do not undergo hydrolysis. They merely get hydrated when dissolved in water.

Reasons:

- K^+ has a relatively low charge density.
- Cl^- is a weak conjugate base of HCl.

(ii) **A strong acid reacts with a weak base to give an acidic salt.**

Example:

$$HCl(aq) + NH_3(aq) \longrightarrow NH_4Cl(aq).$$

Cl^- does not undergo hydrolysis.

On the other hand, NH_4^+ is a strong conjugate acid of the weak base NH_3. It undergoes hydrolysis wherein it behaves like an acid, donating a proton to H_2O:

$$NH_4^+(aq) + H_2O(l) \rightleftharpoons NH_3(aq) + H_3O^+(aq).$$

The excess H_3O^+ formed results in an overall acidic solution.

Q: Why is there excess H_3O^+? Based on the equilibria above, since $[NH_3] = [H_3O^+]$, wouldn't all the H_3O^+ be neutralised by NH_3?

A: If H_3O^+ were completely neutralised by NH_3, then you would not get a resultant acidic solution. The fact that you get an acidic solution must be because the extent of the forward reaction is greater than the backward reaction. It is really interesting to see that an acid (H_3O^+) and base (NH_3) can actually coexist "peacefully" in the same solution. Then again, this is only possible if the particles do not meet.

(iii) **A weak acid reacts with a strong base to give an alkaline salt.**

Example:

$$CH_3COOH(aq) + NaOH(aq) \longrightarrow CH_3COONa(aq) + H_2O(l).$$

CH_3COO^- is a strong conjugate base of CH_3COOH. It undergoes hydrolysis wherein it behaves like a base, accepting a proton from H_2O:

$$CH_3COO^-(aq) + H_2O(l) \rightleftharpoons CH_3COOH(aq) + OH^-(aq).$$

The excess OH^- formed results in an alkaline solution.

Q: Actually, I do not quite understand why CH_3COO^- undergoes hydrolysis?

A: Well, if you understand what makes a weak acid weak is the difficulty in cleaving an O–H bond to release H^+, then you should be able to appreciate that the CH_3COO^- anion that is formed has high affinity to accept an H^+, and hence the ability to break up a water molecule.

(iv) **A weak acid reacts with a weak base to give one of the three types of salt stated.**

Example:

$$CH_3COOH(aq) + NH_3(aq) \longrightarrow CH_3COONH_4(aq).$$

In this case, we have both the cation and anion, i.e., strong conjugate acid and base, respectively, undergoing hydrolysis:

$$\text{Acidic hydrolysis: } NH_4^+(aq) + H_2O(l) \rightleftharpoons NH_3(aq) + H_3O^+(aq).$$

$$\text{Alkaline hydrolysis: } CH_3COO^-(aq) + H_2O(l) \rightleftharpoons CH_3COOH(aq) + OH^-(aq).$$

The pH of the resultant solution depends on the relative strength of the conjugate acid NH_4^+ and that of the conjugate base CH_3COO^-.

Since the K_a of NH_4^+ is slightly greater than the K_b of CH_3COO^-, $[H_3O^+]$ is slightly greater than $[OH^-]$ and the solution is therefore slightly acidic.

In general,

- if K_a of the conjugate acid > K_b of the conjugate base, the extent of the hydrolysis of the conjugate acid is relatively greater and the resultant solution is acidic;
- if K_a of the conjugate acid = K_b of the conjugate base, the extents of the hydrolysis of both species are the same and the resultant solution is neutral;
- if K_a of the conjugate acid < K_b of the conjugate base, the extent of the hydrolysis of the conjugate base is relatively greater and the resultant solution is alkaline.

7.10.2 *Hydrolysis of high charge density cations*

For a salt that contains a cation of high charge density (such as Al^{3+}, Cr^{3+} and Fe^{3+}), it will undergo hydrolysis to form an acidic solution.

In aqueous solution, the Fe^{3+} ion exists as an aqua complex ion $[Fe(H_2O)_6]^{3+}$, with six water molecules forming dative covalent bonds with the cation. Due to its high charge and small size, the Fe^{3+} ion has a **high charge density** and hence **high polarising power**. It distorts the electron cloud of H_2O molecules bonded to it, weakening the O–H bonds and enabling these H_2O molecules to become better proton donors. The free water molecules in the solution act as bases and the following equilibrium is established:

$$[Fe(H_2O)_6]^{3+}(aq) + H_2O(l) \rightleftharpoons [Fe(OH)(H_2O)_5]^{2+}(aq) + H_3O^+(aq).$$

The Fe^{3+} ion is said to undergo **appreciable hydrolysis** in aqueous solution. The slight excess of H_3O^+ ions in the solution renders the solution **acidic.** The acidity of the solution is so high that if you added carbonate or hydrogencarbonate to a solution containing Fe^{3+} ions, CO_2(g) would be liberated. This basically means that one cannot prepare the $Fe_2(CO_3)_3$ compound.

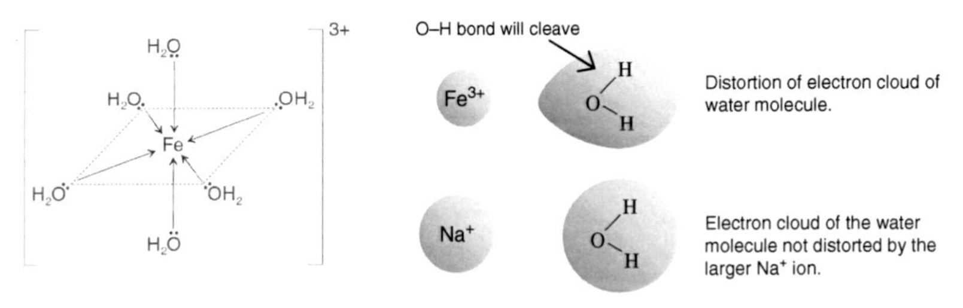

7.11 Buffer Solutions

A buffer solution is able to resist changes in pH when a small amount of acid or base is added. To behave as a buffer, a solution must contain both an acidic species and a basic species that will react with the small amount of base or acid added, respectively.

There are two types of buffer solutions:

(i) Acidic buffer

- It has a pH in the acidic range, i.e., pH < 7.
- It comprises a solution of weak acid with a **K_a value greater than the K_b** value of its conjugate base.
- Example: a solution of ethanoic acid and sodium ethanoate in substantial amounts.

(ii) Alkaline buffer

- It has a pH in the alkaline range, i.e., pH > 7.
- It comprises a solution of weak base with a **K_b value greater than the K_a** value of its conjugate acid.
- Example: a solution of aqueous ammonia and ammonium chloride in substantial amounts.

Q: Does a weak acid and its conjugate base always result in an acidic buffer?

A: No. The nature of the buffer solution (acidic or alkaline) depends on the relative values of the K_a of acid and K_b of its conjugate base — as simple as that. Take, for instance, the NH_4^+/NH_3 buffer system. This is an alkaline buffer because the K_a of NH_4^+ (5.7×10^{-10} mol dm^{-3}) is smaller than the K_b of NH_3 (1.75×10^{-5} mol dm^{-3}); the extent of basic hydrolysis is greater than that of the acidic hydrolysis. So overall, we get an alkaline buffer.

7.11.1 *How does an acidic buffer work*

Consider an acidic buffer solution that contains both ethanoic acid and sodium ethanoate:

- When in solution, ethanoic acid undergoes partial dissociation, since it is a weak acid:

$$CH_3COOH(aq) \rightleftharpoons CH_3COO^-(aq) + H^+(aq). \quad (7.6)$$

- When in solution, sodium ethanoate undergoes complete dissociation, since it is a strong electrolyte:

$$CH_3COO^-Na^+(aq) \longrightarrow CH_3COO^-(aq) + Na^+(aq). \quad (7.7)$$

The supply of the large number of CH_3COO^- ions from reaction (7.7) actually suppresses the dissociation of the weak acid CH_3COOH.

If you recall Le Chatelier's Principle, Eq. (7.6) represents a system that is in a state of dynamic equilibrium. When the introduction of $CH_3COO^-Na^+$ contributes to an increase in $[CH_3COO^-]$, the equilibrium position of Eq. (7.6) shifts to the left in a bid to remove some of the CH_3COO^- ions. The backward reaction is favoured and, as a result, the dissociation of CH_3COOH is further suppressed. (Refer to Chap. 6 on "Chemical Equilibria" for the kinetics' viewpoint on the suppression effect.)

The resultant solution, termed the buffer solution, thus contains high concentrations of both undissociated CH_3COOH molecules (acid) and CH_3COO^- ions (base) which enable the solution to resist changes in pH as follows:

- When a small amount of acid (any source of H^+ ions) is added to the buffer:
  - the ethanoate ions will react with the added H^+ ions in the neutralisation reaction:

$$CH_3COO^-(aq) + H^+(aq) \longrightarrow CH_3COOH(aq);$$

  - the presence of a large reservoir of CH_3COO^- ions helps to ensure that the majority of the added H^+ ions are removed;
  - hence, $[H^+]$ does not change much and the pH of the buffer remains *approximately* constant.

Q: Can I write the equation as $H^+(aq) + OH^-(aq) \rightarrow H_2O(l)$ instead of $CH_3COO^-(aq) + H^+(aq) \rightarrow CH_3COOH(aq)$?

A: If the added H^+ ions encounter OH^- ions, there is nothing to stop them from reacting. But statistically, it is not as probable as encountering CH_3COO^- ions as these are present in larger quantity. Thus, writing the equation as $CH_3COO^-(aq) + H^+(aq) \rightarrow CH_3COOH(aq)$ is more appropriate.

- When a small amount of base (any source of OH^- ions) is added to the buffer:
  - the ethanoic acid molecules will react with the added OH^- ions:

$$CH_3COOH(aq) + OH^-(aq) \longrightarrow CH_3COO^-(aq) + H_2O(l);$$

  - the presence of a large reservoir of undissociated CH_3COOH molecules helps to ensure that the majority of the added OH^- ions are removed;
  - hence, the pH of the buffer remains approximately constant.

Q: If CH_3COOH is a weak acid which partially dissociates, why is it inappropriate to write the equation for its reaction with OH^- as follows?

$$CH_3COOH(aq) + OH^-(aq) \rightleftharpoons CH_3COO^-(aq) + H_2O(l).$$

A: A buffer resists changes in pH. This means that the majority of the OH^- ions added has been removed. Henceforth, it would be more appropriate to use a single arrow ($\rightarrow$) in the equation to depict a completed reaction.

Q: Why can't a buffer solution be made with a strong acid and its conjugate base?

A: Remember from Section 7.10.1: a strong acid dissociates fully in the aqueous medium whereas its conjugate base does not undergo hydrolysis. So such a solution is essentially just an acidic solution. Now, when a small amount of OH^- ions is added, these react with the H^+ ions present. The result is that $[H^+]$ decreases drastically but there is no reservoir of undissociated acid to replenish the lost of H^+. Thus, it cannot act as a buffer.

7.11.2 *How does an alkaline buffer work*

Consider the alkaline buffer solution that contains both aqueous ammonia and ammonium nitrate.

- When in solution, aqueous ammonia undergoes partial ionisation, since it is a weak base:

$$NH_3(aq) + H_2O(l) \rightleftharpoons NH_4{}^+(aq) + OH^-(aq). \qquad (7.8)$$

- When in solution, ammonium nitrate undergoes complete dissociation, since it is a strong electrolyte:

$$NH_4NO_3(aq) \longrightarrow NH_4{}^+(aq) + NO_3{}^-(aq). \qquad (7.9)$$

The supply of the large amount of $NH_4{}^+(aq)$ from reaction (7.9) actually suppresses the ionisation of the weak base NH_3.

The introduction of $NH_4NO_3(aq)$ contributes to an increase in $[NH_4{}^+]$. This causes the equilibrium position in Eq. (7.8) to shift to the left in a bid to remove some of the $NH_4{}^+$ ions. The backward reaction is favoured and, as a result, the ionisation of NH_3 is further suppressed.

The resultant buffer solution contains high concentrations of both unionised NH_3 molecules (base) and $NH_4{}^+$ ions (acid) which enable the solution to resist changes in pH as follows:

- When a small amount of acid is added to the buffer:
  - the NH_3 molecules will react with the added H^+ ions:

$$NH_3(aq) + H^+(aq) \longrightarrow NH_4{}^+(aq);$$

  - the presence of a large reservoir of unionised NH_3 molecules helps to ensure that the majority of the added H^+ ions are removed;
  - hence, the pH of the buffer remains approximately constant.
- When a small amount of base is added to the buffer:
  - the $NH_4{}^+$ ions will react with the added OH^- ions:

$$NH_4{}^+(aq) + OH^-(aq) \longrightarrow NH_3(aq) + H_2O(l);$$

  - the presence of a large reservoir of $NH_4{}^+$ ions helps to ensure that the majority of the added OH^- ions are removed;
  - hence, the pH of the buffer remains approximately constant.

In addition to knowing the different types of buffer, it is also important to know how to select an appropriate buffer to control the pH of a solution.

The selected buffer should have a ($pK_a \pm 1$) range coinciding with the pH that you want to maintain. For instance, if you want to keep the pH constant at 4.5, then you should select a weak acid such that its ($pK_a \pm 1$) value coincides with pH = 4.5. This ($pK_a \pm 1$) range is known as the maximum buffering capacity zone, which means that out of this range, the buffer solution is not able to effectively maintain the pH range.

7.11.3 *The role of buffer in controlling pH in blood*

Enzymes work well over a narrow range of pH. In a healthy person, the pH of blood never departs more than perhaps 0.2 pH units from the average value. Death may result if the pH falls below 6.8 or rises above 7.8.

Although the cells in the human body respire and produce CO_2, which dissolves in the blood producing carbonic acid,

$$CO_2 + H_2O \rightleftharpoons H_2CO_3,$$

the pH of human blood is maintained between 7.35 and 7.45. This constant pH, which is crucial for optimal enzymatic activity, is achieved using different biological buffers:

- the H_2CO_3/HCO_3^- buffer,
- the $H_2PO_4^-/HPO_4^{2-}$ buffer, and
- plasma proteins.

Consider the buffer system that contains carbonic acid (H_2CO_3) and the hydrogencarbonate ion (HCO_3^-), which plays an important role in maintaining the pH of blood:

$$\underset{\text{weak acid}}{H_2CO_3(aq)} \rightleftharpoons \underset{\text{conjugate base}}{HCO_3^-(aq)} + H^+(aq).$$

- When a small amount of acid is introduced, the HCO_3^- ions will react with the added H^+ ions:

$$HCO_3^-(aq) + H^+(aq) \longrightarrow H_2CO_3(aq).$$

The presence of a large reservoir of HCO_3^- ions helps to ensure that most of the added H^+ ions are removed. Hence, the pH of the blood remains approximately constant.

- When a small amount of alkali is introduced, the H_2CO_3 molecules will react with the added OH^- ions:

$$H_2CO_3(aq) + OH^-(aq) \longrightarrow HCO_3^-(aq) + H_2O(l).$$

The presence of a large reservoir of undissociated H_2CO_3 molecules helps to ensure that most of the added OH^- ions are removed. Hence, the pH of the blood remains approximately constant.

7.11.4 *Calculating pH of buffer solutions*

Example 7.8: Calculating the pH of an acidic buffer solution

Calculate the pH of a solution that contains 0.05 mol dm^{-3} of the weak acid HA and 0.15 mol dm^{-3} of the sodium salt NaA (given K_a of HA $= 7.5 \times 10^{-4}$ mol dm^{-3}).

Approach: In order to get the pH value, we must first determine $[H^+]$ at equilibrium.

Method 1: Using the I.C.E. table

Solution: Let equilibrium $[H^+] = w$ mol dm^{-3}.

	HA(aq)	$\rightleftharpoons$	**A^-(aq)**	+	**H^+(aq)**
Initial conc. (mol dm^{-3})	0.05		0.15		—
Change in conc. (mol dm^{-3})	$-w$		$+w$		$+w$
Equilibrium conc. (mol dm^{-3})	$(0.05 - w)$		$(0.15 + w)$		w

The supply of A^-(aq) from the complete ionisation of Na^+A^- suppresses the dissociation of the weak acid HA to such a large extent that the following can be assumed:

- equilibrium [HA] $\approx$ initial [HA], i.e., $(0.05 - w) \approx 0.05$, and
- equilibrium $[A^-] \approx$ source of $[A^-]$ from the complete ionisation of the salt, i.e., $(0.15 + w) \approx 0.15$.

Thus,

$$K_a = \frac{[A^-][H^+]}{[HA]} = \frac{0.15 \times w}{0.05} = 7.5 \times 10^{-4}\,\text{mol dm}^{-3}$$

$$[H^+] = 2.5 \times 10^{-4}\,\text{mol dm}^{-3},$$

$$\text{pH} = -\lg[H^+] = -\lg(2.5 \times 10^{-4}) = 3.60.$$

Method 2: Using the Henderson–Hasselbalch Equation for buffer solutions

Method 2 is not something that is very special. It actually incorporates similar mathematical steps and assumptions as in Method 1 into what is known as the Henderson–Hasselbalch Equation.

Thus, Method 2 can be considered the "short-cut" of Method 1.

The Henderson–Hasselbalch Equation is derived from the K_a expression by taking logarithms on both sides of the expression:

$$\lg K_a = \lg[H^+] + \lg\frac{[A^-]}{[HA]}. \tag{7.10}$$

Rearranging Eq. (7.10) will give us Eq. (7.11):

$$\text{pH} = \text{p}K_a + \lg\frac{[A^-]}{[HA]} \quad (\text{recall: pH } = -\lg[H^+],\ \text{p}K_a = -\lg K_a). \tag{7.11}$$

Since the supply of $A^-(aq)$ from the complete ionisation of Na^+A^- suppresses the dissociation of the weak acid HA, it is assumed that:

- equilibrium [HA] $\approx$ initial [HA], and
- equilibrium $[A^-]$ $\approx$ initial [NaA].

Based on the assumptions made, Eq. (7.11) can then be re-expressed as Eq. (7.12), which is the **Henderson–Hasselbalch Equation**:

$$\text{pH} = \text{p}K_a + \lg\frac{[\text{conjugate base}]}{[\text{conjugate acid}]}. \tag{7.12}$$

Solution: Substituting the initial concentrations of NaA and HA into Eq. (7.12):

$$\begin{aligned}\text{pH} &= \text{p}K_\text{a} + \lg\frac{[\text{conjugate base}]}{[\text{conjugate acid}]}\\ &= -\lg K_\text{a} + \lg\frac{[\text{conjugate base}]}{[\text{conjugate acid}]}\\ &= -\lg(7.5\times10^{-4}) + \lg\left(\frac{0.15}{0.05}\right)\\ &= 3.60 \text{ (same answer obtained in Method 1).}\end{aligned}$$

Note that at equilibrium, $[H^+] \neq [A^-]$ for a buffer since their main sources are different. This is unlike the case when pure HA is added to water whereby at equilibrium, $[H^+] = [A^-]$.

Example 7.9: Calculating the pH of an alkaline buffer solution

Calculate the pH of a buffer solution that contains 0.10 mol of ammonium chloride and 0.05 mol of aqueous ammonia dissolved in 1 dm^3 of water at 25°C (given K_b of NH_3 is 1.75×10^{-5} mol dm^{-3} at 25°C).

Approach:

(i) Calculate K_a of conjugate acid NH_4^+ using the relationship: $K_w = K_a \times K_b$.
(ii) Then calculate pH using the Henderson–Hasselbalch Equation.

Solution:

$$K_\text{a} \text{ of } NH_4^+ = \frac{K_\text{w}}{K_\text{b} \text{ of } NH_3} = \frac{1.0\times10^{-14}}{1.75\times10^{-5}} = 5.7\times10^{-10}\,\text{mol dm}^{-3},$$

$$\begin{aligned}\text{pH} &= \text{p}K_\text{a} + \lg\frac{[\text{conjugate base}]}{[\text{conjugate acid}]}\\ &= -\lg K_\text{a} + \lg\frac{[\text{conjugate base}]}{[\text{conjugate acid}]}\\ &= -\lg(5.7\times10^{-10}) + \lg\left(\frac{0.05}{0.1}\right)\\ &= 8.94.\end{aligned}$$

Q: Why is the maximum buffering capacity occurred at pH = $\mathrm{p}K_\mathrm{a} \pm 1$?

A: When pH = $\mathrm{p}K_\mathrm{a}$, [conjugate base] = [conjugate acid]. Thus, at pH = $\mathrm{p}K_\mathrm{a} \pm 1$, both the concentrations of the conjugate base and conjugate acid are almost equivalent. This would provide a large reservoir of conjugate base and conjugate acid to resist pH changes. If the [conjugate acid] > [conjugate base], the conjugate acid can only resist pH changes when some base is added. But the conjugate base cannot resist pH changes when some acid is added due to its lower concentration value. In addition, when the conjugate acid is consumed by the added base, in order to make sure that the pH does not fluctuate drastically, it needs the conjugate base to replenish the "lost" in the conjugate acid. Thus, a lower concentration of the conjugate base than the conjugate acid would not be able to do the "replenishment" effectively.

7.12 Acid–Base Indicators

An indicator is a substance that changes colour in response to pH changes of the solution it is added to. An indicator is usually a weak acid or a weak base. Different colours will be observed for both the conjugate acid and its conjugate base, depending on the pH of the solution.

For instance, an indicator HIn (weak acid) dissociates partially in solution and the following equilibrium is attained:

$$\underset{\substack{\text{weak acid}\\ \text{(colour A)}}}{\mathrm{HIn(aq)}} \rightleftharpoons \mathrm{H^+(aq)} + \underset{\substack{\text{conjugate base}\\ \text{(colour B)}}}{\mathrm{In^-(aq)}}\ . \qquad (7.13)$$

Similar to the idea of K_a and $\mathrm{p}K_\mathrm{a}$, the equilibrium constant K_In and its pH are expressed as:

$$K_\mathrm{In} = \frac{[\mathrm{H^+}][\mathrm{In^-}]}{[\mathrm{HIn}]} \text{ and } \mathrm{pH} = \mathrm{p}K_\mathrm{In} + \lg\frac{[\mathrm{In^-}]}{[\mathrm{HIn}]}.$$

In acidic conditions, with an abundance of $\mathrm{H^+(aq)}$ ions, the equilibrium position of Eq. (7.13) lies to the left, and colour A is observed.

In alkaline conditions, as large amounts of $H^+(aq)$ ions are consumed, the equilibrium position shifts to the right, and colour B is observed.

Indicators are used to determine end-points of acid–base titrations.

Q: What is the end-point of a titration?

A: The end-point of a titration is the volume of acid or base added in order for the indicator to change colour. Different indicators change colour within different pH ranges. We term this range the "working range" of the indicator. Some indicators change colour in the alkaline pH range (e.g., phenolphthalein), while others do so in the acidic pH range (e.g., methyl orange and screened methyl orange). Refer to Table 7.1 for more details.

The end-point of a titration is detected when there is a distinct change in colour from A to B, or *vice versa*. This happens when the indicator has a colour halfway between its extreme colours, i.e., when both the weak acid (HIn) and its conjugate base (In^-) are present in equal amounts. At this point,

$$\text{pH} = \text{p}K_{\text{In}} \text{ since } \lg \frac{[\text{conjugate base}]}{[\text{conjugate acid}]} = \lg 1 = 0.$$

The colour observed at the end-point will be a mixture of colour A and colour B. As the human eye can detect colour changes when the proportion of the two colours is in the ratio 1:10, the pH range for which we will notice that the indicator changes colour is about 2 pH units (between $\text{pH} = \text{p}K_{\text{In}} - 1$ to $\text{pH} = \text{p}K_{\text{In}} + 1$).

Q: Does the end-point indicate that all the acid has been neutralised by the added base, and *vice versa*?

A: No. When the indicator changes colour during titration, it simply means that *most* of the acid or base has been neutralised.

We know that different indicators change colour within different pH ranges. Thus, when an indicator changes colour, the resultant solution may still be slightly acidic or alkaline. This is especially so if salt hydrolysis sets in (refer to Sec. 7.10).

It is the **equivalence point** of the titration that signifies a complete reaction between the acid and base. For example, if there is

Table 7.1 Working Ranges of Some Indicators

Indicator	pK_{In}	pH Range/ Working Range	Colour in "Acidic" Solution	Colour in "Alkaline" Solution
Methyl orange	3.7	3.1–4.4	red	yellow
Screened methyl orange	3.7	3.1–4.4	violet	green
Methyl red	5.1	4.2–6.3	red	yellow
Litmus	6.5	6.0–8.0	red	blue
Bromothymol blue	7.0	6.0–7.6	yellow	blue
Phenolphthalein	9.3	8.2–10.0	colourless	pink

1 mol of H^+ ions present, then it would need 1 mol of OH^- ions to completely neutralise it. That is, the amount of base needed is precisely what is specified by the stoichiometric equation to react with the amount of acid.

A universal indicator, comprising a mixture of several indicators, provides distinct colours for pH ranging from 1 to 14:

pH Range	Colour
0–3	red
3–6	orange/yellow
7	green
8–11	blue
11–14	purple

Colour of universal indicator at various pH values

When an appropriate indicator is chosen, the end-point will be close to the equivalence point. That is, the pH range of the indicator must fall within the region of marked pH change in the titration

Table 7.2 Types of Acid–Base Titrations and Suitable Indicators

Type of Titration	Region of Marked pH Change	Examples of Suitable Indicators
Strong acid–strong base	4–10	methyl orange screened methyl orange phenolphthalein; bromothymol blue
Weak acid–strong base	7.5–10.5	phenolphthalein; bromothymol blue
Strong acid–weak base	3.5–6.5	methyl orange screened methyl orange
Weak acid–weak base	nil	no suitable indicator

curve. Thus, it is important to choose an indicator wisely (refer to Table 7.2).

The choice of indicator depends on:

- its working pH range, and
- the type of titration involved.

Ideally, it is good to use an indicator that enables the end-point of the titration to coincide exactly with the equivalence point. This can be fulfilled if the chosen indicator has a pK_{In} value that is identical to the pH of the resultant solution at the equivalence point.

7.13 Acid–Base Titrations

Titrations are useful in determining the unknown concentration of one reactant by allowing it to react with a known concentration of another reactant in a controlled manner. Normally, a fixed volume of the reactant of unknown concentration is placed in a conical flask while the other reactant is placed in the burette (titrant).

It is important to control the volume of titrant allowed to react with the solution pipetted so that we know when exactly the reaction is completed. This equivalence point is determined by using a suitable indicator, as discussed earlier.

Table 7.2 shows the different types of acid–base titrations along with suitable choices of indicators.

Q: Why is the region of marked pH change for strong acid–weak base titration in the range of 3.5–6.5?

A: When the weak base reacts with the strong acid, at the equivalence point, the strong conjugate acid of the weak base exists. The hydrolysis of this conjugate acid results in an acidic solution at the equivalence point. Suitable indicators include methyl orange and screened methyl orange.

Q: What happens if one uses phenolphthalein as the indicator for the strong acid–weak base titration?

A: The working range of phenolphthalein is in the alkaline range. When phenolphthalein changes colour from colourless to pink, we have probably added to the acid far more base than is required. The titration reading would be higher than the theoretical amount needed for a complete reaction.

Q: Why is there no suitable indicator for a weak acid–weak base titration?

A: For the equivalence point of an acid–base titration to be identifiable using an indicator, the pH near the equivalence point must change sharply by several units. This does not occur for a weak acid–weak base titration. This is due to the simultaneous hydrolysis brought about by both the strong conjugate base of the weak acid and the strong conjugate acid of the weak base.

Throughout an acid–base titration, there will be variations in pH value that correspond to different stages of the neutralisation process. This change in pH during the course of a titration depends largely upon the type and strength of acid and base used.

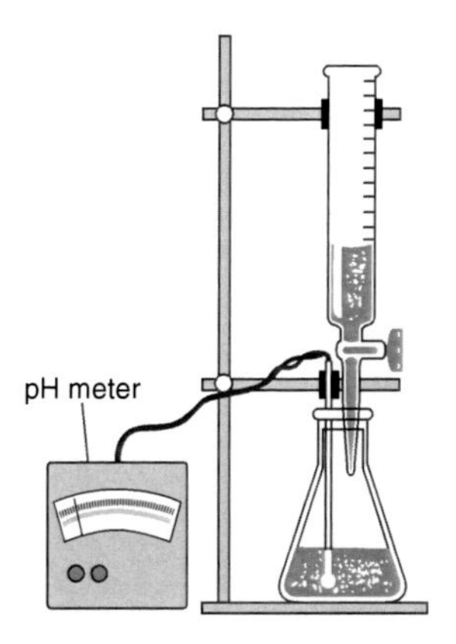

If we were to graphically plot the pH changes during the progress of titration, we would expect different titration curves for the four types of acid–base titrations.

A titration curve is usually a plot of pH against volume of titrant (acid or base). To get the profile of the curve, the pH of the solution is measured using a pH meter at each step-wise volume of titrant added.

7.13.1 *Titration curve of a strong acid–strong base titration*

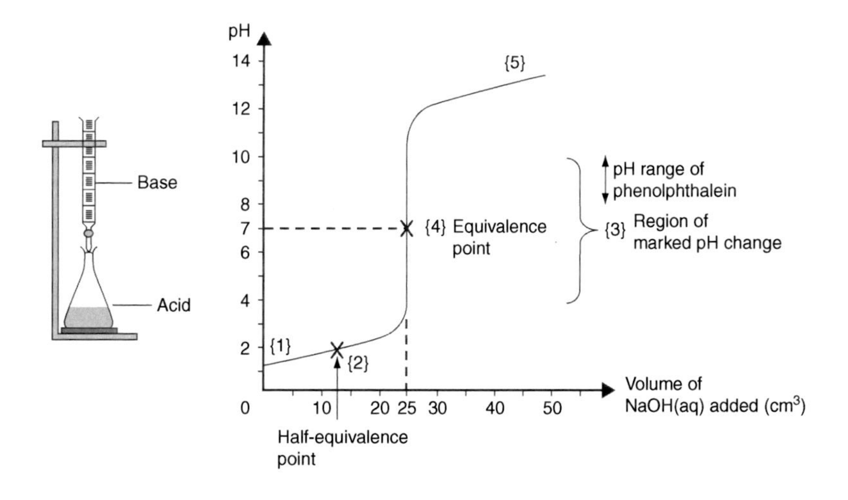

The figure above shows the titration curve when $25.0\,\text{cm}^3$ of $0.50\,\text{mol dm}^{-3}$ HCl(aq) is titrated against $0.50\,\text{mol dm}^{-3}$ NaOH(aq):

$$\text{HCl(aq)} + \text{NaOH(aq)} \longrightarrow \text{NaCl(aq)} + \text{H}_2\text{O(l)}.$$

Features of the titration curve:

(1) Initial pH is low since HCl is a strong acid that fully dissociates in water.

(2) As NaOH(aq) is added, neutralisation and dilution leads to a decrease in $[\text{H}^+]$ and hence an increase in pH. The change in pH is significant.

At the half-equivalence point, exactly half the initial $[\text{H}^+]$ is neutralised when $12.5\,\text{cm}^3$ of NaOH(aq) is added.

(3) A large rapid change in pH (from pH 4 to 10) indicates the approach of the equivalence point.

Suitable indicators whose working range coincides with this region of marked pH change include phenolphthalein and methyl orange.

(4) At the equivalence point, the acid is completely neutralised when $25.0\,\text{cm}^3$ of NaOH(aq) is added. The amount of H^+ present is equal to the amount of OH^- added and the resultant solution contains only NaCl(aq). pH $= 7$ since NaCl(aq) is a neutral salt that does not undergo hydrolysis.

(5) After the equivalence point, the pH continues to rise to a high value typical of a strong base that fully ionises in water.

Titration can also be carried out with the base sitting in the conical flask while the acid is added from the burette. The titration curve is shown below:

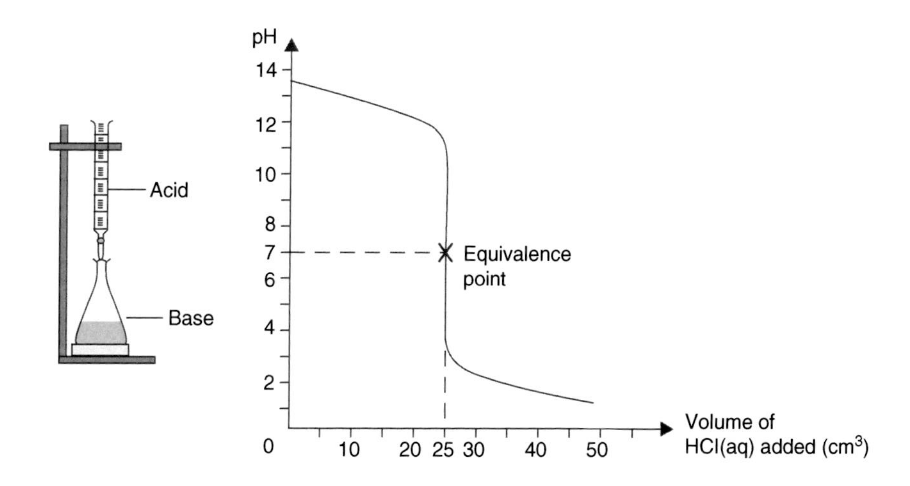

7.13.2 *Titration curve of a weak acid–strong base titration*

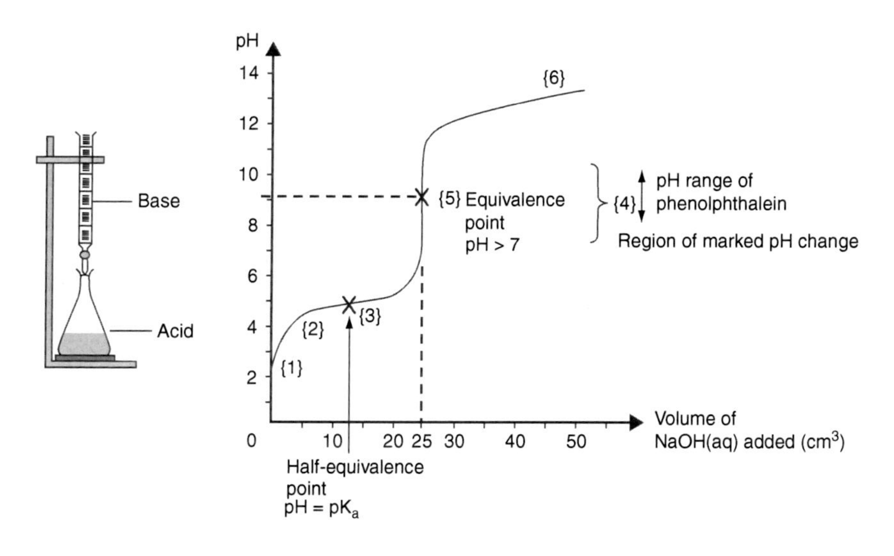

The figure above shows the titration curve when $25.0\,\text{cm}^3$ of $0.50\,\text{mol dm}^{-3}$ $CH_3COOH(aq)$ is titrated against $0.50\,\text{mol dm}^{-3}$ $NaOH(aq)$:

$$CH_3COOH(aq) + NaOH(aq) \longrightarrow CH_3COO^-Na^+(aq) + H_2O(l).$$

Features of the titration curve:

(1) The curve initially begins at an acidic pH since CH_3COOH partially dissociates in water.

(2) After an initial rapid rise in pH, the pH changes more slowly when $NaOH(aq)$ is continuously added. This is because the solution contains both the weak acid and its conjugate base, whereby the buffering effect kicks in! The pH range of $(pK_a \pm 1)$ of the conjugate acid is the buffering region whereby the solution functions as an effective buffer that resists changes in pH when more base is added.

(3) At the **half-equivalence point**, exactly half the initial $[CH_3COOH]$ is neutralised when $12.5\,\text{cm}^3$ of $NaOH(aq)$ is added. Here, $\mathbf{[CH_3COOH] = [CH_3COO^-Na^+]}$ and $\mathbf{pH = p\boldsymbol{K}_a}$ since

$$\text{pH} = \text{p}K_a + \lg\frac{[CH_3COO^-Na^+]}{[CH_3COOH]} = \text{p}K_a + \lg 1 \quad (\text{and } \lg 1 = 0).$$

The buffering efficiency is greatest when [conjugate base]/[conjugate acid] = 1. Thus, the maximum buffering capacity is at pH = $\text{p}K_a$, i.e., at the half-equivalence point.

(4) pH increases drastically (from pH 7.5 to 10.5) near the equivalence point. The marked pH change begins at a higher pH and its range is shorter than that for a strong acid–strong base titration. A suitable indicator is phenolphthalein or bromothymol blue.

(5) At the equivalence point, the acid is completely neutralised when $25.0\,\text{cm}^3$ of $NaOH(aq)$ is added. The amount of CH_3COOH present is equal to the amount of OH^- added and the resultant solution contains only $CH_3COO^-Na^+(aq)$. **pH is greater than 7** since $CH_3COO^-(aq)$ undergoes hydrolysis and produces a slightly alkaline solution:

$$CH_3COO^-(aq) + H_2O(l) \rightleftharpoons CH_3COOH(aq) + OH^-(aq).$$

(6) After the equivalence point, the pH continues to rise to a high value typical of a strong base that fully ionises in water. At this point, the alkaline hydrolysis of $CH_3COO^-(aq)$ is suppressed by the addition of excess strong base!

Titration can also be carried out with the strong base sitting in the conical flask while the weak acid is added from the burette. The titration curve is shown below.

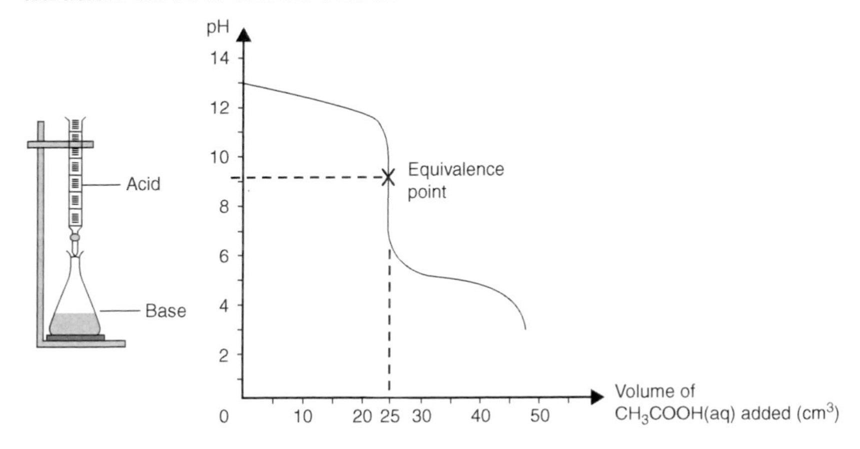

Take note that the buffering zone for this titration is at the region when excess $CH_3COOH(aq)$ has been added after the equivalence point. This is unlike the titration in which NaOH(aq) is added to a conical flask containing $CH_3COOH(aq)$ whereby the buffering zone is before the equivalence point is reached.

Example 7.10: Calculating pH values on a titration curve

Calculate the pH of the resultant solution when the respective volumes of 0.50 mol dm^{-3} NaOH(aq) are added to 25.0 cm^3 of 0.50 mol dm^{-3} $CH_3COOH(aq)$: (i) 0.00 cm^3, (ii) 15.00 cm^3, (iii) 25.00 cm^3 and (iv) 35.00 cm^3.

(Take K_a of ethanoic acid to be $1.8 \times 10^{-5}\,\text{mol}\,\text{dm}^{-3}$ at 298 K.)

Approach: The various pH calculations can be classified into the following types:

(i) calculate the pH of a weak acid given the K_a and its initial concentration (refer to Example 7.3);

(ii) calculate the pH of buffer solution using the Henderson–Hasselbalch Equation (refer to Example 7.8);
(iii) calculate the pH of solution given the K_b of the conjugate base (refer to Example 7.6); and
(iv) calculate the pH of the strong base and ignore the hydrolysis of the weaker base (refer to Example 7.2).

Solution:

(i) Let equilibrium $[H^+] = y\,\text{mol dm}^{-3}$. Ignoring the H^+ contribution from the autoionisation of water, $[CH_3COO^-] = [H^+] = y\,\text{mol dm}^{-3}$.

	$CH_3COOH(aq)$	$\rightleftharpoons$	$CH_3COO^-(aq)$	+	$H^+(aq)$
Initial conc. (mol dm^{-3})	0.50		—		—
Change in conc. (mol dm^{-3})	$-y$		$+y$		$+y$
Equilibrium conc. (mol dm^{-3})	$(0.50 - y)$		y		y

$$K_a = \frac{[CH_3COO^-][H^+]}{[CH_3COOH]}$$

$$1.8 \times 10^{-5} = \frac{y^2}{(0.50 - y)}.$$

Since CH_3COOH is a weak acid, assume $y \ll 0.50$ such that $(0.50 - y) \approx 0.50$. Thus,

$$1.8 \times 10^{-5} = \frac{y^2}{0.50}$$

$$y = 3.0 \times 10^{-3}.$$

Thus, equilibrium $[H^+] = 3.0 \times 10^{-3}\,\text{mol dm}^{-3}$. Then,

$$\begin{aligned} \text{pH} &= -\lg[H^+] \\ &= -\lg(3.0 \times 10^{-3}) \\ &= 2.52. \end{aligned}$$

(ii) Amount of CH_3COOH neutralised
$= \text{amount of NaOH added}$
$= \frac{15}{1000} \times 0.5$
$= 7.5 \times 10^{-3}\,\text{mol}.$
Therefore, amount of CH_3COONa formed $= 7.5 \times 10^{-3}\,\text{mol}$. Thus,

$$\begin{aligned}\text{Amount of } CH_3COOH \text{ unreacted} \\ &= \text{initial amount } CH_3COOH - \text{amount of } CH_3COOH \\ &\quad \text{reacted} \\ &= \left(\frac{25}{1000} \times 0.5\right) - 7.5 \times 10^{-3} \\ &= 5.0 \times 10^{-3}\,\text{mol}.\end{aligned}$$

Upon adding base, the total volume of solution $= 25 + 15 = 40\,\text{cm}^3$. Thus,

$$[CH_3COOH] \text{ unreacted} = \frac{5.0 \times 10^{-3}}{40} \times 1000 = 0.125\,\text{mol dm}^{-3},$$

$$[CH_3COONa] \text{ formed} = \frac{7.5 \times 10^{-3}}{40} \times 1000 = 0.188\,\text{mol dm}^{-3}.$$

Since the supply of CH_3COO^- from the complete ionisation of $CH_3COO^-Na^+$ suppresses the dissociation of the weak acid CH_3COOH, it is assumed that:

- equilibrium $[CH_3COOH] \approx [CH_3COOH]$ unreacted, and
- equilibrium $[CH_3COO^-] \approx [CH_3COOH]$ reacted $=$ $[CH_3COONa]$ formed.

Hence,

$$\begin{aligned}\text{pH} &= \text{p}K_\text{a} + \lg \frac{[CH_3COO^-Na^+]}{[CH_3COOH]} \\ &= -\lg(1.8 \times 10^{-5}) + \lg\left(\frac{0.188}{0.125}\right) = 4.92.\end{aligned}$$

(iii) Amount of CH_3COOH neutralised
$= $ amount of NaOH added
$= \frac{25}{1000} \times 0.5$
$= 1.25 \times 10^{-2}$ mol.

Therefore, amount of CH_3COONa formed $= 1.25 \times 10^{-2}$ mol.

Since there is a complete reaction, the resultant solution contains only $CH_3COONa(aq)$. $CH_3COO^-(aq)$ undergoes hydrolysis and produces a slightly alkaline solution.

Upon adding base, the total volume of solution $= 25 + 25 = 50\,cm^3$. Thus,

$$[CH_3COO^-] = \frac{1.25 \times 10^{-2}}{50} \times 1000 = 0.25\,mol\,dm^{-3}.$$

Let equilibrium $[OH^-] = y\,mol\,dm^{-3}$. Ignoring the OH^- contribution from the autoionisation of water, $[OH^-] = [CH_3COOH] = y\,mol\,dm^{-3}$.

	$CH_3COO^-(aq)$ + $H_2O(l)$ $\rightleftharpoons$	$CH_3COOH(aq)$ +	$OH^-(aq)$
Initial conc. ($mol\,dm^{-3}$)	0.25	—	—
Change in conc. ($mol\,dm^{-3}$)	$-y$	$+y$	$+y$
Equilibrium conc. ($mol\,dm^{-3}$)	$(0.25 - y)$	y	y

$$K_w = K_a \times K_b = 1.0 \times 10^{-14}\,mol^2\,dm^{-6}.$$

Hence,

$$K_b = \frac{1.0 \times 10^{-14}}{1.8 \times 10^{-5}} = 5.56 \times 10^{-10}\,mol\,dm^{-3}$$

and

$$K_b = \frac{[CH_3COOH][OH^-]}{[CH_3COO^-]} = 5.56 \times 10^{-10}\,mol\,dm^{-3}.$$

Thus,

$$5.56 \times 10^{-10} = \frac{y^2}{(0.25 - y)}.$$

Since CH_3COO^- is a weak base, assume $y \ll 0.25$ such that $(0.25 - y) \approx 0.25$. Therefore,

$$5.56 \times 10^{-10} = \frac{y^2}{0.25}$$

$$y = 1.18 \times 10^{-5}.$$

Thus, equilibrium $[OH^-] = 1.18 \times 10^{-5}\,\text{mol}\,\text{dm}^{-3}$. Then,

$$\begin{aligned} \text{pOH} &= -\lg[OH^-] \\ &= -\lg(1.18 \times 10^{-5}) \\ &= 4.93. \end{aligned}$$

Hence,

$$\text{pH} = 14 - \text{pOH} = 14 - 4.93 = 9.07.$$

(iv) The pH after the equivalence point is attributed to the presence of the strong base NaOH. The hydrolysis of the conjugate base CH_3COO^- can be ignored as it is suppressed due to the presence of the strong base.

Amount of NaOH reacted with acid

$$= \text{amount of } CH_3COOH = \frac{25}{1000} \times 0.5 = 1.25 \times 10^{-2}\,\text{mol}.$$

Therefore,

$$\begin{aligned} \text{amount of excess NaOH} &= \text{total amount of NaOH added} \\ &\quad - \text{amount of NaOH reacted} \\ &= \left(\frac{35}{1000} \times 0.5\right) - 1.25 \times 10^{-2} \\ &= 5.0 \times 10^{-3}\,\text{mol}. \end{aligned}$$

Upon adding base, the total volume of solution $= 25 + 35 = 60\,\text{cm}^3$.
Thus,

$$[OH^-] = [\text{NaOH}] = \frac{5.0 \times 10^{-3}}{60} \times 1000 = 8.33 \times 10^{-2}\,\text{mol}\,\text{dm}^{-3}$$

and

$$\begin{aligned} \text{pOH} &= -\lg[\text{OH}^-] \\ &= -\lg(8.33 \times 10^{-2}) \\ &= 1.08. \end{aligned}$$

Hence,

$$\text{pH} = 14 - \text{pOH} = 14 - 1.08 = 12.92.$$

7.13.3 *Titration curve of a strong acid–weak base titration*

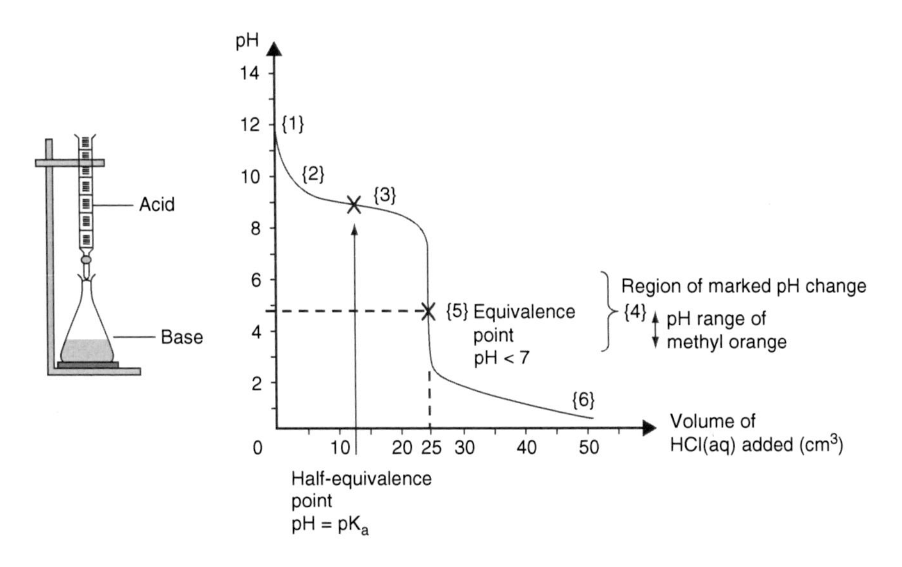

The figure above shows the titration curve when $25.0\,\text{cm}^3$ of $0.50\,\text{mol}\,\text{dm}^{-3}$ $NH_3(aq)$ is titrated against $0.50\,\text{mol}\,\text{dm}^{-3}$ HCl(aq):

$$\text{HCl(aq)} + \text{NH}_3\text{(aq)} \longrightarrow \text{NH}_4{}^+\text{(aq)} + \text{Cl}^-\text{(aq)}.$$

Features of the titration curve:

(1) The curve initially begins at an alkaline pH since NH_3 partially ionises in water.

(2) After an initial rapid drop in pH, the pH changes more slowly when HCl(aq) is continuously added. This is because the solution contains both the weak base and its conjugate acid that can act as a buffer. The pH range of ($pK_a \pm 1$) of the conjugate acid is the buffering region.

(3) At the **half-equivalence point**, exactly half the initial $[NH_3]$ is neutralised when $12.5\,cm^3$ of HCl(aq) is added. The solution is at its maximum buffering capacity since $[\mathbf{NH_3}] = [\mathbf{NH_4}^+]$ and $\mathbf{pH = p}\boldsymbol{K}_\mathbf{a}$, where

$$pH = pK_a + \lg\frac{[\text{conjugate base}]}{[\text{conjugate acid}]} = pK_a + \lg\frac{[NH_3]}{[NH_4{}^+]}.$$

(4) pH decreases drastically (from pH 6.5 to 3.5) near the equivalence point. Suitable indicators include methyl orange and screened methyl orange.

(5) At the equivalence point, the base is completely neutralised when $25.0\,cm^3$ of HCl(aq) is added. The amount of NH_3 present is equal to the amount of H^+ added and resultant solution contains only NH_4Cl(aq).

pH is lower than 7 since $NH_4{}^+$(aq) undergoes hydrolysis and produces a slightly acidic solution:

$$NH_4{}^+(aq) + H_2O(l) \rightleftharpoons NH_3(aq) + H_3O^+(aq).$$

(6) After the equivalence point, the pH continues to fall to a low value typical of a strong acid that fully dissociates in water. At this point, the acidic hydrolysis of $NH_4{}^+$(aq) is suppressed by the addition of excess strong acid!

Titration can also be carried out with the strong acid sitting in the conical flask while the weak base is added from the burette. The titration curve is shown below.

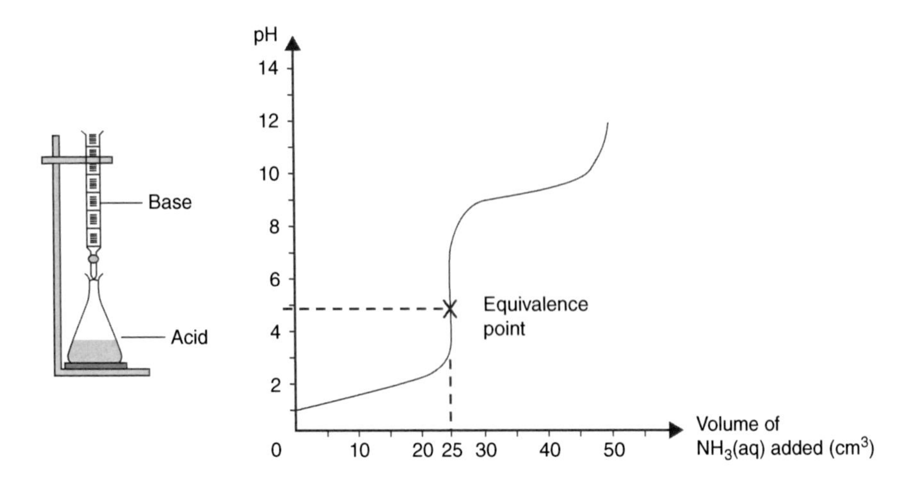

Take note that the buffering zone for this titration is at the region when excess NH_3(aq) has been added after the equivalence point. This is unlike the titration in which HCl(aq) is added to a conical flask containing NH_3(aq) whereby the buffering zone is before the equivalence point is reached.

Exercise: Calculating pH values on a titration curve

(1) Calculate the pH of the resultant solution when the respective volumes of 0.50 mol dm^{-3} HCl(aq) are added to 25.0 cm^3 of 0.50 mol dm^{-3} NH_3(aq):
(i) 0.00 cm^3, (ii) 12.50 cm^3, (iii) 25.00 cm^3, and (iv) 40.00 cm^3.
(Take K_b of NH_3 to be 1.75×10^{-5} mol dm^{-3} at 298 K.)

(2) Hence, sketch the titration curve for this strong acid–weak base titration. Indicate on your graph the region within which the solution acts as a buffer.

Answer:

(i) 11.47,
(ii) 9.24,
(iii) 4.92, and
(iv) 0.94.

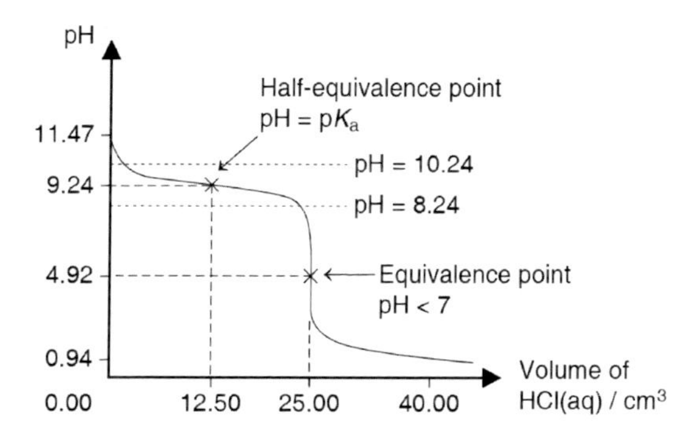

7.13.4 *Titration curve of a weak acid–weak base titration*

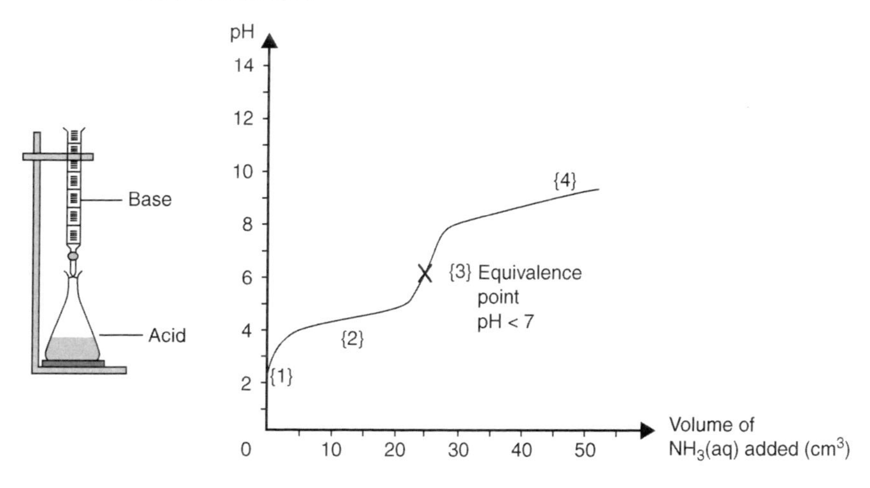

The figure above shows the titration curve when $25.0\,\text{cm}^3$ of $0.50\,\text{mol}\,\text{dm}^{-3}$ $CH_3COOH(aq)$ is titrated against $0.50\,\text{mol}\,\text{dm}^{-3}$ $NH_3(aq)$:

$$CH_3COOH(aq) + NH_3(aq) \longrightarrow CH_3COONH_4(aq).$$

Features of the titration curve:

(1) The curve initially begins at an acidic pH since CH_3COOH partially dissociates in water.

(2) The increase in pH is gradual when the weak base is added. This is due to the buffering effect of the weak acid and its strong

conjugate base. The hydrolysis of the conjugate acid generated from the added weak base is ignored.

(3) There is no straight, vertical section on the graph. This shows that the change in pH at the equivalence point is rather gradual and much less sharp than in any of the previous titrations. This is because upon neutralisation, the weak acid NH_4^+ (a strong conjugate acid of the weak base NH_3) is generated and it does not fully dissociate:

$$NH_4^+(aq) + H_2O(l) \rightleftharpoons NH_3(aq) + H_3O^+(aq).$$

In addition, the weak base CH_3COO^- (a strong conjugate base of the weak acid CH_3COOH) is also produced. The hydrolysis of CH_3COO^- creates an opposing effect to the dissociation of the weak acid NH_4^+:

$$CH_3COO^-(aq) + H_2O(l) \rightleftharpoons CH_3COOH(aq) + OH^-(aq).$$

The pH of the resultant solution depends on the relative strength of the conjugate acid NH_4^+ and that of the conjugate base CH_3COO^-. Since the K_a of NH_4^+ is slightly greater than the K_b of CH_3COO^-, the solution is slightly acidic. There are no suitable indicators whose working range coincides with this region of gradual pH change.

(4) Soon after the equivalence point, the titration curve flattens out at a fairly low alkaline pH since the excess base is a weak one that does not undergo extensive ionisation.

Q: Why do we ignore the hydrolysis effect of the strong conjugate acid of the weak base at the buffering region (at Point 2)?

A: First of all, take note that the weak base that is being used for titration must be a stronger base than the conjugate base of the weak acid being titrated. Otherwise, if it is a weak base, it would not be able to accept a proton from the weak acid, and there would be no point carrying out the titration! This would mean that the conjugate acid of the weak base added must be weaker than the weak acid that is being titrated. Hence, its dissociation would be suppressed by the dissociation of the weak acid being titrated. Thus, its dissociation can be ignored.

7.13.5 *Titration curve of a polybasic acid–strong base titration*

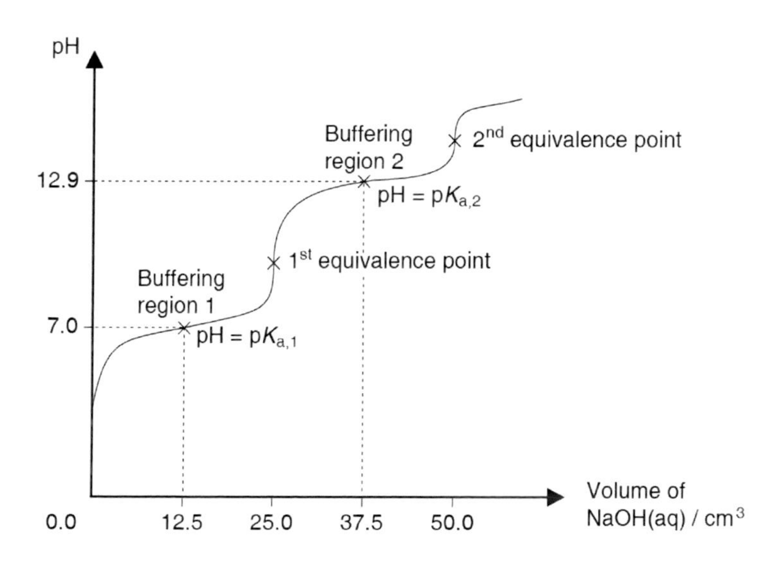

The figure above shows the titration curve when $25.0\,\text{cm}^3$ of $0.50\,\text{mol dm}^{-3}$ $H_2S(aq)$ is titrated against $0.50\,\text{mol dm}^{-3}$ NaOH(aq):

$$H_2S(aq) + 2NaOH(aq) \longrightarrow Na_2S(aq) + 2H_2O(l).$$

H_2S is a weak dibasic acid whose step-wise dissociation is represented by Eqs. (7.14) and (7.15):

$$H_2S(aq) + H_2O(l) \rightleftharpoons H_3O^+(aq) + HS^-(aq),$$

$$K_{a,1} = 1.0 \times 10^{-7}\,\text{mol dm}^{-3}; \quad pK_{a,1} = 7, \tag{7.14}$$

$$HS^-(aq) + H_2O(l) \rightleftharpoons H_3O^+(aq) + S^{2-}(aq),$$

$$K_{a,2} = 1.3 \times 10^{-13}\,\text{mol dm}^{-3}; \quad pK_{a,2} = 12.9. \tag{7.15}$$

There is a large difference in the ease with which a H_2S molecule loses its first and second protons, as reflected in the magnitude of the K_a values. The first dissociation proceeds more readily than the second since it is easier to remove the positively charged proton from the neutral H_2S molecule than from the negatively charged HS^- ion with which it experiences greater attraction. In addition, the first dissociation has to proceed before the second step can go ahead. By then, the H^+ that is produced from the first dissociation suppresses the second dissociation step.

When $H_2S(aq)$ is titrated against a standard solution of NaOH(aq), the titration curve shows two equivalence points that correspond to the step-wise completion of the following reactions:

(i) $H_2S(aq) + OH^-(aq) \longrightarrow HS^-(aq) + H_2O(l)$,
(ii) $HS^-(aq) + OH^-(aq) \longrightarrow S^{2-}(aq) + H_2O(l)$.

Associated with these acids and their conjugate bases are two buffering regions with the following composition:

- Buffer 1 comprises weak acid $H_2S(aq)$ and its conjugate base $HS^-(aq)$, ignoring the further dissociation of $HS^-(aq)$. Maximum buffering capacity is at $pH = pK_{a,1} = 7$.
- Buffer 2 comprises weak acid $HS^-(aq)$ and its conjugate base $S^{2-}(aq)$. Maximum buffering capacity is at $pH = pK_{a,2} = 12.9$.

7.13.6 *Titration curve of a carbonate–strong acid titration (Double-Indicator Method)*

Sodium carbonate is a diprotic base. The following two equations represent the step-wise reaction of sodium carbonate with hydrochloric acid:

	Progress of reaction	*Reaction is marked complete when*
(a) $CO_3{}^{2-}(aq) + H^+(aq) \longrightarrow HCO_3{}^-(aq)$	Na_2CO_3 is only half-neutralised	phenolphthalein turns from pink to colourless
(b) $HCO_3{}^-(aq) + H^+(aq) \longrightarrow CO_2(g) + H_2O(l)$	Na_2CO_3 is completely neutralised	methyl orange turns from yellow to orange

The sum of (a) and (b) gives the overall reaction:

$$CO_3{}^{2-}(aq) + 2H^+(aq) \longrightarrow CO_2(g) + H_2O(l).$$

Titration can be carried out with:

(i) The use of two indicators to signify the end of each equivalence point.

When phenolphthalein, used as the indicator, changes colour from pink to colourless, then the volume of HCl at this end-point (V_1) corresponds to the amount of H^+ required to convert CO_3^{2-} to HCO_3^-.

If now a few drops of methyl orange is added to the solution and titration is continued, then the additional volume of HCl at the second end-point ($V_2 - V_1$) corresponds to the amount of H^+ required to convert HCO_3^- to CO_2.

The two volumes, V_1 and ($V_2 - V_1$), should be the same since the amount of CO_3^{2-} is equal to the amount of HCO_3^-.

(ii) The use of only one indicator to signify the end of the second equivalence point.

If methyl orange is used at the very beginning instead of adding phenolphthalein, then the volume of HCl at this end-point (V_2) corresponds to the amount of H^+ required to convert CO_3^{2-} to CO_2. This volume is twice the volume reading (V_1) if phenolphthalein is used.

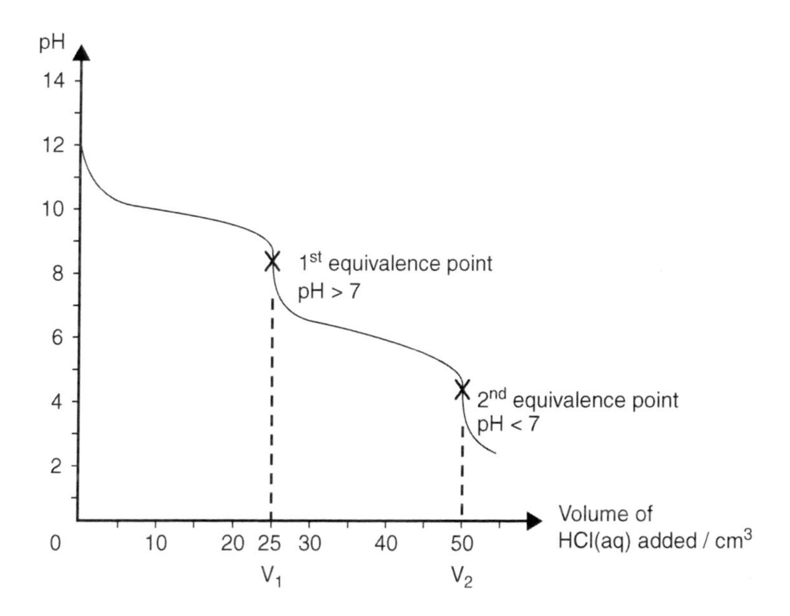

Titration curve when 25 cm^3 of 0.10 mol dm^{-3} $Na_2CO_3(aq)$ is titrated against 0.10 mol dm^{-3} HCl(aq).

Q: What is the use of the Double-Indicator Method?
A: The Double-Indicator Method is useful for determining the following quantities:

(1) amount of carbonate in a carbonate/hydrogencarbonate mixture,
(2) amount of carbonate in a carbonate/hydroxide mixture, and
(3) amount of hydrogencarbonate in a hydrogencarbonate/hydroxide mixture.

The following examples will explain the use of the method and the calculations involved in determining the said quantities.

Example 7.11: Analysing a mixture of Na_2CO_3 and $NaHCO_3$

A 25.0 cm^3 aliquot of a mixture of Na_2CO_3 and $NaHCO_3$ is titrated against 0.5 mol dm^{-3} HCl solution. 17.0 cm^3 of acid is used when phenolphthalein changes from pink to colourless. Upon further titration using methyl orange as indicator, 35.0 cm^3 of acid is used to reach the second end-point. Determine the initial concentrations of Na_2CO_3 and $NaHCO_3$ in the mixture.

Approach:

- The first end-point using phenolphthalein signifies the volume of HCl(aq), V_1 cm^3, used in Reaction 1:
  Reaction 1: $CO_3{}^{2-}(aq) + H^+(aq) \longrightarrow HCO_3{}^-(aq)$.
- The second end-point using methyl orange signifies the volume of HCl(aq), V_2 cm^3, used in Reaction 2:
  Reaction 2: $HCO_3{}^-(aq) + H^+(aq) \longrightarrow CO_2(g) + H_2O(l)$.
- There are two sources of $HCO_3{}^-$: those that are initially present in the mixture and those that are generated from Reaction 1. The difference between the end-points $(V_2 - V_1)$ signifies the amount of $HCO_3{}^-$ that is initially present in the mixture.

Solution:

At the first end-point:

$$Na_2CO_3(aq) + HCl(aq) \longrightarrow NaHCO_3(aq) + NaCl(aq).$$

Amount of HCl reacted $= \frac{17}{1000} \times 0.5 = 8.5 \times 10^{-3}$ mol.

Amount of Na_2CO_3 present
$=$ Amount of HCl reacted
$= 8.5 \times 10^{-3}$ mol.

Therefore,

Concentration of Na_2CO_3 in the mixture
$= 8.5 \times 10^{-3} \times \frac{1000}{25} = 0.340\,\text{mol}\,\text{dm}^{-3}$.

At the second end-point:

$$NaHCO_3(aq) + HCl(aq) \longrightarrow NaCl(aq) + CO_2(g) + H_2O(l).$$

Amount of HCl reacted $= \frac{35}{1000} \times 0.5 = 1.75 \times 10^{-2}$ mol.

Amount of HCl which reacted with $NaHCO_3$ from Reaction 1
$=$ amount of Na_2CO_3 present
$= 8.5 \times 10^{-3}$ mol.

Therefore,
Amount of HCl that reacted with $NaHCO_3$ initially present in mixture
$= 1.75 \times 10^{-2} - 8.5 \times 10^{-3} = 9.0 \times 10^{-3}$ mol.

Concentration of $NaHCO_3$ in the mixture
$= 9.0 \times 10^{-3} \times \frac{1000}{25} = 0.360\,\text{mol}\,\text{dm}^{-3}$.

Or
Volume of HCl required to react with $NaHCO_3$ initially present in mixture $= 35.0 - 17.0 = 18.0\,\text{cm}^3$.

Amount of HCl that reacted with $NaHCO_3$ initially present in mixture $= \frac{18}{1000} \times 0.5 = 9.0 \times 10^{-3}$ mol (same as above).

Example 7.12: Analysing a mixture of Na_2CO_3 and NaOH
A 25.0 cm^3 aliquot of a mixture of Na_2CO_3 and NaOH is titrated against 0.5 mol dm^{-3} HCl solution. 38.0 cm^3 of acid is used when

phenolphthalein changes from pink to colourless. Upon further titration using methyl orange as indicator, $21.0\,cm^3$ of acid is used to reach the second end-point. Determine the initial concentrations of Na_2CO_3 and NaOH in the mixture.

Approach:

- The first end-point using phenolphthalein signifies the volume of HCl(aq), $V_1\,cm^3$, used in Reactions 1 and 2:
  Reaction 1: $CO_3^{2-}(aq) + H^+(aq) \longrightarrow HCO_3^-(aq)$,
  Reaction 2: $OH^-(aq) + H^+(aq) \longrightarrow H_2O(l)$.
- The second end-point using methyl orange signifies the volume of HCl(aq), $V_2\,cm^3$, used in Reaction 3:
  Reaction 3: $HCO_3^-(aq) + H^+(aq) \longrightarrow CO_2(g) + H_2O(l)$.
- The difference between the end-points $(V_1 - V_2)$ signifies the amount of OH^- initially present in the mixture.

Solution:

At the second end-point:

Amount of HCl reacted with $NaHCO_3 = \frac{21}{1000} \times 0.5 = 1.05 \times 10^{-2}$ mol.

Amount of Na_2CO_3 present
= Amount of $NaHCO_3$
= Amount of HCl
$= 1.05 \times 10^{-2}$ mol.

Therefore,

Concentration of Na_2CO_3 in the mixture
$= 1.05 \times 10^{-2} \times \frac{1000}{25} = 0.420\,mol\,dm^{-3}$.

At the first end-point:

Total amount of HCl reacted with both Na_2CO_3 and NaOH $= \frac{38}{1000} \times 0.5 = 1.9 \times 10^{-2}$ mol.

Amount of NaOH present

$$= \text{Total amount of HCl} - \text{amount of HCl reacted with } Na_2CO_3$$

$$= 1.9 \times 10^{-2} - 1.05 \times 10^{-2} = 8.5 \times 10^{-3}\text{ mol.}$$

Concentration of NaOH in the mixture
$= 8.5 \times 10^{-3} \times \frac{1000}{25} = 0.340\,\text{mol}\,\text{dm}^{-3}$.

Or

Volume of HCl required to react with NaOH initially present in mixture $= 38.0 - 21.0 = 17.0\,\text{cm}^3$.

Amount of HCl that reacted with NaOH initially present in mixture
$= \frac{17}{1000} \times 0.5 = 8.5 \times 10^{-3}\,\text{mol}$ (same as above).

Exercise: Analysing a mixture of $NaHCO_3$ and NaOH

A 25.0 cm^3 aliquot of a mixture of $NaHCO_3$ and NaOH is titrated against 0.5 mol dm^{-3} HCl solution. 21.50 cm^3 of acid is used when phenolphthalein changes from pink to colourless. Upon further titration using methyl orange as indicator, 15.0 cm^3 of acid is used to reach the second end-point. Determine the initial concentrations of $NaHCO_3$ and NaOH in the mixture. (Answer: [NaOH] = 0.430 mol dm^{-3}; [$NaHCO_3$] = 0.300 mol dm^{-3}.)

Approach:

- The first end-point using phenolphthalein signifies the volume of HCl(aq) used in Reaction 1:
  Reaction 1: $OH^-(aq) + H^+(aq) \longrightarrow H_2O(l)$.
- The second end-point using methyl orange signifies the volume of HCl(aq) used in Reaction 2:
  Reaction 2: $HCO_3^-(aq) + H^+(aq) \longrightarrow CO_2(g) + H_2O(l)$.

7.13.7 *Titration curve of a mixture of weak acids — strong base titration*

If a mixture of two weak acids is present in a system, then a titration with a strong base results in a two-step titration profile as shown below. This titration profile is actually an inversion of the carbonate-strong acid titration curve.

The first step of titration corresponds to the reaction between the strong base and the stronger of the two weak acids. Thus, from this volume of strong base used, one can determine the amount of the stronger weak acid.

The second step of titration corresponds to the reaction between the second weak acid and the strong base.

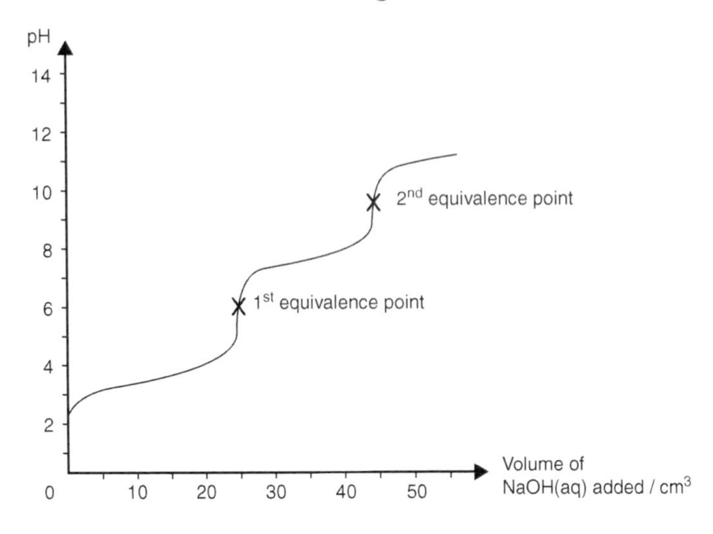

Q: When the base is added to a mixture of weak acids, why does the stronger one react first?

A: You should already know that the stronger weak acid suppresses the dissociation of the weaker weak acid, and hence all H^+ in the solution can be assumed to have come from the stronger weak acid. When OH^- is added, it can react with H^+ (if it encounters it). This causes $[H^+]$ to decrease and according to Le Chatelier's Principle, more of the stronger weak acid dissociates to compensate for the loss of H^+. This happens till all the stronger weak acid has reacted. Now, what happens if OH^- encounters the undissociated stronger weak acid molecule instead? This is not a problem as they can simply react, resulting ultimately in the consumption of the stronger weak acid. But, what if OH^- encounters the undissociated weaker weak acid molecule? This is still not a problem even when the reaction between the undissociated weaker weak acid molecule and OH^- proceeds. When this happens, the conjugate base of the weaker weak acid is generated. This conjugate base of the weaker weak acid does not just stay put. It can accept a proton from the undissociated stronger weak acid molecule as long as the latter is still present or it can also react with a free H^+ that is floating around. Thus, no matter how you see it, the *first volume of base used is solely to neutralise the stronger weak acid* in the mixture!

Example 7.13: Analysing a mixture of two weak acids

A 10.0 cm^3 aliquot of sample W contains two monoprotic (monobasic) weak acids, HA and HB, with pK_a values of 8.86 and 4.20, respectively. Sample W is titrated against 0.050 mol dm^{-3} NaOH solution using a mixture of two indicators, bromothymol blue and phenolphthalein. It is found that 8.6 cm^3 of NaOH is needed to change the colour of the first indicator and a further 7.1 cm^3 is needed to change the colour of the second indicator. Calculate the concentration of each of the two acids in W.

Approach: Since pK_a of HB (4.20) is smaller than that of HA ($pK_a = 8.86$), HB is a stronger acid. Thus, HB reacts with the added base first.

- The first end-point using bromothymol blue signifies the volume of NaOH(aq) used in Reaction 1:
  Reaction 1: $HB(aq) + OH^-(aq) \longrightarrow H_2O(l) + B^-(aq)$.
- The second end-point using phenolphthalein signifies the volume of NaOH(aq) used in Reaction 2:
  Reaction 2: $HA(aq) + OH^-(aq) \longrightarrow H_2O(l) + A^-(aq)$.

Solution:

<u>At the first end-point:</u>

$$\text{Amount of NaOH reacted with HB} = \frac{8.6}{1000} \times 0.05 = 4.3 \times 10^{-4}\,\text{mol}.$$

Therefore,

$$\begin{aligned}\text{Concentration of HB in the mixture} &= 4.3 \times 10^{-4} \times \frac{1000}{10} \\ &= 0.043\,\text{mol}\,\text{dm}^{-3}.\end{aligned}$$

At the second end-point:

$$\text{Amount of NaOH reacted with HA} = \frac{7.1}{1000} \times 0.05 = 3.55 \times 10^{-4}\ \text{mol}.$$

Therefore,

$$\begin{aligned}\text{Concentration of HA in the mixture} &= 3.55 \times 10^{-4} \times \frac{1000}{10} \\ &= 0.0355\,\text{mol}\,\text{dm}^{-3}.\end{aligned}$$

7.13.8 *Back-titration*

There are reactions, especially those involving solid or volatile compounds, for which the determination of these substances cannot be found through direct titration. Possible reasons include the difficulty of detecting the end-point or the lack of accuracy due to the loss of compound or an incomplete reaction. For example, if one needs to determine the amount of ammonia gas released from a reaction, one can bubble the ammonia gas into a known excess amount of aqueous HCl solution. The unreacted HCl is then titrated with a strong base. From the titration results, we can work backwards to calculate the amount of ammonia gas. This is the essence of back-titration.

In another example, the back-titration technique is used to determine the amount of an insoluble carbonate. The following example illustrates how results of back-titration can be used to quantify the amount of $CaCO_3$ present in an impure sample.

Example 7.14: To determine the percentage purity of an impure sample of $CaCO_3$, 0.80 g of the sample is reacted with 100 cm^3 of 0.20 mol dm^{-3} HCl(aq), which is in excess. After the reaction is completed, the remaining unreacted acid is titrated against 0.50 mol dm^{-3} NaOH(aq). It is found that 29.60 cm^3 of base is needed to completely neutralise the acid.

Q: How do we know that 100 cm^3 of this acid is in excess?

A: The sample is assumed to be 100% pure $CaCO_3$. From the stoichiometric equation between $CaCO_3$ and HCl, we can calculate the volume of HCl needed, which is around 80 cm^3. Thus, 100 cm^3 of acid is sufficiently in excess.

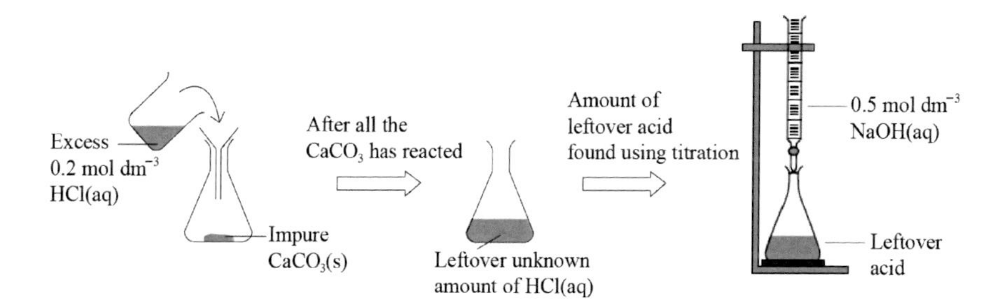

Solution:

$NaOH(aq) + HCl(aq) \longrightarrow NaCl(aq) + H_2O(l)$.
Amount of NaOH reacted $= \frac{29.6}{1000} \times 0.5 = 1.48 \times 10^{-2}$ mol.
Amount of HCl which did not react with $CaCO_3$ = Amount of NaOH used $= 1.48 \times 10^{-2}$ mol.
Total initial amount of HCl $= \frac{100}{1000} \times 0.2 = 0.020$ mol.

Therefore,

Amount of HCl which reacted with $CaCO_3 = 0.020 - 1.48 \times 10^{-2} = 5.2 \times 10^{-3}$ mol.

$CaCO_3(s) + 2HCl(aq) \longrightarrow CaCl_2(aq) + CO_2(g) + H_2O(l)$.
Amount of $CaCO_3$ present = 1/2× Amount of HCl reacted $= 2.6 \times 10^{-3}$ mol.
Mass of $CaCO_3 = 100.1 \times 2.6 \times 10^{-3} = 0.26$ g.

Therefore,

$$\text{Percentage purity of sample} = \frac{\text{Mass of pure substance}}{\text{Mass of sample}} \times 100\%$$
$$= \frac{0.26}{0.80} \times 100\% = 32.5\%.$$

Q: Why is the filter funnel placed over the conical flask when excess acid is added to the impure $CaCO_3$ in the above diagram?

A: As the acid is added to the carbonate, effervescence of CO_2 will cause acid spray. The filter funnel is used to prevent the loss of acid so that the determination of the leftover acid is more accurate.

7.14 Solubility Product K_{sp}

$BaSO_4$ and many other ionic compounds have such low solubilities in water that we often use the term "insoluble salts" to describe them. However, these are more aptly described as sparingly soluble salts. When you keep on adding a sparingly soluble salt to water at a fixed temperature, there will come a point in time when, to the naked eye, no more solute can dissolve.

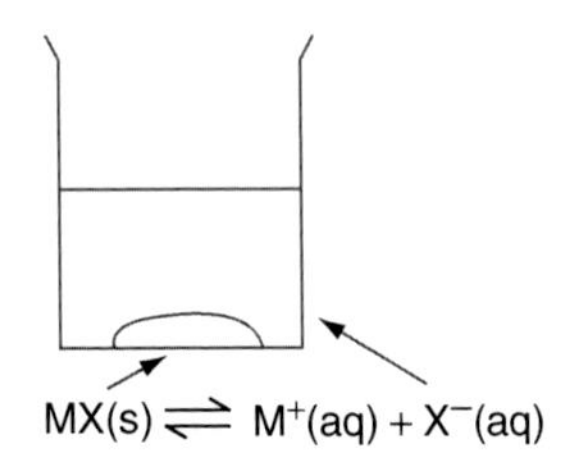

A saturated solution has formed and there exists a dynamic equilibrium between the excess undissolved solute and the ions in the saturated solution:

$$BaSO_4(s) \rightleftharpoons Ba^{2+}(aq) + SO_4{}^{2-}(aq).$$

At equilibrium, the rate of dissolution is equivalent to the rate of precipitation.

The limit to how much solute can dissolve, in a fixed volume of water at a certain temperature, is given by the equilibrium constant called the **solubility product, K_{sp}**.

The K_{sp} expression for the above equilibrium system is written as

$$\boldsymbol{K_{sp}} = [\mathbf{Ba^{2+}(aq)}][\mathbf{SO_4{}^{2-}(aq)}].$$

(Recall: the concentration of the solid is not included in the K_c expression)

The value of K_{sp} differs for each type of salt and is affected by temperature.

In general, K_{sp} is defined as the product of the concentrations of the constituent ions in a saturated solution of the salt raised to the powers as indicated by the stoichiometric coefficients in the balanced equation for the equilibrium:

$$A_nB_m(s) \rightleftharpoons nA^{m+}(aq) + mB^{n-}(aq).$$

The solubility product of A_nB_m is given by:

$$K_{sp} = [A^{m+}]^n[B^{n-}]^m.$$

The units of K_{sp} are $(\text{mol dm}^{-3})^{n+m}$.

7.14.1 *Solubility and* K_{sp}

Solubility of a salt specifies the maximum amount of solute that dissolved in the solvent at equilibrium, i.e., the maximum amount of dissolved solute in a saturated solution.

Solubility of a salt can be measured in terms of

- mass of dissolved solute per unit volume of solution (g cm^{-3}), and
- amount of dissolved solute per unit volume of solution (mol dm^{-3}).

The solubility of a sparingly soluble salt, at a given temperature, is determined experimentally and its value is used to calculate the K_{sp} of the salt.

Q: How do you determine the solubility of a sparingly soluble salt at a given temperature?

A: Dissolve the salt till you can visibly see the undissolved salt left in the solution. Filter the mixture. Collect the filtrate and then pipette a known volume into an evaporating dish of a known mass. Evaporate to a constant mass. Then weigh it again to calculate the mass of the residue. This will give us the solubility (in g cm^{-3}) of the salt in the saturated solution.

Example 7.15: Calculate the K_{sp} of Ag_2CrO_4 given that its solubility is 6.50×10^{-5} mol dm^{-3} at 20°C.

Solution: A solubility of 6.50×10^{-5} mol dm^{-3} indicates that a maximum of 6.50×10^{-5} mol of Ag_2CrO_4 is dissolved in 1 dm^3 of solution:

	$Ag_2CrO_4(s)$ ⇌	$2Ag^+(aq)$	+ CrO_4^{2-} (aq).
Equilibrium conc. ($mol\,dm^{-3}$)		$2 \times (6.50 \times 10^{-5}) = 1.3 \times 10^{-4}$	6.50×10^{-5}

$$\begin{aligned} K_{sp} \text{ of } Ag_2CrO_4 &= [Ag^+]^2[CrO_4{}^{2-}] \\ &= (1.3 \times 10^{-4})^2 \times (6.50 \times 10^{-5}) \\ &= 1.10 \times 10^{-12}\,mol^3\,dm^{-9}. \end{aligned}$$

Likewise, if the K_{sp} value of a salt at a given temperature is known, we can use it to calculate the solubility of the salt at that temperature and the concentration of the ions.

Q: Since $\Delta G^{\ominus} = -RT \ln K_{sp}$, does a small K_{sp} value ($\ll 1$) means that the $\Delta G^{\ominus}$ is positive and hence the reaction is thermodynamically non-spontaneous?

A: You are right here! This also explains why the solubility of the salt is very small as the K_{sp} value is very low. To reiterate, an equilibrium constant that is greater than one would mean a negative $\Delta G^{\ominus}$. Vice versa for an equilibrium constant that is less than one.

Example 7.16: Calculate the K_{sp} of $CaSO_4$ given that its solubility is $0.39\,g\,dm^{-3}$ at 17°C.

Solution: Firstly, convert the units of solubility to $mol\,dm^{-3}$. Solubility of $CaSO_4 = 0.39 \div M_{CaSO_4} = 0.39 \div 136.2 = 2.86 \times 10^{-3}\,mol\,dm^{-3}$.

	$CaSO_4(s)$ ⇌	$Ca^{2+}(aq)$	+ $SO_4^{2-}(aq)$.
Equilibrium conc. ($mol\,dm^{-3}$)		2.86×10^{-3}	2.86×10^{-3}

$$\begin{aligned} K_{sp} \text{ of } CaSO_4 &= [Ca^{2+}][SO_4{}^{2-}] \\ &= (2.86 \times 10^{-3})^2 \\ &= 8.18 \times 10^{-6}\,mol^2\,dm^{-6}. \end{aligned}$$

Example 7.17: The solubility product of lead iodide is $4.2 \times 10^{-9}\,\text{mol}^3\,\text{dm}^{-9}$ at 18°C. Determine its solubility at the same temperature.

Solution: Let the solubility of PbI_2 be $w\,\text{mol}\,\text{dm}^{-3}$:

$$PbI_2(s) \rightleftharpoons Pb^{2+}(aq) + 2I^-(aq).$$

In a saturated solution of PbI_2,
$[Pb^{2+}] = w\,\text{mol}\,\text{dm}^{-3}$,
$[I^-] = 2w\,\text{mol}\,\text{dm}^{-3}$.

$$K_{sp} \text{ of } PbI_2 = [Pb^{2+}][I^-]^2 = w \times (2w)^2 = 4.2 \times 10^{-9}\,\text{mol}^3\,\text{dm}^{-9}$$

$$4w^3 = 4.2 \times 10^{-9}$$

$$w = 1.02 \times 10^{-3}.$$

Thus, the solubility of PbI_2 is $1.02 \times 10^{-3}\,\text{mol}\,\text{dm}^{-3}$ at 18°C.

Exercise:

(1) Calculate the K_{sp} of MgF_2 given that its solubility is $0.12\,\text{g}\,\text{dm}^{-3}$ at 17°C. (Answer: $2.86 \times 10^{-8}\,\text{mol}^3\,\text{dm}^{-9}$.)
(2) K_{sp} of CaF_2 is $3.9 \times 10^{-11}\,\text{mol}^3\,\text{dm}^{-9}$ at 25°C. Determine its solubility at this temperature. (Answer: $2.14 \times 10^{-4}\,\text{mol}\,\text{dm}^{-3}$.)
(3) K_{sp} of $BaCO_3$ is $5.1 \times 10^{-9}\,\text{mol}^2\,\text{dm}^{-6}$ at 25°C. Determine its solubility at this temperature. (Answer: $7.14 \times 10^{-5}\,\text{mol}\,\text{dm}^{-3}$.)

7.14.2 *Ionic product and K_{sp}*

In Chap. 6 on "Chemical Equilibria", we have seen how the value of the reaction quotient Q_c, when compared to the value of K_c, provides us with useful information as to where the position of equilibrium shifts.

When we want to know when a sparingly soluble salt has reached its maximum solubility, i.e., when a saturated solution is formed, we can make use of a term similar to reaction quotient Q_c. In this case, we name the term the "ionic product", which has the same expression as that for K_{sp}. The concentration terms in the ionic product expression correspond to the instantaneous concentration of each of the ions in the solution at any specific point of time. By contrast, the concentration terms in the K_{sp} expression correspond to the instantaneous concentration of each of the ions in the solution at equilibrium!

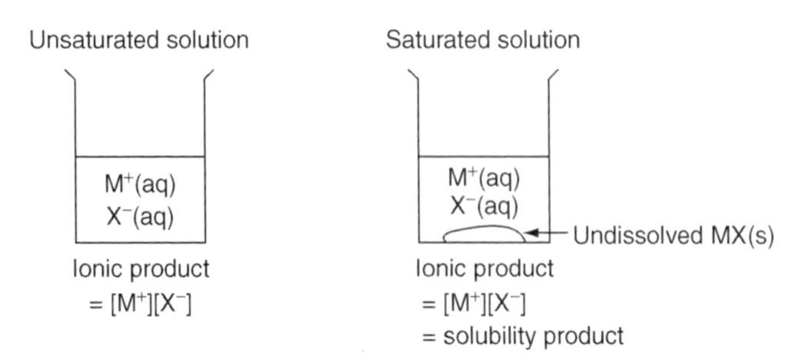

As more salt is added in the process of dissolution, the concentration of the ions increases, leading to an increase in the value of the ionic product. Eventually, the value of the ionic product will be equivalent to that of K_{sp} when a saturated solution is attained.

Thus, with known data of K_{sp} and the ionic product, we can predict whether precipitation can be observed prior to mixing known quantities of two solutions used to form a sparingly soluble salt.

Example 7.18: Will precipitation be observed when 30 cm^3 of 0.05 mol dm^{-3} $AgNO_3$ solution is added to 40 cm^3 of 0.10 mol dm^{-3} NaCl solution? Take K_{sp} of AgCl to be 2.0×10^{-10} mol^2 dm^{-6} at 25°C.

Approach: The precipitate in question is AgCl(s). To determine if precipitation will occur, we need to compare the value of the ionic product against the value of K_{sp}.

In order to do so, we first need to find the concentration of the solvated Ag^+ and Cl^- ions in the mixture.

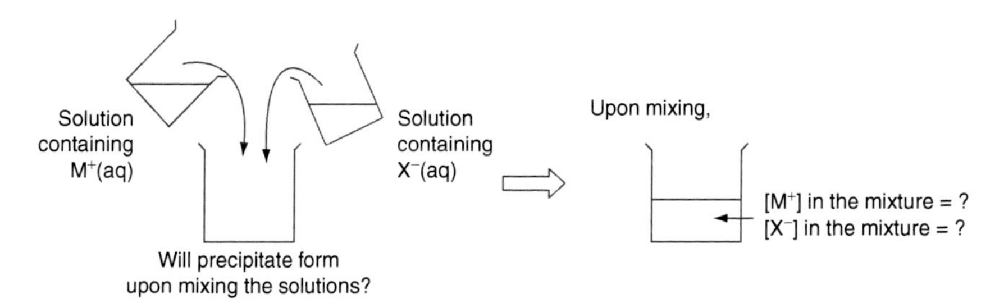

Solution: When the two solutions are mixed,

The total volume of the resultant solution $= 30 + 40 = 70\,\text{cm}^3$.

$$\text{Amount of Ag}^+ \text{ dissolved} = \frac{30}{1000} \times 0.05 = 1.5 \times 10^{-3}\,\text{mol}.$$

$$[\text{Ag}^+] \text{ in } 70\,\text{cm}^3 \text{ resultant solution} = \frac{1.5 \times 10^{-3}}{70} \times 1000$$
$$= 2.14 \times 10^{-2}\,\text{mol dm}^{-3}.$$

Likewise,

$$[\text{Cl}^-] \text{ in } 70\,\text{cm}^3 \text{ resultant solution} = \frac{40}{70} \times 0.10 = 5.71 \times 10^{-2}\,\text{mol dm}^{-3}.$$

Therefore,

$$\text{Ionic product for AgCl} = [\text{Ag}^+(\text{aq})][\text{Cl}^-(\text{aq})]$$
$$= (2.14 \times 10^{-2})(5.71 \times 10^{-2})$$
$$= 1.22 \times 10^{-3}\,\text{mol}^2\,\text{dm}^{-6}.$$

Since the value of the ionic product is greater than the K_{sp} value of 2.0×10^{-10}, precipitation of AgCl will be observed.

In summary, for a solution that contains a sparingly soluble salt:

- When the ionic product $< K_{sp}$, no precipitation is observed since the solution is still unsaturated.
- When the ionic product $= K_{sp}$, no precipitation is observed as the solution just became saturated.
- When the ionic product $> K_{sp}$, precipitation forms as no more salt can dissolve in the saturated solution.

Q: Why does precipitation occur when ionic product $> K_{sp}$?

A: At the point when ionic product > K_{sp}, the rate of precipitation is greater than the rate of dissolution. So the net outcome is that you have precipitate being formed continuously. When ionic product = K_{sp} is reached, the rate of precipitation equals the rate of dissolution, and thus there is no further precipitation.

Exercise: Will precipitation occur when 30 cm^3 of 0.02 mol dm^{-3} $MgSO_4(aq)$ is mixed with 50 cm^3 of 0.10 mol dm^{-3} NaOH(aq)? Take K_{sp} of $Mg(OH)_2$ to be 1.8×10^{-11} mol^3 dm^{-9}.

(Answer: Precipitation will occur. Ionic product = 2.93×10^{-5} mol^3 dm^{-9}.)

7.14.3 *Common ion effect*

We have seen in Sec. 7.11, on buffer solutions, how the dissociation of a strong electrolyte such as $CH_3COO^-Na^+$ further suppresses the partial dissociation of its weak acid CH_3COOH:

$$CH_3COO^-Na^+(aq) \longrightarrow Na^+(aq) + CH_3COO^-(aq),$$

$$CH_3COOH(aq) \rightleftharpoons H^+(aq) + CH_3COO^-(aq). \quad (7.16)$$

The introduction of $CH_3COO^-Na^+$ contributes to an increase in $[CH_3COO^-]$. This causes the equilibrium position in Eq. (7.16) to shift to the left in a bid to remove some of the CH_3COO^- ions. The backward reaction is favoured and as a result, the dissociation of CH_3COOH is suppressed. This phenomenon is known as the **common ion effect** and CH_3COO^- is termed the common ion for the dissociation of CH_3COOH.

The common ion effect is a term that describes the reduced solubility of a solute in a solution that already contains the same ion (which comes from another dissolved solute).

Did you know?
The common ion effect also accounts for the suppressed dissociation of weak acid when it is in the same solution as a strong acid. Can you guess what the common ion is in this case?

Example 7.19: AgI has a solubility of 9.11×10^{-9} mol dm^{-3} in water. However, its solubility decreases when the same amount of AgI is added to 1 dm^3 of $AgNO_3$(aq). Explain.

Solution:

$$\text{AgI(s)} \rightleftharpoons \text{Ag}^+\text{(aq)} + \text{I}^-\text{(aq)}. \tag{7.17}$$

The $AgNO_3$ solution contains the common ion Ag^+. When AgI is added to this $AgNO_3$ solution, the presence of a greater $[Ag^+]$ causes the equilibrium position in Eq. (7.17) to shift to the left in a bid to remove some of the Ag^+ ions. Consequently, the dissociation of AgI is further suppressed. The precipitation of AgI observed indicates that its solubility is reduced in $AgNO_3$ solution compared to its solubility in water.

Example 7.20: Calculate the solubility of AgI in 0.15 mol dm^{-3} NaI solution given K_{sp} of AgI is 8.3×10^{-17} mol^2 dm^{-6} at the same temperature.

Approach: To calculate the solubility of a salt means that you have to calculate the maximum concentration of salt that can be dissolved in the given solution. The value of K_{sp} can provide us with the maximum concentration of the solvated ions.

Solution: Let the solubility of AgI in the NaI solution be y mol dm^{-3}:

$$\text{AgI(s)} \rightleftharpoons \text{Ag}^+\text{(aq)} + \text{I}^-\text{(aq)},$$

$$\text{NaI(aq)} \longrightarrow \text{Na}^+\text{(aq)} + \text{I}^-\text{(aq)}.$$

In the saturated solution,

$$[\text{Ag}^+] = y\,\text{mol dm}^{-3}$$

$$[\text{I}^-] = (y + 0.15)\,\text{mol dm}^{-3} \text{ (recall that there are two sources of I}^-\text{)}.$$

Thus,

$$K_{sp} \text{ of AgI} = [\text{Ag}^+][\text{I}^-] = y(y+0.15) = 8.3\times10^{-17}\,\text{mol}^2\,\text{dm}^{-6}.$$

Since the supply of I^- from the complete dissociation of NaI exerts a common ion effect and suppresses the dissociation of AgI, it can be assumed that $y \ll 0.15$ such that $(y + 0.15) \approx 0.15$.

Therefore,

$$0.15y = 8.3 \times 10^{-17}$$
$$y = 5.53 \times 10^{-16}.$$

Thus, the solubility of AgI in $0.15\,\text{mol}\,\text{dm}^{-3}$ NaI solution is $5.53 \times 10^{-16}\,\text{mol}\,\text{dm}^{-3}$.

Exercise: Calculate the solubility of AgI in $0.15\,\text{mol}\,\text{dm}^{-3}$ MgI_2 solution, given K_{sp} of AgI is $8.3 \times 10^{-17}\,\text{mol}^2\,\text{dm}^{-6}$.

(Answer: $2.77 \times 10^{-16}\,\text{mol}\,\text{dm}^{-3}$.)

Hint: What is $[I^-]$ obtained from the complete dissociation of MgI_2?

7.14.4 *Solubility in qualitative analysis*

The solubility of a sparingly soluble salt may be enhanced with the addition of a reagent that consumes either the cation or the anion of the salt. When the cation or anion is being removed from the equilibria, according to Le Chatelier's Principle, more of the salt has to dissolve to make up for the loss of the cation or anion:

$$A_nB_m(s) \rightleftharpoons nA^{m+}(aq) + mB^{n-}(aq).$$

From the kinetics perspective, the removal of the cation or anion simply decreases the backward rate of reaction, i.e., the rate of precipitation, whereas the forward rate is still the same. Accordingly, this results in dissolution of the sparingly soluble salt.

Examples of solubility being enhanced by the removal of cations:

- The solubility of AgCl with the addition of aqueous ammonia is due to the removal of the Ag^+ ion through the formation of the $[Ag(NH_3)_2]^+$ complex.
- The solubility of $Cu(OH)_2$ with the addition of aqueous ammonia is due to the removal of the Cu^{2+} ion through the formation of the $[Cu(NH_3)_4]^{2+}$ complex.
- The solubility of $Zn(OH)_2$ with the addition of excess NaOH(aq) is due to the removal of the Zn^{2+} ion through the formation of the $[Zn(OH)_4]^{2-}$ complex.

- The solubility of $PbCl_2$ with the addition of concentrated HCl is due to the removal of the Pb^{2+} ion through the formation of the $[PbCl_4]^{2-}$ complex.

Examples of solubility being enhanced by the removal of anions:

- The solubility of $BaCrO_4$ with the addition of dilute HNO_3 is due to the removal of the $CrO_4{}^{2-}$ ion through the formation of $Cr_2O_7{}^{2-}$.
- The solubility of PbI_2 with the addition of H_2O_2 is due to the removal of the I^- ion through the formation of I_2.
- The solubility of $Ag_2C_2O_4$ with the addition of dilute HNO_3 is due to the removal of the $C_2O_4{}^{2-}$ ion through the formation of $H_2C_2O_4$.
- The solubility of $Mg(OH)_2$ with the addition of solid NH_4Cl is due to the removal of the OH^- ion through the reaction of OH^- with the acidic $NH_4{}^+$.
- The solubility of $Pb_3(PO_4)_2$ with the addition of dilute HNO_3 is due to the removal of the $PO_4{}^{3-}$ ion through the formation of $HPO_4{}^{2-}$.

Did you know?

FeS ppt only forms with H_2S in an alkaline medium but not in an acidic medium. Why?

FeS has a relatively large K_{sp} value. H_2S is a weak acid and in an acidic medium, the dissociation of H_2S to form the S^{2-} ion is further suppressed. Thus, the ionic product of FeS is less than its K_{sp} value, and precipitation does not occur. But in an alkaline medium, a sufficient amount of S^{2-} is generated for the ionic product to be greater than the K_{sp} value, and precipitation can thus occur. There are other compounds, such as SnS and CuS, which can form in an acidic medium of H_2S because of their low K_{sp} values.

My Tutorial (Chapter 7)

1. It is now generally accepted that the acidic gas sulfur dioxide is one of the principal causes of acid rain. Atmospheric sulfur dioxide dissolves in water droplets in the air to form an acidic solution of low pH, which damages vegetation and contaminates water bodies when it falls as rain.

(a) Define

(i) acid; and
(ii) pH.

(b) Sulfur dioxide reacts with water to produce sulfuric(IV) acid (H_2SO_3), a weak acid that partially ionises according to:

$$H_2SO_3(aq) + H_2O(l) \rightleftharpoons H_3O^+(aq) + HSO_3^-(aq).$$

(i) Write an expression in terms of concentrations for the acidity constant (K_a) for sulfuric(IV) acid.
(ii) The value of K_a for this reaction at 25°C is 1.5×10^{-2} mol dm^{-3}. Calculate a value for the concentration of dissolved hydrogen ions in a 0.10 mol dm^{-3} solution of sulfuric(IV) acid at 25°C, stating clearly any assumptions make.
(iii) With reference to your answer in part (b)(ii), calculate the pH of a 0.10 mol dm^{-3} solution of sulfuric(IV) acid at 25°C.

2. Acids can be differentiated by the number of hydrogen ions that can be liberated from one molecule of the undissociated acid. Hydrochloric acid is a strong monobasic, or monoprotic, acid, liberating one hydrogen ion per molecule. Sulfuric(VI) acid is a dibasic, or diprotic, acid, with its ionisation in aqueous solution as follows:

$$H_2SO_4(aq) + H_2O(l) \rightleftharpoons H_3O^+(aq) + HSO_4^-(aq),$$
$$K_a = \text{very large}; \quad \text{(I)}$$
$$HSO_4^-(aq) + H_2O(l) \rightleftharpoons H_3O^+(aq) + SO_4^{2-}(aq),$$
$$K_a = 0.01\,\text{mol}\,\text{dm}^{-3}. \quad \text{(II)}$$

(K_a values are quoted for 25°C.)

(a) (i) State the hydrogen ion concentration in 0.02 mol dm^{-3} hydrochloric acid, which is a strong acid, and hence find the pH of this solution.

(ii) State the hydrogen ion concentration of 0.1 mol dm^{-3} sulfuric acid arising from the first stage of ionization, (I).

(iii) A solution of sodium hydrogensulfate, $NaHSO_4$, of concentration 0.1 mol dm^{-3}, which ionises according to Eq. (II) above, has a pH of 1.57. Find the hydrogen ion concentration in this solution, and hence state what you would expect the hydrogen ion concentration in 0.1 mol dm^{-3} sulfuric acid to be.

(iv) In fact, the pH of 0.1 mol dm^{-3} sulfuric(VI) acid is about 0.98. This indicates a hydrogen ion concentration of 0.105 mol dm^{-3}. Considering that reactions (I) and (II) coexist simultaneously, explain the difference in the hydrogen ion concentrations.

(v) If K_a for ionisation (II) has a value of 0.02 mol dm^{-3} at 80°C, state with reasons whether the ionisation is endothermic or exothermic.

(vi) Explain the effect of such an increase in temperature on the pH of this solution.

(b) Pure sulfuric(VI) acid boils at 338°C and it mixes with water in all proportions. Hydrogen chloride boils at –85°C and is extremely soluble in water.

(i) Suggest reasons, in terms of the structure and bonding of H_2SO_4 and HCl, for the large difference between their boiling points.

(ii) Suggest why both compounds are so soluble in water.

(iii) The K_{sp} of magnesium sulfate is smaller than that of barium sulfate; use the following data to suggest why the solubility differs considerably: $\Delta H_{\text{hydration}}$ of $Mg^{2+} = -1920$ kJ mol^{-1}, $Ba^{2+} = -1360$ kJ mol^{-1}.

3. A carboxylic acid W contains 40.0% carbon, 6.70% hydrogen and 53.3% oxygen by mass. When 10.0 cm^3 of an aqueous solution of W, with concentration 4.65 g dm^{-3}, is titrated against 0.050 mol dm^{-3} sodium hydroxide, the following pH readings are obtained:

Volume of NaOH (cm^3)	**0.0**	**2.5**	**5.0**	**7.5**	**10.0**	**14.0**	**15.0**	**16.0**	**17.5**	**20.0**	**22.5**
pH	2.5	3.2	3.5	3.8	4.1	4.7	5.2	9.1	11.5	11.8	12.0

(a) Plot a graph of pH against volume of NaOH used in the titration. Use the graph to determine the end-point of the titration. Hence calculate the relative molecular mass of W.
(b) Calculate the value of K_a for W and state its units.
(c) Calculate the molecular formula of W.

4. (a) When ethanoic acid is dissolved in water, the following equilibrium is established:

$$CH_3CO_2H + H_2O \rightleftharpoons CH_3CO_2{}^- + H_3O^+. \quad \text{(I)}$$

When hydrogen chloride dissolves in ethanoic acid, the equilibrium established is:

$$CH_3CO_2H + HCl \rightleftharpoons CH_3CO_2H_2{}^+ + Cl^-. \quad \text{(II)}$$

What is the role of the ethanoic acid in:

(i) equilibrium (I), and
(ii) equilibrium (II).

(b) What is the relationship between the species $CH_3CO_2H_2{}^+$ and CH_3CO_2H?
(c) The value of K_a for ethanoic acid at 25°C is 1.80×10^{-5} mol dm^{-3} and for methanoic acid, HCO_2H it is 1.56×10^{-4} mol dm^{-3}.

(i) Give the K_a expression for CH_3CO_2H.
(ii) Hence calculate the pH of a 0.100 mol dm^{-3} solution of CH_3CO_2H at 25°C.

(d) The pH of a 0.050 mol dm^{-3} solution of methanoic acid is 2.55. Using this, together with the data in (c) and your answer to (c)(ii):

(i) state which of the two acids is the stronger acid; and
(ii) comment on the relative pH values of the two acids.

(e) (i) Sketch a graph showing how the pH changes during the titration of 20.0 cm^3 of a 0.100 mol dm^{-3} solution of methanoic acid with 0.050 mol dm^{-3} sodium hydroxide solution.

(ii) Select from below a suitable indicator for this titration. Give a brief reason for your choice based on the curve drawn in (e)(i).

Indicator	pH Range
Bromocresol green	3.5–5.4
Bromothymol blue	6.0–7.6
Phenol red	6.8–8.4

(iii) There is no suitable indicator for the titration of methanoic acid with ammonia. Why is this?

5. A buffer solution of pH = 3.87 contains 7.40 g dm^{-3} of propanoic acid ($CH_3CH_2CO_2H$) together with a quantity of sodium propanoate ($CH_3CH_2CO_2Na$). K_a for propanoic acid = 1.35×10^{-5} mol dm^{-3} at 298 K.

   (a) Explain what a buffer solution is and how this particular solution achieves its buffer function.
   (b) Calculate the concentration (in g dm^{-3}) of sodium propanoate in the solution, stating any assumptions made.
   (c) If the sodium propanoate were to be replaced by anhydrous magnesium propanoate, calculate the concentration of magnesium propanoate (in g dm^{-3}), required to give a buffer of the same pH.

6. Chromium(III) hydroxide is a compound commonly found on the surface of chromed metal to act as a protective layer. The layer of $Cr(OH)_3$ is usually formed through anodic oxidation in an alkaline solution. The solubility of $Cr(OH)_3$ in pure water is 1.3×10^{-8} mol dm^{-3}.

   (a) Determine the K_{sp} of $Cr(OH)_3$.
   (b) Sketch a diagram to show the changes in the concentration of the ions in pure water as $Cr(OH)_3$ dissolves.
   (c) Determine the solubility of $Cr(OH)_3$ in 0.10 mol dm^{-3} $CrCl_3$.

(d) Indicate the changes in solubility of $Cr(OH)_3$ when 0.10 mol of solid $CrCl_3$ is added to 1 dm^3 of a saturated solution of $Cr(OH)_3$ on the same diagram in part (b).

(e) When $Cr(OH)_3$ dissolves in water, it is found that $[Cr^{3+}]$:$[OH^-]$ is actually greater than 1:3. Give a reasonable explanation for this observation.

(f) When a solution of H_2O_2 is added to a saturated solution containing the undissolved $Cr(OH)_3$, a yellow solution containing $CrO_4{}^{2-}$ is formed. All the insoluble $Cr(OH)_3$ becomes soluble. With the aid of appropriate equations, explain this observation.

(g) $Al(OH)_3$ is another hydroxide that is insoluble in water. Briefly describe how you would differentiate which compound, $Al(OH)_3$ or $Cr(OH)_3$, is more soluble in water.

(h) The ΔG° of the dissolution of $Cr(OH)_3$ can be determined from the following equation:

$$\Delta G^{\circ} = -RT \ln K_{sp}.$$

With reference to ΔG°, explain why $Cr(OH)_3$ is sparingly soluble in water at 25°C .

(i) The solubility of $Cr(OH)_3$ increases when temperature is increased. With reference to ΔG°, explain why the solubility of $Cr(OH)_3$ changes with temperature.

CHAPTER 8

Redox Chemistry and Electrochemical Cells

Recalling the Brønsted–Lowry theory of acids and bases, an acid–base reaction involves the acid transferring a proton to the base. The acid is a proton donor and the base is a proton acceptor.

A redox reaction is quite similar to an acid–base reaction in the sense that it involves the transfer of particles from one substance to another. Specifically, a redox reaction involves electron transfer between a pair of substances. In the process of electron transfer, one species will lose electron(s) to the other, which will gain the electron(s).

- The species that loses electron(s) is said to be oxidised, i.e., it undergoes oxidation.
- The species that accepts the electron(s) is said to be reduced, i.e., it undergoes reduction.

Take, for instance, the redox reaction represented by Eq. (8.1):

$$\mathrm{Zn(s) + CuSO_4(aq) \longrightarrow ZnSO_4(aq) + Cu(s).} \qquad (8.1)$$

When the reaction is carried out by placing a Zn rod into a beaker containing $CuSO_4(aq)$ solution (see Fig. 8.1), the reaction is observed to proceed spontaneously and heat is evolved. After some time, the Zn rod is slowly eroded and reddish brown deposits of Cu(s) are found.

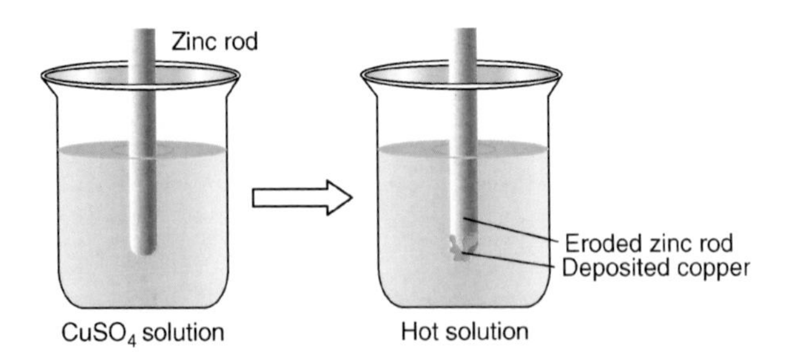

Fig. 8.1.

Based on these observations, it can be concluded that:

- Cu^{2+} is **reduced** to Cu. **Reduction** involves a **gain of electrons**, which is indicated by a **decrease in oxidation number**. Before the reaction, copper has an oxidation state of +2. After the reaction, copper has an oxidation state of zero.

  Since Cu^{2+} gains electrons from Zn, causing it to be oxidised to Zn^{2+}, we say that Cu^{2+} acts as an **oxidising agent (or oxidant)**. Equation (8.2) represents the reduction half-reaction for the overall redox reaction between Cu^{2+} and Zn.

  $$\textbf{Reduction half-equation}: \mathrm{Cu}^{2+} + 2\mathrm{e}^- \longrightarrow \mathrm{Cu}. \tag{8.2}$$

- Zn is **oxidised** to Zn^{2+}. **Oxidation** involves a **loss of electrons**. Before the reaction, zinc has an oxidation state of zero because it is a pure element. After the reaction, zinc has an oxidation state of +2. The **increase in oxidation number** indicates a loss of electrons (it cannot be due to an increase in the number of protons as protons sit "firmly" in the nucleus during a chemical reaction) and thus implies oxidation has occurred. Since Zn loses electrons to Cu^{2+}, helping it to be reduced to Cu, we say that Zn acts as a **reducing agent (or reductant)**. Equation (8.3) represents the oxidation half-reaction for the overall redox reaction between Cu^{2+} and Zn.

  $$\textbf{Oxidation half-equation}: \mathrm{Zn} \longrightarrow \mathrm{Zn}^{2+} + 2\mathrm{e}^-. \tag{8.3}$$

Q: What is meant by oxidation state?

A: An oxidation state indicates the ability of the atom in a species (which can be an atom, ion, or molecule) to undergo oxidation.

A positive oxidation state indicates oxidation whereas a negative oxidation state indicates reduction. The magnitude of this oxidation state is represented by what we term oxidation number, which again is either negative or positive.

Q: So, if an atom has an oxidation number of +1, does that mean that it has lost an electron?

A: No, not necessarily. The oxidation number of Na^+ is +1 and indeed the Na atom has lost an electron to become Na^+. As for the covalent molecule HCl, the H atom also has an oxidation number of +1 yet it has *not* lost any electrons.

Thus, remember that oxidation number indicates the ability of the atom to lose electrons but it does not necessarily mean that the atom *must* lose electrons. Thus, **oxidation number may not be equivalent to the formal charge** that an atom carries.

Q: So, how does one determine the oxidation number of an atom in a species?

A: If it is a cation or anion, the oxidation number is equivalent to the formal charge.

If the atom is covalently bonded to other atoms, one needs to analyse each of the covalent bonds, in turn, and determine which of the two bonding atoms has higher electronegativity. The one that is more electronegative polarises the shared electron cloud more towards itself, and hence this atom has a negative oxidation number. This means that it has the potential to undergo reduction as compared to the other atom that is covalently bonded to it. The oxidation number is determined by the number of electrons that is being contributed by this atom in the sharing process. More on this is discussed in Sec. 8.1.

Q: What if this atom is bonded to two other different atoms, like the C atom in H–C≡N?

A: The oxidation number of the C atom is −1 with respect to the H atom (as the shared electron cloud is more polarised towards C) and +3 with respect to the N atom. Thus, the overall oxidation number of the C atom in HCN is +2.

Q: Why are oxidation and reduction processes known as half-reactions? Why are Eqs. (8.2) and (8.3) known as half-equations?

A: The oxidation of one substance cannot occur without the reduction of the other, and *vice versa*. A redox reaction involves the simultaneous occurrence of both processes (just as its name implies – "red-" stands for reduction and "-ox" stands for oxidation). Adding up both half-equations will give us the overall balanced equation for the redox reaction. In fact, writing half-equations is one useful approach to form an overall balanced equation (see Section 8.2).

Q: What is the role of the SO_4^{2-} ions?

A: Since the oxidation state of the sulfate ion remains the same after the reaction, we say that it is neither reduced nor oxidised and it is just a spectator ion, an ion that does not take part in the reaction of concern. Spectator ions are not written in half-equations. The purpose of the sulfate ion is just to maintain electrical neutrality. You cannot have a solution that contains ONLY Cu^{2+} ions, right?

Q: Is the reaction between hydrogen and oxygen a redox reaction even though there is no complete transfer of electrons from one species to the other?

$$2H_2(g) + O_2(g) \longrightarrow 2H_2O(l).$$

A: Yes, it is a redox reaction that involves a re-sharing of electrons as the reaction proceeds. **A redox reaction need not necessarily only involve a complete transfer of electrons** as in the formation of ionic compounds. To identify a redox reaction, you first have to deduce the oxidation state of each atom that appears in the equation. Next, check for atoms that have changes in their oxidation state before and after the reaction. There must be one atom whose oxidation state has increased after the reaction and one other whose oxidation state has decreased after the reaction.

After assigning oxidation states to all atoms, we can now confirm that it is a redox reaction. There is an increase in oxidation state for hydrogen (i.e., it is oxidised) while a decrease is observed for oxygen, indicating it has undergone reduction:

$$\underset{0}{2H_2}(g) + \underset{0}{O_2}(g) \longrightarrow 2\,\underset{+1-2}{H_2O}(l).$$

Example 8.1: Which of these equations represent redox reactions?

(1) $Al^{3+}(aq) + 3OH^-(aq) \longrightarrow Al(OH)_3(s)$,
(2) $NH_3(aq) + HCl(aq) \longrightarrow NH_4Cl(aq)$,
(3) $Cu_2O(s) + H_2SO_4(aq) \longrightarrow Cu(s) + CuSO_4(aq) + H_2O(l)$.

Solution: Refer to Sec. 8.1 on how to assign oxidation states.

(1) $\underset{+3}{Al^{3+}}(aq) + \underset{-2\ +1}{3OH^-}(aq) \longrightarrow \underset{+3\ -2\ +1}{Al(OH)_3}(s)$.
Since all species have the same oxidation state before and after the reaction, this reaction is not a redox reaction. It is actually a precipitation reaction.

(2) $\underset{-3\ +1}{NH_3}(aq) + \underset{+1\ -1}{HCl}(aq) \longrightarrow \underset{-3\ +1\ -1}{NH_4Cl}(aq)$.
There is no change in oxidation state for all species. This is an acid–base reaction and not a redox reaction.

(3) $\underset{+1\ -2}{Cu_2O}(s) + \underset{+1\ +6\ -2}{H_2SO_4}(aq) \longrightarrow \underset{0}{Cu}(s) + \underset{+2\ +6\ -2}{CuSO_4}(aq) + \underset{+1\ -2}{H_2O}(l)$.
The oxidation state of copper changes from +1, before the reaction, to 0 and +2 after the reaction. This is a redox reaction. More specifically, if a species, like copper in this case, is simultaneously oxidised and reduced, it is said to have undergone disproportionation.

A **disproportionation** reaction is a redox reaction in which a single substance is simultaneously oxidised and reduced. With this, we are saying that one particle undergoes oxidation while the other particle of the very same substance undergoes reduction. Take note that the same particle cannot undergo oxidation and reduction simultaneously.

Another example is the decomposition of H_2O_2 wherein it is simultaneously oxidised and reduced:

$$\underset{+1\ -1}{2H_2O_2}(aq) \longrightarrow 2\,\underset{+1\ -2}{H_2O}\,(l) + \underset{0}{O_2}(g).$$

We have seen how oxidation states are useful in understanding what goes on in a redox reaction. In the next sections, we will learn how to assign oxidation states and construct redox equations.

8.1 Rules for Assignment of Oxidation States

(1) Oxidation states must be written accompanied by either a "+" or a "−" sign:
- A "+" sign implies the atom has lost or might lose its hold on electrons.
- A "−" sign implies the atom has gained or might gain a hold on electrons.

(2) The oxidation state of atoms in the elemental form is assigned zero, e.g., Ca and the F atoms in F_2 have oxidation state of 0.
(3) In all compounds, fluorine has the oxidation state of −1.
(4) In all compounds, hydrogen has the oxidation state of +1. An exception is in the case of metal hydrides such as NaH, where the oxidation state of H is −1.
(5) In all compounds, oxygen has the oxidation state of −2. An exception is in the case of peroxides such as H_2O_2, where the oxidation state of O is −1 and in OF_2, where it is +2.
(6) For a pair of atoms bonded by a single covalent bond, the more electronegative atom is assigned an oxidation state of −1 and the less electronegative atom has an oxidation state of +1.

> Recall that the electronegativity of elements
> - increases across the period, and
> - decreases down the group.

(7) The sum of the oxidation states of all atoms in a neutral compound (e.g., KCl, CO_2) is zero.
(8) For a monoatomic ion, its oxidation state corresponds to the net charge on the ion.

(9) For a polyatomic ion, the sum of the oxidation states of all atoms corresponds to the net charge on the ion.

Example 8.2: Determine the oxidation state of Cr in the $Cr_2O_7^{2-}$ ion.

Solution: In general, oxidation states can be calculated by applying the rules above.

Let the oxidation state of Cr be x. The oxidation state of each O atom is -2. The net charge of this polyatomic ion is -2. Thus, we have

$$2x + 7(-2) = -2$$
$$x = +6.$$

Hence, the oxidation state of Cr is $+6$.

Exercise 8.1: Determine the oxidation state of the underlined atom in each of the following compounds:

(1) $\underline{Mn}O_4^-$, (2) $\underline{N}H_3$, (3) $\underline{Pb}O_2$, (4) $H\underline{N}O_3$, (5) $Na\underline{Cl}O_3$, and (6) $\underline{S}O_3^{2-}$.

[Answer: (1) +7, (2) −3, (3) +4, (4) +5, (5) +5, and (6) +4.]

Exercise 8.2: For each of the following unbalanced equations, identify whether the underlined species acts as a reducing or oxidising agent:

(1) $\underline{AsH_3} + KClO_3 \longrightarrow H_3AsO_4 + KCl$,
(2) $H_2SO_4 + \underline{HBr} \longrightarrow SO_2 + Br_2 + H_2O$, and
(3) $\underline{Cr_2O_3} + Na_2CO_3 + KNO_3 \longrightarrow Na_2CrO_4 + CO_2 + KNO_2$.

[Answer: (1) reducing agent, (2) reducing agent, and (3) reducing agent.]

Q: Based on the rules, the calculated oxidation state of S in the peroxodisulfate ion $S_2O_8^{2-}$ is +7. Sulfur, being in Group 6, has only six valence electrons. How is it possible for S to have an oxidation state of +7?

A: Remember Rule (6)? We shall examine the structure of the ion and apply this golden rule.

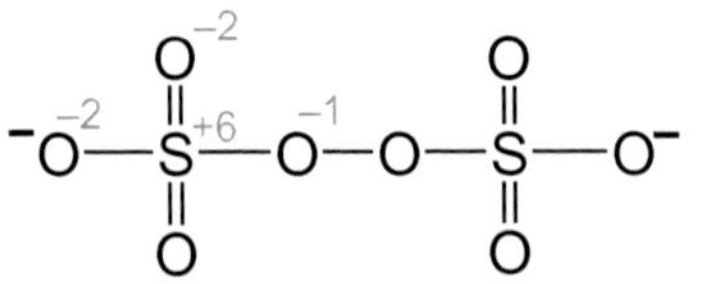

Each S atom is covalently bonded to four O atoms. For each of these bonds, the shared electrons are not equally distributed between the bonding atoms S and O. According to Rule (6), the more electronegative O atom, is assigned an oxidation state of −1. The less electronegative S atom is assigned an oxidation state of +1.

What do these numbers reflect? They actually reflect the unequal distribution of the bonding electrons between the two bonding atoms. O, being more electronegative than S, has a stronger hold on the bonding electrons. Thus the "−1" oxidation state indicates O is relatively more "electron-rich" compared to S, and the latter's oxidation state of "+1" implies it is "electron poor." Thus, with each S acquiring a "+1" for every electron pair that is shared with O, we have a total of "+6" as the oxidation state for S. Each of the peripheral O atoms carry an oxidation state of −2 while the two bridging O atoms each carry an oxidation state of −1. At this point, we have learnt that the maximum oxidation number that an atom could acquire corresponds to the number of valence electrons it has.

Below are more examples on the assignment of oxidation states of S atoms using the structure of the ion:

thiosulfate ion, $S_2O_3^{2-}$

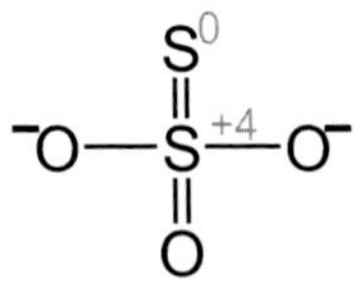

tetrathionate ion, $S_4O_6^{2-}$

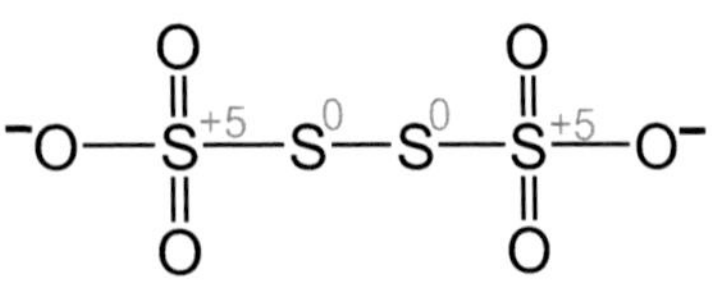

We have seen how the oxidation state of an atom in a compound can be obtained from one of two general methods. One method is to apply the set of rules which gives us the average oxidation states, and the other method requires the use of the molecular structure which depicts actual oxidation states.

Just as how oxidation states help us in the identification of redox reactions, we can also use quantitative data of redox reactions to determine oxidation states.

Example 8.3: A solution containing $40\,cm^3$ of $0.5\,mol\,dm^{-3}$ $VO_2^+(aq)$ was found to require $25\,cm^3$ of $0.8\,mol\,dm^{-3}$ $FeSO_4(aq)$ for a complete reaction. Given that Fe^{2+} is oxidised to Fe^{3+}, which of the following ions could VO_2^+ be reduced to?

(a) VO_3^-; (b) VO^{2+}; (c) V^{3+}; (d) V^{2+}.

Solution: The answer is (b).

$$Fe^{2+} \longrightarrow Fe^{3+} + e^-.$$

Amount of Fe^{2+} reacted $= \frac{25}{1000} \times 0.8 = 0.02\,mol$,
Amount of electrons lost by $0.02\,mol$ of $Fe^{2+} = 0.02\,mol$,
Amount of VO_2^+ that reacted $= \frac{40}{1000} \times 0.5 = 0.02\,mol$,
Amount of electrons gained by $1\,mol$ of $VO_2^+ = \frac{0.02}{0.02} = 1\,mol$.

Thus, the oxidation state of vanadium decreases by 1 unit from +5 to +4 after reduction, i.e., VO_2^+ is reduced to VO^{2+}.

$$VO_2^+(aq) + Fe^{2+}(aq) + 2H^+(aq) \longrightarrow VO^{2+}(aq) + Fe^{3+}(aq) + H_2O(l).$$

8.2 Balancing Redox Equations

A balanced equation must fulfill two criteria:

- The law of conservation of mass. Since matter can neither be created nor destroyed, the number of each type of element must be the same on both sides of the arrow in an equation.
- Conservation of charges. The sum of the charges on one side of the equation must be equal to the sum of the charges on the other side of the equation.

Table 8.1 illustrates the step-by-step approach to constructing the balanced equation for the reaction between $MnO_4^-(aq)$ and $Fe^{2+}(aq)$

Table 8.1 Constructing Equations for a Reaction in an Acidic Medium

Steps Involved	Illustration
(1) Assign oxidation states to determine which reactants undergo reduction and oxidation.	$\underset{+7\ -2}{MnO_4^-} + \underset{+2}{Fe^{2+}} \longrightarrow \underset{+2}{Mn^{2+}} + \underset{+3}{Fe^{3+}}$
(2) Construct two half-equations that show the specific species that are reduced or oxidised to the corresponding products.	Reduction half-equation: $MnO_4^- \longrightarrow Mn^{2+}$ Oxidation half-equation: $Fe^{2+} \longrightarrow Fe^{3+}$
(3) Balance each half-equation by following the simple rules:	
• balance the element that undergoes oxidation or reduction first;	$MnO_4^- \longrightarrow Mn^{2+}$
• balance O atoms by adding the same number of H_2O molecules to the other side of the equation;	$MnO_4^- \longrightarrow Mn^{2+} + 4H_2O$
• balance H atoms by adding H^+ ions to the other side of the equation;	$MnO_4^- + 8H^+ \longrightarrow Mn^{2+} + 4H_2O$
• lastly, balance charges by adding electrons.	$\underbrace{MnO_4^- + 8H^+}_{-1\ +8(+1)=+7} \longrightarrow \underbrace{Mn^{2+} + 4H_2O}_{+2\ +4(0)=+2}$ $\underbrace{MnO_4^- + 8H^+ + 5\ e^-}_{-1+8(+1)+5(-1)=+2} \longrightarrow \underbrace{Mn^{2+} + 4H_2O}_{+2\ +4(0)=+2}$
(4) Repeat Step (3) for the other half-equation.	$Fe^{2+} \longrightarrow Fe^{3+} + e^-$
(5) Ensure that the number of electrons for each half-equation is the same by scaling one or both of these equations by appropriate multiples, i.e., number of e^- lost = number of e^- gained.	$MnO_4^- + 8H^+ + 5e^- \longrightarrow Mn^{2+} + 4H_2O$ (i) $Fe^{2+} \longrightarrow Fe^{3+} + e^-$ (ii)
(6) Add the two half-equations together to obtain the overall balanced equation.	Multiply (ii) by 5, $5Fe^{2+} \longrightarrow 5Fe^{3+} + 5e^-$ (iii)
• Simplify the equation by removing common terms that appear on both sides of the equation. • Double-check that the numbers of atoms and charges are balanced on both sides of the equation.	Add half-equations (i) and (iii), $MnO_4^- + 8H^+ + 5e^- + 5Fe^{2+} \longrightarrow Mn^{2+} + 4H_2O + 5Fe^{3+} + 5e^-$ Balanced redox equation: $5Fe^{2+}(aq) + MnO_4^-(aq) + 8H^+(aq) \longrightarrow 5Fe^{3+}(aq) + Mn^{2+}(aq) + 4H_2O(l)$
• Reduce coefficients to the simplest ratio.	Note: Electrons should not appear in the overall equation.

that takes place in an acidic medium. The products include $Mn^{2+}(aq)$ and $Fe^{3+}(aq)$.

For reactions that occur in a basic medium, additional steps are needed to "neutralise" the hydrogen ions that appear in the equation. Table 8.2 illustrates the step-by-step approach to constructing the balanced equation for the disproportionation of chlorine that takes place in a basic medium. The products include $Cl^-(aq)$ and $ClO_3^-(aq)$.

Exercise:

1. Balance the equations for the following reactions that occur in an acidic medium:

   (a) $Cr_2O_7^{2-}(aq) + HNO_2(aq) \longrightarrow Cr^{3+}(aq) + NO_3^-(aq)$, and
   (b) $I^-(aq) + NO_2^-(aq) \longrightarrow I_2(s) + NO(g)$.

2. Balance the equations for the following reactions that occur in a basic medium:

   (a) $Pb(OH)_4^{2-}(aq) + ClO^-(aq) \longrightarrow PbO_2(s) + Cl^-(aq)$, and
   (b) $MnO_4^-(aq) + H_2O_2(aq) \longrightarrow MnO_2(s) + O_2(g)$.

Solution:

1(a) $Cr_2O_7^{2-}(aq) + 3HNO_2(aq) + 5H^+(aq) \longrightarrow 2Cr^{3+}(aq) + 3NO_3^-(aq) + 4H_2O(l)$.

1(b) $2I^-(aq) + 2NO_2^-(aq) + 4H^+(aq) \longrightarrow I_2(s) + 2NO(g) + 2H_2O(l)$.

2(a) $Pb(OH)_4^{2-}(aq) + ClO^-(aq) \longrightarrow PbO_2(s) + Cl^-(aq) + 2OH^-(aq) + H_2O(l)$.

2(b) $2MnO_4^-(aq) + 3H_2O_2(aq) \longrightarrow 2MnO_2(s) + 3O_2(g) + 2OH^-(aq) + 2H_2O(l)$.

8.3 Redox Titrations

A redox titration is an important titration technique used to quantify the unknown amount of a reactant which takes part in a redox

Table 8.2 Constructing Equations for a Reaction in a Basic Medium

Steps Involved	Illustration
(1) Assign oxidation states to determine which reactants undergo reduction and oxidation.	$\underset{0}{Cl_2} \longrightarrow \underset{-1}{Cl^-} + \underset{+5\ -2}{ClO_3^-}$
(2) Construct two half-equations that show the specific species that are reduced or oxidised to the corresponding products.	Reduction half-equation: $Cl_2 \longrightarrow Cl^-$ Oxidation half-equation: $Cl_2 \longrightarrow ClO_3^-$
(3) Balance each half-equation by following the simple rules:	
• balance the element that undergoes oxidation or reduction first;	$Cl_2 \longrightarrow 2ClO_3^-$
• balance O atoms by adding the same number of H_2O molecules to the other side of the equation;	$Cl_2 + 6H_2O \longrightarrow 2ClO_3^-$
• balance H atoms by adding H^+ ions to the other side of the equation;	$Cl_2 + 6H_2O \longrightarrow 2ClO_3^- + 12H^+$
• "neutralise" H^+ ions, by adding OH^- ions to *both* sides of the equation;	$Cl_2 + 6H_2O + 12OH^- \longrightarrow 2ClO_3^- + 12H^+ + 12OH^-$
• combine H^+ and OH^- ions, on the same side of the equation, into H_2O molecules;	$Cl_2 + 6H_2O + 12OH^- \longrightarrow 2ClO_3^- + 12H_2O$
• simplify the equation by removing common terms that appear on both sides of the equation;	$Cl_2 + 12OH^- \longrightarrow 2ClO_3^- + 6H_2O$
• lastly, balance charges by adding electrons.	$\underbrace{Cl_2 + 12OH^-}_{0+12(-1)=-12} \longrightarrow \underbrace{2ClO_3^- + 6H_2O}_{2(-1)+6(0)=-2}$ $\underbrace{Cl_2 + 12OH^-}_{0+12(-1)=-12} \longrightarrow \underbrace{2ClO_3^- + 6H_2O + 10e^-}_{2(-1)+6(0)+10(-1)=-12}$ $Cl_2 + 2e^- \longrightarrow 2Cl^-$ $Cl_2 + 12OH^- \longrightarrow 2ClO_3^- + 6H_2O + 10e^-$ (i) $Cl_2 + 2e^- \longrightarrow 2Cl^-$ (ii)

(*Continued*)

Table 8.2 (*Continued*)

Steps Involved	Illustration
(4) Repeat Step (3) for the other half-equation.	Multiply (ii) by 5, $5Cl_2 + 10e^- \longrightarrow 10Cl^-$ (iii)
(5) Ensure that the number of electrons for each half-equation is the same by scaling one or both of these equations by appropriate multiples, i.e., number of e^- lost = number of e^- gained.	Add half-equations (i) and (iii), $Cl_2 + 12OH^- \longrightarrow 2ClO_3^- + 6H_2O + 10e^-$ $+5Cl_2 + 10e^- \quad + 10Cl^-$ Balanced redox equation: $6Cl_2 + 12OH^- \longrightarrow 2ClO_3^- + 6H_2O + 10Cl^-$
(6) Add the two half-equations together to obtain the overall balanced equation. • Simplify the equation by removing common terms that appear on both sides of the equation. • Double-check that the numbers of atoms and charges are balanced on both sides of the equation. • Reduce coefficients to the simplest ratio.	Simplified, balanced redox equation: $3Cl_2(g) + 6OH^-(aq) \longrightarrow 5Cl^-(aq) + ClO_3^-(aq) + 3H_2O(l)$

reaction. The following are some important half-equations to take note of:

Reduction reactions:

$MnO_4^- + 8H^+ + 5e^- \longrightarrow Mn^{2+} + 4H_2O$

$Cr_2O_7^{2-} + 14H^+ + 6e^- \longrightarrow 2Cr^{3+} + 7H_2O$

$I_2 + 2e^- \longrightarrow 2I^-$

$Fe^{3+} + e^- \longrightarrow Fe^{2+}$

$H_2O_2 + 2H^+ + 2e^- \longrightarrow 2H_2O$

Oxidation reactions:

$C_2O_4^{2-} \longrightarrow 2CO_2 + 2e^-$

$2S_2O_3^{2-} \longrightarrow S_4O_6^{2-} + 2e^-$

$2I^- \longrightarrow I_2 + 2e^-$

$Fe^{2+} \longrightarrow Fe^{3+} + e^-$

$H_2O_2 \longrightarrow O_2 + 2H^+ + 2e^-$

Examples of redox titrations include the iodine-thiosulfate titration and those involving the use of manganate(VII) ions, dichromate(VI) ions and hydrogen peroxide.

8.3.1 *Manganate(VII) titrations*

Purple potassium manganate(VII), $KMnO_4$, is a common powerful oxidising agent. It is used for the estimation of a wide range of reducing agents such as iron(II) salts, ethanedioates and hydrogen peroxide.

Take note that in an acidic medium, the manganate(VII) ion is reduced as follows:

$$\underset{\text{purple}}{MnO_4^-(aq)} + 8H^+(aq) + 5e^- \longrightarrow \underset{\text{faint pink}}{Mn^{2+}(aq)} + 4H_2O(l).$$

The action of potassium manganate(VII) in a redox titration requires a mineral acid. Sulfuric (VI) acid is the most suitable mineral acid used to provide the acidic medium. Nitric(V) acid (HNO_3) is not employed for acidification as it is itself an oxidising agent while hydrochloric acid undergoes a redox reaction with potassium manganate(VII) to yield chlorine gas, i.e., Cl^- is oxidised to Cl_2.

Neutral or alkaline potassium manganate(VII) is normally not used in redox titration because the manganate(VII) ion is reduced to $MnO_2(s)$, a dark brown/black solid, which inhibits the clear detection of the end-point of titration.

Since $KMnO_4(aq)$ is purple in colour while its reduced product $Mn^{2+}(aq)$ in an acidified medium is essentially colourless, an external indicator is not required to be added in the course of the titration.

Although it is difficult to read off the volume of potassium manganate(VII) when it is placed in a burette due to its dark colour, it is still commonly used because of the ease with which end-point colour change can be detected. The end-point is taken when the first permanent pink colouration is formed because an excess drop of potassium manganate(VII) is added. But if potassium manganate(VII) is placed in the conical flask during a titration, the end-point is taken when the solution changes from purple to colourless. Such a change is not easy to detect with the naked eye.

Example: Solution Q contains a certain amount of Fe^{2+} ions. When $25.0\,cm^3$ of solution Q is titrated with $0.05\,mol\,dm^{-3}$ acidified $KMnO_4$(aq), $27.00\,cm^3$ of the latter is needed for a complete reaction. What is the concentration of Fe^{2+} ions present in solution Q?

Solution:

- First of all, construct the half-equations to get the overall balanced equation:

$$5Fe^{2+}(aq) + MnO_4^-(aq) + 8H^+(aq) \longrightarrow 5Fe^{3+}(aq) + Mn^{2+}(aq) + 4H_2O(l).$$

- Calculate the amount of MnO_4^-(aq) used in the titration:

$$\text{Amount of } MnO_4^-(aq) \text{ reacted} = 0.05 \times \frac{27}{1000} = 1.35 \times 10^{-3}\,\text{mol}.$$

- Calculate the amount of Fe^{2+}(aq) that reacts with the above amount of MnO_4^-(aq):

  Based on the balanced equation, MnO_4^- reacts with Fe^{2+} in the ratio 1:5. Thus,

$$\begin{aligned}&\text{Amount of } Fe^{2+}(aq) \text{ present in } 25.0\,cm^3 \text{ of solution Q}\\ &= 5 \times \text{Amount of } MnO_4^-(aq) \text{ reacted}\\ &= 5 \times 1.35 \times 10^{-3}\\ &= 6.75 \times 10^{-3}\,\text{mol}.\end{aligned}$$

- Calculate the total amount of Fe^{2+}(aq) present in $1000\,cm^3$ of solution Q, i.e., calculate its concentration:

$$\begin{aligned}&\text{Amount of } Fe^{2+}(aq) \text{ present in } 1\,dm^3 \text{ of solution Q}\\ &= \frac{1000}{25} \times 6.75 \times 10^{-3}\\ &= 0.270\,\text{mol}.\end{aligned}$$

  Hence, the concentration of Fe^{2+}(aq) in solution Q is $0.270\,mol\,dm^{-3}$.

8.3.2 *Dichromate(VI) titrations*

Another commonly used oxidising agent in redox titration is potassium dichromate(VI), $K_2Cr_2O_7$. It undergoes reduction in an acidic medium according to the following half-equation:

$$\underset{\text{orange}}{Cr_2O_7^{2-}}(aq) + 14H^+(aq) + 6e^- \longrightarrow \underset{\text{green}}{2Cr^{3+}}(aq) + 7H_2O(l).$$

But unlike potassium manganate(VII), titration using potassium dichromate(VI) is not self-indicating. This is because upon reduction in an acidic medium, the orange dichromate(VI) ion changes to the green chromium(III) ions. The green chromium(III) ions are formed the moment redox reaction starts. Thus, the combination of the green chromium(III) ions (which are continuously being formed before the end-point is reached), and the extra drop of the orange dichromate(VI) solution at the end-point, makes an accurate detection of the end-point colour change virtually impossible. Therefore, an indicator is necessary. The common indicators used are:

(a) barium or sodium diphenylamine p-sulfonate (green to blue-violet),
(b) N-phenylanthranilic acid (green to reddish-violet), and
(c) diphenylamine (green to blue-violet).

Q: How does the indicator work?
A: The organic molecules in the indicator form a complex that has a distinct colour from the green chromium(III) ions when an excess of dichromate(VI) ions is added.

Example: If $25.0\,cm^3$ of solution Q is titrated with $0.05\,mol\,dm^{-3}$ acidified $K_2Cr_2O_7(aq)$, what volume of the latter is needed for a complete reaction? Take the concentration of $Fe^{2+}(aq)$ in solution Q to be $0.270\,mol\,dm^{-3}$.

Solution:

$$6Fe^{2+}(aq) + Cr_2O_7^{2-}(aq) + 14H^+(aq) \longrightarrow 6Fe^{3+}(aq) + 2Cr^{3+}(aq) + 7H_2O(l).$$

Amount of $Cr_2O_7^{2-}(aq)$ needed

$$= \frac{1}{6} \times \text{Amount of } Fe^{2+}(aq) \text{ present in } 25.0\,\text{cm}^3 \text{ of solution Q}$$

$$= \frac{1}{6} \times \left(\frac{25}{1000} \times 0.270\right)$$

$$= 1.125 \times 10^{-3}\,\text{mol}.$$

$$\text{Volume of } Cr_2O_7^{2-}(aq) \text{ needed} = \frac{1.125 \times 10^{-3}}{0.05} = 0.0225\,\text{dm}^3$$

$$= 22.5\,\text{cm}^3.$$

8.3.3 *Iodine-thiosulfate titrations*

Thiosulfate reacts with iodine as represented by the equation:

$$\underset{\text{colourless}}{2S_2O_3^{2-}(aq)} + \underset{\text{brown}}{I_2(aq)} \longrightarrow \underset{\text{colourless}}{S_4O_6^{2-}(aq)} + \underset{\text{colourless}}{2I^-(aq)}\,.$$

Such a titration is known as an iodometric titration. It is extremely useful for the determination of iodine present in a solution. Iodine in the presence of excess iodide ions forms the brown I_3^- complex:

$$I_2(aq) + I^-(aq) \rightleftharpoons I_3^-(aq) \text{ (brown coloured complex).}$$

As thiosulfate ions are added, it reduces the iodine, resulting in a decrease in the concentration of iodine. This causes the brown I_3^- complex to break down due to equilibrium shift (refer to Chap. 6 on Chemical Equilibria). As the concentration of the I_3^- complex diminishes, the resultant solution approaches a pale yellow colouration. In order to allow a more accurate detection of the end-point, a few drops of starch solution are normally added when the solution turns pale yellow:

$$I_2(aq) + \text{starch molecules} \rightleftharpoons I_2\text{-starch complex (deep blue colour).}$$

The starch molecules, which are none other than polysaccharides, form a deep blue water-insoluble complex. The deep blue colouration arises due to iodine molecules being trapped within the polysaccharides. The formation of the iodine-starch complex is a reversible reaction. Thus, when more thiosulfate solution is added, more iodine reacts with the thiosulfate ions. This causes an equilibrium shift to the left, whereby the iodine-starch complex breaks down to release the trapped iodine molecules. Hence, a distinct colour change from deep blue to colourless is observed. This should be the point where the volume of titrant used is noted. This is because if the titrated solution is left to stand in air, the deep blue colouration may be restored. This should be ignored, as it is due to the atmospheric oxidation of the excess iodide in the reaction mixture back to iodine.

Q: Why can't we add starch solution at the very beginning of the titration process?

A: Since iodine is being trapped in the starch molecules, adding starch solution at the start may cause lots of iodine molecules to be embedded in the polysaccharides. The release of iodine from the starch molecules takes time, and as such, it might decrease the accuracy of the titration measurement.

Q: How do the iodine molecules adsorb onto the starch molecules?

A: Iodine is a non-polar diatomic molecule whereas, since the starch molecule is made up of glucose molecules, it is polar in nature. The interaction between iodine and starch molecules are van der Waals' forces of the instantaneous dipole–induced dipole (id–id) type.

Example: 20 cm^3 of copper(II) nitrate solution is reacted with excess potassium iodide to liberate iodine and copper(I) iodide. The amount of iodine is determined by titration with sodium thiosulfate. Given that 22.50 cm^3 of 0.750 mol dm^{-3} thiosulfate solution is required, what is the concentration of Cu^{2+} ions in the copper(II) nitrate solution?

Solution:

$$I_2(aq) + 2S_2O_3{}^{2-}(aq) \longrightarrow S_4O_6{}^{2-}(aq) + 2I^-(aq).$$

$$\text{Amount of } S_2O_3{}^{2-}(aq) \text{ reacted} = \frac{22.5}{1000} \times 0.75 = 1.688 \times 10^{-2}\ \text{mol}.$$

$$\text{Amount of } I_2(aq) = 1/2 \times \text{Amount of } S_2O_3^{2-}(aq)$$
$$= 1/2 \times 1.688 \times 10^{-2}$$
$$= 8.44 \times 10^{-3}\ \text{mol.}$$

$$4I^-(aq) + 2Cu^{2+}(aq) \longrightarrow I_2(aq) + 2CuI(s).$$

$$\text{Amount of } Cu^{2+}(aq) = 2 \times \text{Amount of } I_2(aq) = 2 \times 8.44 \times 10^{-3}$$
$$= 1.688 \times 10^{-2}\ \text{mol.}$$

Hence,

$$\text{Concentration of } Cu^{2+}(aq) = \frac{1.688 \times 10^{-2}}{20} \times 1000$$
$$= 0.844\ \text{mol dm}^{-3}.$$

8.4 Redox Reactions and Electricity

Since the reaction between zinc metal and copper(II) ion is exothermic, wouldn't it be nice to convert the chemical energy from the redox reaction to some other useful energy instead of heat energy?

$$Zn(s) + CuSO_4(aq) \longrightarrow ZnSO_4(aq) + Cu(s) \quad \text{(exothermic reaction).}$$

Chemical energy can be converted to useful electrical energy if we employ a set-up in which the redox reaction is allowed to proceed and we are able to tap into the electrical energy that is produced. And *vice versa*, electrical energy can be converted to chemical energy. Such a system that involves the inter-conversion between electrical and chemical energies is known as an electrochemical cell.

There are two types of electrochemical cells that are of interest here, namely, voltaic cell (or galvanic cell) and electrolytic cell. In order to have a good understanding of what an electrochemical cell is, let us understand some fundamental concepts first.

8.4.1 *Electrode potential*

If a piece of zinc metal is allowed to come in contact with a solution containing aqueous $ZnSO_4$ (Fig. 8.2), two reactions happen:

(i) The zinc atoms have a certain tendency to lose electrons. Hence, some atoms will go into solution as Zn^{2+} ions:

$$\mathrm{Zn(s)} \longrightarrow \mathrm{Zn^{2+}(aq)} + 2\mathrm{e^-} \text{ (oxidation).}$$

The electrons will be left behind on the metal. As time passes, there will be a build-up of electrons on the zinc surface. Thus, the surface area of the metal inside the solution will be surrounded by a layer of positive ions which are attracted to the electrons that have been deposited on the metal surface.

(ii) Meanwhile, the positive ions in the solution may gain electrons, resulting in a reformation of the metal:

$$\mathrm{Zn^{2+}(aq)} + 2\mathrm{e^-} \longrightarrow \mathrm{Zn(s)} \text{ (reduction).}$$

When the rate of oxidation is equal to the rate of reduction, a dynamic equilibrium is established:

$$\mathrm{Zn^{2+}(aq)} + 2\mathrm{e^-} \rightleftharpoons \mathrm{Zn(s)}.$$

Generalising, for a metal M(s) in an aqueous solution of its ions:

$$\mathrm{M}^{n+}\mathrm{(aq)} + n\mathrm{e^-} \rightleftharpoons \mathrm{M(s)}.$$

When an equilibrium is established (at a particular temperature), there will be a *constant* amount of negative charge on the metal

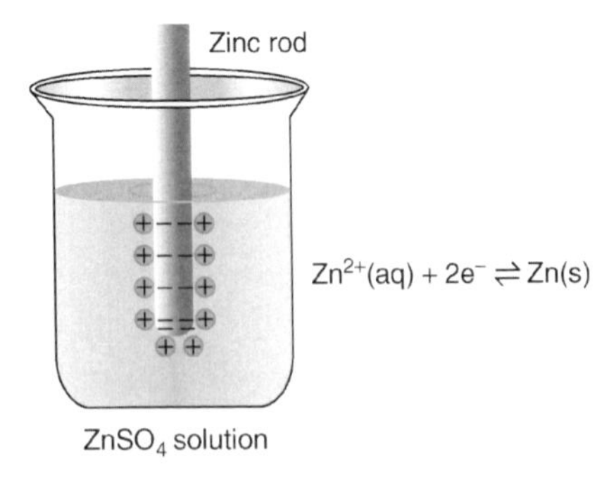

Fig. 8.2. A potential difference develops at the electrode surface.

surface, balanced by a *constant* amount of positive ions adsorbed on the metal surface. This separation of charges across the metal–solution interface creates an electrical potential known as **absolute electrode potential, $\boldsymbol{E}$**, of the metal. The set-up where the metal is in equilibrium with its own metal ions is known as a **half-cell.**

It is not possible to take a voltmeter and connect it across the metal–solution interface to measure the amount of electrode potential that is generated simply because the interface is too narrow for viable electrical contacts to be made. In addition, the moment electrical contact is made, this may cause electrons to be "drained" out.

A way to solve the problem is to hook up this half-cell to another half-cell and physically determine the *difference* in potential (or e.m.f.) between these two electrode systems. In this way, we can generate a list of measurements of the electrode potential of any half-cell with respect to a chosen reference system and from here obtain the e.m.f. value when any other two half-cells are connected. But then, what is an appropriate reference system to be used?

In this case, the chosen reference is the **standard hydrogen electrode (S.H.E.)**.

The standard hydrogen electrode comprising of the $H^+ \mid H_2$ half-cell revolves around the equilibrium between H^+ and H_2:

$$2H^+(aq) + 2e^- \rightleftharpoons H_2(g), \quad \boldsymbol{E} = 0.00 \text{ V (arbitrarily assigned value).}$$

Since the absolute electrode potential for the $H^+ \mid H_2$ half-cell cannot be determined (just like for any other half-cell), it is assigned an arbitrary value of 0.00 V.

The $H^+ \mid H_2$ half-cell does not consist of a metal electrode as hydrogen itself is not electrically conducting. We need an inert conducting electrode which is not reactive to either H^+ or H_2. The electrode of choice is an inert electrode made of platinum.

The inert Pt(s) electrode acts as the interface, allowing $H_2(g)$ molecules to be adsorbed onto its surface. An equilibrium between H^+ and H_2 is established. Figure 8.3 shows the set-up for the S.H.E.

The measurement with the reference electrode has to be done under standard conditions since the **electrode potential varies with changes such as that in temperature, partial pressure of gases and concentration of ions.**

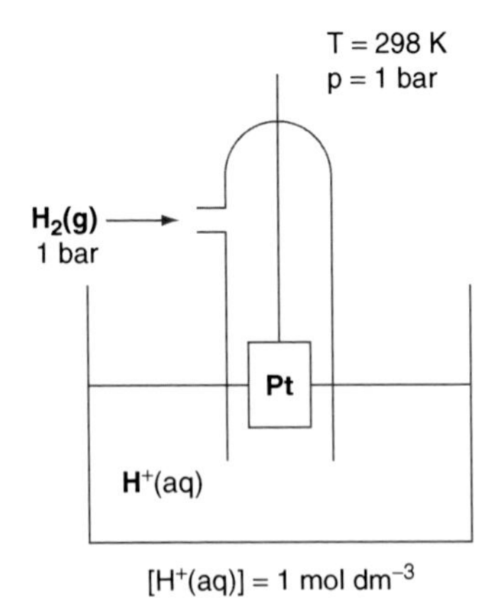

Fig. 8.3. Set-up of the standard hydrogen electrode.

The standard conditions are:

- temperature of 298 K;
- pressure of gases at 1.00 bar (e.g., pressure of H_2 is 1.00 bar for the $H^+ \mid H_2$ half-cell);
- concentration of ions at 1.00 mol dm^{-3} (e.g., $[H^+]$ is 1.0 mol dm^{-3} for the $H^+ \mid H_2$ half-cell).

When a half-cell is connected to the S.H.E., the potential difference (which arises due to the difference in the ability of each half-cell to undergo oxidation or reduction) between the two connected electrodes will generate electric current.

The electricity generated is measured in terms of the electromotive force (abbreviated as e.m.f.), which, in turn, is measured using a voltmeter. The actual e.m.f. of the system is the first reading shown on the voltmeter, the moment the two half-cells are connected. This is because once the circuit is completed, current flows. The current that is generated is a result of the chemical reaction that occurs and this leads to a decrease in the concentration of the reactants which inevitably affects the subsequent e.m.f. measured. When the system reaches equilibrium, the cell goes flat.

Since we are measuring the e.m.f. of a half-cell with respect to a reference half-cell (S.H.E.), the measured e.m.f. is not the absolute electrode potential, but is known as the **standard electrode potential**. By convention, we always represent it in the reduction form; it is thus also known as the **standard reduction electrode potential**. It is denoted by $E^{\ominus}$ (the plimsoll symbol "$\ominus$" represents standard conditions).

> **Definition:** The *standard electrode potential* of a half-cell is the electromotive force, measured at 298 K and 1 bar, between the half-cell and the standard hydrogen electrode, in which the reacting species in solution are at molar concentrations and gaseous species are at a pressure of 1 bar.

Q: Why not named it "standard electrode potential difference" as what we are measuring in indeed a potential difference with respect to the standard hydrogen electrode?

A: Well, you are right that we are actually measuring the potential difference rather than a potential. But because the S.H.E. is assigned an arbitrary value of 0.00 V, it is simply called the standard electrode potential.

8.4.2 *Experimental set-up of half-cells*

We can classify half-cells into three main categories that consist of one of the following sets of components:

(i) A metal and its ion. Example: Cu^{2+}(aq)|Cu(s) electrode system.
(ii) A non-metal and its ion. Example: Pt(s)|Cl_2(g)|Cl^-(aq) electrode system.
(iii) Two ions of the same element but in different oxidation states. Example: Pt(s)|Fe^{3+}(aq), Fe^{2+}(aq) electrode system.

Each of these requires specific types of set-up, which are described in Examples 8.5, 8.6 and 8.7.

Example 8.5: Measuring standard electrode potential $E^{\ominus}$ for the Cu^{2+}(aq)|Cu(s) half-cell

The standard electrode potential $E^{\ominus}$ of the Cu^{2+}(aq)|Cu(s) half-cell (the term "electrode" is also used to refer to the whole half-cell set-up) is determined by combining the system with a standard hydrogen electrode and measuring the e.m.f. of the resultant cell.

The standard copper electrode comprises Cu(s) dipped into $1.00\,\text{mol dm}^{-3}$ Cu^{2+}(aq) solution at 25°C. The standard hydrogen electrode comprises H_2(g) at 1 bar bubbled over a platinum electrode immersed in a $1.00\,\text{mol dm}^{-3}$ H^+(aq) solution [$1.00\,\text{mol dm}^{-3}$ HCl(aq) solution or $0.5\,\text{mol dm}^{-3}$ H_2SO_4(aq) solution].

These two electrodes are connected to a voltmeter via an electrical wire. The **standard electrode potential** $\boldsymbol{E^\circ}$ of the $Cu^{2+}(aq)|Cu(s)$ half-cell is the measured e.m.f. or potential difference of the cell, i.e.,

$$Cu^{2+}(aq) + 2e^- \rightleftharpoons Cu(s), \quad E^\circ = +0.34\,\text{V}.$$

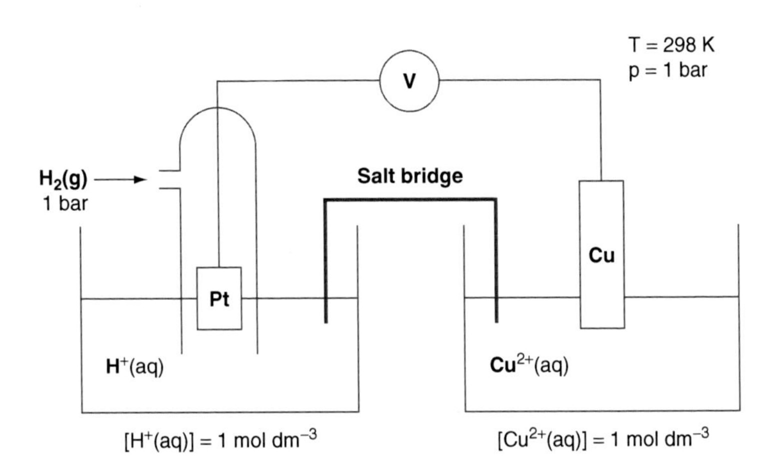

Fig. 8.4.

Example 8.6: Measuring standard electrode potential $\boldsymbol{E^\circ}$ for the $Cl_2(g)|Cl^-(aq)$ half-cell.

The E° value for the $Cl_2(g)|Cl^-(aq)$ half-cell can be measured by replacing the $Cu^{2+}(aq)|Cu(s)$ half-cell in Fig. 8.4 with a $Cl_2(g)|Cl^-(aq)$ half-cell. A piece of platinum metal is used as a metal electrode, creating a surface for the reactions of the Cl_2 and Cl^- species. Chlorine gas at 1 bar is bubbled over the platinum electrode immersed in a solution of $1.00\,\text{mol dm}^{-3}$ Cl^-(aq) ions at 25°C.

The e.m.f. measured by the voltmeter is the standard electrode potential of the $Cl_2(g)|Cl^-(aq)$ half-cell, i.e.,

$$Cl_2(g) + 2e^- \rightleftharpoons 2Cl^-(aq), \quad E^\ominus = +1.36\,V.$$

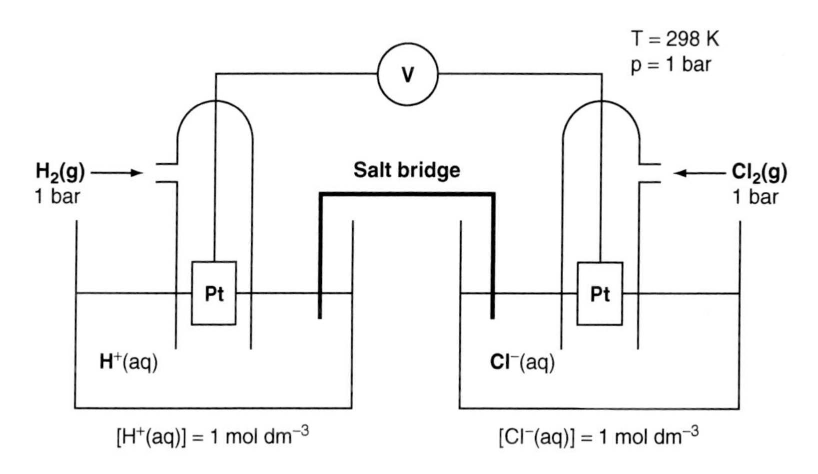

Fig. 8.5.

Example 8.7: Measuring standard electrode potential $E^\ominus$ for the $Fe^{3+}(aq)|Fe^{2+}(aq)$ half-cell

The $E^\ominus$ value for the $Fe^{3+}(aq)|Fe^{2+}(aq)$ half-cell can be measured by replacing the $Cu^{2+}(aq)|Cu(s)$ half-cell in Fig. 8.4 with a half-cell consisting of a platinum electrode dipped into a solution where $[Fe^{2+}(aq)] = [Fe^{3+}(aq)] = 1.00\,mol\,dm^{-3}$ at 25°C, i.e.,

$$Fe^{3+}(aq) + e^- \rightleftharpoons Fe^{2+}(aq), \quad E^\ominus = +0.77\,V.$$

Q: Why should there be a salt bridge?

A: Consider the cell set-up in Example 8.5 without a salt bridge (see Fig. 8.7): as Cu^{2+} from the solution is being reduced to Cu, the solution becomes increasingly negatively charged, which in this case is caused by the presence of the spectator anions [remember we use $CuSO_4(aq)$ solution?]. This prevents further reduction of Cu^{2+} from taking place. Likewise, in the other half-cell, as more H_2 gas is being oxidised to form H^+ ions, the solution becomes increasingly positively charged. As time passes, there is

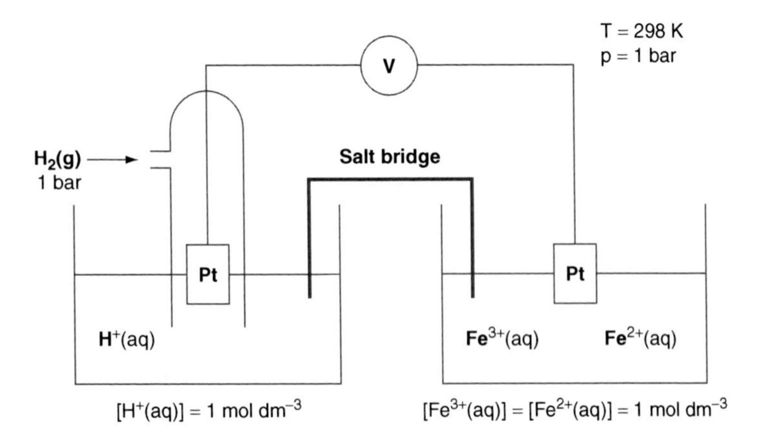

Fig. 8.6.

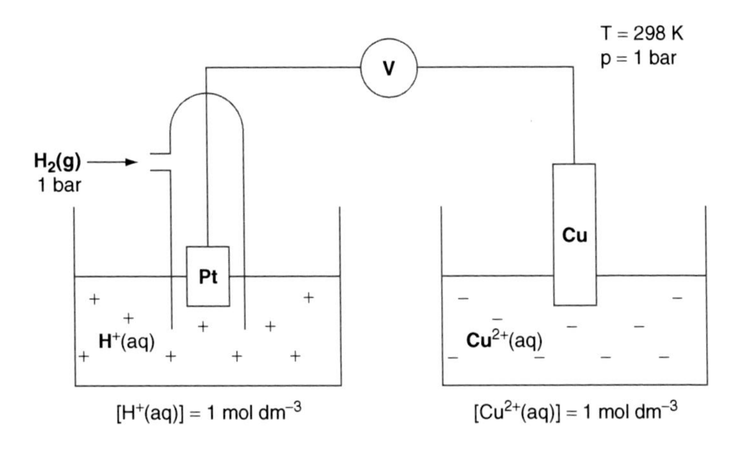

Fig. 8.7.

an opposite potential difference built up, which prevents further redox reaction from occurring.

However, when a salt bridge containing a relatively inert compound, such as concentrated KNO_3 solution, is put in place, both the K^+ and NO_3^- ions can move into each of the half-cells to maintain electrical neutrality. The K^+ ions move into the $Cu^{2+}(aq)|Cu(s)$ half-cell while the NO_3^- ions move into the other half-cell. This migration ensures the continuous smooth operation of the electrochemical cell until the reactants are depleted.

8.4.3 *Information obtained from the standard electrode potential*

Standard electrode potentials ($E^{\ominus}$) are relative values since all electrode potential values are defined with reference to the standard hydrogen electrode potential. $E^{\ominus}$ values are measured in volts, which is actually J C^{-1}. What do $E^{\ominus}$ values tell us?

When the standard electrode potential of the $Cu^{2+}|Cu$ electrode is measured, an $E^{\ominus}$ of $+0.34$ V is obtained:

$$2H^+(aq) + 2e^- \rightleftharpoons H_2(g), \quad E^{\ominus} = 0.00\,V,$$

$$Cu^{2+}(aq) + 2e^- \rightleftharpoons Cu(s), \quad E^{\ominus} = +0.34\,V.$$

This positive value indicates that the reduction of Cu^{2+} to Cu proceeds more favourably over the reduction of H^+ to H_2. In a redox reaction, you cannot possibly have two half-reactions that are reductions or two half-reactions that are oxidations. It is not possible for electrons to be gained without a corresponding oxidation reaction that supplies the electrons to be gained!

Thus, if reduction of Cu^{2+} occurs in the $Cu^{2+}|Cu$ half-cell, then in the $H^+|H_2$ half-cell, we expect oxidation to occur, i.e., H_2 is oxidised to H^+.

The $Cu^{2+}|Cu$ half-cell is the **cathode** (a reduction electrode) with a **positive** polarity. The $H^+|H_2$ half-cell is the **anode** (an oxidation electrode) which is **negatively charged**.

For all types of half-cells, their standard electrode potentials will either be a negative value or a positive value.

- A positive $E^{\ominus}$ indicates that reduction takes place in the half-cell of interest while at the reference S.H.E., $H_2(g)$ is oxidised to $H^+(aq)$.
- A negative $E^{\ominus}$ indicates that oxidation takes place in the half-cell of interest while at the reference S.H.E., $H^+(aq)$ is reduced to $H_2(g)$.

Standard electrode potential is therefore also known as standard reduction potential, since its magnitude represents the relative tendency of reduction taking place in the half-cell of interest, with respect to the reference $H^+|H_2$ half-cell.

A list of standard reduction potentials of various elements is listed in Table 8.3.

We can also make use of standard reduction potentials to determine:

- **The relative oxidising power of elements.**
For instance, between F_2 and Cl_2, which will be a better oxidising agent? To answer this, we refer to the following standard reduction potentials:

$$F_2(g) + 2e^- \rightleftharpoons 2F^-(aq), \quad E^\ominus = +2.87\,V,$$
$$Cl_2(g) + 2e^- \rightleftharpoons 2Cl^-(aq), \quad E^\ominus = +1.36\,V.$$

The $F_2(g)|F^-(aq)$ system has a **more positive** $E^\ominus$, which indicates that the equilibrium position lies further to the right as compared to that of the $Cl_2(g)|Cl^-(aq)$ system. This means that F_2 has a greater tendency to gain electrons and be **reduced**, i.e., F_2 is a **stronger oxidising agent** than Cl_2.

- **The relative reactivity of metals.**

$$Ag^+(aq) + e^- \rightleftharpoons Ag(s), \quad E^\ominus = +0.80\,V,$$
$$Ca^{2+}(aq) + 2e^- \rightleftharpoons Ca(s), \quad E^\ominus = -2.87\,V,$$
$$K^+(aq) + e^- \rightleftharpoons K(s), \quad E^\ominus = -2.92\,V.$$

The relative reactivity of metals is related to the relative ease with which they lose electrons (i.e., the backward reaction).

Just as how a more positive $E^\ominus$ indicates that equilibrium position lies further to the right, a more negative $E^\ominus$ indicates an equilibrium position lying further to the left.

The $K^+(aq)|K(s)$ system has the **most negative** $E^\ominus$ value followed by the $Ca^{2+}(aq)|Ca(s)$ system and lastly the $Ag^+(aq)|Ag(s)$ system. This means that K loses electrons most easily and is therefore the **greatest reducing agent**, followed by Ca, and lastly Ag.

- **The reaction, if any, that will take place for any pair of substances.**
For instance, if we were to connect a $Cu^{2+}(aq)|Cu(s)$ half-cell and a $K^+(aq)|K(s)$ half-cell together, what reaction will occur?

Table 8.3 Standard Reduction Potentials at 25°C

					E° (V)
Ag^{+}	+	e^{-}	$\rightleftharpoons$	Ag	+0.80
Al^{3+}	+	$3e^{-}$	$\rightleftharpoons$	Al	−1.66
Ba^{2+}	+	$2e^{-}$	$\rightleftharpoons$	Ba	−2.90
Br_2	+	$2e^{-}$	$\rightleftharpoons$	$2Br^{-}$	+1.07
Ca^{2+}	+	$2e^{-}$	$\rightleftharpoons$	Ca	−2.87
Cl_2	+	$2e^{-}$	$\rightleftharpoons$	$2Cl^{-}$	+1.36
Co^{2+}	+	$2e^{-}$	$\rightleftharpoons$	Co	−0.28
Co^{3+}	+	e^{-}	$\rightleftharpoons$	Co^{2+}	+1.82
$[Co(NH_3)_6]^{2+}$	+	$2e^{-}$	$\rightleftharpoons$	$Co + 6NH_3$	−0.43
Cr^{2+}	+	$2e^{-}$	$\rightleftharpoons$	Cr	−0.91
Cr^{3+}	+	$3e^{-}$	$\rightleftharpoons$	Cr	−0.74
Cr^{3+}	+	e^{-}	$\rightleftharpoons$	Cr^{2+}	−0.41
$Cr_2O_7^{2-} + 14H^{+}$	+	$6e^{-}$	$\rightleftharpoons$	$2Cr^{3+} + 7H_2O$	+1.33
Cu^{+}	+	e^{-}	$\rightleftharpoons$	Cu	+0.52
Cu^{2+}	+	$2e^{-}$	$\rightleftharpoons$	Cu	+0.34
Cu^{2+}	+	e^{-}	$\rightleftharpoons$	Cu^{+}	+0.15
$[Cu(NH_3)_4]^{2+}$	+	$2e^{-}$	$\rightleftharpoons$	$Cu + 4NH_3$	−0.05
F_2	+	$2e^{-}$	$\rightleftharpoons$	$2F^{-}$	+2.87
Fe^{2+}	+	$2e^{-}$	$\rightleftharpoons$	Fe	−0.44
Fe^{3+}	+	$3e^{-}$	$\rightleftharpoons$	Fe	−0.04
Fe^{3+}	+	e^{-}	$\rightleftharpoons$	Fe^{2+}	+0.77
$[Fe(CN)_6]^{3-}$	+	e^{-}	$\rightleftharpoons$	$[Fe(CN)_6]^{4-}$	+0.36
$Fe(OH)_3$	+	e^{-}	$\rightleftharpoons$	$Fe(OH)_2 + OH^{-}$	−0.56
$2H^{+}$	+	$2e^{-}$	$\rightleftharpoons$	H_2	0.00
I_2	+	$2e^{-}$	$\rightleftharpoons$	$2I^{-}$	+0.54
K^{+}	+	e^{-}	$\rightleftharpoons$	K	−2.92
Li^{+}	+	e^{-}	$\rightleftharpoons$	Li	−3.04
Mg^{2+}	+	$2e^{-}$	$\rightleftharpoons$	Mg	−2.38
Mn^{2+}	+	$2e^{-}$	$\rightleftharpoons$	Mn	−1.18
Mn^{3+}	+	e^{-}	$\rightleftharpoons$	Mn^{2+}	+1.49
$MnO_2 + 4H^{+}$	+	$2e^{-}$	$\rightleftharpoons$	$Mn^{2+} + 2H_2O$	+1.23
MnO_4^{-}	+	e^{-}	$\rightleftharpoons$	MnO_4^{2-}	+0.56
$MnO_4^{-} + 4H^{+}$	+	$3e^{-}$	$\rightleftharpoons$	$MnO_2 + 2H_2O$	+1.67
$MnO_4^{-} + 8H^{+}$	+	$5e^{-}$	$\rightleftharpoons$	$Mn^{2+} + 4H_2O$	+1.52
$NO_3^{-} + 2H^{+}$	+	e^{-}	$\rightleftharpoons$	$NO_2 + H_2O$	+0.81
$NO_3^{-} + 3H^{+}$	+	$2e^{-}$	$\rightleftharpoons$	$HNO_2 + H_2O$	+0.94
$NO_3^{-} + 10H^{+}$	+	$8e^{-}$	$\rightleftharpoons$	$NH_4^{+} + 3H_2O$	+0.87
Na^{+}	+	e^{-}	$\rightleftharpoons$	Na	−2.71
Ni^{2+}	+	$2e^{-}$	$\rightleftharpoons$	Ni	−0.25
$H_2O_2 + 2H^{+}$	+	$2e^{-}$	$\rightleftharpoons$	$2H_2O$	+1.77

(*Continued*)

Table 8.3 (*Continued*)

					$E^{\ominus}$ (V)
$O_2 + 4H^+$	+	$4e^-$	⇌	$2H_2O$	+1.23
$O_2 + 2H_2O$	+	$4e^-$	⇌	$4OH^-$	+0.40
$O_2 + 2H^+$	+	$2e^-$	⇌	H_2O_2	+0.68
$2H_2O$	+	$2e^-$	⇌	$H_2 + 2OH^-$	−0.83
$SO_4^{2-} + 4H^+$	+	$2e^-$	⇌	$SO_2 + 2H_2O$	+0.17
$S_2O_8^{2-}$	+	$2e^-$	⇌	$2SO_4^{2-}$	+2.01
$S_4O_6^{2-}$	+	$2e^-$	⇌	$2S_2O_3^{2-}$	+0.09
V^{2+}	+	$2e^-$	⇌	V	−1.20
V^{3+}	+	e^-	⇌	V^{2+}	−0.26
$VO^{2+} + 2H^+$	+	e^-	⇌	$V^{3+} + H_2O$	+0.34
$VO_2^+ + 2H^+$	+	e^-	⇌	$VO^{2+} + H_2O$	+1.00
$VO_3^- + 4H^+$	+	e^-	⇌	$VO^{2+} + 2H_2O$	+1.00
Zn^{2+}	+	$2e^-$	⇌	Zn	−0.76

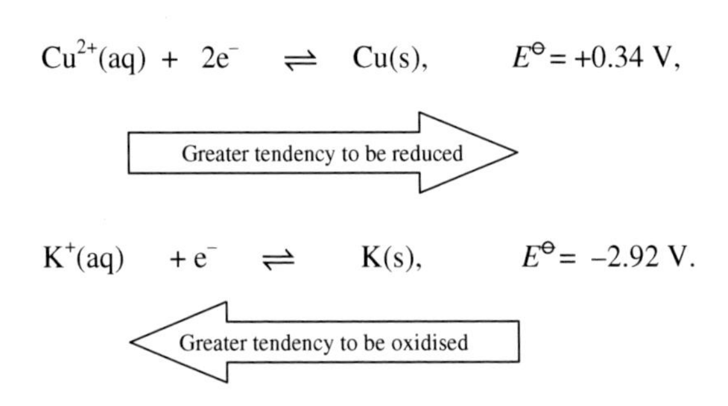

The Cu^{2+}(aq)|Cu(s) system has a more positive $E^{\ominus}$ value compared to the K^+(aq)|K(s) system. This means that Cu^{2+}(aq) tends to be reduced, and as such, K(s) is oxidised. We thus have a redox reaction that can generate electricity. But how much electricity is generated? This question is especially important if we are to construct batteries with maximum e.m.f.. In Section 8.4.5, we will learn to calculate the e.m.f. of a cell.

8.4.4 *Describing cell components using conventional notation*

Describing each of the componential half-cells of the electrochemical cell and the redox reaction that happens seems to require lengthy details of the half-cell processes that occur.

For instance, for the electrochemical reaction between Zn and Cu^{2+}, we have:

- the overall cell reaction: $Cu^{2+}(aq) + Zn(s) \rightarrow Cu(s) + Zn^{2+}(aq)$,
- oxidation taking place at the Zn anode: $Zn(s) \rightarrow Zn^{2+}(aq) + 2e^-$, and
- reduction occurring at the Cu cathode: $Cu^{2+}(aq) + 2e^- \rightarrow Cu(s)$.

There is actually a shorthand system with the following conventions involved:

- The oxidation reaction is written first in the shorthand notation. Since Zn is oxidised to Zn^{2+}, the first part of the notation reads as "$Zn(s)|Zn^{2+}(aq)$". The vertical line represents a phase (refer to Chap. 4 on Chemical Thermodynamics) boundary, which separates the solid Zn and aqueous Zn^{2+} solution.
- The reduction reaction is written second. Since Cu^{2+} is reduced to Cu, the second part of the notation reads as "$Cu^{2+}(aq)|Cu(s)$". The vertical line divides the components in different phases.
- Combining both parts of the notation gives us the convention used to describe the overall cell reaction:

$$Zn(s)|Zn^{2+}(aq) \vdots\vdots\, Cu^{2+}(aq)|Cu(s).$$

  The pair of dashed lines represents the salt bridge.
- In some cases, when there is no conducting solid species involved in the electrochemical reaction, platinum is used as the inert electrode.
- When there is more than one species in the same phase, these are separated by a comma in the cell notation.

  For instance, the oxidation half-cell involving Fe^{3+} and Fe^{2+} in the absence of a conducting solid, has the following notation: $Pt(s)|Fe^{2+}(aq), Fe^{3+}(aq)$.

In summary, the conventional notation is read as follows:

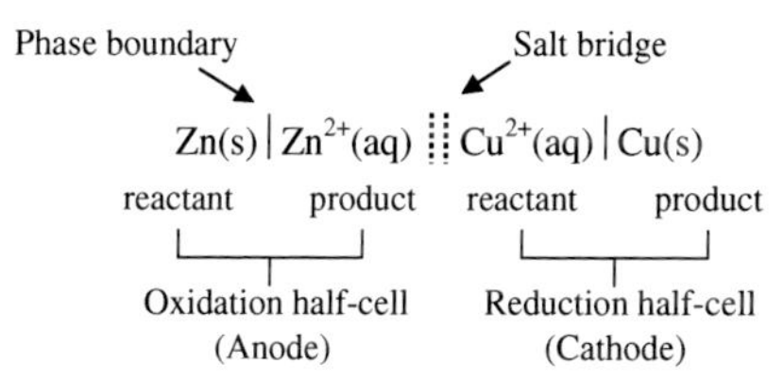

Half-cells can also be described using such notations. For instance, the half-cell containing Zn electrode and an aqueous solution of its ions can simply be represented by the $Zn^{2+}(aq)|Zn(s)$ half-cell. Similarly, $Cu^{2+}(aq)|Cu(s)$ half-cell implies that we are talking about the half-cell containing an Cu(s) electrode dipped into a solution containing $Cu^{2+}(aq)$ ions.

However, take note that for the half-cell shorthand notations, we always write the chemical formula for the oxidised species on the left and the reduced species on the right. This is because the standard electrode potential of a half-cell is always written in the reduction form. The vertical line denotes a phase boundary.

8.4.5 *Calculating standard cell potential*

Example: Calculate the e.m.f. for a cell that consists of the $Cu^{2+}(aq)|Cu(s)$ half-cell and the $K^+(aq)|K(s)$ half-cell.

Approach:

(1) Write the overall cell equation by constructing the balanced half-equations:

Reduction half-equation:

$$Cu^{2+}(aq) + 2e^- \longrightarrow Cu(s), \text{ (i)}$$

Oxidation half-equation:

$$2K(s) \longrightarrow 2K^+(aq) + 2e^-. \text{ (ii)}$$

> In writing the half-equations, use a "single-headed" arrow since the reaction occurs in the specific direction.

Make sure the number of electrons lost in the oxidation reaction is equal to that gained in the reduction reaction.

Adding the two half-equations will give a balanced redox equation, i.e., adding Eqs. (i) and (ii) gives the overall cell equation:

$$Cu^{2+}(aq) + 2K(s) \longrightarrow 2K^+(aq) + Cu(s).$$

(2) The standard cell potential, which is the difference between the standard reduction potentials of the half-cells, is calculated using

Eq. (iii):

$$E^{\ominus}{}_{\text{cell}} = E^{\ominus}{}_{\text{Red}} - E^{\ominus}{}_{\text{Ox}}, \tag{iii}$$

where $E^{\ominus}{}_{\text{Red}}$ denotes the standard reduction potential of the half-cell wherein a species undergoes reduction, and $E^{\ominus}{}_{\text{Ox}}$ denotes the standard reduction potential of the half-cell wherein a species undergoes oxidation.

The e.m.f. of the cell generated from the redox reaction between $Cu^{2+}(aq)$ and K(s) is calculated below:

$$Cu^{2+}(aq) + 2e^- \rightleftharpoons Cu(s), \quad E^{\ominus}{}_{\text{Red}} = +0.34\,V,$$

$$K^+(aq) + e^- \rightleftharpoons K(s), \quad E^{\ominus}{}_{\text{Ox}} = -2.92\,V.$$

Thus,

$$E^{\ominus}{}_{\text{cell}} = +0.34 - (-2.92) = +3.26\,V.$$

Q: Why isn't the computed $E^{\ominus}{}_{\text{cell}} = +0.34 - \mathbf{2}(-2.92)$ since the equation for the $K^+(aq)|K(s)$ half-cell is multiplied by a factor of 2 and hence $E^{\ominus}{}_{\text{Ox}} = 2(-2.92)$ V?

A: When equations are multiplied by a certain factor, its corresponding standard electrode potential remains unchanged. This is because half-cell reduction potentials and voltaic cell potentials both have a unit of Volt, which is actually Joule per Coulomb — the amount of energy that one would get when one Coulomb of charges passed through an object, such as a resistor.

When the half-cell equation is multiplied by any number, the number of electrons is indeed greater, but at the same time, the amount of energy one gets also increases proportionately. Hence, when one calculates the voltage, which is normalised as per Coulomb, one still gets the same voltage. This essentially means that voltage is independent of the number of electrons in the equation. We term such a quantity "intensive property".

Other intensive properties include density and temperature, which are not affected by the amount of substance present. In contrast, thermodynamic quantities such as enthalpy and entropy are extensive properties affected by the actual amount of substance.

Example: Write the overall cell equation and calculate the e.m.f. based on the cell notation given:

$$\mathrm{Pt|Fe^{2+}(aq), Fe^{3+}(aq) :: MnO_4^-(aq), Mn^{2+}(aq)|Pt.}$$

Approach:

(1) Based on the cell notation:
Oxidation half-equation: $Fe^{2+} \longrightarrow Fe^{3+} + e^-$,
Reduction half-equation: $MnO_4^- + 8H^+ + 5e^- \longrightarrow Mn^{2+} + 4H_2O$.
Thus, the overall cell equation is:

$$\mathrm{MnO_4^- + 8H^+ + 5Fe^{2+} \longrightarrow Mn^{2+} + 4H_2O + 5Fe^{3+}}.$$

(2) Select the relevant $E^\ominus$ data in Table 8.3.
From Table 8.3, we will have to find relevant equations for the redox processes.
For instance, there are a few equations that contain MnO_4^- but only Eq. (vi) is appropriate.

$$\mathrm{MnO_4^- + e^- \rightleftharpoons MnO_4^{2-}}, \qquad E^\ominus = +0.56\,\mathrm{V}, \quad \text{(iv)}$$

$$\mathrm{MnO_4^- + 4H^+ + 3e^- \rightleftharpoons MnO_2 + 2H_2O}, \qquad E^\ominus = +1.67\,\mathrm{V}, \quad \text{(v)}$$

$$\mathrm{MnO_4^- + 8H^+ + 5e^- \rightleftharpoons Mn^{2+} + 4H_2O}, \qquad E^\ominus = +1.52\,\mathrm{V}. \quad \text{(vi)}$$

Q: Why are Eqs. (iv) and (v) not suitable?
A: Both equations do not depict the correct product, Mn^{2+}, as given in the cell notation. The relevant equations needed are

$$\mathrm{MnO_4^- + 8H^+ + 5e^- \rightleftharpoons Mn^{2+} + 4H_2O}, \qquad E^\ominus{}_{\mathrm{Red}} = +1.52\,\mathrm{V},$$

$$\mathrm{Fe^{3+} + e^- \rightleftharpoons Fe^{2+}}, \qquad E^\ominus{}_{\mathrm{Ox}} = +0.77\,\mathrm{V}.$$

(3) Calculate the e.m.f. of the cell.

$$E^\ominus{}_{\mathrm{cell}} = E^\ominus{}_{\mathrm{Red}} - E^\ominus{}_{\mathrm{Ox}} = +1.52 - (+0.77) = +0.75\,\mathrm{V}.$$

Exercise: Calculate the e.m.f. of the cell: $\mathrm{Ni(s)|Ni^{2+}(aq) :: Ag^+(aq)|Ag(s)}$. (Answer: +1.05 V.)

8.4.6 *Using* $E^{\ominus}_{cell}$ *to predict spontaneity of a reaction*

Just as how it is possible to use the sign of $E^{\ominus}$ values to predict what reactions (oxidation or reduction with reference to S.H.E.) are likely to occur, we can also use $E^{\ominus}{}_{cell}$ values to predict the spontaneity of a redox reaction.

- $E^{\ominus}{}_{cell} > 0$ implies that the redox reaction is thermodynamically spontaneous under standard conditions.
- $E^{\ominus}{}_{cell} = 0$ implies that there is no net redox reaction occurring as the system is in equilibrium.
- $E^{\ominus}{}_{cell} < 0$ implies that the reaction is not thermodynamically spontaneous under standard conditions. However, the reaction in the reverse direction is thermodynamically spontaneous under standard conditions.

Q: How is $E^{\ominus}{}_{cell}$ link to thermodynamic spontaneity of the redox reaction which is actually predicted by $\Delta G^{\ominus}$?

A: The relationship between $E^{\ominus}{}_{cell}$ or $E^{\ominus}$ value and $\Delta G^{\ominus}$ is $\Delta G^{\ominus} = -\mathrm{n}\mathrm{F}E^{\ominus}{}_{cell}$, where n is the number of moles of electrons involved in the balanced redox equation and F is the Faraday's constant ($96{,}500\,\mathrm{C\,mol^{-1}}$). As can be seen, a positive $E^{\ominus}{}_{cell}$ or $E^{\ominus}$ value gives a negative $\Delta G^{\ominus}{}_{cell}$, which thus means that the redox reaction is thermodynamically spontaneous under standard conditions.

Q: Since $\Delta G^{\ominus}$ is dependent on the stoichiometric ratio of the balanced chemical equation, does that mean that the value of n, which is the number of moles of electrons involved in the balanced redox equation would also be dependent on the stoichiometric ratio of the equation?

A: Certainly! For examples, the $\Delta G^{\ominus}$ for the following two equations would differ:

$$\mathrm{Zn(s) + 2Fe^{3+}(aq) \rightarrow Zn^{2+}(aq) + 2Fe^{2+}(aq)}$$
$$\mathrm{1/2Zn(s) + Fe^{3+}(aq) \rightarrow 1/2Zn^{2+}(aq) + Fe^{2+}(aq)}$$

Hence, the values of n for the first equation would be 2 (2 moles of Fe^{3+} needs 2 moles of electrons to become 2 moles of Fe^{2+} or

1 mole of Zn releases 2 moles of electrons to become 1 mole of Zn^{2+}). While for the second equation, the value of n would be 1 (1 mole of Fe^{3+} needs 1 mole of electrons to become 1 moles of Fe^{2+} or 1/2 mole of Zn releases 1 mole of electrons to become 1/2 mole of Zn^{2+}).

Q: So, the value of $E^{\ominus}_{\text{cell}}$ is not dependent on the stoichiometric ratio of the balanced chemical equation?

A: Yes, you are right that the $E^{\ominus}_{\text{cell}}$ is independent on the stoichiometric ratio because the unit for $E^{\ominus}_{\text{cell}}$ is volt, which fundamentally is joules of energy per coulomb of charges (J C^{-1}). The per coulomb of charges would have already "taken care" of the amount of charges that the balanced equation has used.

Q: Why is the system at equilibrium when the $E^{\ominus}_{\text{cell}}$ is zero? That is, why is the redox system at equilibrium when the battery runs out of power?

A: When a battery still has an $E^{\ominus}_{\text{cell}}$ value, this would mean the two half-cells that made up the battery, each has a different potential to undergo reduction or oxidation. As the battery is used, the redox reaction proceeds and the reactants are used up. This would affect the potential of each of the two half-cells to undergo oxidation or reduction. At the point when the battery is flat, the potential of each of the two half-cells to undergo oxidation or reduction is the same. Hence, an equilibrium is said to have established in which there would be no nett redox reaction.

Example: Use $E^{\ominus}$ data to predict whether the reaction between F^- and H_2O_2 is thermodynamically spontaneous under standard conditions when they are mixed in an acidic medium.

Approach:

(1) Select the relevant $E^{\ominus}$ data from Table 8.3.
Between Eqs. (vii) and (viii) containing H_2O_2, Eq. (vii) is the correct choice.

$$H_2O_2 + 2H^+ + 2e^- \rightleftharpoons 2H_2O, \quad E^{\ominus} = +1.77\,\text{V}, \qquad \text{(vii)}$$

$$O_2 + 2H^+ + 2e^- \rightleftharpoons H_2O_2, \quad E^{\ominus} = +0.68\,\text{V}. \qquad \text{(viii)}$$

From prior knowledge, we know that F^- tends to get oxidised rather than reduced. Thus, if we expect F^- to be oxidised, then the other reactant, H_2O_2, is expected to get reduced. Hence, we choose the equation wherein H_2O_2 appears on the left-hand side, i.e., Eq. (vii).

Tips: If one reactant is found on the right-hand side of a reduction equation, i.e., it is expected to be oxidised, then the other reactant is expected to be reduced, i.e., it should be located on the left-hand side of the reduction equation, and *vice versa.* (Recall that oxidation and reduction go hand-in-hand.)

The relevant equations are:

$$F_2 + 2e^- \rightleftharpoons \mathbf{2F^-}, \qquad E^\ominus = +2.87\,V,$$

$$\mathbf{H_2O_2} + 2H^+ + 2e^- \rightleftharpoons 2H_2O, \quad E^\ominus = +1.77\,V.$$

(2) Calculate $E^\ominus{}_{cell}$ to find out if a reaction between F^- and H_2O_2 is feasible.
To calculate the e.m.f. of the cell, we use the same formula:

$$\mathbf{E^\ominus{}_{cell} = E^\ominus{}_{Red} - E^\ominus{}_{Ox}},$$

but, in this case, $\mathbf{E^\ominus{}_{Red}}$ denotes the standard reduction potential of the half-cell wherein a species *is expected* to undergo reduction, and $\mathbf{E^\ominus{}_{Ox}}$ denotes the standard reduction potential of the half-cell wherein a species *is expected* to undergo oxidation.

Thus,

$$E^\ominus{}_{cell} = +1.77 - (+2.87) = -1.10\,V.$$

Since $E^\ominus{}_{cell}$ is less than zero, the reaction between F^- and H_2O_2 is not thermodynamically spontaneous under standard conditions and is thus not likely to occur under standard conditions.

If the reaction between F^- and H_2O_2 is not thermodynamically spontaneous under standard conditions, we expect the reverse reaction to be thermodynamically spontaneous under standard conditions, i.e., reaction between F_2 and H_2O is thermodynamically spontaneous under standard conditions since the $E^\ominus{}_{cell}$ calculated is greater than 0 (i.e., $E^\ominus{}_{cell} = +1.10\,V$).

Example: Use $E^\ominus$ data to predict whether the reactions between the given pairs of reagents are thermodynamically spontaneous under standard conditions. State the observations for those reactions predicted to occur:

(i) $I^-(aq)$ and $Fe^{2+}(aq)$,
(ii) $SO_2(g)$ and acidified $Cr_2O_7^{2-}(aq)$.

Solution:

(i) Relevant data:

$$I_2(aq) + 2e^- \rightleftharpoons \mathbf{2I^-(aq)}, \qquad E^\ominus{}_{Ox} = +0.54\,V,$$

$$\mathbf{Fe^{2+}(aq)} + 2e^- \rightleftharpoons Fe(s), \qquad E^\ominus{}_{Red} = -0.44\,V.$$

$$E^\ominus{}_{cell} = -0.44 - (+0.54) = -0.98\,V.$$

Since $E^\ominus{}_{cell} < 0\,V$, the reaction is not thermodynamically spontaneous under standard conditions and is thus not likely to occur under standard conditions.

(ii) Relevant data:

$$\mathbf{Cr_2O_7^{2-}(aq)} + 14H^+(aq) + 6e^- \rightleftharpoons 2Cr^{3+}(aq) + 7H_2O(l),$$

$$E^\ominus{}_{Red} = +1.33\,V,$$

$$SO_4^{2-}(aq) + 4H^+(aq) + 2e^- \rightleftharpoons \mathbf{SO_2(g)} + 2H_2O(l),$$

$$E^\ominus{}_{Ox} = +0.17\,V.$$

$$E^\ominus{}_{cell} = +1.33 - (+0.17) = +1.16\,V.$$

Since $E^\ominus{}_{cell} > 0\,V$, the reaction is thermodynamically spontaneous under standard conditions and the following reaction is likely to occur:

$$Cr_2O_7^{2-}(aq) + 2H^+(aq) + 3SO_2(g) \rightarrow 2Cr^{3+}(aq) + H_2O(l) + 3SO_4^{2-}(aq).$$

Orange $Cr_2O_7^{2-}(aq)$ will be reduced to green $Cr^{3+}(aq)$. Thus, this redox reaction can be used to confirm the presence of sulfur dioxide gas.

Example: Use $E^\ominus$ data to account for the following observations: When fluorine gas is bubbled into an acidified pink solution of $CoSO_4(aq)$, a blue solution is formed. Upon standing in air, the blue solution changes to pink and a colourless gas is evolved.

Solution: Relevant data:

$$Co^{3+}(aq) + e^- \rightleftharpoons Co^{2+}(aq), \qquad E^\ominus = +1.82\,V,$$

$$F_2(g) + 2e^- \rightleftharpoons 2F^-(aq), \qquad E^\ominus = +2.87\,V,$$

$$O_2(g) + 4H^+(aq) + 4e^- \rightleftharpoons 2H_2O(l), \qquad E^\ominus = +1.23\,V.$$

$$2Co^{2+}(aq) + F_2(g) \longrightarrow 2Co^{3+}(aq) + 2F^-(aq).$$

$$E^\ominus{}_{cell} = +2.87 - (+1.82) = +1.05\,V.$$

Since $E^\ominus{}_{cell} > 0$, the reaction is thermodynamically spontaneous under standard conditions. $F_2(g)$ oxidises pink $Co^{2+}(aq)$ to blue $Co^{3+}(aq)$.

$$2Co^{3+}(aq) + H_2O(l) \longrightarrow 2Co^{2+}(aq) + 1/2O_2(g) + 2H^+(aq).$$

$$E^\ominus{}_{cell} = +1.82 - 1.23 = +0.59\,V$$

Since $E^\ominus{}_{cell} > 0$, the reaction is thermodynamically spontaneous under standard conditions. Upon standing in air, blue $Co^{3+}(aq)$ oxidises water to $O_2(g)$ and is itself reduced to pink $Co^{2+}(aq)$.

The use of $E^\ominus{}_{cell}$ to predict the thermodynamic spontaneity of a reaction has its limitations. There are cases when a reaction does not occur although it was predicted to be thermodynamically spontaneous under standard conditions. Possible explanations include the fact that the reaction is conducted under non-standard conditions whereas $E^\ominus$ values are measured under standard conditions. Another explanation stems from the fact that $E^\ominus{}_{cell}$ predicts the thermodynamic spontaneity of a reaction but not its kinetic feasibility because the reaction has high activation energy which needs to be overcome at a higher temperature.

For example, when hydrogen gas is bubbled into copper(II) sulfate solution, the expected reaction is:

$$Cu^{2+}(aq) + H_2(g) \longrightarrow Cu(s) + 2H^+(aq), \quad E^\ominus = +0.34\,V > 0.$$

However, no reaction occurs. Although the reaction is thermodynamically spontaneous under standard conditions, it is kinetically not feasible due to the high activation energy involved that causes the reaction to be non-observable under standard conditions.

In addition, if a side reaction occurs, a thermodynamically non-spontaneous reaction may become spontaneous. Consider the following reaction between Cu^{2+} and I^-:

$$I_2(aq) + 2e^- \rightleftharpoons 2I^-(aq), \qquad E^\ominus = +0.54\,V$$

$$Cu^{2+}(aq) + e^- \rightleftharpoons Cu^+(aq), \quad E^\ominus = +0.15\,V$$

$E^\ominus{}_{cell} = +0.15 - (+0.54) = -0.39$ V < 0 (redox is not thermodynamically spontaneous under standard conditions).

Since $E^\ominus{}_{cell} < 0$, no reaction is expected when aqueous solutions of Cu^{2+} and I^- are mixed. However, a reaction does occur: $2Cu^{2+}(aq) + 4I^-(aq) \rightarrow 2CuI(s) + I_2(aq)$, i.e., A cream precipitate (CuI) in brown solution (I_2) is observed when blue $Cu^{2+}(aq)$ is mixed with colourless $I^-(aq)$. Why? Copper(I) iodide is insoluble and the side reaction of precipitation of CuI occurs between Cu^+ ions and I^- ions. Due to this side reaction occurring, $[Cu^+(aq)]$ becomes very low, so that E_{Red} is more positive than $+0.15$ V. Thus, it is possible for $E_{cell} = E_{Red} - E_{Ox}$ to be greater than zero and the reaction to become spontaneous.

Q: Why is the "$\ominus$" missing in the $E_{cell} = E_{Red} - E_{Ox}$ equation?
A: Under non-standard conditions, we have to drop the "$\ominus$" from all the $E^\ominus$ expression.

8.4.7 *Effect of concentration changes on* $E^\ominus{}_{cell}$ *value*

The e.m.f. of a cell depends on variables such as temperature and concentration of reactants and products.

The effect of changes in concentration can be accounted for by using Le Chatelier's Principle. $E^\ominus$ values measure the relative tendency for a reduction reaction to occur. If a change is imposed in one half-cell that is in equilibrium, according to Le Chatelier's Principle, the position of the equilibrium shifts to nullify the change. This in turn affects the e.m.f. of the half-cell and hence the e.m.f. of the cell.

Example: A cell is set up between the $Cu^{2+}(aq)|Cu(s)$ and acidified $Cr_2O_7{}^{2-}(aq)|Cr^{3+}(aq)$ systems.

(a) Calculate the e.m.f. of the cell and write the balanced equation for the reaction that occurs.

(b) Explain the effect on the cell's e.m.f. when

(i) KOH(aq) is added to the $Cu^{2+}(aq)|Cu(s)$ half-cell;

(ii) water is added to the $Cu^{2+}(aq)|Cu(s)$ half-cell;

(iii) $H_2SO_4(aq)$ is added to the $Cr_2O_7{}^{2-}(aq)|Cr^{3+}(aq)$ half-cell.

Solution:

(a) $Cu^{2+}(aq) + 2e^- \rightleftharpoons Cu(s), \quad E^\ominus{}_{Ox} = +0.34\,V,$
$Cr_2O_7{}^{2-}(aq) + 14H^+(aq) + 6e^- \rightleftharpoons 2Cr^{3+}(aq) + 7H_2O(l),$
$E^\ominus{}_{Red} = +1.33\,V.$
$E^\ominus{}_{cell} = E^\ominus{}_{Red} - E^\ominus{}_{Ox} = +1.33 - (+0.34) = +0.99\,V.$
Since $E^\ominus{}_{cell} > 0\,V$, the reaction is thermodynamically spontaneous under standard conditions and the following reaction occurs:

$$Cr_2O_7{}^{2-}(aq) + 14H^+(aq) + 3Cu(s) \longrightarrow 2Cr^{3+}(aq) + 7H_2O(l) + 3Cu^{2+}(aq).$$

(b)(i) $Cu^{2+}(aq) + 2e^- \rightleftharpoons Cu(s), \quad E^\ominus{}_{Ox} = +0.34\,V.$
Cu^{2+} forms a blue precipitate, $Cu(OH)_2$, with OH^- added. This causes $[Cu^{2+}]$ to decrease. According to Le Chatelier's Principle, the backward reaction will be favoured and the equilibrium position shifts to the left. The e.m.f. of the $Cu^{2+}(aq)|Cu(s)$ half-cell becomes less positive, and hence the e.m.f. of the cell becomes more positive. (refrain from referring to the "$E^\ominus$ of the $Cu^{2+}(aq)|Cu(s)$ half-cell" because

the conditions are no longer standard; so you should drop the $^{\ominus}$ notation.)

(b)(ii) $Cu^{2+}(aq) + 2e^- \rightleftharpoons Cu(s)$, $E^{\ominus}{}_{Ox} = +0.34\,V$.
Adding water causes $[Cu^{2+}]$ to decrease. According to Le Chatelier's Principle, the backward reaction will be favoured and the equilibrium position shifts to the left. The e.m.f. of the $Cu^{2+}(aq)|Cu(s)$ half-cell becomes less positive, and hence the e.m.f. of the cell becomes more positive.

(b)(iii) $Cr_2O_7{}^{2-}(aq) + 14H^+(aq) + 6e^- \rightleftharpoons 2Cr^{3+}(aq) + 7H_2O(l)$,
$E^{\ominus}{}_{Red} = +1.33\,V$.
Adding $H_2SO_4(aq)$ causes $[H^+]$ to increase. According to Le Chatelier's Principle, the forward reaction will be favoured and the equilibrium position shifts to the right. The e.m.f. of the $Cr_2O_7{}^{2-}(aq)|Cr^{3+}(aq)$ half-cell becomes more positive, and hence the e.m.f. of the cell becomes more positive.

This also explains why dichromates are weak oxidising agents in neutral solution but are stronger oxidising agents in acidic solution due to higher $[H^+]$.

Q: If we use a bigger piece of copper electrode, would it affect the the $E^{\ominus}$ value?

A: No! This is because the concentration of a piece of pure metal or pure liquid is a constant value.

8.4.8 *Effect of ligands on* $E^{\ominus}{}_{cell}$ *value*

Transition metal ions form stable complexes with other ions or molecules called ligands through dative covalent bond formation. (The concept of ligand is discussed in Chap. 11 on Introduction to Transition Metals and their Chemistry.)

Metal ions are present in water as aqua-complex ions. The presence of ligands other than water molecules affects the standard electrode potentials. It can be seen that when Co^{2+} is co-ordinated by a more strongly electron-releasing ligand than water (i.e., NH_3), its tendency to gain electrons is reduced and $E^{\ominus}$ becomes more negative:

$$[Co(H_2O)_6]^{2+} + 2e^- \rightleftharpoons Co + 6H_2O,\ E^{\ominus} = -0.28\,V,$$

$$[Co(NH_3)_6]^{2+} + 2e^- \rightleftharpoons Co + 6NH_3,\ E^{\ominus} = -0.43\,V,$$

$$[\mathrm{Fe(H_2O)_6}]^{3+} + \mathrm{e}^- \rightleftharpoons [\mathrm{Fe(H_2O)_6}]^{2+}, \quad E^\ominus = +0.77\,\mathrm{V},$$
$$[\mathrm{Fe(CN)_6}]^{3-} + \mathrm{e}^- \rightleftharpoons [\mathrm{Fe(CN)_6}]^{4-}, \quad E^\ominus = +0.36\,\mathrm{V}.$$

In the case of Fe^{3+}, the formation of a negatively charged complex with CN^- ligands decreases the tendency of the complex to undergo reduction as compared to the $[Fe(H_2O)_6]^{3+}$ complex. Thus, $E^\ominus$ becomes less positive.

Q: Why does the presence of CN^- ligands decrease the tendency of Fe^{3+} to be reduced?

A: It is thermodynamically more demanding to add an electron to the negatively charged ion $[Fe(CN)_6]^{3-}$ than to the positively charged $[Fe(H_2O)_6]^{3+}$ ion.

A simple experiment that can be used to demonstrate the weaker oxidising power of $[Fe(CN)_6]^{3-}$ compared to $[Fe(H_2O)_6]^{3+}$ involves the use of KI(aq). Add KI(aq), followed by starch, to a test tube containing $[Fe(CN)_6]^{3-}$ and another containing $[Fe(H_2O)_6]^{3+}$. A blue-black colouration will only be observed in the test tube containing $[Fe(H_2O)_6]^{3+}$. Being a stronger oxidising agent, $[Fe(H_2O)_6]^{3+}$ oxidises I^- to I_2 that forms the coloured iodine-starch complex. But not $[Fe(CN)_6]^{3-}$!

8.5 Types of Electrochemical Cells

A **battery** or **voltaic cell** consists of one or more electrochemical cells in which stored chemical energy can be converted into electrical energy. There are two broad categories of batteries, namely, primary batteries and secondary batteries.

Primary batteries consist of electrochemical cells in which the redox reactions that happen in the cell are irreversible. The zinc-carbon dry cell and alkaline batteries are some examples of primary batteries. Thus, because the electrochemical reaction is irreversible in a primary battery, this results in the use-and-dispose nature of the cell, which poses many environmental problems due to the chemicals present in the cell.

Secondary batteries are just the opposite of primary batteries. The electrochemical reaction that generates electric current in secondary batteries can simply be reversed by electrical recharging. In the recharging process, electrical energy is converted to stored chemical energy, which can be released once again when the redox reaction occurs. Some examples of secondary batteries include the lead-acid accumulator, which is made of a few lead-acid unit cells connected in series. Others include the nickel-cadmium (Ni-Cd), nickel-metal hydride (NiMH) and lithium-ion (Li-ion) cells, which are currently commonly used as energy storage for mobile phones and laptop computers. Secondary batteries cannot be recharged an infinite number of times. After each recharging, there is inevitably some irreversible changes to the cell, especially if there is overcharging. Some changes are accumulative in nature and shorten the lifespan of such batteries.

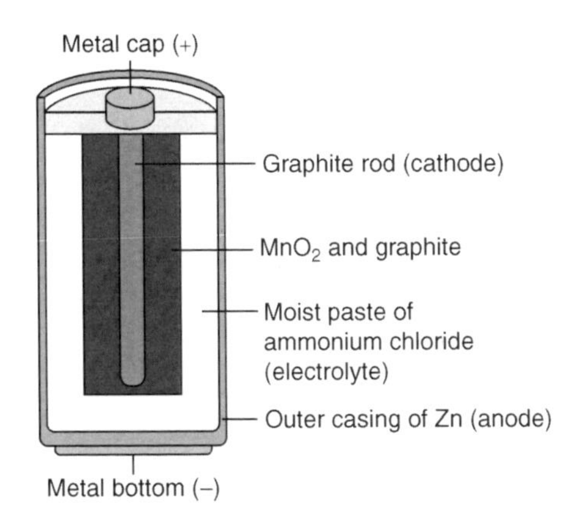

Fig. 8.8. A zinc-carbon dry cell.

(a) **Zinc-carbon dry cell** (Leclanche Cell)

The zinc-carbon dry cell consists of a graphite rod and an outer casing of zinc, which act as electrodes for the following half-reactions to occur:

$$\text{Anode:} \quad Zn(s) \longrightarrow Zn^{2+}(aq) + 2e^-,$$

$$\text{Cathode:} \quad 2MnO_2(s) + 2NH_4{}^+(aq) + 2e^- \longrightarrow Mn_2O_3(s) + 2NH_3(aq) + H_2O(l),$$

$$\text{Overall cell reaction : } Zn(s) + 2MnO_2(s) + 2NH_4^+(aq) \longrightarrow Zn^{2+}(aq) + Mn_2O_3(s) + 2NH_3(aq) + H_2O(l).$$

This cell is commonly used in flashlights, clocks, toy cars, etc. because of its low production cost. The battery has a very limited shelf life even without usage because of the possible reaction between the zinc metal and the acidic NH_4^+ ion. Once the zinc casing has been corroded, chemical leakages can occur.

(b) Alkaline dry cell

The electrodes that are involved in the electrochemical reaction of an alkaline dry cell are similar to the zinc-carbon dry cell, except that the acidic NH_4Cl electrolyte is replaced by alkaline KOH:

$$\text{Anode: } Zn(s) + 2OH^-(aq) \longrightarrow ZnO(s) + H_2O(l) + 2e^-,$$

$$\text{Cathode: } 2MnO_2(s) + H_2O(l) + 2e^- \longrightarrow Mn_2O_3(s) + 2OH^-(aq),$$

$$\text{Overall cell reaction: } Zn(s) + 2MnO_2(s) \longrightarrow ZnO(s) + Mn_2O_3(s).$$

The alkaline dry cell is much more durable than the zinc-carbon dry cell because there is no corrosion of Zn.

(c) Ni-Cd Rechargeable Cell

As compared to the dry cell, the economic cost of the Ni-Cd rechargeable cell is higher. But the multiple recharges possible with this cell have made its usage worthwhile. The electrochemical reactions that occur are as follows:

$$\text{Anode: } Cd(s) + 2OH^-(aq) \longrightarrow Cd(OH)_2(s) + 2e^-,$$

$$\text{Cathode: } NiO(OH)(s) + H_2O(l) + e^- \longrightarrow Ni(OH)_2(s) + OH^-(aq),$$

Overall cell reaction:

$$Cd(s) + 2NiO(OH)(s) + 2H_2O(l) \longrightarrow Cd(OH)_2(s) + 2Ni(OH)_2(s).$$

Other types of cells

As science and technology advances, new inventions in electrochemical cells have been made to meet the demand for smaller size, lower mass, higher voltage, longer lifespan, greater reliability, and quicker recharging. These new electrochemical cells are now used in various applications such as in electric vehicles, mobile phones, laptops, cardiac pacemakers, etc. Two such environmentally friendly inventions that have increased in popularity in recent years during the current era of energy conservation are fuel cells and solar cells.

(d) Fuel cells

The biggest difference between a fuel cell and other electrochemical cells that have been discussed so far is that the fuel cell does not deplete its supply of electrical energy as long as reactants are continuously being "fed" into the cell. The oxidising agent, oxygen gas, is supplied to the cathode compartment whereas the fuel, which can be hydrogen gas, hydrazine ($H_2N\text{-}NH_2$), methanol (CH_3OH), sugar ($C_6H_{12}O_6$) and other organic compounds, is fed to the anode compartment:

$$\begin{aligned}
\text{Anode: } & 2H_2 + 4OH^- \longrightarrow 4H_2O + 4e^-, \\
& H_2N\text{-}NH_2 + 4OH^- \longrightarrow 4H_2O + N_2 + 4e^-, \\
& CH_3OH + 6OH^- \longrightarrow 5H_2O + CO_2 + 6e^-, \\
& C_6H_{12}O_6 + 24OH^- \longrightarrow 18H_2O + 6CO_2 + 24e^-, \\
\text{Cathode: } & O_2 + 2H_2O + 4e^- \longrightarrow 4OH^-.
\end{aligned}$$

The greater the number of electrons generated in the anodic process, the greater the amount of electrical energy generated. Besides sodium hydroxide, sulfuric acid can also be used. Both electrolytes are known for causing high temperature corrosion to the steel materials used to construct fuel cells.

The fuel cell is popular due to the non-polluting products that are generated and its high energy efficiency. The main drawbacks arise from expensive noble metals such as platinum and palladium being used as electrodes and the high temperature that must be maintained to increase the electrical mobility of the sodium hydroxide electrolyte.

(e) Solar cell or photovoltaic cell

The solar cell is basically made of semiconductive materials that can convert solar energy into more useful electrical energy. When

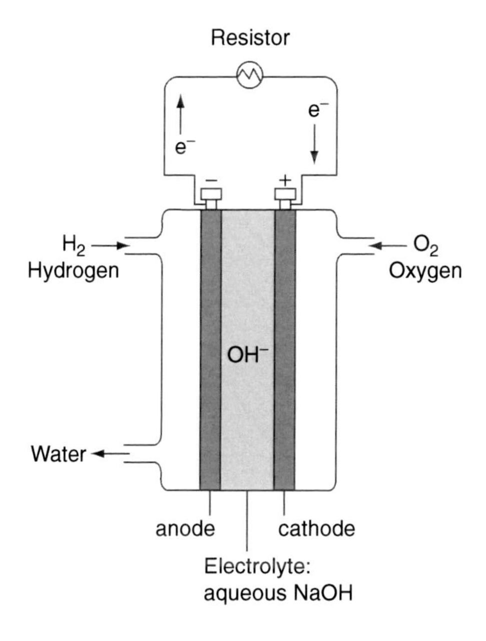

Fig. 8.9. A hydrogen-oxygen fuel cell.

light falls on certain materials, such as specially treated silicon, it generates a flow of electricity which can do useful work. Solar powered cars are environmentally friendly and use a sustainable energy source. The main drawbacks are the cost of construction and the lack of functionality in the absence of sunlight.

8.6 Electrolysis

Electrochemical cells generate electricity from the spontaneous redox reactions that take place. We can also make a non-spontaneous electrochemical change occur, but for this, we need to input electricity. This process of passing an electric current to force an otherwise non-spontaneous redox reaction to occur is known as *electrolysis*, and the set-up is termed the electrolytic cell (see Fig. 8.10).

The main component of an electrolytic cell is an electrolyte, which is usually an aqueous solution with dissolved ions or molten salts such as sodium chloride. Reduction takes place at the cathode, which

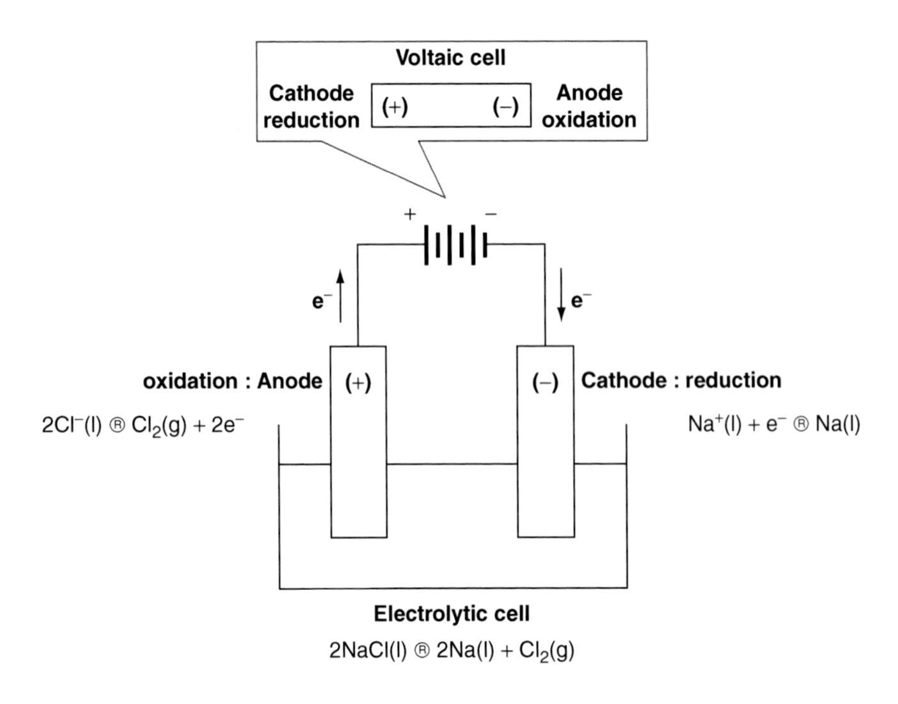

Fig. 8.10.

is negatively charged, and oxidation occurs at the anode, which is positively charged. When electricity is applied to the electrodes, the electrolyte provides ions that migrate to one of the two electrodes based on the concept of "opposites attract". Positive ions (cations) are attracted to the negative electrode (cathode) and move towards it. Since the cathode is electron-rich, the cations accept these electrons and are reduced. Negative ions (anions) are attracted to the positive electrode (anode) and move towards it. At the anode, the anions lose electrons and are oxidised.

For both the voltaic cell and the electrolytic cell, the anode is always where oxidation occurs and the cathode is always where reduction occurs. The phrase "*An* *ox* *c*harges *r*ed" may help you to remember this. Electrons flow from the anode to the cathode. However, the polarities of the electrodes in each cell are different. In the voltaic cell, the anode is the negative electrode and the cathode is the positive electrode. The polarities are reversed in the electrolytic cell.

8.6.1 *Faraday's laws of electrolysis*

Faraday's laws of electrolysis provide the quantitative relationship between electricity and chemical change.

Faraday's First Law of Electrolysis: The mass of the substance and/or the volume of gas liberated during electrolysis is directly proportional to the amount of charge that passes through the cell.

Faraday's Second Law of Electrolysis: The amount of charge required to discharge one mole of an element depends on the charge z on the ion.

The amount of charge Q in coulombs (C) is the product of current I in amperes (A or Cs^{-1}) and time t in seconds (s), i.e.,

$$Q = I \times t.$$

1 C of charge is the electric charge passed by a current of 1 A in 1 s.

One faraday F is the charge (in C mol^{-1}) carried by one mole of electrons:

$$F = Le,$$

where

L = Avogadro constant = $6.02 \times 10^{23}\,\text{mol}^{-1}$,
e = charge on an electron = $-1.60 \times 10^{-19}\,\text{C}$,
F = $96{,}500\,\text{C mol}^{-1}$.

In other words,

$$\text{Amount of e}^- = \frac{Q}{96500}.$$

The amount of charge required to discharge one mole of the following ions has been found experimentally to be:

$$1\,\text{mol Ag}^+ : 1 \times 9.65 \times 10^4\,\text{C},$$

$$1\,\text{mol Cu}^{2+} : 2 \times 9.65 \times 10^4\,\text{C},$$

$$1\,\text{mol Al}^{3+} : 3 \times 9.65 \times 10^4\,\text{C},$$

since

$$\text{Ag}^+ + 1\text{e}^- \rightleftharpoons \text{Ag},$$

$$\text{Cu}^{2+} + 2\text{e}^- \rightleftharpoons \text{Cu},$$

$$\text{Al}^{3+} + 3\text{e}^- \rightleftharpoons \text{Al}.$$

Hence one mole of Ag^{+}, Cu^{2+} and Al^{3+} requires 1, 2 and 3 faradays, respectively, to discharge at the cathode.

Example: How many grams of copper are deposited at the cathode of an electrolytic cell if an electric current of 2.00 A is run through an electrolytic cell comprising a solution of copper(II) sulfate, $CuSO_4(aq)$, for a period of 20.0 minutes?

Solution:

$$\begin{aligned}\text{Amount of charge passed } Q &= I \cdot t \\ &= 2.00 \times 20.0 \times 60 \\ &= 2400\,\text{C}. \\ \text{The amount of electrons passed through} &= \frac{2400}{96500}\ \text{mol}.\end{aligned}$$

From the reduction half-equation $Cu^{2+}(aq) + 2e^{-} \longrightarrow Cu(s)$, two moles of e^{-} deposits one mole of Cu(s). Therefore,

$$\frac{2400}{96500}\ \text{mol of e}^{-}\text{deposits}\left(\frac{2400}{96500}\right) \div 2\,\text{mol of Cu(s)}.$$

$$\begin{aligned}\text{Hence, mass of Cu(s) deposited} &= \left(\frac{2400}{96500}\right) \div 2 \times 63.5 \\ &= 0.790\,\text{g}.\end{aligned}$$

The electrolysis will deposit 0.790 g of copper at the cathode.

Example: How much time will it take to deposit 1.00 g of chromium when an electric current of 0.120 A flows through an electrolytic cell containing a solution of chromium(III) sulfate, $Cr_2(SO_4)_3(aq)$?

Solution:

$$Cr^{3+}(aq) + 3e^{-} \longrightarrow Cr(s).$$

1 mol of Cr requires 3 mol of electrons.

1 mol of electrons carries 96,500 C of charges.

Therefore,

$$1.00\,\text{g of Cr requires } \frac{1.00}{52.0} \times 3 \times 96{,}500\,\text{C of charges.}$$

$$Q = I \cdot t$$

$$\frac{1.00}{52.0} \times 3 \times 96,500 = 0.120 \times t$$

$$t = 46{,}394\,\text{s} \approx 12.9\,\text{hrs.}$$

8.6.2 *Selective discharge of ions*

When an electrical potential is applied across the two electrodes in an electrolytic cell, not all species undergo simultaneous discharge, i.e., are oxidised or reduced. There is preferential discharge of certain species over others and this is influenced by

- the electrode potentials of the species,
- the relative concentrations of the species (refer to Sec. 8.6.3.4 on The Electrolysis of Brine),
- the nature of the electrolyte, and
- the nature of the electrode.

Example: The use of molten versus aqueous electrolyte

- When molten sodium chloride is electrolysed using graphite electrodes, the following reactions occur:

$$\text{At the cathode: } Na^+(l) + e^- \longrightarrow Na(l),$$

$$\text{At the anode: } 2Cl^-(l) \longrightarrow Cl_2(g) + 2e^-.$$

However, if the electrolyte is aqueous sodium chloride, there are three types of species, Na^+(aq), H^+(aq) and H_2O molecules, competing for the reaction at the cathode. The electrode potentials can be used to predict the result of the competition.

- Competing reactions:
  Reduction can occur for both Na^+ ions and H_2O molecules. However, based on the $E^\ominus$ data below, H_2O has a greater tendency to be reduced and is thus preferentially discharged:

$$2H^+(aq) + 2e^- \rightleftharpoons H_2(g), \qquad E^\ominus = 0.00\,V,$$

$$Na^+(aq) + e^- \rightleftharpoons Na(s), \qquad E^\ominus = -2.71\,V,$$

$$2H_2O(l) + 2e^- \rightleftharpoons H_2(g) + 2OH^-(aq), \quad E^\ominus = -0.83\,V.$$

  The first reaction will go forward more readily than the second (since its $E^\ominus$ is more positive). But it is unlikely to happen as the concentration of H^+ is extremely low ($10^{-7}\,mol\,dm^{-3}$)!

  H_2O is more easily discharged (accepts electrons more readily) than Na^+. Hence, the cathode reaction is:

$$2H_2O(l) + 2e^- \longrightarrow H_2(g) + 2OH^-(aq).$$

  Likewise at the anode,

$$O_2(g) + 2H_2O(l) + 4e^- \rightleftharpoons 4OH^-(aq), \quad E^\ominus = +0.40\,V,$$

$$Cl_2(g) + 2e^- \rightleftharpoons 2Cl^-(aq), \qquad E^\ominus = +1.36\,V,$$

$$O_2(g) + 4H^+(aq) + 4e^- \rightleftharpoons 2H_2O(l), \qquad E^\ominus = +1.23\,V.$$

  OH^- is preferentially discharged, but it is unlikely to happen as the concentration of OH^- is extremely low ($10^{-7}\,mol\,dm^{-3}$)! Based on the $E^\ominus$ data above, H_2O has a greater tendency to be oxidised and is preferentially discharged. Hence, the anode reaction is:

$$2H_2O(l) \longrightarrow O_2(g) + 4H^+(aq) + 4e^-.$$

  Overall, the electrolysis of NaCl(aq) results in the electrolysis of water!

Example: The use of active versus inert electrodes

- When aqueous copper(II) sulfate is electrolysed using active copper electrodes, there are competing reactions.

  Reduction can occur for H^+, Cu^{2+}, SO_4^{2-} ions and H_2O molecules. But due to the low concentration of H^+ ($10^{-7}\,mol\,dm^{-3}$), it is not reduced. Now, based on the $E^\ominus$ data below, Cu^{2+} has the greatest tendency to be reduced and is thus preferentially discharged:

$$Cu^{2+}(aq) + 2e^- \rightleftharpoons Cu(s), \qquad E^\ominus = +0.34\,V,$$

$$2H_2O(l) + 2e^- \rightleftharpoons H_2(g) + 2OH^-(aq), \qquad E^\ominus = -0.83\,V,$$

$$SO_4^{\ 2-}(aq) + 4H^+(aq) + 2e^- \rightleftharpoons SO_2(g) + 2H_2O(l), \qquad E^\ominus = +0.17\,V.$$

Therefore at the cathode:

$$Cu^{2+}(aq) + 2e^- \longrightarrow Cu(s).$$

Oxidation can occur for Cu metal, $SO_4^{\ 2-}$ ions and H_2O molecules (the concentration of OH^- is too low for it to be oxidised). Based on the $E^\ominus$ data below, Cu has the greatest tendency to be oxidised and is thus preferentially discharged:

$$Cu^{2+}(aq) + 2e^- \rightleftharpoons Cu(s), \qquad E^\ominus = +0.34\,V,$$

$$O_2(g) + 4H^+(aq) + 4e^- \rightleftharpoons 2H_2O(l), \qquad E^\ominus = +1.23\,V,$$

$$S_2O_8^{\ 2-}(aq) + 2e^- \rightleftharpoons 2SO_4^{\ 2-}(aq), \qquad E^\ominus = +2.01\,V$$

Therefore, at the anode:

$$Cu(s) \longrightarrow Cu^{2+}(aq) + 2e^-.$$

- When aqueous copper(II) sulfate is electrolysed using inert graphite electrodes, the competing reduction processes are similar to the case when active electrodes are used. Cu^{2+} is preferentially discharged.

$$\text{At the cathode: } Cu^{2+}(aq) + 2e^- \longrightarrow Cu(s).$$

As for oxidation, there are only $SO_4^{\ 2-}$ ions and H_2O molecules competing with each other. Based on the $E^\ominus$ data below, H_2O has a greater tendency to be oxidised and is thus preferentially discharged:

$$O_2(g) + 4H^+(aq) + 4e^- \rightleftharpoons 2H_2O(l), \qquad E^\ominus = +1.23\,V,$$

$$S_2O_8^{\ 2-}(aq) + 2e^- \rightleftharpoons 2SO_4^{\ 2-}(aq), \qquad E^\ominus = +2.01\,V.$$

Therefore, at the anode:

$$2H_2O(l) \longrightarrow O_2(g) + 4H^+(aq) + 4e^-.$$

8.6.3 *Industrial uses of electrolysis*

8.6.3.1 *Electroplating*

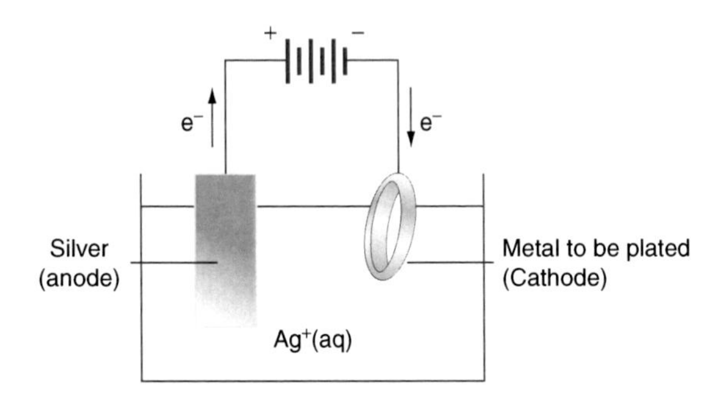

Fig. 8.11.

Electroplating is primarily used to coat a thin layer of material (of desirable properties) onto another material (which lacks the desired property). For instance, a popular use is in the electroplating of jewellery. Inexpensive jewellery is often coated with a thin layer of a precious metal such as silver or gold.

The piece of metal to be coated with silver is made the cathode. This cathode is placed into an electrolytic solution that contains the ions of the coating material, i.e., Ag^+ ions.

Q: Why do we make the metal to be plated be the cathode?
A: The objective is to have the metal ions be reduced and get deposited as a coat. This can only occur at the cathode as it is a reduction electrode.

The coating material, i.e., Ag(s), is made the anode.

When electricity is passed through, the Ag^+ ions in the solution will migrate to the cathode and undergo reduction to form Ag(s) on the surface of the cathode.

8.6.3.2 *The anodising of aluminium*

Anodising refers to the growing of an oxide film on certain metals, like aluminium, using an electrolytic process. This layer of aluminium oxide is of substantial commercial and technological importance as it prevents the corrosion of automobile and aerospace structures and it provides electrical insulation.

Q: Why is aluminium highly corrosion resistant although it has a very negative $E^\ominus$ value?

$$Al^{3+}(aq) + 3e^- \rightleftharpoons Al(s), \quad E^\ominus = -1.66\,V.$$

A: The corrosion resistance of aluminium is due to the surface coating of aluminium oxide that is impervious and hence protects the underlying metal from further chemical attack by air and water.

Although such a protective coating can be formed naturally in air, the layer is not thick enough. Therefore, the process of anodisation is used to increase the protective surface coating on aluminium objects.

Anodisation of aluminium is usually carried out in a sulfuric acid electrolyte where the aluminium workpiece to be anodised is used as the anode. During the electrolytic process of anodisation, oxygen gas is discharged at the anode, which reacts with the unoxidised aluminium metal to form a thick aluminium(III) oxide layer. The freshly formed film can then further be dyed to give colour-anodised aluminium. The cathode can be graphite, stainless steel or other electrical conductors that are inert in the anodising bath.

Q: If the electrolyte is sulfuric acid, doesn't the amphoteric Al_2O_3 react with it?

A: Yes indeed, it does react. But the rate of reaction is slower than its formation. In addition, we are using dilute sulfuric acid, so the reaction does not pose a serious problem. In fact, we are interested in creating a porous layer of aluminium oxide through the reaction of the oxide and sulfuric acid so that we can adhere other chemicals such as dye to the surface later. In addition, the anode being positively charged would repel the H^+ ions away. This thus decreases the amount of reaction between the amphoteric Al_2O_3 and H^+ ion.

Half-cell reactions:

At the anode (the aluminium object):

$$2H_2O(l) \longrightarrow O_2(g) + 4H^+(aq) + 4e^-,$$

$$4Al(s) + 3O_2(g) \longrightarrow 2Al_2O_3(s).$$

At the graphite cathode: $2H^+(aq) + 2e^- \longrightarrow H_2(g)$.

8.6.3.3 *Purification of copper (refining of copper)*

The purity of copper is about 99% when it is first obtained from its ore. The presence of main impurities such as silver, platinum, iron, gold and zinc, decreases the electrical conductivity of the metal. So to further enhance the electrical conductivity, impure copper must be purified before being used.

A piece of pure copper is made the cathode. The piece of impure copper to be refined is made the anode. Both electrodes are placed into the electrolytic solution of $CuSO_4(aq)$. When electricity is passed through, the following reactions will occur at the electrodes:

$$\text{At the cathode: } Cu^{2+}(aq) + 2e^- \longrightarrow Cu(s),$$

$$\text{At the anode: } Cu(s) \longrightarrow Cu^{2+}(aq) + 2e^-.$$

Reddish-brown Cu(s) will be deposited at the cathode. As for the anode, it will dissolve over a period of time and its mass reduced.

Q: So, the increase in mass of the copper cathode is not equal to the decrease in mass of the impure copper anode?

A: Indeed! Although when one Cu atom is oxidised at the anode, a corresponding Cu^{2+} is reduced at the cathode, the decreased in mass of the anode is much greater than the increased in mass of the cathode because of the additional "loss" of impurities at the anode. In addition, take note that the blue $CuSO_4$ solution will not fade as long as there is still copper at the anode to supplement the "consumption" of Cu^{2+} at the cathode.

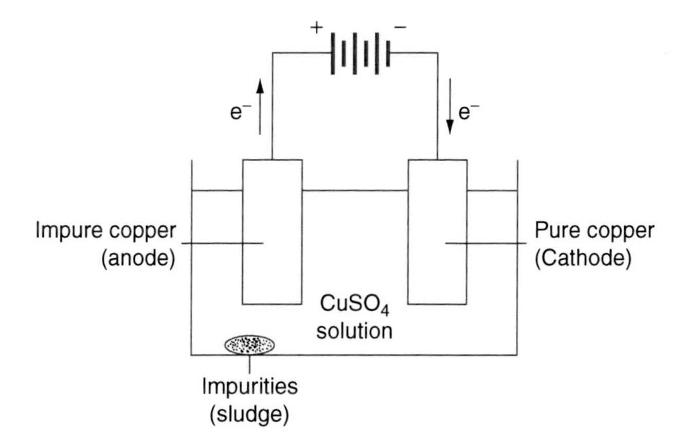

Fig. 8.12. Purification of copper by electrolysis.

Anode sludge

At the anode, other than copper, metals such as zinc and iron, which are more easily oxidised than copper, are also oxidised to form the dissolved ions. Once formed, these ions do not undergo reduction at the cathode because they are less preferentially discharged than the Cu^{2+} ion. Metal impurities that are less reactive than copper, such as silver and gold, simply drop to the bottom of the electrolytic bath as anode sludge. The recovery of silver, gold and platinum from the anode sludge is an important revenue in carrying out the purification process.

8.6.3.4 *The electrolysis of brine*

Brine, which is simply saturated sodium chloride solution, is an important starting material for the production of hydrogen, chlorine gases and sodium hydroxide. The reaction of chlorine gas and sodium hydroxide is important for the production of sodium chlorate(I) (NaClO) a powerful oxidising agent, which acts as the active ingredient in bleaching agents. A diaphragm cell is normally used in the electrolytic production of sodium hydroxide and chlorine, by introducing purified brine into the anode compartment. The anode compartment is separated from the cathode compartment by a permeable diaphragm made of asbestos fibre.

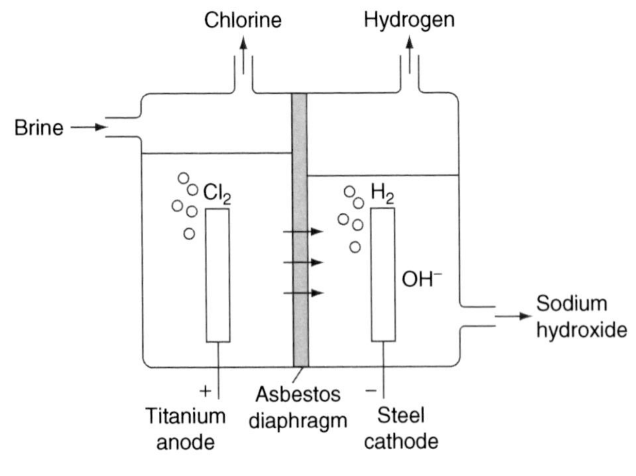

Fig. 8.13. A diaphragm cell.

At the cathode, reduction can occur for H^+, Na^+ ions and H_2O molecules. But due to the low concentration of H^+ ($10^{-7}\,\text{mol dm}^{-3}$), it would not be reduced. Thus, water is much more easily reduced than Na^+, forming hydrogen gas:

$$2H_2O(l) + 2e^- \longrightarrow H_2(g) + 2OH^-(aq).$$

At the anode, OH^- is preferentially discharged but this is unlikely to happen as the concentration of OH^- is extremely low ($10^{-7}\,\text{mol dm}^{-3}$)! Based on the $E^\ominus$ data, H_2O has a greater tendency to be oxidised (as compared with Sec. 8.6.2) but it would not undergo oxidation because of the high concentration of Cl^- ions:

$$Cl_2(g) + 2e^- \rightleftharpoons 2Cl^-(aq), \qquad E^\ominus = +1.36\,\text{V},$$

$$O_2(g) + 4H^+(aq) + 4e^- \rightleftharpoons 2H_2O(l), \quad E^\ominus = +1.23\,\text{V}.$$

Hence, under such non-standard conditions where the prediction using the $E^\ominus$ value does not work, the anode reaction involves $Cl^-(aq)$ being discharged to form $Cl_2(g)$:

$$2Cl^-(aq) \longrightarrow Cl_2(g) + 2e^-.$$

Due to the corrosive nature of chlorine gas, the anode metal is made of titanium.

Overall reaction:

$$\underbrace{2Na^+(aq) + 2Cl^-(aq)}_{2NaCl(aq)} + 2H_2O(l) \longrightarrow H_2(g) + Cl_2(g)$$

$$+ \underbrace{2Na^+(aq) + 2OH^-(aq)}_{2NaOH(aq)}.$$

Q: Why do the two gaseous products, hydrogen and chlorine, have to be prevented from mixing?

A: Mixtures of H_2 and Cl_2 are highly reactive when they come into contact; thus they need to be separated. In addition, this allows for H_2 and Cl_2 to be sold as separate chemicals.

The asbestos diaphragm also prevents the mixing of sodium hydroxide and chlorine gas, where the chlorine can disproportionate to form sodium chlorate(I) solution:

$$2NaOH(aq) + Cl_2(g) \longrightarrow NaClO(aq) + NaCl(aq) + H_2O(l).$$

Hence, one cannot say that sodium chlorate(I) is a side-product of the electrolysis of brine but NaOH is indeed the side-product that is being produced. In addition, the level of liquid in the anode compartment is kept at a higher level than that in the cathode compartment so that the brine will seep through the diaphragm in one direction. Back flow from the cathode compartment to the anode compartment is prevented to avoid the mixing of products formed from the two compartments.

Since not all the chloride ions are oxidised in the anode compartment, some of it would seep through the asbestos into the cathode compartment. Thus, the solution of sodium hydroxide which is formed in the cathode compartment also contains about 15% sodium chloride. This is evaporated; solid NaCl crystallises and is removed leaving a solution containing 50% by mass sodium hydroxide and only 1% sodium chloride.

My Tutorial (Chapter 8)

1. (a) An aqueous solution of hydrogen peroxide, H_2O_2, decomposes in the presence of a catalyst according to the equation:

$$2H_2O_2(aq) \rightarrow 2H_2O(l) + O_2(g).$$

(i) Calculate the number of moles of H_2O_2 required to produce $10\,dm^3$ of oxygen gas measured at room temperature and pressure.

(ii) The number of moles of H_2O_2 calculated in (a)(i) is present in $1\,dm^3$ of H_2O_2 solution. Calculate the volume of this solution required to make $250\,cm^3$ of a $0.200\,mol\,dm^{-3}$ solution, by dilution with water.

(b) Calculate the mass of potassium manganate(VII), $KMnO_4$, required to make $200\,cm^3$ of solution having a concentration of $0.040\,mol\,dm^{-3}$.

(c) When $20.0\,cm^3$ of the $0.200\,mol\,dm^{-3}$ solution of H_2O_2 is acidified with sulfuric(VI) acid and titrated against a $0.040\,mol\,dm^{-3}$ solution of potassium manganate(VII), $KMnO_4$, $40.0\,cm^3$ of the latter is required for a complete reaction.

(i) Calculate the number of moles of $KMnO_4$ in $40.0\,cm^3$ of the $0.040\,mol\,dm^{-3}$ solution.

(ii) Calculate the number of moles of H_2O_2 in $20.0\,cm^3$ of the $0.200\,mol\,dm^{-3}$ solution.

(iii) Hence, deduce the number of moles of H_2O_2 which reacts with 1 mole of $KMnO_4$.

(iv) Give a balanced equation for the reaction taking place in the titration.

(v) Explain why dilute hydrochloric acid is not used for acidification.

(vi) Why is potassium manganate(VII) usually placed in a burette, despite the difficulties it presents in reading the burette?

(d) The solution at the end of this reaction contains potassium sulfate(VI) and manganese(II) sulfate(VI) only.

(i) Write formulas for the cations present in this aqueous solution.

(ii) Treatment of the solution with dilute sodium hydroxide gives a precipitate that does not dissolve in excess sodium hydroxide solution. Identify the precipitate by name or formula.

2. (a) Explain the meaning of the term *oxidation* in terms of

    (i) electron transfer, and
    (ii) change in oxidation number (oxidation state).

   (b) What is the oxidation number of nitrogen in hydroxylamine, NH_2OH?
   (c) (i) Write down the half-equations for the oxidation of iron(II) to iron(III) ions and the reduction of manganate(VII) to manganese(II) ions under acidic conditions.
    (ii) Deduce the ionic equation for the reaction between iron(II) ions and manganate(VII) ions under acidic conditions.
   (d) The following experiment is used to determine the equation for the reaction between hydroxylamine and iron(III) ions. 0.074g of hydroxylamine was dissolved in water and made up to $50.0\,cm^3$. The solution was reacted with an excess solution of an acidified iron(III) salt. When the reaction was completed, the iron(II) produced required $44.8\,cm^3$ of $0.0200\,mol\,dm^{-3}$ potassium manganate(VII) solution to oxidise the iron(II) back to iron(III).

    (i) Calculate the amount of hydroxylamine used in the reaction.
    (ii) Calculate the amount of iron(II) formed in the reaction.
    (iii) Determine the molar ratio of iron(III) to hydroxylamine reacting together.
    (iv) Using both parts (b) and (d)(iii), deduce the oxidation number of nitrogen in the product.
    (v) Which of the following possible nitrogen containing compounds, NO, N_2O, N_2O_4, N_2 and NH_3, is the most likely product of the reaction?
    (vi) Write the equation for the reaction between hydroxylamine and iron(III) ions.

3. A possibility for the future is to use electric cars that generate their own electricity. Instead of burning methanol in an internal combustion engine, methanol could be used to power a *fuel cell* in the car. The fuel cell generates an electric current which drives an electric motor. The fuel and oxygen are fed continuously to two electrodes immersed in an acidic electrolyte solution. The electrodes are made of platinum dispersed onto a porous carbon support.

   (a) Write half-equations for the reactions that take place at the anode and the cathode of the fuel cell, and combine these to give an overall equation for the cell reaction.
   (b) Explain where the energy which powers the electric motor comes from.
   (c) Explain why the fuel cell is much more environmentally sound than a conventional internal combustion engine.
   (d) One refinement of the fuel cell design is to replace the acidic electrolyte solution with a film of solid H^+ ion-conducting electrolyte. Explain why this would be an improvement.

4. Tin cans made of tin-plated iron are used to preserve food. Tin has the advantages that it corrodes much less readily than iron and that it forms a protective layer protecting the iron from rusting. When the coating is scratched, however, the iron rusts faster when it is in contact with tin. Fortunately, neither Fe^{2+} nor Sn^{2+} ions are toxic.

   (a) State *two* substances that are necessary for iron to rust and from which iron is protected by the tin layer.
   (b) Write a balanced equation for the reaction where the presence of tin ions encourages the iron to corrode.
   (c) Write a balanced equation for the corrosion of iron.
   (d) Rust is often given the formula $Fe_2O_3.xH_2O$, where x has a variable non-integral values. Calculate the value of x for a sample of rust that loses 22% mass (as steam) when heated to constant mass.

(e) An underground iron pipe is less likely to corrode if bonded at intervals to magnesium stakes. Give a reason for this. Explain why aluminium would be a poor substitute for magnesium.

5. This question concerns the lead-acid battery. The following data will be required:

	$E^{\ominus}$(V)
$PbO_2(s) + 4H^+(aq) + SO_4{}^{2-}(aq) + 2e^- \rightleftharpoons PbSO_4(s) + 2H_2O(l)$	+1.69
$PbSO_4(s) + 2e^- \rightleftharpoons Pb(s) + SO_4{}^{2-}(aq)$	−0.36

(a) The lead-acid battery is one form of storage cell. What substance is used for:

(i) the negative pole,
(ii) the positive pole, and
(iii) the electrolyte.

(b) Give the equation for the overall cell reaction during discharge.
(c) Calculate the e.m.f. of the cell.
(d) A storage cell, as used in the lead-acid battery, is a simple cell in which the reactions are reversible, i.e., once the chemicals have been used up they can be re-formed. Write an equation for the chemical reaction that occurs on charging.
(e) Give one disadvantage of such batteries used in cars.

CHAPTER 9

The Periodic Table — Chemical Periodicity

In previous chapters, we have seen how atoms participate in reactions, rearranging themselves to form new entities. We have studied these reactions from various angles. Thermodynamically and kinetically, we try to account for the spontaneity of such reactions in terms of structure and bonding. We have also attributed their chemical properties, such as equilibria and redox, in terms of the physical properties of substances.

In this chapter, and in fact the following two chapters, we will apply the theories we have learnt in the prior Physical Chemistry section to understand both the physical and chemical properties of some of the elements in the periodic table. We will dwell on the basics of Inorganic Chemistry, which deals with the study of elements and compounds that are not carbon based with the exception of organometallic compounds [e.g., $B(CH_3)_3$], carbonates and oxides.

There seem to be a lot of elements to cover in the periodic table. This task is made simpler by recognising the periodicity, in other words, the patterns, that lie within the periodic table which allow us to study the properties of the elements, as collective groups, and even to predict the properties of an element that we are not familiar with just by analysing its location in the Periodic Table.

First of all elements are placed in order of increasing atomic number in the Periodic Table.

An element's position is defined by two numbers: the Period Number and the Group Number.

Period numbers, assigned to the horizontal rows, indicate two features for an atom of a specific element:

- the valence (outermost) principal quantum shell that contains electrons, and
- the total number of principal quantum shells that contain electrons.

An element in Period 3 will have a $n = 3$ valence principal quantum shell and correspondingly a total of three principal quantum shells that contain electrons.

Group numbers, assigned to the vertical columns, represent the number of valence electrons in the valence principal quantum shell after we have subtracted away ten units from the group number.

The combination of these two numbers reflects the element's electronic configuration. It can therefore be inferred for an atom such as $_{15}P$, which is located in Period 3 and Group 15, that it has three principal quantum shells of electrons and the valence principal quantum shell contains five electrons $(15 - 10 = 5)$. This means that the valence shell electronic configuration of P is $3s^2 3p^3$ and the electronic configuration of P is written as $1s^2 2s^2 2p^6 3s^2 3p^3$.

Q: Why do we need to subtract ten units from the group number in order to determine the number of valence electrons?

A: The group number is assigned in accordance to the number of elements that are present in Period 4, i.e. a total of 18 elements. For periods 1 to 3, although there are no *d* block elements (Groups 3 to 12) between Groups 2 to 13, the group number is still assigned according to the presence of the *d* block elements in Period 4. Thus, in order to determine the number of valence electrons for elements that come from Groups 13 to 18, we need to subtract ten units from the group number.

Take note that the outermost principal quantum shell may not necessarily be completely filled with electrons.

Since chemical reactions involve the valence shell electrons, an element's chemical properties are determined by its valence electrons. Elements with similar chemical properties are grouped into specific columns and also blocks, with each block denoting the type of valence subshell:

- *s* block: Consists of hydrogen, helium, elements in Group 1 (alkali metals) and Group 2 (alkaline earth metals), all with the *s* subshell filled with electrons. With the exception of hydrogen and helium, these elements are highly reactive and are powerful reducing agents. Unlike the metals in the *d* block, Group 1 and 2 metals are soft with low melting points.
- *p* block: Consists of elements in Groups 13 to 18, all with the *p* subshell as the highest energy subshell that is filled with electron(s). This group contains the non-metals, metalloids (semi-metals with properties of both metals and non-metals) and certain metals such as lead.
- *d* block: Consists of elements from Groups 3 to 12, that have partially or fully filled *d* subshell (d^1 to d^{10}). These elements are hard metals with high melting points. The chemistry behind these elements will be discussed in Chap. 11 on Transition Metal Chemistry.
- *f* block: Consists of the lanthanides (elements with atomic number 57–71) and the actinides (elements with atomic number 89–103). These elements have partially or fully filled *f* subshell.

s block　　d block　　p block

H																	He
Li	Be											B	C	N	O	F	Ne
Na	Mg											Al	Si	P	S	Cl	Ar
K	Ca	Sc	Ti	V	Cr	Mn	Fe	Co	Ni	Cu	Zn	Ga	Ge	As	Se	Br	Kr
Rb	Sr	Y	Zr	Nb	Mo	Tc	Ru	Rh	Pd	Ag	Cd	In	Sn	Sb	Te	I	Xe
Cs	Ba	La*	Hf	Ta	W	Re	Os	Ir	Pt	Au	Hg	Tl	Pb	Bi	Po	At	Rn
Fr	Ra	Ac**	Rf	Db	Sg	Bh	Hs	Mt	Ds	Rg	Cn						

f block

*Lanthanides	Ce	Pr	Nd	Pm	Sm	Eu	Gd	Tb	Dy	Ho	Er	Tm	Yb	Lu
** Actinides	Th	Pa	U	Np	Pu	Am	Cm	Bk	Cf	Es	Fm	Md	No	Lr

9.1 Atomic Structure and Period 3 Elements

In Chap. 1, we have learnt that the variations in physical properties, such as ionisation energy, atomic size and electronegativity, of elements are primarily attributed to the degree of effective nuclear charge (ENC) experienced by the valence electrons.

These fundamental concepts are revisited in this chapter to explain the trend in physical properties of elements in Period 3, excluding in some cases, the noble gases. All noble gases have very stable electronic arrangements, as evidenced by their **high ionisation energy**, **low affinity for additional electrons**, and **general lack of reactivity**.

Understanding the same set of concepts will allow you to account for trends among elements in other periods as well.

The properties covered in this section are:

(i) atomic radius,
(ii) ionic radius,
(iii) first ionisation energy, and
(iv) electronegativity.

9.1.1 *Trend in atomic radius*

Atomic radius ***generally*** *decreases across a period because of greater effective nuclear charge.*

From Na to Ar, there is an increase in nuclear charge (due to the increasing number of protons) but the shielding effect is relatively constant since the number of inner core electrons is the same across a period. Overall, the effective nuclear charge increases across the period, resulting in electrons being pulled closer towards the nucleus.

This brings about two phenomena:

- a general decrease in atomic radius of elements, and
- a general increase in first ionisation energies of the elements across the period.

Q: Based on the above explanation, shouldn't the Group 18 elements have smaller radii than those in Group 17? Why does the graph in Fig. 9.1 indicate the contrary?

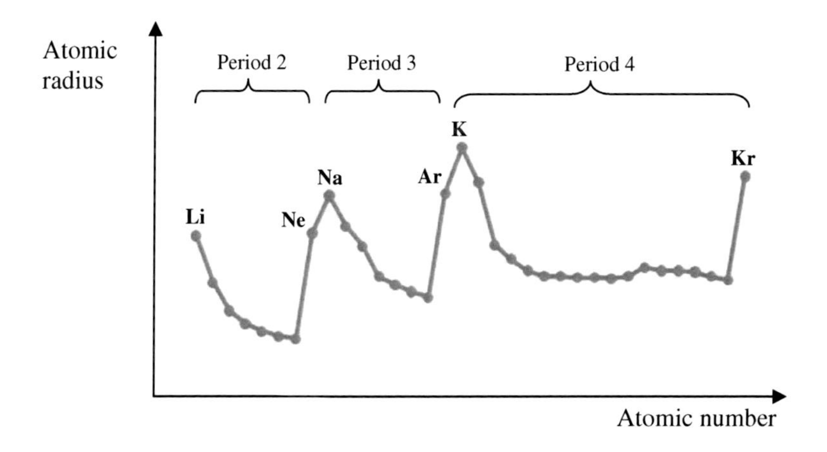

Fig. 9.1. Trend in atomic radius.

A: There are altogether three different types of measurement for atomic radius which depend on the nature of the elemental state of the substance.

For elements that are metallic in nature, such as Na, Mg and Al, the atoms are packed together in a giant metallic lattice structure. Thus, when we measure the atomic radius, we are actually taking the distance from the centre of a nucleus of one atom to its immediate neighbour.

For elements that exist as either simple discrete molecules (such as Cl_2, P_4, S_8) or macromolecules (such as Si, B), the atomic radius is actually what we call the "covalent radius." This is because when two atoms form a covalent bond, the electron clouds overlap, resulting in a smaller measurement of atomic radius as compared to if the atom exists alone in the metallic lattice.

As for Group 18 elements, they are all in the monatomic form. When we solidify them to measure the atomic radii, the weak van der Waals' forces pull the atoms closer to one another. The proximity of the atoms is not as great as that in metals as there is actually a large space in between the atoms. The radius that is being determined is known as the van der Waals' radius, which is a relatively large value, thus accounting for the plots in Fig. 9.1.

9.1.2 *Trend in ionic radius*

The ionic radius refers to the size of a spherical ion in a crystal lattice.

The cationic radius is smaller than the atomic radius.

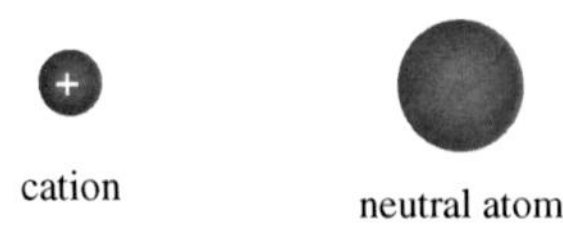

When the valence electrons are removed, the resulting cation has fewer electrons than the neutral atom. The inter-electronic repulsion experienced by the electron cloud of the cation is less than in the neutral atom, and since both species have the same amount of nuclear charge, the net electrostatic attractive force on the electron cloud in the cation is greater than the neutral atom, resulting in smaller cationic size.

The cationic radius decreases from Na^+ to Mg^{2+} to Al^{3+}.

These three cations are iso-electronic, i.e., they have the same number of electrons. The inter-electronic repulsion experienced by the electron cloud of the three cations is the same but the amount of nuclear charge increases from Na^+ to Mg^{2+} to Al^{3+}. Thus, the net electrostatic attractive force on the electron cloud in the cation increases, resulting in smaller cationic size.

The anionic radius is greater than the atomic radius.

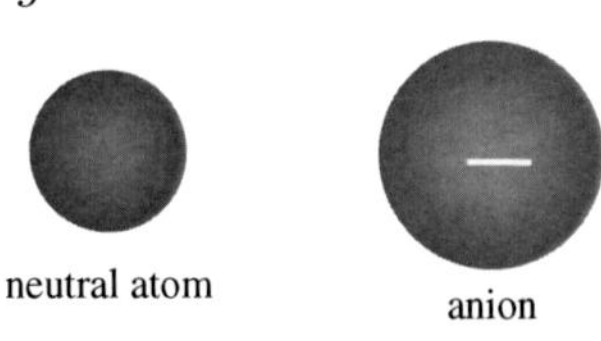

When electrons are added to the valence shell, the resulting anion has more electrons than the neutral atom. The inter-electronic repulsion experienced by the electron cloud of the anion is greater than in the neutral atom, and since both species have the same amount of nuclear charge, the net electrostatic attractive force on the electron cloud in the anion is smaller than the neutral atom, resulting in greater anionic size.

The anionic radius decreases from P^{3-} to S^{2-} to Cl^{-}.

These three anions are iso-electronic. The inter-electronic repulsion experienced by the electron cloud of the three anions is the same but the amount of nuclear charge increases from P^{3-} to S^{2-} to Cl^{-}. Thus, the net electrostatic attractive force on the electron cloud in the anion increases, resulting in smaller anionic size.

Atoms	Na	Mg	Al	Si	P	S	Cl
Atomic radii (pm)	186	160	143	117	110	104	99

Ions	Na^{+}	Mg^{2+}	Al^{3+}		P^{3-}	S^{2-}	Cl^{-}
Ionic radii (pm)	95	65	50		212	184	181

In general, Group 14 elements do not have the tendency to form ions but rather share their electrons in forming covalent compounds.

9.1.3 *Trend in first ionisation energy (1st I.E.)*

1st ionisation energy ***generally*** *increases across a period because of greater effective nuclear charge.*

> From Na to Ar, nuclear charge increases but the shielding effect is relatively constant. Overall, the effective nuclear charge increases, resulting in electrons being pulled closer towards the nucleus. It becomes increasingly difficult to remove the valence electron, i.e., more energy is required, and thus 1st I.E. increases.

However, elements from the **same period** may have actual I.E. different from those predicted by the general trend. There are two possible attributes (more details in Chap. 1):

(i) removal of an electron from different subshells results in the 1st I.E. of Al being lower than Mg;

(ii) removal of an electron facilitated by inter-electronic repulsion results in the 1st I.E. of S being lower than P.

- **Inter-electronic repulsion**:

  Example 1: Why does sulfur have a lower 1st I.E. than phosphorus?

  P: $1s^2 2s^2 2p^6 3s^2 3p^3$
  S: $1s^2 2s^2 2p^6 3s^2 3p^4$

  S has a lower 1st I.E. as less energy is required to remove one of the paired $3p$ electrons since there is inter-electronic repulsion between these electrons in the same orbital.

  Example 2: Why does chlorine have a lower 2nd I.E. than sulfur?

  Cl^+: $1s^2 2s^2 2p^6 3s^2 3p^4$
  S^+: $1s^2 2s^2 2p^6 3s^2 3p^3$

  Cl has a lower 2nd I.E. as less energy is required to remove one of the paired $3p$ electrons since there is inter-electronic repulsion between these electrons in the same orbital.

- **Removal of an electron further away from the nucleus**

  Example: Why does aluminium have a lower 1st I.E. than magnesium?

  Mg: $1s^2 2s^2 2p^6 3s^2$
  Al: $1s^2 2s^2 2p^6 3s^2 3p^1$

  Al has a lower 1st I.E. since less energy is needed to remove its $3p$ electron, which is at a higher energy level (i.e., further away from the nucleus) than the $3s$ electron of Mg.

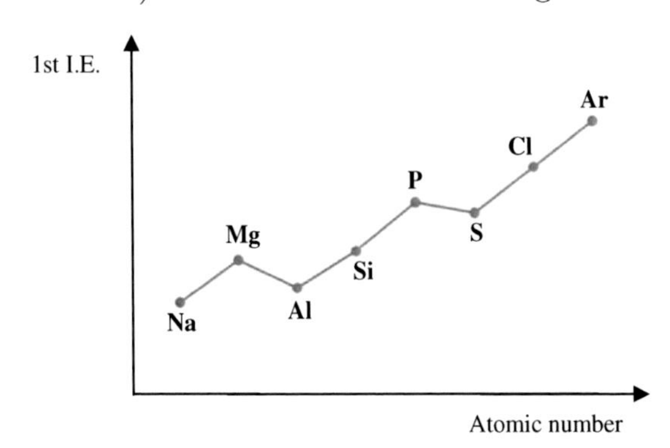

9.1.4 *Trend in electronegativity*

Electronegativity refers to the ability of an atom *in a molecule* to attract shared electrons. The higher the effective nuclear charge, the higher the electronegativity. Thus, we expect electronegativity to increase across a period (here again, with the exception of the noble gases).

Note the term "electronegativity" is used when considering a bonding atom's ability to attract shared electrons. The term "electropositivity" has a directly opposite meaning to electronegativity; it refers to the ability of an atom to lose electrons.

9.2 Structure, Bonding and Period 3 Elements

In this section, we will see how structure and bonding influence the physical properties of the elements. We will touch mainly on the following two properties:

- melting or boiling point, and
- electrical conductivity.

9.2.1 *Variation in melting points and boiling points*

Melting point refers to the temperature at which a pure solid is in equilibrium with its pure liquid at atmospheric pressure. It measures the amount of energy required to break down the regular arrangement of atoms/ions/molecules in a crystal lattice.

The variations in melting point of Period 3 elements are accounted for by the type of structure and the type of bonding or intermolecular forces that need to be overcome for the phase change to occur:

- Group 1, 2 and 13: metals with giant metallic lattice structures;
- Group 14: macromolecules with giant molecular lattice structures;
- Group 15–17: discrete molecules with simple molecular structures.

Na, Mg and Al are metals with **giant metallic lattice structures**. They have high melting points as a great amount of energy is needed to overcome the strong metallic bonds between the positively charged ions and the "sea of delocalised valence electrons." The increasing melting point from Na to Al reflects the increasing strength of metallic bonds from Na to Al due to the **greater number of delocalised electrons** available for metallic bonding.

Si has a **macromolecular structure**. Its melting point is very high as a large amount of heat energy is needed to overcome the **strong Si–Si covalent bonds**.

Elements P, S and Cl exist as discrete molecules with **simple molecular structures**, i.e., P_4, S_8 and Cl_2. Ar exists in the monatomic form. These elements have low melting points since less heat energy is required to overcome the **weak instantaneous dipole–induced dipole (id–id)** interactions between the molecules and, in the case of Ar, atoms.

The strength of id–id interactions depends on the polarisability of the electron cloud and hence on the **number of electrons per molecule**. Since the number of electrons decreases in the order $S_8 > P_4 > Cl_2 > Ar$, the strength of the id–id interactions and hence the melting point decreases in the same order.

The greater the number of electrons present, the more polarisable the electron cloud.

Similar arguments can be used to account for variations in boiling point of these elements.

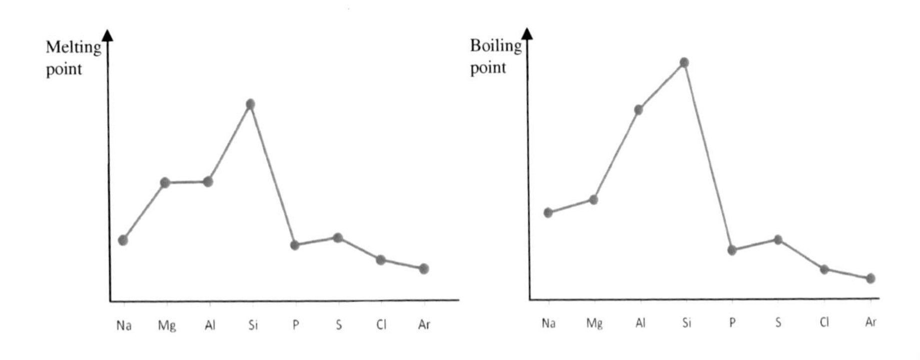

9.2.2 *Variation in electrical conductivity*

Variation in electrical conductivity can also be explained in terms of the structure and bonding involved for the elements.

Na, Mg and Al are metals with **giant metallic lattice structures**. They are good conductors of electricity as there exist delocalised electrons that can act as mobile charge carriers when a potential difference is applied. Electrical conductivity increases from Na to Al due to the **greater number of delocalised electrons** in the metallic lattice structure.

Si has a giant molecular lattice structure that comprises **strong covalent Si–Si bonds**. There are **neither delocalised electrons nor ions** that can act as charge carriers. Thus, Si does not conduct electricity.

Elements P, S and Cl exist as discrete molecules with **simple molecular structures** whereas Ar exists as atoms. For these four elements, there are **neither delocalised electrons nor ions** that can act as charge carriers. Thus, these elements do not conduct electricity.

The periodic trends of Period 3 are summarised as follows:

Group	1	2	13	14	15	16	17
Period 3	Na	Mg	Al	Si	P	S	Cl
Type of element	Metal			Metalloid	Non-metal* Discrete molecules (P_4, S_8, Cl_2)		
Structure	Giant metallic lattice			Giant molecular lattice	Simple molecular		
Bonding	Metallic bonding: Strong electrostatic forces of attraction between **cations** and **sea of delocalised valence electrons**			Covalent bonding: Strong covalent bonds between **atoms** in the molecule. These bonds extend throughout the entire lattice.	Covalent bonding: Strong covalent bonds between **atoms** in the molecule. Weak van der Waals' forces of attraction between **molecules.**		
Melting/ boiling points	High			Very high	Low		
Conductivity	Good			Poor	Nil		

*Group 18 elements, such as Ar and Ne, exist as atoms in all phases. Only weak id–id interactions exist between the atoms. We do not speak of molecules or covalent bonding for these elements.

9.3 Oxides and Chlorides of Period 3 Elements

When asked to describe what metals essentially are, you would most probably list down characteristics that are typical of metals. Physical properties of metals include their good electrical conductivity, malleability and ductility. Chemical properties of metals include their tendency to form **cations, basic oxides, and act as strong reducing agents**.

Why do metals behave this way in reactions? The underlying reason is due to the relative ease of removing electrons from metal atoms. This fact can be *inferred* from the low values of ionisation energy, electron affinity and electronegativity and also the negative standard reduction potentials. All these physical properties simply tell us that metals form cations with ease!

When subjected to reaction with oxygen, metal atoms give up electrons, which are accepted by the oxygen atoms, forming oxide ions, e.g.,

$$2\text{Na(s)} + \frac{1}{2}\text{O}_2\text{(g)} \longrightarrow \text{Na}_2\text{O(s)}.$$

On the other hand, as electronegativity values increase across the period, elements such as P, S, O and Cl have rather small differences in electronegativity values to the point that when they combine with each other, it is more energetically favourable to share electrons forming covalent bonds than to transfer electrons forming ionic bonds.

Thus, the reason ionic compounds are formed between a non-metal and a metal is primarily the large difference in their ability to accept electrons or to lose electrons.

As expected, non-metals readily accept electrons to form **anions and acidic oxides, and behave as oxidising agents**. They exhibit high values of ionisation energy, electron affinity, electronegativity, and also positive standard reduction potentials.

Of course, we may always quote counter examples, such as $AlCl_3$ being a covalent compound, that disagree with this general phenomenon. It is nonetheless useful in helping us to predict the type of compounds that are formed when elements come together in a reaction.

Thus, when the elements of Period 3 react with oxygen and chlorine (refer to subsequent sections), you will find that the metals form ionic compounds with these reactants (except $AlCl_3$), whereas non-metals form simple molecules. Of interest in this chapter are the oxides of Na to S and the chlorides of Na to P.

Element	**Reaction with Oxygen**	**Reaction with Chlorine**	**Possible Oxidation State**
Na	Reacts vigorously to form basic oxide Na_2O: $2Na(s) + \frac{1}{2}O_2(g) \rightarrow Na_2O(s)$	Reacts vigorously to form NaCl: $2Na(s) + Cl_2(g) \rightarrow 2NaCl(s)$	+1
Mg	Reacts vigorously to form basic oxide MgO: $Mg(s) + \frac{1}{2}O_2(g) \rightarrow MgO(s)$	Reacts vigorously to form $MgCl_2$: $Mg(s) + Cl_2(g) \rightarrow MgCl_2(s)$	+2
Al	Vigorous reaction initially but the oxide layer formed soon prevents further reaction: $2Al(s) + \frac{3}{2}O_2(g) \rightarrow Al_2O_3(s)$	Reacts vigorously to form $AlCl_3$: $Al(s) + \frac{3}{2}Cl_2(g) \rightarrow AlCl_3(s)$	+3
Si	Reacts slowly to form SiO_2: $Si(s) + O_2(g) \rightarrow SiO_2(s)$	Reacts slowly to form $SiCl_4$: $Si(s) + 2Cl_2(g) \rightarrow SiCl_4(l)$	+4
P	Reacts vigorously to form P_4O_6 and P_4O_{10} depending on reaction conditions: $4P(s) + 3O_2(g) \rightarrow P_4O_6(s)$, $4P(s) + 5O_2(g) \rightarrow P_4O_{10}(s)$	Reacts slowly to form PCl_3 and PCl_5: $P(s) + \frac{3}{2}Cl_2(g) \rightarrow PCl_3(l)$, $P(s) + \frac{5}{2}Cl_2(g) \rightarrow PCl_5(s)$	+3, +5
S	Reacts slowly to form SO_2, which oxidises very slowly to SO_3 without a catalyst: $S(s) + O_2(g) \rightarrow SO_2(g)$, $2SO_2(g) + O_2(g) \rightarrow 2SO_3(g)$	—	+4, +6

One important phenomenon to note is that the maximum oxidation state possible for an element is limited by the number of valence electrons the element has. The higher the oxidation state, the more likely the compound is covalent in nature.

When we talk about the degree of metallic character of an element, we are actually referring to the extent to which the element exhibits chemical properties pertaining to metals. Thus, moving from left to right across Period 3, there is a decreasing degree of metallic character. This trend is a result of the increasing difficulty of removing valence electrons from the atoms since effective nuclear charge increases across the period. We then expect the strongest reducing agent to be francium (Fr), the bottommost member of Group 1, and the strongest oxidising agent to be fluorine (F), the topmost member of Group 17.

Group	**1**	**2**	**13**	**14**	**15**	**16**	**17**
Period 3	Na	Mg	Al	Si	P	S	Cl
Metallic character	Decreases from left to right						
Reactivity nature of elements	Reducing agents				Oxidising agents		

9.3.1 *Oxides of Period 3 elements*

9.3.1.1 *Variation in melting points and boiling points*

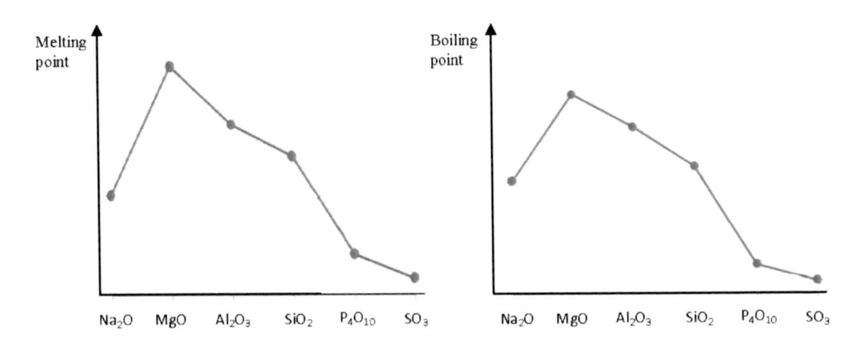

Na_2O and MgO have giant ionic lattice structures with strong electrostatic forces of attraction between the oppositely charged ions in the ionic lattice. Mg^{2+} has a greater charge and smaller ionic radius than Na^+.

These two factors result in a more exothermic lattice energy for MgO and consequently its higher melting point compared to Na_2O. The presence of covalent character in Al_2O_3 diminishes the charge density of the Al^{3+} ion, resulting in a weaker ionic bond and hence a lower melting point than MgO.

> Strength of ionic bond measured by Lattice energy $\propto \frac{q_+ \times q_-}{r_+ + r_-}$.

SiO_2 has a giant covalent lattice structure and its melting point is high as a large amount of heat energy is needed to overcome the strong Si–O covalent bonds.

P_4O_{10} and SO_3 have simple molecular structures. These elements have relatively low melting points since less energy is required to overcome the weak van der Waals' forces of attraction that exist between the molecules. For these covalent oxides, the melting point is determined by the strength of the id–id interactions, which increases with an increase in the number of electrons per molecule. The melting point for P_4O_{10} is thus higher than SO_3.

Similar arguments can be made for the boiling points of the oxides, or any other groups of compounds. The general idea is to account for the data given using your knowledge about structure and bonding for the four types of compounds: metals, ionic compounds, macromolecules and simple covalent compounds.

9.3.1.2 *Acid–base nature of the oxides*

Generally, in crossing the period, we move from the ionic oxides of metals, which are basic, to the oxides of metalloids with giant covalent lattice structure, which are weakly basic, weakly acidic or amphoteric, and finally to the simple molecular oxides of non-metals, which are acidic.

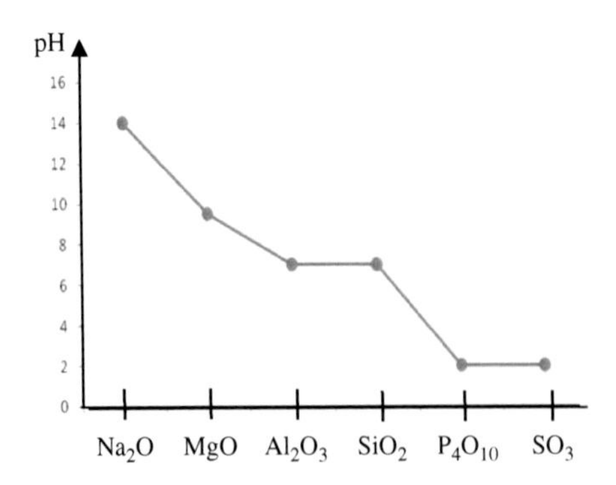

Basic oxides: Na_2O and MgO
Type of structure: Giant ionic lattice structure
Type of bonding: Strong electrostatic forces between cations and anions

	Reaction with H_2O	Reaction with Acid
Na_2O	Na_2O reacts vigorously with H_2O, forming a strongly alkaline solution of pH 13–14: $Na_2O(s) + H_2O(l) \rightarrow 2Na^+(aq) + 2OH^-(aq)$.	$Na_2O(s) + 2H^+(aq) \rightarrow 2Na^+(aq) + H_2O(l)$
MgO	MgO is only slightly soluble in H_2O, forming a weakly alkaline solution of pH 9.5–10.5: $MgO(s) + H_2O(l) \rightleftharpoons Mg^{2+}(aq) + 2OH^-(aq)$. Its low solubility is attributed to the rather strong ionic bond, which holds ions more tightly.	$MgO(s) + 2H^+(aq) \rightarrow Mg^{2+}(aq) + H_2O(l)$

Amphoteric oxide: Al_2O_3
Type of structure: Giant ionic lattice structure
Type of bonding: Strong electrostatic forces between cations and anions

	Reaction with H_2O	Reaction with Acid and Base
Al_2O_3	Al_2O_3 does not react with water due to the extremely strong ionic bond. The pH of solution remains neutral, i.e., pH = 7 at 25°C.	$Al_2O_3(s) + 6H^+(aq) \rightarrow 2Al^{3+}(aq) + 3H_2O(l)$ $Al_2O_3(s) + 2OH^-(aq) + 3H_2O(l) \rightarrow 2[Al(OH)_4]^-(aq)$

Acidic oxide: SiO_2
Type of structure: Giant covalent lattice structure
Type of bonding: Strong covalent bonds between Si and O atoms throughout the entire lattice

	Reaction with H_2O	Reaction with Base
SiO_2	SiO_2 does not react with water due to the extremely strong covalent bond. The pH of solution remains neutral, i.e., pH = 7 at 25°C.	It only reacts with *concentrated* alkalis forming silicate(IV) ions: $SiO_2(s) + 2OH^-(aq) \rightarrow SiO_3^{2-}(aq) + H_2O(l)$

Acidic oxides: P_4O_{10}, P_4O_6, SO_3 and SO_2
Type of structure: Simple molecular structure
Type of bonding: Strong covalent bonds between atoms within each discrete molecule, and weak van der Waals' forces of attraction between molecules

	Reaction with H_2O	Reaction with Base
P_4O_{10}	P_4O_{10} reacts readily with water, forming a strongly acidic solution of pH 2: $P_4O_{10}(s) + 6H_2O(l) \rightarrow 4H_3PO_4(aq)$ phosphoric (V) acid	$P_4O_{10}(s) + 12OH^-(aq) \rightarrow 4PO_4^{3-}(aq) + 6H_2O(l)$
P_4O_6	P_4O_6 reacts readily with water, forming a strongly acidic solution of pH 2: $P_4O_6(s) + 6H_2O(l) \rightarrow 4\ H_3PO_3(aq)$ phosphoric (III) acid	$P_4O_6(s) + 12OH^-(aq) \rightarrow 4PO_3^{3-}(aq) + 6H_2O(l)$
SO_3	SO_3 reacts readily with water, forming a strongly acidic solution of pH 2: $SO_3(g) + H_2O(l) \rightarrow H_2SO_4(aq)$ Sulfuric acid	$SO_3(g) + 2OH^-(aq) \rightarrow SO_4^{2-}(aq) + H_2O(l)$

	Reaction with H_2O	Reaction with Base
SO_2	SO_2 dissolves in water and an equilibrium is established: $SO_2(g) + H_2O(l) \rightleftharpoons H_2SO_3(aq)$ Sulfurous acid	$SO_2(g) + 2OH^-(aq) \rightarrow SO_3^{2-}(aq) + H_2O(l)$

Q: Why are metal oxides known as basic oxides whereas non-metal oxides are acidic oxides?

A: When one dissolves a metal oxide (provided it dissolves) such as Na_2O in water, there are two components in water, namely Na^+ and O^{2-}. Na^+ ions form ion–dipole interactions with water molecules but the charge density is not high enough for them to hydrolyse water molecules. On the other hand, the highly electron rich O^{2-} ion has such a strong affinity for H^+ that it actually abstracts a proton from an H_2O molecule (i.e., O^{2-} acts as the base and H_2O is the acid, from the Brønsted–Lowry definition of acid and base):

$$O^{2-} + H_2O \longrightarrow 2OH^-.$$

Unlike metal oxides, non-metal oxides contain the element covalently bonded to oxygen. The highly electronegative nature of the oxygen atom causes the non-metal element to be highly electron deficient. Since it carries a partial positive charge ($\delta+$), it "welcomes" the attack by the lone pair of electrons from the water molecule. After the attack, H^+ is released, resulting in the formation of oxo-acids (molecular acids that contain oxygen).

Q: Why is Al_2O_3 amphoteric, but Na_2O and MgO are basic?

A: All three oxides are ionic, and the presence of the O^{2-} ions make them basic, i.e., able to react with acid. However, Al^{3+} has a higher charge density than the other two cations and this allows the compound to react with a base (which is electron rich), hence giving rise to its acidic property. Other examples of amphoteric oxides include zinc oxide, tin oxide and water.

Q: Why is MgO partially soluble in water and both Al_2O_3 and SiO_2 are insoluble in water?

A: If we want to know the reason as to why a particular ionic compound dissolves in water or why it does not dissolve, we need to consider two important energies: the amount of energy that is required to break up the lattice and the amount of energy that is released when the ions are hydrated. Thus, we can account for the low solubility of MgO and Al_2O_3, simply because the energy that is released during hydration cannot compensate for the larger amount of energy that is required to break up the lattice (refer to Chap. 2). On the other hand, SiO_2 is a macromolecular compound and it is not energetically feasible to break its strong covalent bonds.

9.3.2 *Chlorides of Period 3 elements*

From Na to P, the chloride varies from ionic to covalent in character. This is because across the period, the electronegativity of the element increases, resulting in a reduced ability for the element to lose electrons. Therefore, forming a covalent compound is a more "preferred" choice!

9.3.2.1 *Variation in melting points and boiling points*

We will now see how the same principles regarding structure and bonding can be applied to account for the boiling point trend of the chlorides of Period 3 elements.

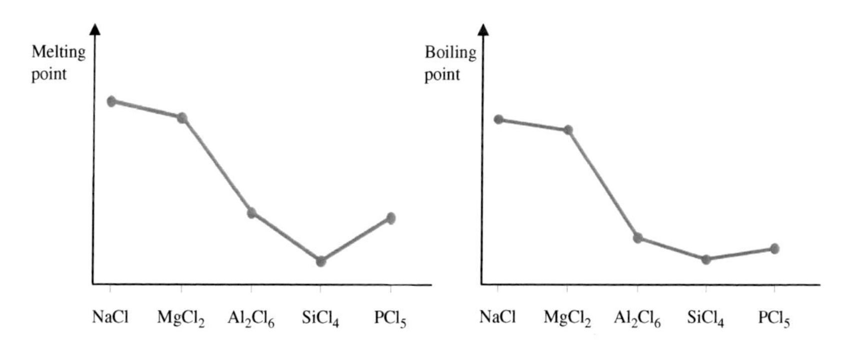

Both NaCl and $MgCl_2$ have giant ionic lattice structures with strong electrostatic forces of attraction between the oppositely charged ions in the ionic lattice. The presence of covalent character in $MgCl_2$ diminishes the charge density of the Mg^{2+} ion, resulting in a weaker ionic bond and hence a lower boiling point than NaCl.

On the other hand, Al_2Cl_6, $SiCl_4$ and PCl_5 are discrete molecules with simple molecular structures. These elements have low boiling points since less heat energy is required to overcome the weak van der Waals' forces of attraction of the id–id type that exist between the molecules. For these covalent chlorides, the boiling point is determined by the strength of the id–id interactions, which increases with an increase in the number of electrons per molecule. Thus, Al_2Cl_6 has the highest boiling point followed by PCl_5 and lastly $SiCl_4$.

Take note that if $AlCl_3$ is used to plot the graph, then the boiling point is lower than $SiCl_4$ because the id–id interaction is weaker due to the smaller number of electrons for $AlCl_3$. Similarly, if PCl_3 is used, the boiling point of PCl_3 should be lower than $SiCl_4$ for the same reason.

9.3.2.2 *Acid–base nature of the chlorides*

We have seen in the previous section that it is the O^{2-} ions that render ionic oxides basic. When it comes to ionic chlorides, the Cl^- ion is unfortunately not electron-rich enough to abstract a H^+ from a water molecule. Thus, when NaCl(s) is added to water, the Na^+ and Cl^- ions merely get hydrated.

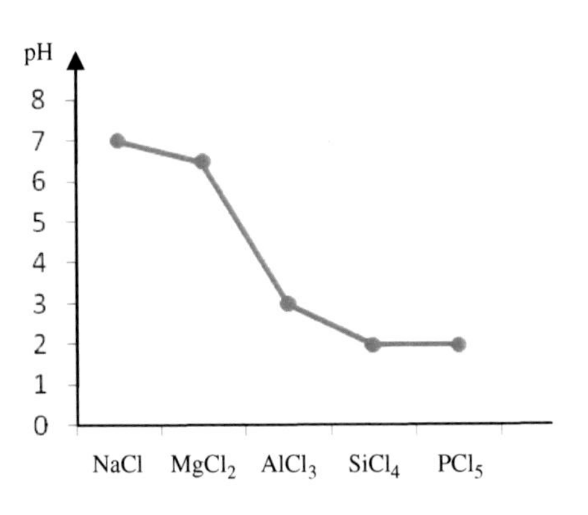

Period 3 Chloride	Type of Structure and Bonding	Reaction with Water
NaCl	Giant ionic lattice structure: Strong electrostatic forces of attraction between cations and anions	Dissolves readily in water, forming a neutral solution of pH 7: $NaCl(s) + aq \rightarrow Na^+(aq) + Cl^-(aq)$
$MgCl_2$		Dissolves readily in water with a small extent of salt hydrolysis, forming a weakly acidic solution of pH 6.5: $MgCl_2(s) + 6H_2O(l) \rightarrow [Mg(H_2O)_6]^{2+}(aq) + 2Cl^-(aq)$, $[Mg(H_2O)_6]^{2+}(aq) + H_2O(l) \rightleftharpoons [Mg(OH)(H_2O)_5]^+(aq) + H_3O^+(aq)$
$AlCl_3$ (or Al_2Cl_6)	Simple molecular structure: Strong covalent bonding between atoms in the discrete molecule.	Undergoes appreciable hydrolysis, producing HCl fumes; the resultant solution has pH = 3: $AlCl_3(s) + 6H_2O(l) \rightarrow [Al(H_2O)_6]^{3+}(aq) + 3Cl^-(aq)$, $[Al(H_2O)_6]^{3+}(aq) + H_2O(l) \rightleftharpoons [Al(OH)(H_2O)_5]^{2+}(aq) + H_3O^+(aq)$ with limited amount of water $AlCl_3(s) + 3H_2O(l) \rightarrow Al(OH)_3(s) + 3HCl(g)$
$SiCl_4$		Readily hydrolyses in water, producing HCl fumes; the resultant solution has pH = 2: $SiCl_4(l) + 2H_2O(l) \rightarrow SiO_2(s) + 4HCl(aq)$
PCl_3	Weak van der Waals' forces of attraction between molecules.	Readily hydrolyses in water, producing HCl fumes; the resultant solution has pH = 2: $PCl_3(l) + 3H_2O(l) \rightarrow H_3PO_3(aq) + 3HCl(aq)$
PCl_5		Readily hydrolyses in water, producing HCl fumes; the resultant solution has pH = 2. The reaction of PCl_5 is highly exothermic! Cold water or a limited amount of water present can limit further hydrolysis of $POCl_3$. When the mole ratio of $PCl_5:H_2O$ = 1:1 **or** when the water added is cold or in a limited amount: $PCl_5(s) + H_2O(l) \rightarrow POCl_3(aq) + 2HCl(aq)$. (i)

(*Continued*)

Period 3 Chloride	Type of Structure and Bonding	Reaction with Water
		When <u>more water</u> is added: $POCl_3(aq) + 3H_2O(l) \rightarrow H_3PO_4(aq) + 3HCl(aq)$. (ii) When <u>excess</u> water is added **AND** water is <u>not cold</u>: $PCl_5(s) + 4H_2O(l) \rightarrow H_3PO_4(aq) + 5HCl(aq)$ (iii)=(i)+(ii)

Q: Why is $AlCl_3$ regarded as a covalent and not an ionic compound?

A: We would expect aluminium chloride to be an ionic compound. However, the high charge density of the Al^{3+} ion and the highly polarisable Cl^- ion result in an accumulation of electron density in the inter-nuclei region, causing the formation of covalent bonds. For the compound AlF_3, the charge density of Al^{3+} is still high but the polarisability of F^- is not high. Thus, overall the compound is ionic in nature.

Q: Why is NaCl neutral in water but both $MgCl_2$ and $AlCl_3$ give acidic solutions when dissolved in water?

A: Both Na^+ and Cl^- ions are not able to break up the water molecule. But for Mg^{2+} and Al^{3+} ions, their charge densities are relatively high, with Al^{3+} much greater than Mg^{2+}. These ions undergo hydrolysis in water, polarising the electron cloud of the surrounding water molecules to the extent that the HO–H bond in water is cleaved, and causing them to give up H^+ ions which contribute to the acidity of the resultant solution. The hydrolysis for the $AlCl_3$ solution is so appreciable that when a carbonate or hydrogencarbonate is added to the solution, CO_2 is evolved. Thus, it is not possible to isolate $Al_2(CO_3)_3$. There are other cations such as Cr^{3+} and Fe^{3+} that behave similarly to Al^{3+}.

Q: Why is Al^{3+}(aq) not form when limited amount of water is added to $AlCl_3$?

A: With limited amount of water, the HCl fumes form do not have a "chance" to dissolve and hence it escapes. Thus, acid-base reaction between $Al(OH)_3$ and HCl(aq) does not take place, unlike when excess water is used.

The reaction of ionic compounds in water is mainly attributed to the reactivity of the ions that go into the aqueous phase. For non-metal chlorides, their effect on water is the same as that of non-metal oxides, i.e., producing acidic solutions. The acidity of both non-metal oxides and chlorides arises due to the electron-deficient non-metal centre.

The elements Si, P and S all have lower electronegativity values compared to Cl or O. Thus, when covalently bonded to Cl or O, each of these elements acquires a partial positive charge ($\delta+$) and is electron deficient. These covalent molecules behave as Lewis acids, accepting the lone pair of electrons from an H_2O molecule. The mechanism of the hydrolysis of $SiCl_4$ is believed to proceed as follows:

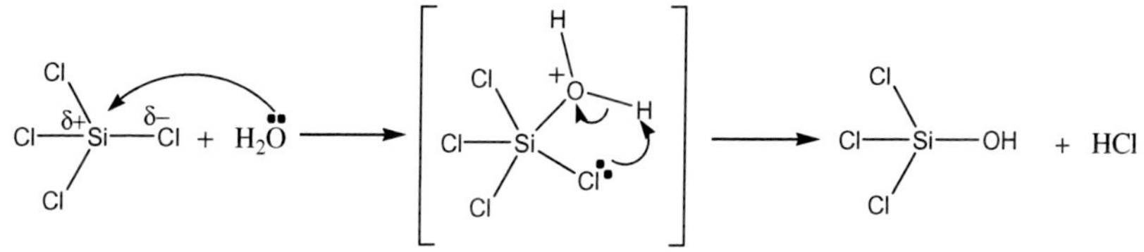

Subsequent attacks on the electron-deficient Si centre will eventually yield $Si(OH)_4$, which will spontaneously lose water to form hydrated SiO_2.

Although carbon is in the same group as silicon, CCl_4 does not hydrolyse in water because of the much smaller size of the C atom relative to the large Cl atoms which hinder the approach of the water molecule. In addition, C cannot accommodate the lone pair of electrons from a water molecule since it does not have vacant low-lying orbitals for expansion of the octet. See Fig. 9.2.

In summary:

- An acidic solution may be accounted for by the presence of
  - a non-metal oxide or chloride,
  - a highly charged metal cation, or
  - a salt containing the conjugate acid of a weak base (e.g., NH_4^+, $CH_3NH_3^+$).
- A basic solution may be accounted for by the presence of
  - a soluble metal oxide, or
  - a salt containing the conjugate base of a weak acid (e.g., CO_3^{2-}, CH_3COO^-).

My Tutorial (Chapter 9)

1. (a) Elements in the *p* block of the periodic table show great variation in physical and chemical properties. Explain this variation, with reference to the properties of aluminium, silicon, phosphorus, and chlorine.
   (b) For any two of these elements explain how their large-scale uses are determined by their physical and chemical properties.
   (c) Normal electric wiring consists of copper wire surrounded by polyvinyl chloride (PVC). In one type of electric wiring used in fire alarm systems, a copper wire is surrounded by solid magnesium oxide acting as an insulator, and then encased in a copper tube covered with PVC.
      (i) What type of bonding is present in magnesium oxide? Explain it can act as an insulator.
      (ii) Suggest why magnesium oxide is preferred to PVC as an insulator in fire alarm systems.

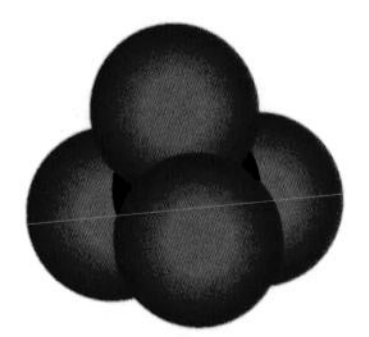 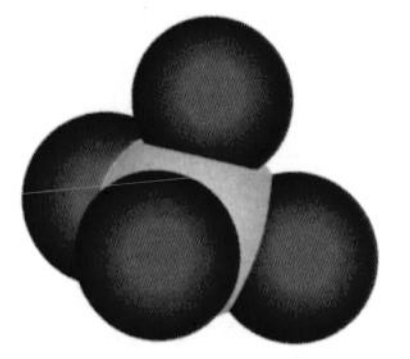

Fig. 9.2. Space filling models of CCl_4 (left) and $SiCl_4$ (right).

2. Aluminium chloride occurs in both the anhydrous state and in the hydrated state. The structure of the anhydrous state may be perceived as having the formula $AlCl_3$. When water is added to solid anhydrous aluminium chloride, white acidic fumes are seen.

   (a) Name the white acidic fumes.
   (b) Write an equation for the reaction occurring.
   (c) Explain with reference to the structure of $AlCl_3$ how the first step of this reaction occurs.
   (d) When water is added to hydrated aluminium chloride, no white fumes are observed. Explain this in terms of the bonding in hydrated aluminium chloride.
   (e) Hydrated aluminium chloride dissolves in water to give an acidic solution. Explain why the solution is acidic with an appropriate equation.

3. NO_2 reacts with aqueous sodium hydroxide according to the following equation:

$$2OH^- + 2NO_2 \longrightarrow NO_2{}^- + NO_3{}^- + H_2O.$$

   (a) What type of reaction is this? Justify your answer.
   (b) Deduce the ionic half-equations for this reaction.

4. (a) Sodium hydroxide is manufactured by an electrolytic process using a diaphragm cell.

      (i) What is used as the electrolyte?
      (ii) What material is each of the anode and cathode made of, respectively?
      (iii) Give an equation for the reaction occurring at each of the electrodes.
      (iv) Give one reason why it is necessary to separate the two electrodes into two compartments.
      (v) Write an equation for the overall cell reaction.

   (b) Give one large-scale industrial use for each of the following:

      (i) chlorine; and
      (ii) hydrogen.

(c) Iron(III) oxide is a basic oxide. What type of oxide is:

(i) aluminium oxide; and
(ii) silicon dioxide?

(d) Bauxite is an ore containing hydrated aluminium oxide, iron(III) oxide and silicon dioxide. In order to obtain a purer form of aluminium oxide, bauxite is heated with a 10% solution of sodium hydroxide in which the aluminium oxide dissolves.

(i) Write an equation for the reaction of aluminium oxide with sodium hydroxide.
(ii) Why does iron(III) oxide not dissolve in sodium hydroxide?
(iii) Why does silicon dioxide not dissolve in a 10% solution of sodium hydroxide?

5. (a) Give the formulas of the chlorides of the elements of Period 3.
(b) Calculate the percentage by mass of chlorine in the chloride of silicon.
(c) (i) Draw a dot-and-cross diagram to show the bonding in the chloride of silicon.
(ii) Draw the shape of this molecule. Explain your answer in terms of the Valence Shell Electron Pair Repulsion (VSEPR) theory.
(iii) State the shape of a molecule of $AlCl_3$ and explain why it is different from that of the chloride of silicon.
(iv) Give an equation for the reaction of the chloride of silicon with cold water.
(v) How does the behaviour of carbon tetrachloride with cold water compare with that in part (iv)? Explain any differences.

6. (a) Study the table of ionisation energies below and answer the questions which follow.

Ionisation Energy ($kJ\,mol^{-1}$)	1st	2nd	3rd	4th
Sodium	494	4560	6940	9540
Magnesium	736	1450	7740	10500
Aluminium	577	1820	2740	11600

Explain the *relative* magnitudes of the following:

(i) the 1st ionisation energies of sodium and magnesium;
(ii) the 1st ionisation energies of magnesium and aluminium;
(iii) the 2nd ionisation energies of sodium and magnesium;
(iv) the 2nd ionisation energies of magnesium and aluminium;
(v) the 3rd and 4th ionisation energies of aluminium.

(b) Consider the electron affinities for oxygen given below:

Electron Affinity ($kJ\,mol^{-1}$)	1st	2nd
	−142	+844

(i) Write equations representing the changes to which the 1st and 2nd electron affinities of oxygen relate.
(ii) Explain the relative magnitudes of the 1st and 2nd electron affinities of oxygen.
(iii) Given the endothermic nature of the 2nd electron affinity of oxygen, comment briefly on the thermodynamic stability of ionic metal oxides.

CHAPTER 10

Chemistry of Groups 2 and 17

In this chapter, we will deal closely with Group 2 and Group 17 elements, looking at group trends with regard to their properties and the reactions of their compounds. We will find patterns and similarities for elements that fall within the group as they have the same valence shell electronic configuration. This is one of the most important attributes of the properties displayed, as the valence electrons are the actual "participants" in a chemical reaction.

Group 2 elements are commonly called alkaline earth metals. The term is derived from the fact that Group 2 oxides react with water to produce alkalis and most of the compounds are found in the earth's crust. The properties of magnesium and elements below it are discussed as a group with the exclusion of beryllium, which has distinctively different properties from the others. The chemistry of radioactive radium will not be discussed here.

Group 17 elements are commonly known as the halogens, which has its roots in the Greek for salt generators. As you might have guessed by now, halogens are known to form salts, when they react with metals. Group 17 elements exist as discrete diatomic molecules — molecules which comprise, as the prefix *di-* suggests, two atoms (e.g., F_2 and Cl_2). See Table 10.1.

Table 10.1 Elements in Groups 2 and 17

Group 2	Group 17
Beryllium (Be)	Fluorine (F_2)
Magnesium (Mg)	Chlorine (Cl_2)
Calcium (Ca)	Bromine (Br_2)
Strontium (Sr)	Iodine (I_2)
Barium (Ba)	Astatine (At_2)
Radium (Ra)	

10.1 Atomic Structure and Group Trends

10.1.1 *Trend in atomic radius*

As we move down a group, nuclear charge increases but electrons are being added to a higher energy principal quantum shell which is further away from the nucleus. This results in the valence electrons experiencing weaker attractive forces from the nucleus, which gives rise to a greater atomic radius.

10.1.2 *Trend in ionic radius*

Cationic radius increases down a group. The cations of the elements in the same group have the same charge, e.g., +2 for cations of Group 2 elements and +1 for cations of Group 1 elements. As we move down a group, the nuclear charge of the cation increases but its valence electrons are further away from the nucleus. This results in the valence electrons experiencing weaker attractive forces, which gives rise to a greater cationic size. This is also the explanation for the increase in the ionic radius of the anions, of Group 17 elements, down the group.

10.1.3 *Trend in 1st I.E.*

In general, 1st I.E. decreases down a group. As we move down a group, nuclear charge increases but electrons are being added to a higher energy principal quantum shell which is further away from the nucleus. This results in the valence electrons experiencing weaker attractive forces, and they are consequently more easily removed. This is also the explanation for a decrease in the nth I.E. down the group of elements.

10.1.4 *Trend in electron affinity*

Electron affinity generally decreases (i.e., becomes less exothermic) down a group for similar reasons to those that applied to the trend in ionisation energy — both group trends are attributed to the increase in the distance of the valence shell from the nucleus.

Overall, the distance of the valence shell from the nucleus increases down a group, resulting in weaker attractive forces that lead to the valence electrons being less tightly held by the nucleus. Since it becomes easier to remove a valence electron from atoms of increasing sizes, this also implies that it is more difficult for atoms of increasing atomic sizes to attract electrons from an external source.

10.1.5 *Trend in electronegativity*

Just as in the case of electron affinity, the greater the atomic size, the smaller the electronegativity value. Thus, we expect electronegativity to decrease down a group.

10.1.6 *Trend in metallic character*

As we go down a group, it becomes easier to remove electrons from the atoms and thus metallic character increases down a group. In other words, electropositivity increases down the group.

Example 10.1: State the equation that represents the first electron affinity of an atom. Account for the trend in electron affinity of Group 17 elements.

	F	**Cl**	**Br**	**I**
1st electron affinity ($kJ\,mol^{-1}$)	−333	−348	−340	−297

Solution:

$$X(g) + e^- \longrightarrow X^-(g).$$

From Cl to I, the attraction of the nucleus for the valence electrons gets weaker because of the increased distance of the valence shell from the nucleus. Any added electron will be less strongly attracted by the nucleus and this is indicated by the less exothermic electron affinity from Cl to I. The electron affinity of fluorine is less exothermic than that of chlorine because of the very small size of the fluorine atom, which results in the added electron being strongly repelled by the electrons that are already present in the valence shell.

Example 10.2: Account for the decrease in the 1st I.E. down Group 2.

Solution: Moving down Group 2, electrons are added to a higher energy principal quantum shell. Nuclear charge increases but the valence electron is further away from the nucleus. This results in weaker electrostatic attractive forces experienced by the valence electrons. Removal of a valence electron is easier and this leads to a decrease in the 1st I.E. down the group.

Note: The same answer is used to account for the increase in reducing power of Group 2 elements down the group.

Example 10.3: Account for the trend in electronegativity of Group 17 elements.

Solution: Electronegativity refers to the ability of the atom to distort shared electron cloud and valence electrons are used in sharing. Moving down Group 17, nuclear charge increases but the valence electrons are further away from the nucleus. This results in weaker electrostatic attractive forces experienced by the shared electrons, and thus less distortion of the shared electron cloud.

10.2 Physical Properties of Group 2 Elements

Melting and boiling for Group 2 elements (Table 10.2) involve breaking of the metallic bond. Generally, metallic bond strength decreases down the group. Although the number of valence electrons available for delocalisation is the same down the group, atomic size increases down the group. This results in weaker electrostatic attractive forces experienced by the valence electrons, and thus weaker metallic bond strength. Anomalies that are observed may arise from the specific packing of atoms in the giant metallic crystal lattice.

Table 10.2 Melting and Boiling Points of Group 2 Elements

	Be	Mg	Ca	Sr	Ba
Melting point (°C)	1278	649	839	769	725
Boiling point (°C)	2970	1107	1484	1384	1643

10.3 Chemical Properties of Group 2 Elements

As implied by the group number "2", the valence shell electronic configuration is ns^2 for these elements. They are readily oxidised, losing the two valence electrons to form cations of +2 charge. Group 2 elements are therefore strong reducing agents.

Since ionisation energy decreases down the group, it is expected that as we go down Group 2, the atom loses electrons more readily. In other words, the reducing power (the element itself is oxidised, i.e., loses electrons) of the Group 2 elements increases down the group. The strength of the reducing power is reflected by the increasing *negative* $E^\ominus$ value down the group.

The more negative $E^\ominus$ value indicates the increasing difficulty of reducing the metal cation back to the metal. This means that it is easier to oxidise the metal atom and thus the reducing power of the metal is stronger, the more negative the $E^{\ominus}$ value:

$$Mg^{2+} + 2e^- \rightleftharpoons Mg, \quad E^{\ominus} = -2.38\,V,$$
$$Ca^{2+} + 2e^- \rightleftharpoons Ca, \quad E^{\ominus} = -2.87\,V,$$
$$Sr^{2+} + 2e^- \rightleftharpoons Sr, \quad E^{\ominus} = -2.89\,V,$$
$$Ba^{2+} + 2e^- \rightleftharpoons Ba, \quad E^{\ominus} = -2.90\,V.$$

Group 2 elements, with the exception of Be (see Sec. 10.6 on Be chemistry), generally undergo similar reactions, although in varying conditions, to form ionic compounds. When subjected to the same reactants such as water and oxygen, the reaction occurs more readily and the enthalpy change of reaction is more exothermic down the group. This is due to the increasing reactivity from Mg to Ba.

What best describes this reactivity? The reactivity depends on the ease with which they lose electrons. The easier it is to lose electrons, the higher the degree of spontaneity of a reaction occurring. Therefore, **the reactivity of Group 2 elements in forming other compounds actually mirrors their reducing power.**

With respect to their reactions with acid and water, Group 2 elements are similar to Group 1 elements:

$$2K(s) + 2HCl(aq) \longrightarrow 2KCl(aq) + H_2(g),$$
$$Ca(s) + 2HCl(aq) \longrightarrow CaCl_2(aq) + H_2(g),$$

$$2K(s) + 2H_2O(l) \longrightarrow 2KOH(aq) + H_2(g),$$
$$Ca(s) + 2H_2O(l) \longrightarrow Ca(OH)_2(aq) + H_2(g).$$

Q: Should one expect Group 2 elements to react more vigorously with water than Group 1 elements from the same period?

A: The relative reactivity of metals is related to the relative ease with which they lose electrons. A more negative standard reduction potential indicates that the oxidation of metal occurs more readily:

$$Ca^{2+} + 2e^- \rightleftharpoons Ca, \quad E^{\ominus} = -2.87\,V,$$
$$K^+ + e^- \rightleftharpoons K, \quad E^{\ominus} = -2.92\,V.$$

Hence, potassium metal is expected to react more vigorously with water than calcium metal. This stems from the fact that it is less energetically demanding to remove one electron from the potassium atom than to remove two electrons from the calcium atom.

Q: Since the enthalpy change of hydration of Ca^{2+} is more exothermic than that of K^+, shouldn't the oxidation of Ca, and hence its reaction with water, be more favourable?

A: The more exothermic enthalpy change of hydration of Ca^{2+} is due to the higher charge density of Ca^{2+} than that of K^+, which leads to stronger ion-dipole interactions formed between Ca^{2+} ions and water molecules. Unfortunately, the energy released from the hydration of Ca^{2+} is insufficient to compensate for the large amount of energy needed to ionise Ca. The net result is a less negative standard reduction potential for Ca than K.

10.3.1 *Reaction with water*

Mg reacts very slowly with cold water since it is less reactive than Ca, Sr and Ba. The reaction becomes faster at high temperatures with Mg reacting rapidly with steam to form the metal oxide and hydrogen gas:

$$\mathrm{Mg(s) + H_2O(g) \longrightarrow MgO(s) + H_2(g).} \qquad (10.1)$$

On the other hand, for Ca, Sr and Ba, their greater reducing strength renders their reaction with water spontaneous even at low temperatures. They need only react with cold water to readily form the metal oxide. The resultant solution is alkaline as the oxide formed further reacts with water to produce the respective hydroxide, e.g.,

$$\mathrm{CaO(s) + H_2O(l) \longrightarrow Ca(OH)_2(aq).}$$

Thus, for Ca, Sr and Ba, the overall equation for the reaction with water is as follows:

$$M\mathrm{(s) + 2H_2O(l) \longrightarrow} M\mathrm{(OH)_2(aq) + H_2(g),} \qquad (10.2)$$

where M = Ca, Sr or Ba.

The different products in Eqs. (10.1) and (10.2) reflect the relative difference in solubility of the Group 2 hydroxides. These hydroxides have low solubility in water with $Mg(OH)_2$ being the least soluble (see Sec. 10.7 for more details).

10.3.2 *Reaction with oxygen*

As with the Group 1 elements, the highly reactive nature of Group 2 elements with oxygen and water prevents them from naturally existing in the free elemental form. Instead, these metals are extracted by electrolysis from their respective salts. Once formed, they quickly tarnish when exposed to air as the reaction with oxygen produces the metal oxide which coats the metal surface. That is why these metals are normally stored in liquid paraffin to prevent oxidation from occurring, e.g.,

$$2Ca(s) + O_2(g) \longrightarrow 2CaO(s).$$

However, storage in an oxygen-free environment is not crucial for Be and Mg as these metals, once produced, are coated with thin layer of fairly impermeable oxide which is difficult to remove. This so-called "protective" oxide layer prevents the element from further oxidation and accounts for its less reactive nature compared to the other Group 2 metals.

As expected, the reactivity with oxygen increases from Mg to Ba to the extent that finely divided Sr metal and Ba ignite spontaneously in air without heating being involved.

When Group 2 metals are subjected to combustion, they burn with distinctive brilliant coloured light. This property is harnessed in pyrotechnics and in the manufacture of flares. For instance, magnesium metal gives off radiant white light and calcium metal produces intense red light.

10.4 Thermal Stability of Group 2 Compounds

In this section, we will discuss the thermal decomposition of some Group 2 salts, namely, the nitrates, carbonates and hydroxides.

We will seek to understand the driving force behind these decompositions.

Group 2 carbonate decomposes to give the oxide and carbon dioxide, e.g.,

$$MgCO_3(s) \underset{\Delta}{\longrightarrow} MgO(s) + CO_2(g).$$

Group 2 nitrate decomposes to give the oxide, nitrogen dioxide and oxygen, e.g.,

> *Contrast with Group 1 nitrates:*
> $2KNO_3(s) \underset{\Delta}{\longrightarrow} 2KNO_2(s) + O_2(g)$

$$Ca(NO_3)_2(s) \underset{\Delta}{\longrightarrow} CaO(s) + 2NO_2(g) + 1/2O_2(g).$$

Group 2 hydroxide decomposes to give the oxide and steam, e.g.,

$$Ba(OH)_2(s) \underset{\Delta}{\longrightarrow} BaO(s) + H_2O(g).$$

Q: Why does the decomposition of Group 2 carbonate, nitrate and hydroxide all result in the formation of the same product, the oxide?

A: This shows that the oxide formed is relatively more stable than the corresponding Group 2 carbonate, nitrate or hydroxide compounds. This is the driving force for decomposition — the formation of a more stable compound.

Q: How can we know that the oxide is more stable than the other three types of compounds? How is stability measured?

A: Since Group 2 compounds (with the exception of Be compounds) are ionic, the stability of the solid compound is dependent on the strength of the ionic bonds which is indicated by the magnitude of the lattice energy. In general, the more exothermic the lattice energy, the stronger the ionic bonds, the more stable the compound, and the more difficult it is to decompose.

$$\text{Lattice energy} \propto \frac{q_+ \times q_-}{r_+ + r_-}, \tag{10.3}$$

where q is charge and r is ionic radius.

Equation (10.3) shows that an ionic compound comprising ions of a higher charge and a smaller ionic radius will tend to be more stable than a compound comprising ions of a smaller charge and a larger ionic radius.

The underlying reason is that the greater the charges on the ions, the stronger the attraction between the oppositely charged ions will be, and this gives rise to a more stable ionic compound. With smaller sizes, oppositely charged ions can be closer to one another in the solid lattice, forming stronger attractions, and this also leads to a more stable compound.

Example 10.4: Why does $CaCO_3$ decompose to CaO upon heating?

Solution:

$$CaCO_3(s) \xrightarrow[\Delta]{} CaO(s) + CO_2(g).$$

The driving force for the decomposition lies in the formation of an energetically more stable compound CaO compared to $CaCO_3$. The greater stability of CaO is attributed to the smaller anionic radius of O^{2-} as compared to CO_3^{2-}. This allows the O^{2-} ions to be closer to the cations. The ionic bonds between the Ca^{2+} and O^{2-} ions are thus much stronger and this makes CaO more stable than $CaCO_3$. In addition, the decomposition reaction is also driven by an increase in the entropy of the system due to the formation of gaseous CO_2, which has a higher degree of disorder.

Using similar arguments, the decomposition of nitrate and hydroxide to form the oxide can be explained as follows.

Example 10.5: Why does $Ca(NO_3)_2$ decompose to CaO upon heating?

Solution:

$$Ca(NO_3)_2(s) \xrightarrow[\Delta]{} CaO(s) + 2NO_2(g) + 1/2O_2(g).$$

The driving force for the decomposition lies in the formation of an energetically more stable compound CaO compared to $Ca(NO_3)_2$. The greater stability of CaO is attributed to the higher charge and smaller anionic radius of the O^{2-} ion that results in stronger ionic bonds between the Ca^{2+} and O^{2-} ions, and this makes CaO more stable than $Ca(NO_3)_2$. In addition, the decomposition reaction is also driven by an increase in the entropy of the system due to the formation of gaseous NO_2 and O_2, which have a higher degree of disorder.

Since the stability of ionic compounds is dependent on the strength of ionic bonds, the thermal stability of the Group 2 carbonates also differs for the same reason and this is reflected in the temperature at which decomposition occurs.

Based on Table 10.3, $MgCO_3$ decomposes at a relatively low temperature of about 660°C and as we move down the group, an increasing amount of heat energy is needed to initiate decomposition of the same amount of carbonates, i.e., the thermal stability of carbonates increases down the group.

Table 10.3 Decomposition Temperatures of Group 2 Carbonates

	$MgCO_3$	$CaCO_3$	$SrCO_3$	$BaCO_3$
Decomposition temp. (°C)	660	900	1100	1300

As discussed in Chap. 2 on Chemical Bonding, ionic compounds do not have 100% pure ionic bonding with distinct separation of the opposite charges. If the cation has a sufficiently high charge density, it can actually attract the electron density of the anion. Using CO_3^{2-}, NO_3^- and OH^- as examples, this distortion of the electron cloud of the anion actually weakens the intra-molecular bonds, i.e., C–O, N–O and O–H bonds in the anions, respectively. The anion is said to be polarised by the cation. The greater the extent of distortion, the weaker the intra-molecular bonds in the anion and the easier it is to break these bonds. Thus, a lower temperature is needed to decompose the compound.

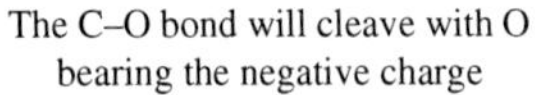

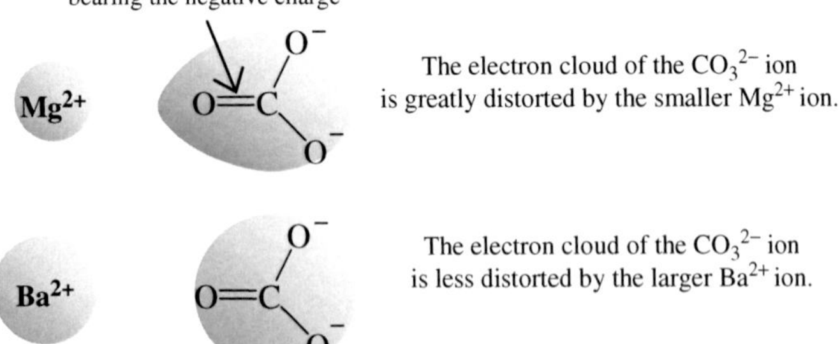

> Recall: charge density $\propto \frac{q_+}{r_+}$.

The extent of the distortion of the electron cloud is dependent on both the polarising power of the cation and the polarisability of the anion (refer to Chap. 2). How strongly the cation can distort the electron cloud of the anion (i.e., its polarising power) is measured by its charge density. This makes plain sense: the more "concentrated" the positive charge, the more "powerful is its attractive power".

Moving down Group 2, the size of the cation increases but the amount of charge is the same at +2. Therefore, charge density decreases from Mg^{2+} to Ba^{2+} and this translates to a decrease in polarising power of the cations down the group.

As for the anions, CO_3^{2-}, NO_3^- and OH^-, they are easily polarised as they are large anions — the greater the anionic size, the further the electron cloud is from the nucleus, which means it will be less tightly attracted to the nucleus and hence more easily distorted.

Example 10.6: Account for the trend in decomposition temperatures of Group 2 carbonates.

Solution: Moving down Group 2, the cationic size increases, resulting in a decrease in charge density and hence a decrease in the polarising power of the cations. The electron cloud of the CO_3^{2-} ion is distorted to a lesser extent and thus a smaller weakening effect is acted upon the intra-molecular C–O bonds by the cations as we move down Group 2. Thermal decomposition becomes increasingly difficult, thus thermal stability increases down the group. This is observed by the increasing temperature needed for decomposition to occur.

The decomposition temperatures of Group 2 nitrates and hydroxides are observed to follow a similar trend and thus we use the same arguments to account for this trend (in the description given above, just replace CO_3^{2-} with NO_3^- or OH^- and replace C–O with N–O or O–H).

10.5 Some Uses of Group 2 Elements and Their Compounds

- Magnesium is used in the manufacture of strong and lightweight alloys, which are widely used in industries such as automobiles and aerospace.
- Magnesium oxide, because of its high melting point and low reactivity, is mainly used as a refractory material in furnace linings needed in the production of materials such as iron and steel.
- Magnesium hydroxide is used in "milk of magnesia," which is used to alleviate constipation and treat acid indigestion.
- Calcium oxide (known as lime or quicklime), when mixed with water, produces calcium hydroxide (slaked lime) whose uses include reducing soil acidity in agriculture and in the production of mortar and plaster.
- Calcium carbonate has primary uses in the construction industry: as building materials and in the building of roads, and in the making of lime and glass.
- Barium sulfate is used as a radiocontrast agent for the X-ray imaging of the gastro-intestinal tract. Given its very low solubility and ease of removal from the body, the patient is protected from absorbing harmful amounts of the heavy metal.

10.6 Properties of Beryllium

Beryllium is less reactive than the other Group 2 elements. The protective oxide layer that coats the Be metal surface prevents the element from further oxidation, and at the same time, accounts for its less reactive nature. This explains why beryllium does not react with water or steam.

Other properties of beryllium which differ from the other Group 2 elements include the following:

- Many beryllium compounds, such as $BeCl_2$, BeO and BeH_2 are covalent and not ionic in nature.
- The oxide and hydroxide of Be are amphoteric whereas those of the other Group 2 elements are basic.

Example 10.7: Explain why $BeCl_2$ (b.p. 547°C) is more volatile than $MgCl_2$ (b.p. 1418°C).

Solution: Unlike $MgCl_2$, which is an ionic compound, $BeCl_2$ exists as discrete non-polar covalent molecules. Less energy is needed to overcome the weak instantaneous dipole-induced dipole interactions between $BeCl_2$ molecules than that needed to overcome the stronger ionic bonds in $MgCl_2$. Thus, the boiling point of $BeCl_2$ is lower and it is more volatile.

The considerably lower boiling point of $BeCl_2$ compared to $MgCl_2$, and in fact to all ionic compounds, is evidence that $BeCl_2$ is not ionic in nature. But what could be the reasons for its lack of the ionic property?

For one, the idea of it being a covalent compound means that there is sharing of electrons between the metal beryllium atom and the chlorine atoms. We can infer from this that it is more energetically favourable to share electrons than for the beryllium atom to lose electrons to form Be^{2+}. This supposition can be substantiated by the fact that the Be atom has a much higher ionisation energy compared to its Group 2 counterparts as the attraction between the nucleus and the valence electrons is substantially stronger due to their close proximity.

Instead of putting an end to the discussion, let us venture further. What if the formation of Be^{2+} were favourable; there would then be electron transfer from the metal atom to the chlorine atoms, resulting in ionic bonding between oppositely charged ions. However, if ionic

$BeCl_2$ were formed this way, the high charge and small size of Be^{2+} would cause it to have such high polarising power that it could distort the electron cloud of the anion and create a covalent bond. This is analogous to the explanation that is used to explain the nature of $AlCl_3$ in Chap. 9 on Periodicity.

The high charge density of the beryllium cation also accounts for its acidic properties. In aqueous solution, the Be^{2+} ion exists as an aqua-complex ion, $[Be(H_2O)_4]^{2+}$. Due to its high charge and small size, the Be^{2+} ion has a high charge density and hence high polarising power. It distorts the electron cloud of the H_2O molecules bonded to it, weakening the O–H bonds and enabling these bonded H_2O molecules to become better proton donors. The free water molecules in the solution act as bases and the following equilibrium is established:

$$[Be(H_2O)_4]^{2+}(aq) + H_2O(l) \rightleftharpoons [Be(OH)(H_2O)_3]^{+}(aq) + H_3O^{+}(aq).$$

The Be^{2+} ion is consequently said to undergo **appreciable hydrolysis** in aqueous solution. The slight excess of H_3O^+ ions in the solution causes the solution to be acidic and enables it to react with strong bases. Again, such behaviour is extremely similar to that of Al^{3+} (refer to Chap. 9).

The acidic nature of BeO and $Be(OH)_2$ can be concluded from observations made when separate solutions, each containing a Group 2 cation, are subjected to drop-wise addition of NaOH until the latter is added in excess (see Fig. 10.1).

Observations:	white precipitate of $Mg(OH)_2$ is insoluble in excess NaOH(aq)
	white precipitate of $Be(OH)_2$ is soluble in excess NaOH(aq)

Q: BeO and $Be(OH)_2$ are amphoteric. We can use the reaction with NaOH(aq) to test for their acidic property. How do we test for their basic property?

A: One just needs to add a mineral acid to it. If it is a base, one gets salt and water being formed.

If you study the properties of beryllium, you will find them somewhat similar to those of aluminium. For instance, both are not reactive

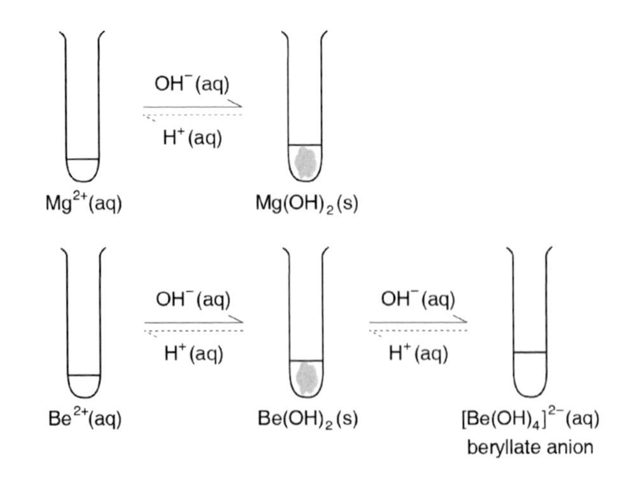

Fig. 10.1. Reaction of Mg^{2+}(aq) and Be^{2+}(aq) with NaOH(aq).

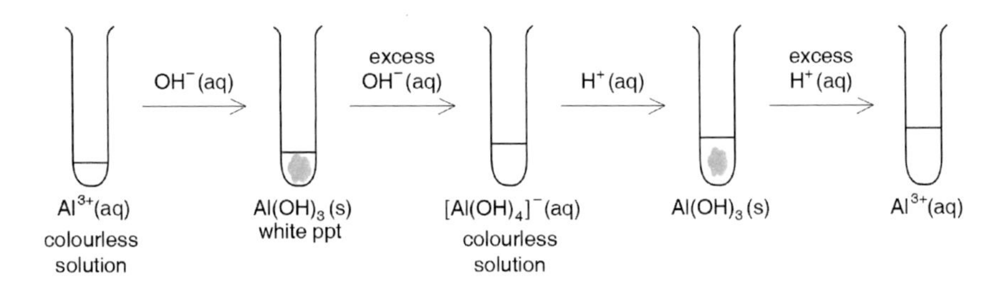

Fig. 10.2. Reactions showing the amphoteric nature of $Al(OH)_3$.

due to the formation of a protective oxide layer and they both form amphoteric oxides (see Fig. 10.2). This is one pair of diagonally adjacent elements which exhibit similar properties to each other — a feature known as the "diagonal relationship" in the Periodic Table.

Example 10.8: Explain why beryllium oxide is amphoteric whereas the other Group 2 oxides are basic.

Solution: Due to the **high charge density** of the beryllium ion, there is **covalent character** in beryllium oxide. The Be atom is highly electron deficient, giving it an **acidic characteristic**, thereby enabling it to react with a stronger base such as NaOH or KOH. The presence of O^{2-} allows BeO to be able to react with acid to form salt and water.

Q: Would you expect beryllium salts to be more or less thermally unstable compared to those of the other Group 2 elements?

A: Beryllium salts are more thermally unstable due to the higher charge density of Be^{2+}, which gives rise to greater polarising power. This makes the electron clouds of the anions that are attracted to the Be^{2+} ion more distorted, causing the intra-molecular bonds to be weakened to a greater extent. Thus, it is easier to decompose beryllium salts.

10.7 Solubility of Group 2 Compounds

In general, solubility decreases down the group for Group 2 compounds that consist of large anions such as SO_4^{2-}, CO_3^{2-} and NO_3^- (see Table 10.4).

On the other hand, solubility increases down the group for Group 2 compounds that consist of small anions such as OH^- and F^-.

Why is $MgSO_4$ soluble but not $BaSO_4$? Why does Mg^{2+} form the white precipitate $Mg(OH)_2$ with $NH_3(aq)$ but not Ba^{2+}?

The opposing trends in solubility of the above compounds may be explained by considering the variations in hydration energy and the energy that is required to break-up the lattice (−L.E.) that arise from the differences in ionic sizes.

Table 10.4 Solubility (mol/g water) at 25°C of some Group 2 Compounds

Cation	Hydroxide, OH^-	Sulfate, SO_4^{2-}	Carbonate, CO_3^{2-}	Nitrate, NO_3^-
Mg^{2+}	0.2×10^{-6}	1830×10^{-6}	1.3×10^{-6}	4.9×10^{-3}
Ca^{2+}	15×10^{-6}	11×10^{-6}	0.13×10^{-6}	6.2×10^{-3}
Sr^{2+}	33.7×10^{-6}	0.71×10^{-6}	0.07×10^{-6}	1.86×10^{-3}
Ba^{2+}	150×10^{-6}	0.009×10^{-6}	0.09×10^{-6}	0.39×10^{-3}

Recall from Chap. 4 on Thermodynamics, the enthalpy change of solution can be represented by the formula

$$\Delta H_{\text{soln}} = \Delta H_{\text{hyd}}[\text{cation}] + \Delta H_{\text{hyd}}[\text{anion}] - \text{L.E.}, \tag{10.4}$$

where ΔH_{hyd} stands for enthalpy change of hydration and L.E. stands for lattice energy. The relationship is illustrated in the energy cycle below:

$MX(s) \xrightarrow{\Delta H_{\text{soln}}} M^{2+}(aq) + X^{2-}(aq)$

$MX(s) \xrightarrow{-\text{L.E.}} M^{2+}(g) + X^{2-}(g)$

$M^{2+}(g) + X^{2-}(g) \xrightarrow{\Delta H_{\text{hyd}}[\text{cation}] + \Delta H_{\text{hyd}}[\text{anion}]} M^{2+}(aq) + X^{2-}(aq)$

M^{2+}	= cation of Group 2 element
X^{2-}	= any anion of -2 charge

Let us now analyse the trend in solubility for Group 2 compounds.

Hydration is concerned with the strength of the ion–dipole interactions between ions and polar water molecules. $\Delta H_{\text{hyd}}[\text{cation}]$ thus measures the amount of energy released upon hydration of the gaseous cation. The stronger the ion–dipole interactions, the more energy will be evolved.

Since we are making a comparison of Group 2 compounds involving the same kind of anion, $\Delta H_{\text{hyd}}[\text{anion}]$ is a fixed constant term that we do not need to consider.

Comparing the Group 2 cations of the same charges, the smaller the cation, the greater is its charge density and hence the stronger the attraction it will have to water molecules. Therefore, going down Group 2, **ΔH_{hyd}[cation] becomes less exothermic as cationic size increases.**

The next term to discuss is L.E., which is dependent on both the size and charge on the ions. We do not need to consider the anion since it is common. As the charges on the Group 2 cations are the same, L.E. varies with the size of the cations. Therefore, going down Group 2, **L.E. becomes less exothermic as cationic size**

increases:

$$\Delta H_{\text{soln}} \propto (\Delta H_{\text{hyd}}[\text{cation}] - \text{L.E.}). \quad (10.5)$$

To sum up, the ΔH_{soln} term is dependent on $\Delta H_{\text{hyd}}[\text{cation}]$ and L.E., both of which become less exothermic as cationic size increases down Group 2. In general, a less exothermic ΔH_{soln} signifies a less soluble salt. Since both $\Delta H_{\text{hyd}}[\text{cation}]$ and L.E. are negative values, their combined effect on ΔH_{soln} depends on the relative extent of the influence the cationic size has on each of the following terms:

$$\Delta H_{\text{hyd}}[\text{cation}] \propto q_+/r_+, \quad (10.6)$$

$$\text{Lattice energy} \propto \frac{q_+ \times q_-}{r_+ + r_-}. \quad (10.7)$$

A less exothermic ΔH_{hyd} decreases solubility but a less exothermic L.E. enhances solubility.

The carbonate ion has a large anionic radius. Even though r_+ increases down Group 2, the increase in $(r_+ + r_-)$ is not very significant since $r_- \gg r_+$, i.e., the denominator in Eq. (10.7) does not change much so L.E. does not change appreciably. Hence, the effect of increasing ionic size on L.E. is less significant than its effect on ΔH_{hyd}. ΔH_{hyd} becomes less exothermic (which decreases solubility) but **L.E. becomes less exothermic to a much smaller extent** (which increases solubility only slightly). Overall, the solubility of carbonate (or other large anions) decreases down Group 2.

Hydroxide (OH^-) has a small ionic radius. As r_+ increases down Group 2, the increase in $(r_+ + r_-)$ is significant, i.e., the denominator in Eq. (10.7) increases significantly so L.E. becomes much less exothermic down Group 2. The effect of increasing ionic size on L.E. is even more significant than its effect on ΔH_{hyd}. ΔH_{hyd} becomes less exothermic (which decreases solubility) but **L.E. becomes less exothermic to a much greater extent** (which increases solubility more significantly). Overall, the solubility of hydroxide (or other small anions) increases down Group 2.

This trend is reflected in the increasing K_{sp} values at 25°C for the solubility equilibria shown below.

	K_{sp} ($mol^3\,dm^{-9}$)
$Mg(OH)_2(s) \rightleftharpoons Mg^{2+}(aq) + 2OH^-(aq)$	1.8×10^{-11}
$Ca(OH)_2(s) \rightleftharpoons Ca^{2+}(aq) + 2OH^-(aq)$	5.5×10^{-6}
$Sr(OH)_2(s) \rightleftharpoons Sr^{2+}(aq) + 2OH^-(aq)$	1.5×10^{-4}
$Ba(OH)_2(s) \rightleftharpoons Ba^{2+}(aq) + 2OH^-(aq)$	5.0×10^{-3}

Recall that the larger the equilibrium constant, the greater the extent of the forward reaction.

Q: Why are Group 2 oxides and hydroxides less soluble in water compared to those of the Group 1 elements from the same period?

A: The ionic bond strength of a Group 2 oxide or hydroxide is stronger than that of Group 1 since a Group 2 cation has a higher charge than a Group 1 cation. In addition, the hydration energy of the Group 2 cation is more exothermic than that of the Group 1 cation. Now, that Group 2 oxides and hydroxides are less soluble than their Group 1 counterparts is an indication that the hydration energy cannot really compensate for the energy required to break up the lattice.

Recall:

$$\Delta H_{\text{hydration}} \propto \frac{q_+}{r_+}.$$

10.8 Physical Properties of Group 17 Elements

10.8.1 *Melting point, boiling point and volatility*

Group 17 elements exist as simple discrete diatomic molecules that are non-polar with only weak instantaneous dipole-induced dipole interactions between their molecules. The presence of such weak intermolecular forces of attraction accounts for the relatively low values of these elements' physical properties such as melting point, boiling point and density.

Moving down the group from fluorine to iodine, there is a gradual increase in the magnitude of these properties. For example, observing that fluorine and chlorine are gases, bromine is a liquid, and iodine is a solid at room temperature, we can easily deduce that down the group,

- both melting point and boiling point increase,
- ΔH_{vap} becomes more endothermic, and
- volatility decreases.

What is volatility? It refers to the changing of phase from a solid or liquid to a gas. Volatility and boiling point have a reciprocal relationship. When we say that ethanol is highly volatile, it goes without saying that it has a low boiling point.

Enthalpy change of vaporisation (ΔH_{vap}) is a measure of volatility. The more endothermic ΔH_{vap}, the greater the energy needed for a substance to be vapourised (converted to the gaseous state) and thus the substance is said to have low volatility.

Example 10.9: Account for the trend in boiling points of the halogens.

Solution: The elements exist as simple discrete diatomic molecules that are non-polar with only weak instantaneous dipole-induced dipole (id–id) interactions between their molecules. The strength of id–id interactions depends on the number of electrons in the molecule. From F_2 to I_2, the number of electrons increases, resulting in a more polarisable electron cloud and leading to stronger id–id interactions. The boiling point therefore increases from F_2 to I_2 as more energy is required to overcome the stronger id–id interactions for a phase change to occur.

The above answer is used to account for trends in melting point and volatility as well.

10.8.2 *Colour*

Even when it comes to colour, there is an increase in colour intensity down the group. The various colours of the elements in their physical state at 20°C are shown in Table 10.5.

Table 10.5 Physical State of Group 17 Elements at 20°C

Group 17 Element	Physical State at 20°C
Fluorine	pale yellow gas
Chlorine	yellowish-green gas
Bromine	reddish-brown liquid
Iodine	dark violet solid (looks black)

Due to the fact that the elements in Group 17 show similar characteristics and there is a steady progression for some of these as we go down the group, we can provide fairly accurate predictions about these properties for those elements that we are not familiar with. Take for example astatine, the element below iodine. Can you predict the physical state of astatine and its colour?

With reference to Table 10.5, it can be predicted that astatine will have a much darker colour than iodine; presumably, it is a black solid. In fact, until recently, most of the properties of radioactive astatine have been inferred from the properties of the other group members.

10.8.3 *Solubility in water*

Fluorine oxidises water vigorously to yield oxygen even at cold temperatures:

$$2F_2(g) + 2H_2O(l) \longrightarrow 4HF(aq) + O_2(g).$$

Chlorine disproportionates (it is reduced and oxidised at the same time) partially in water to form hydrochloric and chloric(I) acids:

$$Cl_2(g) + H_2O(l) \rightleftharpoons HCl(aq) + HClO(aq).$$

Chloric(I) acid is a weak acid and decomposes slowly to give O_2:

$$2HClO(aq) \longrightarrow 2HCl(aq) + O_2(g).$$

Note: This decomposition is accelerated by *sunlight* and catalysts such as *platinum(Pt)*. HClO is responsible for the bleaching and disinfecting action of chlorine water.

Bromine is moderately soluble in water and forms a light brown solution containing mainly aqueous Br_2. The amount of HBrO formed is negligible.

Iodine dissolves only slightly in water giving a pale yellow solution. However, its solubility is greatly enhanced in the presence of some iodide ions due to the formation of a complex ion. The complex ion forms strong ion–dipole interactions with the water molecules, accounting for its high solubility. The resulting solution is brown in colour because of $I_3^-(aq)$:

$$I_2(s) + I^-(aq) \rightleftharpoons I_3^-(aq).$$

10.8.4 *Solubility in organic solvent*

Due to their non-polar nature, the halogens are more soluble in non-polar organic solvents than in polar solvents, of which water is an example. In particular, when bromine and iodine are each dissolved in an organic solvent, they will produce distinct colours as shown in Fig. 10.3. It must be noted that the actual colour of the solution is very much dependent on the concentration of the halogen present. Both fluorine and chlorine, on the other hand, are colourless in water and in organic solvent.

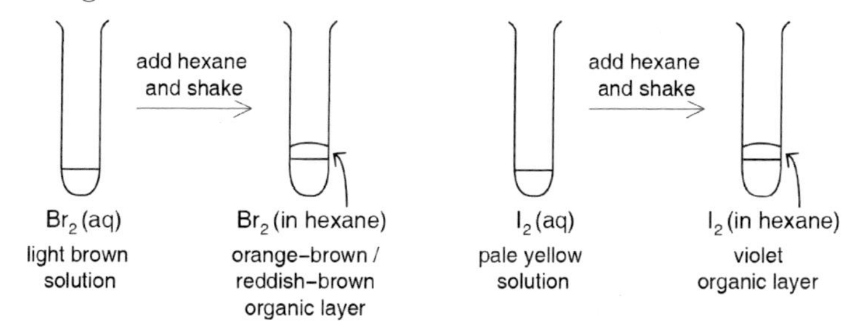

Fig. 10.3. Colours of Br_2 and I_2 in water and in organic solvent.

10.8.5 *Trend in bond energy*

	F–F	Cl–Cl	Br–Br	I–I
Bond energy ($kJ\,mol^{-1}$)	158	244	193	151

The bond energy decreases from chlorine to iodine (with the exception of fluorine) and this trend indicates that the bond strength decreases in the order: Cl–Cl > Br–Br > I–I. This is because as the size of the halogen increases from F to I, the valence orbital used for bonding is larger and more diffuse. As a result, the overlap of the orbitals is less effective. This accounts for the weaker bond strength that is reflected in the bond energies.

Q: Shouldn't the F–F bond be the strongest since the orbital F atom used for bonding is the smallest among all of the halogens?

A: Since the valence orbital of F atom is the smallest, one would expect the overlap of this orbital with that of another F atom to be very effective, i.e., that there would be greater accumulation of electron density between the two bonding F atoms, resulting in stronger attraction to the positively charged nuclei. However, when the two F atoms come into close proximity to form a bond, their non-bonding lone pair of electrons repel each other and this tension weakens the F–F bond. This effect is less prominent as the atomic size increases since the non-bonding electrons have a larger region of space to spread out into.

10.9 Chemical Properties of Group 17 Elements

Group 17 elements are **strong oxidising agents**. As a result of their high electronegativity, they tend to accept electrons readily from other substances and become reduced, forming halide ions that carry a charge of −1.

Since electronegativity decreases down the group, it is expected that the ability to accept electrons decreases. In other words, the oxidising power of Group 17 elements decreases down the group with fluorine being the strongest and iodine being the weakest oxidising agent in the group.

The relative strength of these oxidising agents (i.e., the ability to gain electrons) is reflected by the decreasing *positive* $E^{\ominus}$ values down the group. The higher the positive $E^\ominus$ value, the greater the extent of the reduction process and the stronger the oxidising power of the halogen:

$$\begin{aligned}
&F_2 + 2e^- \rightleftharpoons 2F^-, && E^\ominus = +2.87\,\text{V},\\
&Cl_2 + 2e^- \rightleftharpoons 2Cl^-, && E^\ominus = +1.36\,\text{V},\\
&Br_2 + 2e^- \rightleftharpoons 2Br^-, && E^\ominus = +1.07\,\text{V},\\
&I_2 + 2e^- \rightleftharpoons 2I^-, && E^\ominus = +0.54\,\text{V}.
\end{aligned}$$

Q. How does one account for the decrease in oxidising power of the halogens down the group?

A. Consider the formation of X^-(aq) from X_2 (X= F, Cl, Br or I) as being made up of three steps:

$$\begin{array}{ccc}
\tfrac{1}{2}X_2 + e^- & \xrightarrow{\Delta H_f} & X^-(aq) \\
\Delta H_{at}\big\downarrow & & \big\uparrow \Delta H_{hyd} \\
X(g) + e^- & \xrightarrow{\text{E.A.}} & X^-(g)
\end{array}$$

By Hess' Law, $\Delta H_f = \Delta H_{at} + \text{E.A.} + \Delta H_{hyd}$. The enthalpy change of formation (ΔH_f) of X^-(aq) is calculated based on data given in the table below:

	F	**Cl**	**Br**	**I**
ΔH_{at} (kJ mol^{-1})	79	121	112	107
E.A. (kJ mol^{-1})	−333	−364	−342	−295
ΔH_{hyd} (kJ mol^{-1})	−457	−381	−351	−307
ΔH_f (kJ mol^{-1})	−711	−624	−581	−495

ΔH_f of F^-(aq) is the most exothermic, which is to say that F^- is most readily formed from F_2, indicating that F_2 is the most powerful oxidising agent. The state symbol for X_2 depends on the halogen: Cl_2 is a gas, Br_2 is a liquid, and I_2 is a solid under standard conditions.

The reduction process involves:

- breaking of the X–X bond (reflected as the enthalpy change of atomisation, ΔH_{at}),
- adding an electron to X (electron affinity, E.A.), and
- hydrating the X^- gaseous ion (ΔH_{hyd}).

Hence, the reduction process is favoured by

- a less endothermic bond enthalpy, coupled with
- a more exothermic electron affinity, and
- a more exothermic hydration energy.

From chlorine to iodine, the bond enthalpy becomes less endothermic, but this change is not significant enough to compensate for the greater changes in both the electron affinity and hydration enthalpy. Thus, one can see that the decrease in oxidising power from chlorine to iodine is mainly due to the changes in both E.A. and ΔH_{hyd}.

In short, to account for the decrease in oxidising power down Group 17, one simply needs to mention that it is due to the decrease in electronegativity of the elements. The trend is also reflected by the less positive $E^{\ominus}_{X_2/X^-}$ values.

The halogens generally undergo similar reactions, although under varying conditions, as these depend on their oxidising strengths. Thus, through the following reactions, we can observe the decrease in oxidising power in the order $F_2 > Cl_2 > Br_2 > I_2$:

- reaction with water,
- displacement reaction of halogens,
- reaction with thiosulfate,
- reaction with alkali, and
- reaction with hydrogen.

10.9.1 *Displacement reaction of halogens*

In a displacement reaction, a halogen that is a stronger oxidising agent will be able to oxidise the halide ion of a weaker oxidising agent that is below it in the group. The halide ion is said to be displaced from the solution.

We can make use of this type of reaction to identify the presence of Br^- and I^- in solution. Just add aqueous Cl_2 followed by an organic solvent. Upon shaking the mixture, the presence of Br^- or I^- can be confirmed as Br_2 and I_2 provide distinct colouration in the organic layer:

$$\mathrm{Cl_2(aq) + 2Br^-(aq) \longrightarrow 2Cl^-(aq) + Br_2(aq),}$$
$$\mathrm{Cl_2(aq) + 2I^-(aq) \longrightarrow 2Cl^-(aq) + I_2(aq).}$$

If we try to react I_2 with Cl^-, no reaction will occur. I^- and Cl_2 are not produced. We can use $E^\ominus{}_{\text{cell}}$ values to predict the spontaneity of a given reaction. Calculating the $E^\ominus{}_{\text{cell}}$ of the reaction between I_2 and Cl^-, $E^\ominus{}_{\text{cell}} < 0$ indicates that the reaction is thermodynamically non-spontaneous under standard conditions; $E^\ominus{}_{\text{cell}} = +0.54 - (+1.36) = -0.82\,\text{V}$.

What happens when F_2 and Br^- are allowed to react? Does a displacement reaction occur with products F^- and Br_2 formed? Calculating the $E^\ominus{}_{\text{cell}}$ value, it is predicted that the reaction between F_2 and Br^- is thermodynamically spontaneous under standard conditions since $E^\ominus{}_{\text{cell}} > 0$; $E^\ominus{}_{\text{cell}} = +2.87 - (+1.07) = +1.8\,\text{V}$.

Being the strongest oxidising agent, F_2 is expected to displace the other halide ions, Cl^-, Br^- or I^-, from aqueous solution but in actual fact the displacement reaction does not occur. Why is this so?

Fluorine is strongly oxidising — and this is essentially the answer! It is so strongly oxidising that it readily oxidises water, which is present in greater amount than the halogens or halides introduced into the solvent. There is just not enough fluorine left to react with the minute amount of halide.

Thus, remember that using $E^\ominus{}_{\text{cell}}$ to predict the spontaneity of a reaction is, as mentioned, only a prediction. There are cases when the prediction fails and we should seek answers to account for the limitations of the prediction.

Example 10.10: Describe and explain what is observed when chlorine gas is bubbled into aqueous potassium iodide solution.

Solution:

$$Cl_2(g) + 2I^-(aq) \longrightarrow 2Cl^-(aq) + I_2(aq).$$

Observations: The colourless solution turns brown with some black solid I_2 seen.

$$E^\ominus{}_{cell} = +1.36 - (+0.54) = +0.82\,V.$$

The positive $E^\ominus{}_{cell}$ indicates that the above redox reaction is thermodynamically spontaneous under standard conditions.

Confirmatory test for iodine: Add starch solution.
A blue-black colouration is observed due to the formation of the starch-iodine complex.

10.9.2 *Reaction with thiosulfate*

Having greater oxidising abilities than I_2, both Cl_2 and Br_2 are able to oxidise the thiosulfate ($S_2O_3{}^{2-}$) ion to the sulfate(VI) ion. The change in the average oxidation state of S is from +2 to +6, an increase of 4 units:

$$4Cl_2(aq) + S_2O_3{}^{2-}(aq) + 5H_2O(l) \longrightarrow 8Cl^-(aq) + 2SO_4{}^{2-}(aq) + 10H^+(aq),$$

$$4Br_2(aq) + S_2O_3{}^{2-}(aq) + 5H_2O(l) \longrightarrow 8Br^-(aq) + 2SO_4{}^{2-}(aq) + 10H^+(aq).$$

On the other hand, the weaker oxidising agent I_2 is only able to oxidise the thiosulfate to tetrathionate. The change in the average oxidation state of S in this case is only from +2 to +2.5:

$$I_2(aq) + 2S_2O_3{}^{2-}(aq) \longrightarrow 2I^-(aq) + S_4O_6{}^{2-}(aq).$$

The stronger the oxidising agent, the more readily it accepts electrons. This can be inferred from the greater magnitude of change in the oxidation state of the species oxidised.

Q. Why is the phrase "average oxidation state" used? Can the word "average" be dropped?

A. Recall in Chap. 8 that there are two ways to calculate oxidation states: one method is to apply the set of rules and the other requires the use of the molecular structure.

Through the use of molecular structures, oxidation states are assigned to each of the S atoms based on the difference in electronegativity of the bonding atoms. Thus, when $S_2O_3^{2-}$ is oxidised to SO_4^{2-}, the oxidation state of the terminal S actually increases from 0 to +6 whereas that of the central S increases from +4 to +6.

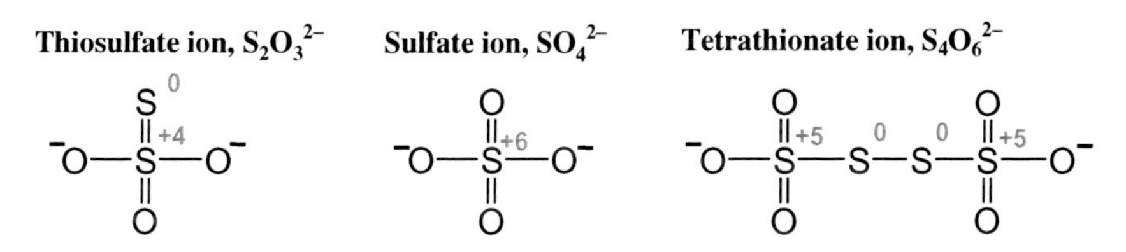

Instead of giving such specific details for each S atom in the various compounds, we can give a clear yet succinct account for the changes in oxidation state by using the concept of "average oxidation state," which is mathematically calculated based on the rules found in Chap. 8.

Take the case of $S_2O_3^{2-}$ for example. Let the oxidation state of S be x. The oxidation state of each O atom is −2. The overall charge of this polyatomic ion is −2.

Thus, we have

$$2x + 3(-2) = -2$$
$$x = +2.$$

The oxidation state of +2 is calculated as an "average" value.

Apply a similar set of calculations for $S_4O_6^{2-}$, letting the oxidation state of S be y:

$$4y + 6(-2) = -2$$
$$y = +2.5.$$

The "average" value calculated is +2.5.

Example 10.11: Discuss the relative oxidising strengths of the halogens using the reaction with $FeSO_4(aq)$. Substantiate your explanation using $E^{\ominus}$ values.

Approach: Since the halogens are oxidising agents, we are interested in their relative ability to oxidise Fe^{2+} to Fe^{3+}. The oxidation of SO_4^{2-} is ruled out since S is already in its highest oxidation state.

To determine if a redox reaction occurs between each halogen and Fe^{2+}, we need to calculate $E^{\ominus}_{cell}$ and a positive value indicates a thermodynamically spontaneous reaction under standard conditions.

The following reduction potentials are thus needed:

	$E^{\ominus}$(V)
$Fe^{3+} + e^- \rightleftharpoons Fe^{2+}$	+0.77
$F_2 + 2e^- \rightleftharpoons 2F^-$	+2.87
$Cl_2 + 2e^- \rightleftharpoons 2Cl^-$	+1.36
$Br_2 + 2e^- \rightleftharpoons 2Br^-$	+1.07
$I_2 + 2e^- \rightleftharpoons 2I^-$	+0.54

Solution:

$$F_2(aq) + 2Fe^{2+}(aq) \longrightarrow 2F^-(aq) + 2Fe^{3+}(aq),$$
$$E^{\ominus}_{cell} = +2.87 - (+0.77) = +2.10\,V,$$
$$Cl_2(aq) + 2Fe^{2+}(aq) \longrightarrow 2Cl^-(aq) + 2Fe^{3+}(aq),$$
$$E^{\ominus}_{cell} = +1.36 - (+0.77) = +0.59\,V,$$
$$Br_2(aq) + 2Fe^{2+}(aq) \longrightarrow 2Br^-(aq) + 2Fe^{3+}(aq),$$
$$E^{\ominus}_{cell} = +1.07 - (+0.77) = +0.30\,V,$$
$$I_2(aq) + 2Fe^{2+}(aq) \longrightarrow 2I^-(aq) + 2Fe^{3+}(aq),$$
$$E^{\ominus}_{cell} = +0.54 - (+0.77) = -0.23\,V.$$

Based on the calculated negative value of $E^{\ominus}_{cell}$, I_2 is not able to oxidise Fe^{2+} to Fe^{3+}.

Based on the respective calculated positive values of $E^{\ominus}_{cell}$, F_2, Cl_2 and Br_2 are able to oxidise Fe^{2+} to Fe^{3+}. More positive $E^{\ominus}_{cell}$ values indicate that reactions are more thermodynamically spontaneous under standard conditions.

Thus, it can be concluded that oxidising strength decreases in the order $F_2 > Cl_2 > Br_2 > I_2$.

10.9.3 *Reaction with alkali*

Chlorine, bromine and iodine undergo **disproportionation** upon treatment with alkali. The composition and type of products formed are influenced by the reaction temperature and the oxidising strength of the halogen.

The following reaction occurs when the halogen is treated with a cold dilute alkali, such as NaOH or KOH, at about 15°C:

$$X_2(aq) + 2OH^-(aq) \longrightarrow \underset{\text{halide ion}}{X^-(aq)} + \underset{\text{halate(I) ion}}{XO^-(aq)} + H_2O(l), \qquad (10.8)$$

where $X_2 = Cl_2$, Br_2 or I_2.

The mixture of halide and halate(I) salts obtained contains the halogen in different oxidation states. But the reaction does not stop here. Once formed, the halate(I) ions may undergo further disproportionation to produce both halide ions and halate(V) ions:

$$\underset{\text{halate(I) ion}}{3XO^-(aq)} \longrightarrow \underset{\text{halide ion}}{2X^-(aq)} + \underset{\text{halate(V) ion}}{XO_3{}^-(aq)}. \qquad (10.9)$$

The reaction represented by Eq. (10.9) occurs at different rates depending on the temperature and also the type of halogen.

When chlorine is treated with cold dilute NaOH(aq), the following reaction occurs:

$$Cl_2(aq) + 2OH^-(aq) \underset{15^\circ C}{\rightarrow} Cl^-(aq) + \underset{\text{chlorate(I) ion}}{ClO^-(aq)} + H_2O(l). \qquad (10.10)$$

At 15°C and below, the disproportionation of ClO^- is too slow to generate substantial $ClO_3{}^-$ ions and thus the main reaction when chlorine reacts with cold dilute NaOH(aq) is shown by Eq. (10.10) with ClO^- as the major halate product formed. Under such conditions, one may still be able to detect the smell of the pungent chlorine gas!

When the reaction temperature is increased, to say around 70°C, the disproportionation of ClO^- proceeds rapidly [see Eq. (10.11)] and the major products generated are Cl^- and $ClO_3{}^-$. The pungent smell

Table 10.6 Types of Halate Product Formed

When Alkali Reacts with:	Major Halate Product Obtained at: 0°C	15°C	70°C
Cl_2	ClO^-	ClO^-	ClO_3^-
Br_2	BrO^-	BrO_3^-	BrO_3^-
I_2	←	mostly IO_3^-	→

of the chlorine gas would totally disappear under such conditions:

$$3ClO^-(aq) \longrightarrow 2Cl^-(aq) + ClO_3^-(aq). \quad (10.11)$$

When chlorine reacts with hot aqueous NaOH, the overall reaction can be represented as:

$$3Cl_2(aq) + 6OH^-(aq) \underset{70°C}{\longrightarrow} 5Cl^-(aq) + ClO_3^-(aq) + 3H_2O(l).$$

Table 10.6 provides a summary of the type of main halate product formed for each halogen when subjected to treatment with NaOH(aq) at various temperatures.

As shown in Table 10.6, the decreasing ease of disproportionation of halate(I) ions is in the order $IO^- > BrO^- > ClO^-$.

For the reaction with bromine, the disproportionation of BrO^- occurs readily at cold temperatures of around 15°C. For the case of iodine, disproportionation of IO^- occurs readily even at 0°C. This makes it difficult to produce IO^- without obtaining IO_3^-.

Q: Why is it that when a more powerful oxidising agent (such as Cl_2) reacts with NaOH, the more likely halate product formed is XO^- rather than XO_3^-? How does the formation of the halate(V) ion demonstrate the decrease in the strength of the oxidising power down Group 17?

A: Oxidising power decreases down the group, which means that it is less spontaneous for the halide ion to form from the halogen. The reverse is also true: it is more spontaneous for the halogen to act as reducing agent and be oxidised. So, the fact that IO_3^- is likely to form even at low temperature is an indication that I_2 is a better reducing agent than Br_2 and Cl_2 — a clear indication of an increase in reducing power down the group or a decrease in oxidising power!

10.9.4 *Reaction with hydrogen*

All the halogens, in the gaseous phase, react with hydrogen to produce hydrogen halides. The stronger the oxidising agent, the more vigorous the reaction is with hydrogen:

$$X_2(g) + H_2(g) \longrightarrow 2HX(g),$$

where X = F, Cl, Br.

Consequently, the reaction conditions become more stringent for a halogen with weaker oxidising strength to react effectively with hydrogen.

Fluorine is so highly oxidising that it reacts explosively with hydrogen in the dark at room temperature and pressure. Chlorine reacts explosively with hydrogen when in the presence of sunlight. For bromine to react with hydrogen, heating at 300°C with a Pt catalyst is required.

As for iodine, heating and a catalyst are required, but even with such drastic reaction conditions, the reaction with hydrogen does not go to completion:

$$H_2(g) + I_2(g) \rightleftharpoons 2HI(g).$$

Q: The breaking of the X–X bond exerts an influence on the reaction energetic, right? If the bond strength is in the decreasing order Cl–Cl > Br–Br > I–I, do we not expect the formation of HX from H_2 and X_2 to proceed more favourably for a halogen whose X–X bond is easier to cleave?

A: A reaction is not merely affected by the breaking of bonds but is also affected by the formation of new bonds, giving rise to the products. Thus, if observations tell us that the reaction is vigorous for Cl_2 but becomes less so down Group 17, this tells us in fact that the dominant factor that influences the formation of the hydrogen halide is not the breaking of the X–X bond but rather it is due to the formation of the H–X bond. The more stable the product, the greater the tendency for it to form.

Looking up the data on bond energy (BE) of the H–X bond, there is a decreasing trend as follows:

$$\text{BE(H−Cl)} > \text{BE(H−Br)} > \text{BE(H−I)}.$$

The H–I bond is the weakest: its formation is least favoured and once it is formed, there is higher tendency for it to cleave. This results in the formation of a dynamic equilibrium.

This same explanation accounts for the relative thermal stability of the hydrogen halides and their reactions, which we will be discussing in greater detail in the next section.

10.10 Hydrogen Halides

The hydrogen halides (or hydrohalic acids), abbreviated here as HX, where X represents a halogen (F, Cl, Br or I), are colourless gases at room temperature and pressure.

There exists hydrogen bonding between the HF molecules. The hydrogen bonding is so extensive that the molecules associate with one another in both the liquid and gas phases, accounting for its relatively high melting and boiling points.

Hydrogen Halide	Melting Point (°C)	Boiling Point (°C)
HF	−83	20
HCl	−114	−85
HBr	−87	−66
HI	−51	−34

For HCl, HBr and HI, only van der Waals' forces of attraction (both pd–pd and id–id interactions) exist between these diatomic molecules. As seen from the table, the boiling point increases from HCl to HI.

Q: Can we compare the strengths of the pd–pd interactions to account for the trend in boiling point?

A: We know that the greater the magnitude of the molecule's net dipole moment, the stronger the pd–pd interactions. If we are to follow this line of argument, the strength of pd–pd interactions is highest for HCl followed by HBr and HI. This would translate to a similar trend in boiling point, which is contrary to experimental data.

Therefore, to account for the observed trend in boiling point, we have to look at the main attribute, which in this case revolves around the id–id interactions.

From HCl to HI, the number of electrons in the molecule increases, resulting in a more polarisable electron cloud and leading to stronger id–id interactions. The boiling point therefore increases from HCl to HI as more energy is required to overcome the stronger id–id interactions for a phase change to occur.

10.10.1 *Thermal stability of hydrogen halides*

HX decomposes, on heating, to form the constituent elements $H_2(g)$ and $X_2(g)$:

$$2HX(g) \longrightarrow H_2(g) + X_2(g),$$

with bond energies decreasing in the order:

$$BE(H{-}Cl) > BE(H{-}Br) > BE(H{-}I).$$

The bond strength also decreases in the same order:

$$H{-}Cl > H{-}Br > H{-}I.$$

Thermal stability thus decreases from HF to HI.

Q: What accounts for the decreasing bond strength and hence thermal stability from H–F to H–I?

A: As the size of the halogens gets bigger, from F to I, the valence orbital used for bonding is larger and more diffuse. As a result, the overlap of the orbital with the 1*s* orbital of the hydrogen atom becomes less effective and this accounts for the weaker bond strength that is reflected in the bond energies.

The following observations help to illustrate the trend in thermal stability: Thermal stability of HI is the weakest of all — when a red-hot needle is inserted into a jar filled with HI(g), violet fumes of I_2 are observed. The heat from such a needle is not sufficient to decompose HBr. Instead, strong heating is required to obtain brown bromine vapour.

Q: How does the decrease in thermal stability of the hydrogen halide down Group 17 tie in with the fact that the oxidising power of the halogen decreases down the group?

A: If the halogen is less likely to be reduced to form the halide as we go down the group, then the halide is more likely to be oxidised back to form the halogen. In short, the oxidising power of the halogens decreases down the group, and the reducing power of the halides increases down the group.

10.10.2 *Acidity of hydrogen halides*

With the exception of HF, the hydrogen halides are strong acids that completely dissociate in aqueous solution:

$$\mathrm{HX(g) + H_2O(l) \longrightarrow H_3O^+(aq) + X^-(aq).}$$

For dissociation to occur, the H–X bond must be broken. The acidity of the hydrogen halides thus depends on the ease of cleavage of the H–X bond. Since it is easiest to cleave the weakest H–I bond followed by the H–Br bond and subsequently the H–Cl bond, the acid strength decreases in the order:

$$\mathrm{HI > HBr > HCl.}$$

Q: Heating is not required when HX is mixed with water. Where do we get the energy to cleave the H–X bond in water?

A: The energy that is required to break the bond comes mainly from the hydration of the proton (H^+) and partially from that of X^-.

Q: Why is HF a weak acid?

A: Before dissociation can occur, the strong hydrogen bonding among the HF molecules needs to be overcome. Add to that the need to break the very strong H–F bond. Another attribute of the weak extent of dissociation is the low electron affinity of the F atom. Once the H–F bond is cleaved, the bonding electrons are acquired by the F atom to form F^-. However, because of the very small size of fluorine, the added electron tends to be strongly repelled by the electrons in the valence shell. All these energetically demanding processes are not compensated for sufficiently, thus resulting in low dissociation of the HF acid.

10.10.3 *Reaction of halides with concentrated acid*

Hydrogen halides can be obtained from the reaction of concentrated H_2SO_4 and solid halides in the form of salts such as sodium halide, NaX (where X represents Cl^-, Br^- or I^-):

$$NaX(s) + H_2SO_4(l) \longrightarrow NaHSO_4(s) + HX(g)$$

The HX product formed may be further oxidised to X_2 by concentrated H_2SO_4, which is oxidising in nature. The ease of oxidation (reducing strength) is in the order HI > HBr > HCl.

Q: Why is HI a stronger reducing agent than HCl or HBr?
A: It is easier for I^- to lose an electron (be oxidised) than Br^- and Cl^- since the valence electron in I^- is less strongly bound due to its further distance from the nucleus.

The oxidising power of concentrated H_2SO_4 is not strong enough to oxidise HCl to Cl_2. Thus, the reaction of NaCl with concentrated H_2SO_4 yields only gaseous HCl:

$$NaCl(s) + H_2SO_4(l) \longrightarrow NaHSO_4(s) + HCl(g).$$

HBr is oxidised by concentrated H_2SO_4 to Br_2. Thus, the reaction of NaBr with concentrated H_2SO_4 produces a mixture of products HBr and Br_2:

$$NaBr(s) + H_2SO_4(l) \longrightarrow NaHSO_4(s) + HBr(g),$$

$$2HBr(g) + H_2SO_4(l) \longrightarrow SO_2(g) + Br_2(l) + 2H_2O(l).$$

HI is easily oxidised by concentrated H_2SO_4 to I_2. Thus, the reaction of NaI with concentrated H_2SO_4 produces a mixture of products, consisting mainly of I_2 and to a lesser extent, HI. Apart from H_2SO_4 being reduced to SO_2, S or H_2S may also be produced, depending on reaction conditions:

$$NaI(s) + H_2SO_4(l) \longrightarrow NaHSO_4(s) + HI(g),$$

$$2HI(g) + H_2SO_4(l) \longrightarrow SO_2(g) + I_2(s) + 2H_2O(l),$$

$$6HI(g) + H_2SO_4(l) \longrightarrow S(s) + 3I_2(s) + 4H_2O(l),$$

$$8HI(g) + H_2SO_4(l) \longrightarrow H_2S(g) + 4I_2(s) + 4H_2O(l).$$

Q: Why is S or even H_2S formed when HI reacts with concentrated H_2SO_4, whereas SO_2 is formed when HBr reacts with concentrated H_2SO_4?

A: The oxidation state of the sulfur-containing product that is formed serves as an indication of the strength of the reducing power of the hydrogen halide. In the presence of the stronger reducing agent HI, the oxidation state of S decreases by 8 units (from +6 in SO_4^{2-} to −2 in H_2S). In the presence of HBr, the oxidation state of S decreases only by 2 units (from +6 in SO_4^{2-} to +4 in SO_2).

10.10.4 *Distinguishing tests for halide ions*

If you are given three test tubes of colourless solutions, each containing one of the halide ions $Cl^-(aq)$, $Br^-(aq)$ and $I^-(aq)$, how can you tell one apart from another?

This sums up the gist of "qualitative inorganic analysis," which is devoted to establishing the identities of constituent ions present in samples through the use of various reagents. The identity of a specific constituent ion is deduced using reactions which produce results characteristic of that particular species, be it a colour change, the formation of a precipitate, or some other type of visible change.

The usual method employed to distinguish between the halides involves reacting each of these with aqueous silver nitrate to form insoluble silver halide salts. Since the precipitates may be similar

Table 10.7 Distinguishing Tests for Halide Ions

	Type of Halide	Cl^-(aq)	Br^-(aq)	I^-(aq)
Step 1	Add $AgNO_3$(aq) to a test tube containing the halide	White precipitate of AgCl formed	Cream precipitate of AgBr formed	Yellow precipitate of AgI formed
Step 2a	To the precipitate in Step 1, add dilute NH_3(aq)	Precipitate soluble in dilute NH_3(aq)	Precipitate insoluble in dilute NH_3(aq)	Precipitate insoluble in dilute NH_3(aq)
Step 2b	To the precipitate in Step 1, add concentrated NH_3	Precipitate soluble in concentrated NH_3	Precipitate soluble in concentrated NH_3	Precipitate insoluble in concentrated NH_3

in colour, a further confirmatory test is done. The second test usually involves either the use of dilute aqueous ammonia or concentrated ammonia to test the solubility of the silver halide in such solvents. The results, characteristic of each halide, are given in Table 10.7.

Q: What is the reason behind the different solubility of the silver halides in aqueous ammonia?

A: Silver halides have very low solubility in water. If the ionic product of AgX exceeds its K_{sp} value, precipitation is observed:

$$\mathrm{AgX(s)} \rightleftharpoons \mathrm{Ag^+(aq) + X^-(aq)}, \quad K_{sp} \text{ of AgX} = [\mathrm{Ag^+}][\mathrm{X^-}]. \tag{10.12}$$

When dilute NH_3(aq) is added, there is complex formation between Ag^+ ions and NH_3 giving rise to the diamminesilver(I) complex, $[Ag(NH_3)_2]^+$(aq). There needs to be a critical concentration of Ag^+ in order for complex formation to take place:

$$\mathrm{Ag^+(aq) + 2NH_3(aq) \longrightarrow [Ag(NH_3)_2]^+(aq)}.$$

The removal of the Ag^+ via complexation causes $[Ag^+]$ in the solution to decrease. Consequently, according to Le Chatelier's Principle, the forward reaction [see Eq. (10.12)] is favoured to produce more Ag^+ and more AgX is expected to dissolve. However, this is not the case since we know that not all silver halides are soluble in $NH_3(aq)$. Why is this so?

	K_{sp} Values at 25°C ($mol^2\,dm^{-6}$)
AgCl	1.8×10^{-10}
AgBr	5.3×10^{-13}
AgI	8.3×10^{-17}

From the above K_{sp} data, AgCl is the most soluble silver halide compound. Thus, when AgCl dissolves in water, it can provide the critical amount of Ag^+ for the $[Ag(NH_3)_2]^+(aq)$ complex to form. As a result, the solubility equilibrium for AgCl is affected. Hence, according to Le Chatelier's Principle, more AgCl dissolves. Now, all this is not possible for AgBr and AgI, simply because their low solubilities do not provide the critical amount of Ag^+ for complexation to take place. However, if concentrated ammonia is used, then the large amount of NH_3 present causes AgBr to dissolve but not AgI!

Q: Why is AgBr insoluble in dilute $NH_3(aq)$ but soluble in concentrated NH_3?

A: Although the concentration of Ag^+ is low due to the smaller K_{sp} value for AgBr, there is a higher concentration of NH_3 in concentrated NH_3 than the dilute one, which causes a substantial amount of $[Ag(NH_3)_2]^+(aq)$ to be formed. Thus, according to Le Chatelier's Principle, equilibrium [see Eq. (10.12)] is affected, causing more AgBr to dissolve.

10.11 Industrial Uses and Environmental Impact of Group 17 Elements and Their Compounds

10.11.1 *Fluorine and its compounds*

Fluorine is used to make chlorofluorocarbons (CFCs), such as CF_2Cl_2, which are widely used as refrigerants and aerosol propellants. But nowadays, their use has been greatly reduced as they are believed to have caused ozone depletion in the stratosphere.

Fluorine is used to manufacture tetrafluoroethene ($CF_2{=}CF_2$), which can be polymerised to give poly(tetrafluoroethene), PTFE. This resists attack by most chemicals and is used in the chemical industry for the manufacture of corrosion-proof valves and seals. In addition, it is also used as a non-stick coating for pans and skis.

Fluoride is added to toothpaste to prevent dental cavities. The source of fluoride includes NaF.

10.11.2 *Chlorine and its compounds*

Chlorine is used as a bleaching agent and in water treatment, e.g., as a disinfecting agent in swimming pools. Chlorine disproportionates in water to give chloric(I) acid (HClO) and hydrochloric acid. HClO is a strong oxidising agent and kills germs by oxidising them.

Chlorine is used in the manufacturing of antiseptic (NaClO and $NaClO_3$); explosives, matches and fireworks; weed killer; HCl and inorganic $AlCl_3$, $FeCl_3$, PCl_3, etc.; solvents such as chloroform ($CHCl_3$); organochlorine products such as chloroethene for the making of poly(vinyl chloride) or PVC plastic. PVC is non-biodegradable and poses disposal problems, while chlorinated insecticides, pesticides and herbicides are toxic and may leave residual toxicity in the birds, fishes and humans who consume them.

10.11.3 *Bromine and its compounds*

Bromine is used in the production of silver bromide used in photographic film. 1,2-dibromoethane is added to petrol to remove lead by generating volatile lead bromide and so prevent it from fouling spark plugs.

My Tutorial (Chapter 10)

1. (a) Outline one industrial process for the manufacture of chlorine, stating the starting materials.
   (b) Name any one commercially important compound that is made directly or indirectly from chlorine and state its use.
   (c) Chlorine reacts slightly with water. The reversible reaction can be represented by the equation:

$$Cl_2 + H_2O \rightleftharpoons HCl + HClO.$$

   (i) What is the oxidation state of chlorine in HClO?
   (ii) Give a name for the reaction.
   (iii) Write two ionic half-equations for this process.

   (d) Chlorine is widely used to disinfect water. The non-ionised acid HClO, formed in the above reaction, is a much better disinfectant than the ClO^- ion. HClO is a weak acid:

$$HClO(aq) \rightleftharpoons H^+(aq) + ClO^-(aq).$$

   (i) In which direction should the pH of the solution be adjusted so as to increase the disinfectant power? Explain your answer.
   (ii) Give one adverse effect that might follow if the pH were adjusted too far in this direction.

   (e) Bromine occurs as bromide ions at low concentration in seawater. It is extracted commercially by the following series of steps.

   (A) Chlorine is passed into acidified seawater and bromine is then removed in a stream of air. The concentration of bromine in the gas phase is too low at this stage for it to be condensed out efficiently.

(B) Therefore, it is converted to hydrogen bromide by reaction with added sulfur dioxide and a small excess of water vapour:

$$Br_2(g) + SO_2(g) + 2H_2O(g) \longrightarrow 2HBr(g) + H_2SO_4(g).$$

A moderately concentrated, aqueous mixture of hydrobromic and sulfuric acids is obtained when the vapour is cooled and condensed.

(C) The concentrated mixture of acids is treated with chlorine and steam, producing a vapour from which liquid bromine is condensed on cooling. The sulfuric acid remains in the solution.

(D) The crude bromine is purified by fractional distillation.

(i) Explain in terms of the relevant standard electrode potentials the reactions taking place in steps A and B. (Assume $SO_2 + H_2O$ is H_2SO_3.)

(ii) Why does liquid bromine vapourise easily?

(iii) Explain why, in step B, a large volume of gaseous hydrogen bromide dissolves in a small volume of water, and give the chemical species present in aqueous hydrobromic acid.

(iv) Describe a test for the detection of bromide ions in aqueous solution.

2. (a) Explain why alkaline solutions result when the oxides of the *s* block elements dissolve in water.

(b) Most of these oxides dissolve readily in water. However, the oxide of beryllium and magnesium have very low solubility in water. Explain in terms of enthalpy changes why this is so.

(c) An aqueous solution of beryllium sulfate is acidic whereas an aqueous solution of magnesium sulfate is almost neutral. Why is there this difference in behaviour of Be^{2+}(aq) and Mg^{2+}(aq)?

(d) Given a sample of solid magnesium chloride contaminated with magnesium carbonate, describe tests you would perform in order to confirm the presence of:

(i) magnesium ions,

(ii) chloride ions, and
(iii) carbonate ions.

3. (a) (i) Write an equation for the reaction of barium with water.
(ii) Would the reaction in (a)(i) occur more vigorously or less vigorously than the reaction of calcium with water? Identify one contributory factor and use it to justify your answer.
(iii) Write an equation for the action of heat on solid barium carbonate.
(iv) At a given high temperature, which of the two carbonates, barium carbonate or calcium carbonate, would decompose more easily? Explain your answer.
(v) How would you distinguish between solutions of barium chloride and calcium chloride? State in each case what you would see as a result of the test on each solution.

(b) 1.71 g of barium reacts with oxygen to form 2.11 g of an oxide W.

(i) Calculate the formula of W.
(ii) Give the formula of the anion present in W.
(iii) What is the oxidation number of oxygen in this anion?
(iv) Sodium forms an oxide, Y, which contains this same anion. Give the formula of Y.

(c) Treatment of either W or Y with dilute sulfuric acid leads to the formation of the sulfate of the metal, together with an aqueous solution of hydrogen peroxide, H_2O_2.

(i) Write an equation for the reaction of Y with dilute sulfuric acid.
(ii) The hydrogen peroxide solution produced may be separated from the other reaction product. Explain briefly why this is easier to achieve if W is used as the initial reagent rather than Y.

(d) (i) Write an expression for K_p for the following equilibrium, giving the units:

$$BaCO_3(s) \rightleftharpoons BaO(s) + CO_2(g).$$

(ii) How would the numerical value of K_p change if $CaCO_3$ were used in place of $BaCO_3$ in (d)(i)? Explain your answer.

4. (a) (i) Define the term *lattice enthalpy*, illustrating your answer by reference to the oxide of a Group 2 metal.
(ii) What are the factors that affect the magnitude of lattice enthalpy?
(iii) Why is lattice enthalpy for a given compound found by use of a Born–Haber cycle rather than being measured by direct experiment?
(iv) Lattice enthalpies can be calculated from a formula based on a purely ionic model. Why do values calculated in this way often differ from those found from the Born–Haber cycle?
(b) (i) How does the solubility of the hydroxides of Group 2 metals change with increasing atomic number?
(ii) Suggest an explanation for this trend.
(c) (i) Explain why there is an increase in metallic character with increase in atomic number in Group 14.
(ii) Explain why compounds of lead in oxidation state +4 are oxidising.
(d) From the position of radium in the periodic table, predict the following:
(i) the formula of radium carbonate;
(ii) the equation for the thermal decomposition of solid radium carbonate; and
(iii) how the decomposition temperature required in (d)(ii) would compare with that required for magnesium carbonate.

5. (a) Concentrated sulfuric acid reacts with sodium chloride as follows:

$$H_2SO_4 + Cl^- \rightleftharpoons HCl + HSO_4^-.$$

(i) Identify the conjugate acid-base pairs in this reaction.
(ii) What would be the observable result of this reaction?

(iii) Explain why this reaction goes almost completely to the right despite the fact that both hydrochloric and sulfuric acids are strong.

(b) When concentrated sulfuric acid reacts with solid sodium iodide, hydrogen iodide, hydrogen sulfide, sodium hydrogen-sulfate, and water are formed together with one other product.

(i) Identify this product and state how you would recognise it.

(ii) Write an ionic half-equation to show the conversion of sulfuric acid into hydrogen sulfide.

(iii) Hence, write the full ionic equation for the reaction between concentrated sulfuric acid and sodium iodide.

(iv) What is the function of the sulfuric acid in this reaction?

(c) Explain concisely why the type of reaction occurring in (b) does not occur with sodium chloride.

CHAPTER 11

Introduction to Transition Metals and Their Chemistry

By definition, d block elements refer to those atoms with electrons occupying the d subshell. The number of d electrons may vary from d^1 to d^{10}. For instance, the electronic configuration of the Br atom is $[Ar]3d^{10}4s^24p^5$. Although it has 10 d electrons, it is not a d block element because its p subshell is partially filled. It is considered a p block element instead.

In the periodic table, the d block elements are found in groups located between Group 2 and Group 13 as shown in Table 11.1.

A transition element (otherwise known as a transition metal) is a d block element that can form at least one stable ion with *a partially filled d subshell.*

Here, we will focus on the first row of transition elements, namely:

Ti	V	Cr	Mn	Fe	Co	Ni	Cu
titanium	vanadium	chromium	manganese	iron	cobalt	nickel	copper

Strictly speaking, both Sc and Zn are not considered transition elements as they do not form stable species with partially filled d subshells:

$$
\begin{array}{llll}
\text{Sc} & [\text{Ar}]3d^14s^2 & \text{Zn} & [\text{Ar}]3d^{10}4s^2 \\
\text{Sc}^{3+} & [\text{Ar}] & \text{Zn}^{2+} & [\text{Ar}]3d^{10}
\end{array}
$$

Table 11.1 The *d* Block Elements

d block (shaded)

H																	He
Li	Be											B	C	N	O	F	Ne
Na	Mg											Al	Si	P	S	Cl	Ar
K	Ca	Sc	Ti	V	Cr	Mn	Fe	Co	Ni	Cu	Zn	Ga	Ge	As	Se	Br	Kr
Rb	Sr	Y	Zr	Nb	Mo	Tc	Ru	Rh	Pd	Ag	Cd	In	Sn	Sb	Te	I	Xe
Cs	Ba	La*	Hf	Ta	W	Re	Os	Ir	Pt	Au	Hg	Tl	Pb	Bi	Po	At	Rn
Fr	Ra	Ac**	Rf	Db	Sg	Bh	Hs	Mt	Ds	Rg	Cn						

*Lanthanides	Ce	Pr	Nd	Pm	Sm	Eu	Gd	Tb	Dy	Ho	Er	Tm	Yb	Lu
** Actinides	Th	Pa	U	Np	Pu	Am	Cm	Bk	Cf	Es	Fm	Md	No	Lr

The common stable species for Sc is Sc^{3+}, which has an empty *d* subshell. As for Zn, its only stable ion Zn^{2+} has a completely filled *d* subshell.

Q: What is so special about the element forming species with a partially filled *d* subshell?

A: We will see later that it is the very presence of a partially filled *d* subshell that gives rise to interesting properties such as the ability to exist in variable oxidation states, the ability to form complexes, catalytic activity, and colour — all of which are features of transition elements.

11.1 Writing Electronic Configuration

Let us do a warm-up exercise by listing the electronic configurations for the first row transition elements and their ions shown below:

(a) ${}_{22}Ti$ and Ti^{2+}, (e) ${}_{26}Fe$ and Fe^{2+},
(b) ${}_{23}V$ and V^{2+}, (f) ${}_{27}Co$ and Co^{2+},
(c) ${}_{24}Cr$ and Cr^{2+}, (g) ${}_{28}Ni$ and Ni^{2+},
(d) ${}_{25}Mn$ and Mn^{2+}, (h) ${}_{29}Cu$ and Cu^{2+}.

Solution:

(a) ${}_{22}Ti$: $1s^2 2s^2 2p^6 3s^2 3p^6 3d^2 4s^2$; Ti^{2+}: $1s^2 2s^2 2p^6 3s^2 3p^6 3d^2$,
(b) ${}_{23}V$: $1s^2 2s^2 2p^6 3s^2 3p^6 3d^3 4s^2$; V^{2+}: $1s^2 2s^2 2p^6 3s^2 3p^6 3d^3$,
(c) ${}_{24}Cr$: $1s^2 2s^2 2p^6 3s^2 3p^6 3d^5 4s^1$; Cr^{2+}: $1s^2 2s^2 2p^6 3s^2 3p^6 3d^4$,

(d)	${}_{25}$Mn:	$1s^22s^22p^63s^23p^63d^54s^2$;	Mn^{2+}:	$1s^22s^22p^63s^23p^63d^5$,
(e)	${}_{26}$Fe:	$1s^22s^22p^63s^23p^63d^64s^2$;	Fe^{2+}:	$1s^22s^22p^63s^23p^63d^6$,
(f)	${}_{27}$Co:	$1s^22s^22p^63s^23p^63d^74s^2$;	Co^{2+}:	$1s^22s^22p^63s^23p^63d^7$,
(g)	${}_{28}$Ni:	$1s^22s^22p^63s^23p^63d^84s^2$;	Ni^{2+}:	$1s^22s^22p^63s^23p^63d^8$,
(h)	${}_{29}$Cu:	$1s^22s^22p^63s^23p^63d^{10}4s^1$;	Cu^{2+}:	$1s^22s^22p^63s^23p^63d^9$.

Approach: Remember the following points, which tend to be overlooked:

- The $4s$ subshell is filled first with electrons before the $3d$ subshell. This is because the empty $4s$ subshell has a lower energy than the $3d$ subshell at that particular atomic number 19.
- However, as the $3d$ subshell is closer to the nucleus (since $n = 3$ principal quantum shell is closer to the nucleus than $n = 4$), once it is occupied by electrons, the $3d$ electrons repel the $4s$ electrons further away from the nucleus and cause the latter to be at a higher energy level. Thus, the electronic configuration of the element is written in the form of increasing principal quantum number.
- When writing the electronic configuration of the cation, write the electronic configuration of the neutral atom first and then remove electrons from the subshell with the highest energy, i.e., remove electrons from the $4s$ subshell before the $3d$ subshell.
- The anomalous cases of Cr and Cu: Apparently, electronic configurations with *half-filled* or *fully filled* $3d$ subshells are unusually stable due to symmetrical distribution of charge around the nucleus. The electronic configuration of Cr is $[Ar]3d^54s^1$ and that of Cu is $[Ar]3d^{10}4s^1$.

Q: Why is the energy of the $4s$ subshell lower than that of the $3d$ subshell? Isn't the $n = 3$ principal quantum shell nearer to the nucleus than that of $n = 4$?

A: Indeed, on average, the $n = 3$ principal quantum shell is closer to the nucleus than the $n = 4$. But within a particular principal quantum shell, different subshells have different energy due to their slightly different distance from the nucleus. We term them as having different "penetrating power". The more penetrating the subshell, the lower the energy it has because of the stronger attractive force experienced due to the closer proximity to the nucleus.

The penetrating power of the subshells in the same principal quantum shell decreases in the order $s > p > d > f$. The corresponding energy of the subshells increases in the order $s < p < d < f$. Across different principal quantum numbers, the relative penetrating power for the same type of sub-shell decreases in the order $1s > 2s > 3s > 4s > 5s$, etc. In addition, the number of protons cause each particular subshell of a particular principal quantum number to change to a different extent. All these complex factors combine in a non-linear relationship, which coincidentally results in the $3d$ subshell's energy being higher than that of the $4s$ subshell at atomic number = 19 (Fig. 11.1).

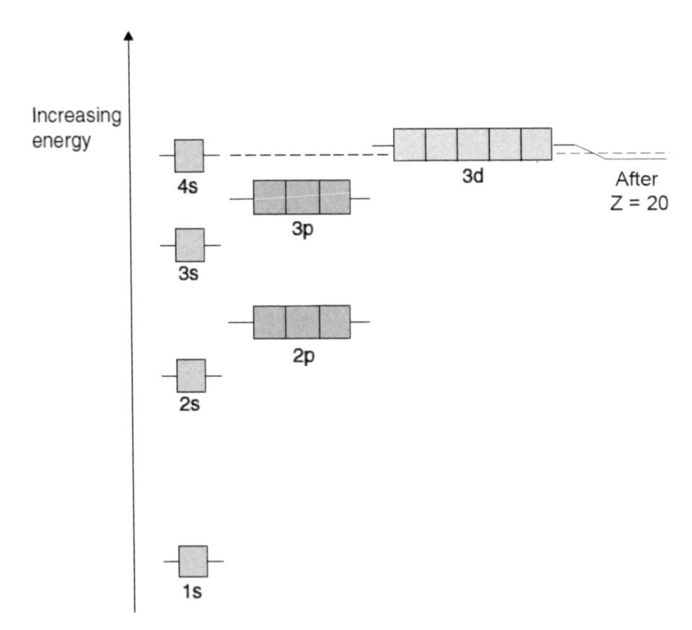

Fig. 11.1. Variation in potential energy of subshells against atomic number Z.

Q: If $3d$ electrons repel $4s$ electrons and raise the energy of the latter, why do the $4s$ electrons not subsequently "come down" to the lower energy $3d$ subshell?

A: This is a case of "chicken or the egg". One needs to recognise that the presence of d electrons does increase the energy of the

$4s$ electrons, and if for this reason the $4s$ electrons "drop down" to the lower energy $3d$ subshell, leaving the $4s$ subshell empty, then the scenario is similar to filling up the $3d$ subshell first instead of filling up the $4s$ subshell. But do not forget that at atomic number $= 19$, the $4s$ subshell has a lower energy than the $3d$ subshell.

Q: Why is a half-filled or completely filled d subshell more preferred than a partially filled d subshell?

A: The five d orbitals when combined, make a complete sphere. Thus, if the d subshell is half-filled or completely filled, the distribution of electron density is symmetrical as compared to an asymmetrical distribution of electron density for a partially filled d subshell. For the latter asymmetrical distribution of electron density, there is a potential difference built up, which means that there is a driving force to decrease this tension to achieve a "tension free" state, which has lower energy. The latter corresponds to a symmetrical distribution of electron density.

Q: If a more symmetrical distribution of electron density around the nucleus results in a more stable atom, then why is the electronic configuration of $_{23}V$ not $1s^2 2s^2 2p^6 3s^2 3p^6 3d^4 4s^1$? Doesn't it constitute a more symmetrical distribution than $1s^2 2s^2 2p^6 3s^2 3p^6 3d^3 4s^2$?

A: Yes, there is indeed an increase in the degree of symmetrical distribution of electron density if the electronic configuration is $1s^2 2s^2 2p^6 3s^2 3p^6 3d^4 4s^1$. But meanwhile, do not forget that the $3d$ subshell occupies a smaller region of space than the $4s$ subshell. Bringing the electron into the $3d$ subshell increases inter-electronic repulsion at the expense of a slight increase in the degree of electron density symmetry, which might not be thermodynamically feasible. So it seems that nature knows when to strike a good balance — it is at Cr and Cu.

Q: Why isn't the electronic configuration of Fe^{2+} be like that of the Cr atom of $[Ar]3d^5 4s^1$ instead of $[Ar]3d^6$?

A: The special electronic configurations of Cr and Cu atom, respectively of $[Ar]3d^5 4s^1$ and $[Ar]3d^{10} 4s^1$, only happened at that particular atomic numbers of 24 and 29, respectively. For Fe^{2+}, although the number of outmost electrons (i.e. 6 electrons in the $3d$ subshell) is the same as that of the Cr atom, the number of protons (atomic number) of Fe^{2+} (i.e. 26) does not "allow" it to have the electronic configuration of $[Ar]3d^5 4s^1$ because it is not the most stable form for this particular atomic number.

Q: So, if we "return" an electron to Fe^{2+}, the electronic configuration of Fe^+ is $[Ar]3d^6 4s^1$ and not $[Ar]3d^7$?

A: Yes, absolutely! The electronic configuration of Fe^{2+} is obtained by removing two electrons from a neutral Fe atom and thus if you put back an electron back to Fe^{2+} to form Fe^+, it is equivalent to the removal of one electron from a neutral Fe atom. The two electronic configurations must be the same.

11.2 Physical Properties of Transition Metals

As with all metals, transition elements are good conductors of heat and electricity. Yet they also show properties different from the metals of Groups 1 and 2. Unlike the latter, transition metals are hard, have higher densities, and higher melting and boiling points.

Recall that variations in physical properties, across a period, are attributed primarily to the degree of effective nuclear charge experienced by valence electrons.

If we move across the first row of transition elements, we expect the effective nuclear charge felt by the valence electrons to increase and consequently lead to a decrease in atomic radii from Ti to Cu. Likewise, we expect an increase in the first ionisation energies as we cross from Ti to Cu. Actual experimental data paints a different picture: the transition elements actually show remarkably similar properties without much variance across the d block.

11.2.1 *Trend in atomic radius*

As we move from Ti to Cu, there is an increase in nuclear charge. Each additional electron enters the penultimate (i.e., second to last) $3d$ subshell. As such, there is an increase in shielding effect caused by the increasing number of d electrons. Overall, the effective nuclear charge increases very gradually, accounting for the relatively constant atomic radii without much variation observed across the *d* block (Fig. 11.2).

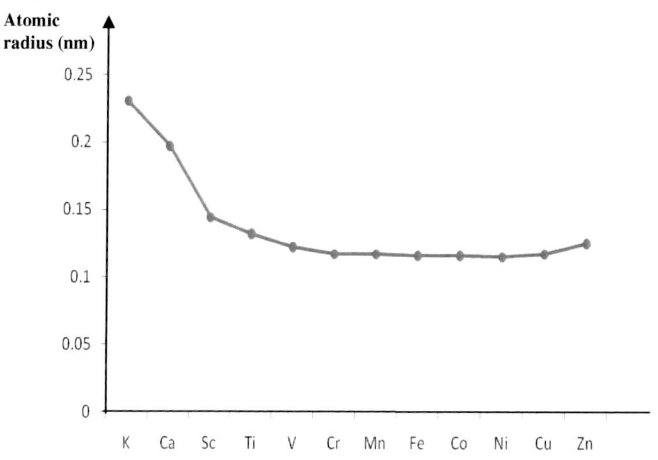

Fig. 11.2. Trend in atomic radius.

Q: Since electrons are being added to the penultimate *d* subshell, the increase in shielding effect should be able to nullify, to a substantial degree, the increase in nuclear charge that the 4*s* electrons experience. But if one compares the atomic radii of the *s* block elements (such as K and Ca) with the *d* block elements, one sees a very drastic difference. How does one account for the much smaller atomic radii of the *d* block elements compared to the *s* block elements?

A: The *d* electron in the penultimate subshell does not provide an effective shielding effect compared to the *s* electrons or *p* electrons. This is easy to rationalise. Remember we have said that the five *d* orbitals, when combined together, form a sphere? This means that each *d* orbital, on average, only occupies about 20% of the space; hence, this drastically decreases the amount of shielding effect it can provide. This results in a sharp increase in the effective nuclear charge, which causes a drastic decrease in atomic size.

In short, the smaller atomic radii of the *d* block elements compared to the *s* block elements arises from the greater effective nuclear charge in the former because of higher nuclear charge, but poor shielding effect, provided by the *d* electrons.

Q: So, in short, if we compare two species, then the one that has a smaller radius must be because of higher effective nuclear charge acting on the outmost electrons?

A: Absolutely spot on! The nett force acting on the outmost electrons is a "tug-of-war" between the nuclear charge and the inter-electronic repulsion, which results in the effective nuclear charge.

11.2.2 *Trend in ionic radius*

In comparing cations of the same charge, there is generally a gradual decrease in ionic radius across the *d* block (Fig. 11.3). The phenomenon is similar to the trend for atomic radii. This is attributed to the small yet gradual increase in effective nuclear charge from Sc to Ni.

On the whole, transition metal ions are considered to have higher charge density than their *s* block counterparts owing to their smaller ionic radii and higher charges. With such higher charge density, the transition metal cations undergo appreciable hydrolysis in aqueous solution, causing the solution to be relatively acidic (refer to Chap. 7).

11.2.3 *Trend in first ionisation energy*

Moving across from Sc to Zn, the first ionisation energies are approximately constant, i.e., there is only a very small increase across the

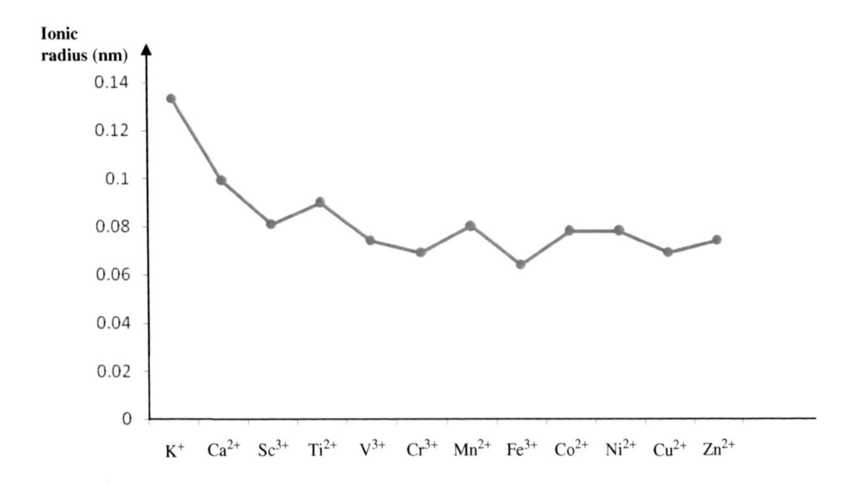

Fig. 11.3. Trend in ionic radius.

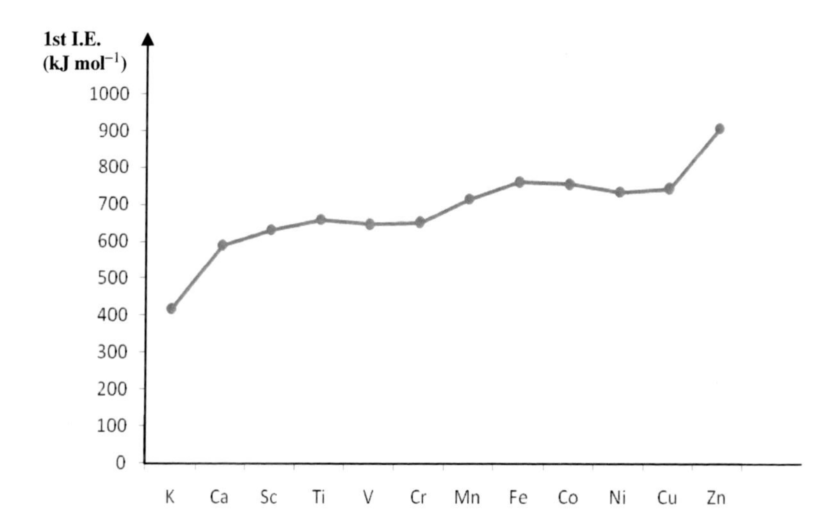

Fig. 11.4. Trend in first ionisation energy.

d block (Fig. 11.4). This can be ascribed to the relatively constant effective nuclear charge across the *d* block.

• Comparison between *d* block and *s* block elements in the same period

Example 11.1: Why do compounds containing Cr^{3+} exist but not those of Ca^{3+}?

Solution: Useful information from the data booklet:
First three ionisation energies of Ca (in kJ mol^{-1}): 590, 1150 and 4940.
First three ionisation energies of Cr (in kJ mol^{-1}): 653, 1590 and 2990.
Electronic configurations of:

$$_{20}Ca^{2+}: 1s^2 2s^2 2p^6 3s^2 3p^6; \quad _{24}Cr^{2+}: 1s^2 2s^2 2p^6 3s^2 3p^6 3d^4.$$

The 3*d* electron is at a higher energy level than the 3*p* electron. Hence, the *d* electron is easier to remove and less energy is required for the ionisation of Cr^{2+} to form Cr^{3+}. On the other hand, it is too energetically demanding to remove the 3*p* electron from Ca^{2+} to form Ca^{3+} as shown by the much higher 3rd ionisation energy of Ca compared to that of Cr.

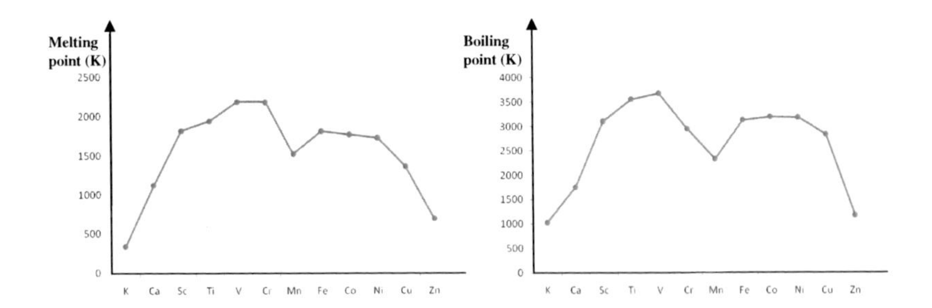

Fig. 11.5. Trends in melting and boiling points.

11.2.4 *Trend in melting and boiling points*

In general, the *d* block elements have high melting and boiling points (Fig. 11.5). These data suggests strong metallic bonding in these elements.

• Comparison between *d* block and *s* block elements in the same period

Example 11.2: Why do first row *d* block elements have higher melting points than *s* block elements?

Solution: The higher melting point of the *d* block elements compared to the *s* block elements indicates stronger metallic bonding in the former. The greater strength of the metallic bond is due to the greater number of electrons available for delocalisation. This is possible as both the 3*d* and 4*s* electrons are similar in energy.

The same argument applies when comparing the boiling points between the *d* block and *s* block elements.

11.2.5 *Trend in electrical conductivity*

The transition elements are metals with giant metallic lattice structures. They are good conductors of electricity as there exist delocalised electrons that can act as mobile charge carriers when a potential difference is applied.

• Comparison between *d* block and *s* block elements in the same period

Example 11.3: How do you expect the electrical conductivity of the first row *d* block elements to differ from those of the *s* block elements?

Solution: Transition metals are better electrical conductors than the *s* block elements. The better electrical conductivity is due to the greater number of delocalised electrons in the metallic lattice structure, i.e., both the 3*d* and 4*s* electrons are available since these have similar energies. In the *s* block elements, only the 4*s* delocalised electrons can act as mobile charge carriers.

11.2.6 *Trend in density*

Moving from Sc to Cu, there is a gradual increase in density, which can be accounted for by the small decrease in atomic size coupled with the increase in relative atomic mass. In general, more atoms of the *d* block elements are packed in a unit volume compared to the *s* block elements within the same period.

• Comparison between *d* block and *s* block elements in the same period

Example 11.4: How do we account for the higher density of the first row *d* block elements compared to that of the *s* block elements?

Solution: Comparing one unit volume of a *d* block and *s* block metal, the higher density of the former is attributed to the smaller atomic size and greater relative atomic mass of the *d* block metal atom compared to that of the *s* block metal atom.

11.3 Chemical Properties of Transition Metals

This section discusses some typical chemical properties unique to transition metals but not the *s* block elements, including:

- displaying variable oxidation states,
- exhibiting catalytic properties,
- forming stable complexes, and
- forming coloured compounds and ions.

11.3.1 *Variable oxidation states*

Exercise: State the oxidation state of Cr in the following compounds: (a) $CrCl_2$, (b) $CrCl_3$, (c) CrO_2, and (d) $K_2Cr_2O_7$.

Solution:

(a) +2, (b) +3, (c) +4, and (d) +6.

As shown in the above example, Cr, like the other transition elements, exhibits a variety of different oxidation states in its compounds. This is not true for K, Ca, Al or any other metals whose oxidation number corresponds solely to its group number.

The ability to display many different oxidation numbers is made possible by the fact that the 4*s* and 3*d* electrons have similar energies (Fig. 11.1) and different numbers of these electrons are available for use in bond formation.

There is a general relationship between the oxidation states and the type of bond formed (be it ionic or covalent):

- the elements exhibit higher oxidation states when they form covalently bonded oxo-anions with oxygen, e.g., $Cr_2O_7^{2-}$, VO_3^-, CrO_4^{2-};
- they exhibit lower oxidation states when they exist as cations.

Q: Is it possible for a transition element to use all its 3*d* and 4*s* electrons in bond formation?

A: Yes and no. It is observed for Sc to Mn. These elements exhibit the maximum oxidation number, which is equivalent to the total number of 3*d* and 4*s* electrons. For example in MnO_4^-, the Mn atom has the maximum oxidation state of +7. However, this is not the case for Fe and elements to the right of it in the periodic table. One reason for this is the limitation of size and space. Cu has a total of 11 electrons in both the 3*d* and 4*s* subshells. If you can, for a moment, picture Cu being surrounded by six O atoms (pretending that it is CuO_6^-), do you see it is a tight squeeze? In addition, since the *d* subshell of Cu is completely filled, the formation of a compound such as CuO_6^- would require the utilisation of the higher energy 4*p* subshell. This becomes thermodynamically non-spontaneous as far as "investment" is concerned. Anyway, the common oxidation states of Cu are +1 and +2 in its compounds.

Table 11.2 Vanadium and Its Various Ions

Oxidation State of V	**+5**	**+4**	**+3**	**+2**
Formula of ion	$VO_2^+(aq)$	$VO^{2+}(aq)$	$V^{3+}(aq)$	$V^{2+}(aq)$
Colour of ion	yellow	blue	green	purple

The reduction of $VO_2^+(aq)$ using the reducing agents $FeSO_4(aq)$ and Zn(s) is a good example to demonstrate the ability of the transition elements in exhibiting various oxidation states. The spontaneity of the redox reaction can be verified by calculating the $E^\ominus_{cell}$ value of the reaction. A positive value indicates that the reaction is thermodynamically spontaneous under standard conditions.

Table 11.2 shows vanadium in various oxidation states, the corresponding ions and their colours.

We will take a look at what happens when $FeSO_4(aq)$ and Zn(s) are each added, until in excess, to a solution of $NH_4VO_3(aq)$ acidified with dilute sulfuric acid.

Firstly, we must note the acid–base reaction that occurs between $VO_3^-(aq)$ and the acid as shown by Eq. (11.1). Hence, the subsequent reaction with $FeSO_4(aq)$ or Zn(s) actually involves the $VO_2^+(aq)$ ion and not the $VO_3^-(aq)$ ion:

$$VO_3^-(aq) + 2H^+(aq) \longrightarrow VO_2^+(aq) + H_2O(l). \qquad (11.1)$$

Relevant $E^\ominus$ values are used to compute $E^\ominus_{cell}$ to show the spontaneity of a particular reaction:

	$E^\ominus$ (V)
$VO_2^+ + 2H^+ + e^- \rightleftharpoons VO^{2+} + H_2O$	+1.00
$VO^{2+} + 2H^+ + e^- \rightleftharpoons V^{3+} + H_2O$	+0.34
$V^{3+} + e^- \rightleftharpoons V^{2+}$	−0.26
$V^{2+} + 2e^- \rightleftharpoons V$	−1.20
$Fe^{3+} + e^- \rightleftharpoons Fe^{2+}$	+0.77
$Zn^{2+} + 2e^- \rightleftharpoons Zn$	−0.76

- *Addition of aqueous iron(II) sulfate*

$$\underset{\text{yellow}}{VO_2^{+}(aq)} + \underset{\text{pale green}}{Fe^{2+}(aq)} + 2H^{+}(aq) \longrightarrow \underset{\text{blue}}{VO^{2+}(aq)} + \underset{\text{yellow-brown}}{Fe^{3+}(aq)} + H_2O(l)$$

$$E^{\ominus}_{cell} = 1.00 - (+0.77) = +0.23\,V.$$

Since $E^{\ominus}_{cell} > 0\,V$, the reaction is considered to be thermodynamically spontaneous under standard conditions. $Fe^{2+}(aq)$ is able to reduce $VO_2^{+}(aq)$ to $VO^{2+}(aq)$.

The original yellow colour of $VO_2^{+}(aq)$ will turn green. The green colour arises from the mixing of colours of the products — blue $VO^{2+}(aq)$ and yellow-brown $Fe^{3+}(aq)$.

Q: Since there is excess $Fe^{2+}(aq)$, can it further reduce the $VO^{2+}(aq)$ that is formed?

A: For this, let us calculate $E^{\ominus}_{cell}$ if the following reaction is to occur:

$$VO^{2+}(aq) + Fe^{2+}(aq) + 2H^{+}(aq) \longrightarrow V^{3+}(aq) + Fe^{3+}(aq) + H_2O(l).$$

$$E^{\ominus}_{cell} = 0.34 - (+0.77) = -0.43\,V.$$

Since $E^{\ominus}_{cell} < 0\,V$, the reaction is thermodynamically non-spontaneous under standard conditions. $Fe^{2+}(aq)$ is not able to reduce $VO^{2+}(aq)$.

Thus, when $FeSO_4(aq)$ is added in excess to a solution containing $VO_2^{+}(aq)$, the reducing power of $Fe^{2+}(aq)$ is only strong enough to reduce $VO_2^{+}(aq)$ to $VO^{2+}(aq)$.

- *Addition of zinc*

$$\underset{\text{yellow}}{2VO_2^{+}(aq)} + Zn(s) + 4H^{+}(aq) \longrightarrow \underset{\text{blue}}{2VO^{2+}(aq)} + \underset{\text{colourless}}{Zn^{2+}(aq)} + 2H_2O(l). \tag{11.2}$$

$$E^{\ominus}_{cell} = 1.00 - (-0.76) = +1.76\,V.$$

Since $E^{\ominus}_{cell} > 0\,V$, the reaction is thermodynamically spontaneous under standard conditions. Zn(s) is able to reduce $VO_2^{+}(aq)$ to

$VO^{2+}(aq)$. However, the reaction does not stop here. Zn(s) can further reduce the $VO^{2+}(aq)$ formed:

$$\underset{\text{blue}}{2VO^{2+}(aq)} + Zn(s) + 4H^+(aq) \longrightarrow \underset{\text{green}}{2V^{3+}(aq)} + Zn^{2+}(aq) + 2H_2O(l). \quad (11.3)$$

$$E^\ominus_{cell} = 0.34 - (-0.76) = +1.10\,V.$$

Since $E^\ominus_{cell} > 0\,V$, the reaction is thermodynamically spontaneous under standard conditions. Zn(s) is able to reduce $VO^{2+}(aq)$ to $V^{3+}(aq)$.

In fact, Zn(s) is able to further reduce the $V^{3+}(aq)$ that is formed:

$$\underset{\text{green}}{2V^{3+}(aq)} + Zn(s) \longrightarrow \underset{\text{purple}}{2V^{2+}(aq)} + Zn^{2+}(aq). \quad (11.4)$$

$$E^\ominus_{cell} = -0.26 - (-0.76) = +0.50\,V.$$

Since $E^\ominus_{cell} > 0\,V$, the reaction is thermodynamically spontaneous under standard conditions.

Q: Can Zn(s) further reduce the $V^{2+}(aq)$ that is formed?
A: For this, let us calculate $E^\ominus_{cell}$ if the following reaction is to occur:

$$V^{2+}(aq) + Zn(s) \longrightarrow V(s) + Zn^{2+}(aq).$$

$$E^\ominus_{cell} = -1.20 - (-0.76) = -0.44\,V.$$

Since $E^\ominus_{cell} < 0\,V$, the reaction is thermodynamically non-spontaneous under standard conditions. Zn(s) is not able to reduce $V^{2+}(aq)$.

Thus, when Zn(s) is added in excess to a solution containing $VO_2^+(aq)$, Zn(s) is such a strong reducing agent that it can reduce $VO_2^+(aq)$ to $V^{2+}(aq)$. The original yellow colour of the $VO_2^+(aq)$ will turn purple.

The overall equation for the reaction between $VO_2^+(aq)$ and Zn(s) is the summation of Eqs. (11.2), (11.3) and (11.4), i.e.,

$$\underset{\text{yellow}}{2VO_2^+(aq)} + 3Zn(s) + 8H^+(aq) \longrightarrow \underset{\text{purple}}{2V^{2+}(aq)} + 3Zn^{2+}(aq) + 4H_2O(l).$$

Based on the examples given, for as long as the intended reducing agent has a more negative $E^\ominus$ value than that of the species to be reduced, the reduction is considered to be thermodynamically spontaneous.

Example 11.5: With reference to $E^\ominus$ values, propose a reducing agent that can reduce manganese(IV) oxide to $Mn^{2+}(aq)$ in an acidic solution.

Solution: The reduction of MnO_2 can be represented by

$$MnO_2 + 4H^+ + 2e^- \rightleftharpoons Mn^{2+} + 2H_2O, \quad E^\ominus = +1.23\,V.$$

The intended reducing agent has to have a more negative $E^\ominus$ value than that of the species to be reduced, i.e., the $E^\ominus$ value has to be more negative than $+1.23\,V$.

Glancing through the list of standard electrode potentials in the Data Booklet, there are many possibilities. Just pick one but it is recommended to know, at the back of your mind, common reducing agents such as $FeSO_4(aq)$, $Zn(s)$, $KI(aq)$ and $SO_2(g)$.

In this case, the proposed reducing agent is $FeSO_4(aq)$:

$$Fe^{3+} + e^- \rightleftharpoons Fe^{2+}, \qquad E^\ominus = +0.77\,V.$$

The reaction between Fe^{2+} and MnO_2 is given as

$$MnO_2(s) + 2Fe^{2+}(aq) + 4H^+(aq) \longrightarrow Mn^{2+}(aq) + 2Fe^{3+}(aq) + 2H_2O(l).$$

$$E^\ominus_{cell} = 1.23 - (+0.77) = +0.46\,V.$$

Since $E^\ominus_{cell} > 0\,V$, the reaction is thermodynamically spontaneous under standard conditions. $Fe^{2+}(aq)$ is able to reduce $MnO_2(s)$ to $Mn^{2+}(aq)$.

11.3.2 *Catalytic properties*

Transition elements and their compounds are useful catalysts. The two types of catalysts are:

- heterogeneous catalysts, and
- homogeneous catalysts.

- Heterogeneous catalysts
  In heterogeneous catalysis, the catalyst is in a different phase from the reactants. How does this catalyst function in its role?

The reactants are usually liquids or gases and the catalyst is normally a solid that provides active sites at which reaction can occur. Reactant molecules are *adsorbed* onto the surface of the catalyst by forming weak bonds with the atoms on the active sites. For such bond formation to occur, the metal catalyst must ***possess partially filled d orbitals*** for the following two possible mechanisms to take place:

- the *d* electrons can be used to form bonds with the reactant molecules, and
- low-lying vacant orbitals can be used to accommodate lone pairs of electrons from reactant molecules, resulting in bond formation.

The increase in reaction rate is achieved through the following:

- The weak bonds that are formed between reactant molecules and the metal atoms at the active sites weaken the intra-molecular bonds in the reactant molecules. This reduces the activation energy for the reaction to occur.
- In addition, there will be a high concentration of reactants on the catalyst surface. Thus, the reactants are brought in close proximity to one another and with the correct orientation for effective reactions to take place.

In essence, the catalyst acts as a "reactant"! The reaction pathway with the involvement of the catalyst has a lower activation energy than the uncatalysed pathway. As a result, at the same temperature that the uncatalysed reaction is carried out, the catalysed reaction has more reactant particles possessing kinetic energy greater than the lower activation energy. This leads to a higher reaction rate.

Examples of heterogeneous catalysts include:

- finely divided Fe powder and Fe_2O_3 used in the Haber process to manufacture ammonia: $N_2(g) + 3H_2(g) \rightleftharpoons 2NH_3(g)$;
- Ni(s) used in the hydrogenation of unsaturated fats to the saturated form in the manufacture of margarine: $R{-}CH{=}CH{-}R'(l) + H_2(g) \longrightarrow R{-}CH_2{-}CH_2{-}R'(l)$.

Refer to Chap. 5 on Reaction Kinetics for details of the above examples.

• Homogeneous catalysts

In homogeneous catalysis, the catalyst is in the same phase as the reactants. How does this catalyst function in its role?

The involvement of the catalyst in the reaction creates an alternative reaction pathway that has lower activation energy (Fig. 11.6). It does so by reacting with one reactant to form an intermediate species that subsequently reacts with the other reactant to form the desired products. The catalyst is regenerated at the end of the reaction.

Therefore, for transition elements to serve the role of a homogeneous catalyst:

- they must be able to exist in different oxidation states, and
- it should be relatively easy to convert from one oxidation state to another.

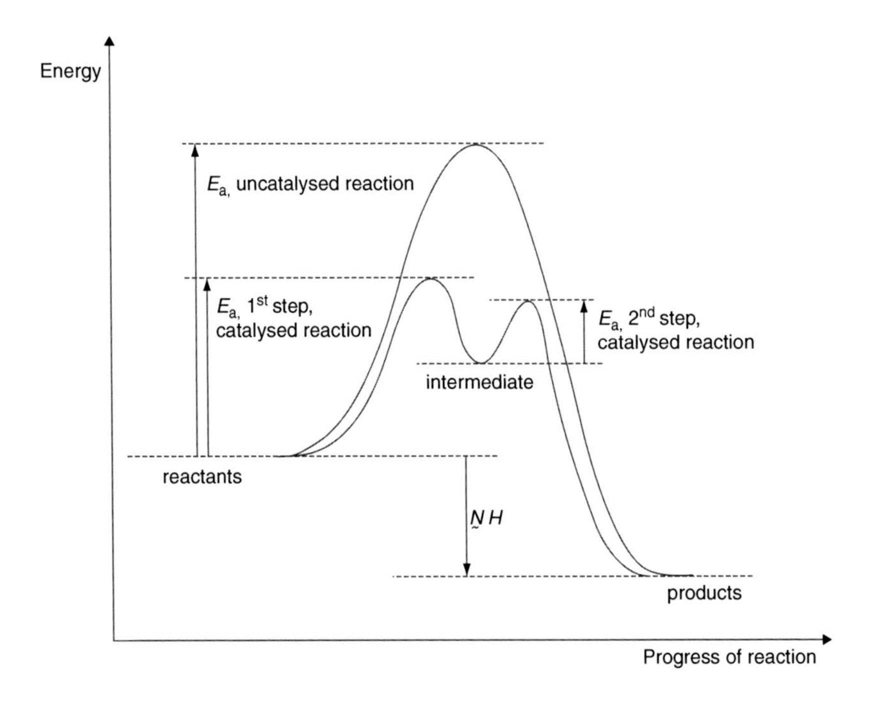

Fig. 11.6. Energy profiles for a catalysed and an uncatalysed reaction.

Transition metal ions are particularly effective at acting as homogeneous catalysts because they can *exist in different oxidation states and they can change their oxidation states easily*, facilitating the formation and breakdown of intermediate compounds between the ions and the reactants.

The following section provides examples of transition elements and their ions functioning as homogeneous catalysts. In addition, we will use $E^{\ominus}{}_{cell}$ calculations to show the feasibility of their usage.

Example 11.6: Account for the use of either Fe^{2+} or Fe^{3+} as catalysts in the reaction between peroxodisulfate ions and iodide ions.

Solution: $E^{\ominus}$ values needed:

$$S_2O_8{}^{2-} + 2e^- \rightleftharpoons 2SO_4{}^{2-}, \quad E^{\ominus} = +2.01\,V,$$
$$Fe^{3+} + e^- \rightleftharpoons Fe^{2+}, \quad E^{\ominus} = +0.77\,V,$$
$$I_2 + 2e^- \rightleftharpoons 2I^-, \quad E^{\ominus} = +0.54\,V.$$

The uncatalysed reaction is

$$S_2O_8{}^{2-}(aq) + 2I^-(aq) \longrightarrow 2SO_4{}^{2-}(aq) + I_2(aq).$$

$$E^{\ominus}{}_{cell} = +2.01 - (+0.54) = +1.47\,V.$$

Since $E^{\ominus}{}_{cell} > 0$, it is predicted that the uncatalysed reaction is thermodynamically spontaneous under standard conditions. However, kinetically, the electrostatic repulsion between the two negatively charged ions partly causes the reaction to have high activation energy. As a result, the rate of reaction is slow.

- When the reaction is catalysed by Fe^{2+}

Step 1: $$\underset{\text{catalyst}}{2Fe^{2+}(aq)} + S_2O_8{}^{2-}(aq) \longrightarrow 2SO_4{}^{2-}(aq) + \underset{\text{intermediate}}{2Fe^{3+}(aq)}.$$

$$E^{\ominus}{}_{cell} = +2.01 - (+0.77) = +1.24\,V.$$

Step 2: $$2I^-(aq) + \underset{\text{intermediate}}{2Fe^{3+}(aq)} \longrightarrow I_2(aq) + \underset{\text{catalyst regenerated}}{2Fe^{2+}(aq)}.$$

$$E^{\ominus}{}_{cell} = +0.77 - (+0.54) = +0.23\,V.$$

When the reaction is catalysed by Fe^{2+}, it proceeds via a two-step pathway that is thermodynamically spontaneous under standard conditions ($E^\ominus_{cell}$ for both steps > 0). These steps are expected to take place more readily than the uncatalysed reaction since they involve the collision between two oppositely charged ions. The activation energy for each of the steps is lower than that of the uncatalysed reaction and as a result, the reaction rate increases. The catalyst Fe^{2+} is regenerated at the end of the reaction.

- When the reaction is catalysed by Fe^{3+}

$$\text{Step 1:}\quad 2Fe^{3+}(aq) + 2I^-(aq) \longrightarrow I_2(aq) + 2Fe^{2+}(aq).$$
$$E^\ominus_{cell} = +0.77 - (+0.54) = +0.23\,V.$$
$$\text{Step 2:}\quad S_2O_8{}^{2-}(aq) + 2Fe^{2+}(aq) \longrightarrow 2SO_4{}^{2-}(aq) + 2Fe^{3+}(aq).$$
$$E^\ominus_{cell} = +2.01 - (+0.77) = +1.24\,V.$$

As shown by the positive $E^\ominus_{cell}$ values for the two steps, Fe^{3+} is also an effective catalyst for the same reasons why Fe^{2+} is a good one.

To select a suitable metal ion as a homogeneous catalyst, $E^\ominus$ values can provide us with a good starting point. The intended catalyst should have a standard reduction potential between those of the reactants, that is in this case, it should be between 0.54 V and 2.01 V for the reaction of peroxodisulfate and iodide ions.

11.3.3 *Formation of complexes*

A complex consists of a central metal atom or ion surrounded by other ions or molecules called ligands bonded to the central atom/ion by dative covalent bonds. We can have neutral complexes such as $Ni(CO)_4$ and also complex ions, which depending on their overall charge, can be either a cation or an anion, e.g., $[Fe(H_2O)_6]^{3+}$ and $[Fe(CN)_6]^{3-}$.

In complex formation, there is dative covalent bond formation between the central atom/ion and each of the ligands. A ligand must contain *at least* one donor atom with a lone pair of electrons to be used in forming a dative covalent bond with the central metal atom or ion. The central atom or ion must possess low-lying vacant orbitals to receive these lone pairs of electrons from the ligands.

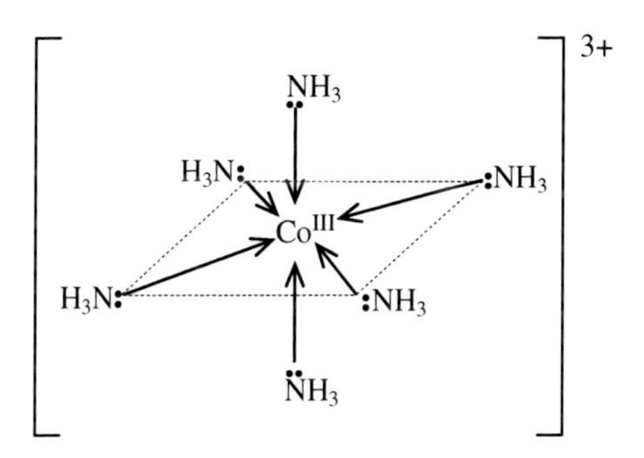

Fig. 11.7. A diagram depicting the complex $[Co(NH_3)_6]^{3+}$.

As shown in Fig. 11.7, the dative covalent bond is represented by an arrow "$\longrightarrow$" showing the direction of electron flow from each of the six NH_3 ligands to the central Co^{3+} ion.

The total number of dative covalent bonds that are associated with the central atom or ion is called the **co-ordination number**. Thus, for $[Co(NH_3)_6]^{3+}$, the co-ordination number is 6.

Example 11.7: State the co-ordination number for the following metal complexes:

(a) $[CuCl_4]^{2-}$, (b) $[Fe(CN)_6]^{4-}$, (c) $Ni(CO)_4$, and (d) $[Ag(NH_3)_2]^+$.

Solution:

(a) 4, (b) 6, (c) 4, and (d) 2.

11.3.3.1 *Nature of ligands*

A ligand can be either a molecule or an ion, as long as it contains at least one atom bearing a lone pair of electrons to be used in dative covalent bond formation. Examples of ligands include the neutral molecules H_2O, NH_3 or CO and the anions Cl^-, OH^-, CN^-or SCN^-.

There are ligands that can donate more than one pair of electrons and hence form more than one dative covalent bond to the same central atom or ion. These ligands are classified as polydentate ligands. More specifically, a *bi*dentate ligand forms two bonds, a *tri*dentate ligand forms three bonds, and so forth. A *mono*dentate ligand is capable of forming, as its prefix implies, only one dative covalent bond with the central metal atom or ion. Polydentate ligands typically have a relatively long skeletal chain length, usually comprising of carbon atoms, which allows the ligand molecule to coil round the metal centre.

Another term for polydentate ligand is *chelating agent*, derived from the Greek word meaning "claw." The resultant complex ion formed is commonly called a chelate or chelated compound. Comparing complexes of the same central metal ion and the same co-ordination number, one with a polydentate ligand is generally more stable than one with many monodentate ligands. This is because in order to dissociate the polydentate ligand from the metal centre, all the dative covalent bonds need to break off simultaneously at one go — a feat statistically less probable than breaking off a single covalent bond formed with a monodentate ligand.

There are many polydentate molecules present naturally in the biological system. They include saccharides, amino acids, proteins, polypeptides, and fatty acids.

Example 11.8: State the oxidation number of the metal in the following complex ions.

(a) $[V(H_2O)_6]^{3+}$, (b) $[Cr(NH_3)_6]Cl_3$, (c) $[Fe(CN)_6]^{4-}$, and (d) $[Fe(SCN)\ (H_2O)_5]^{2+}$.

Solution:

(a) +3, (b) +3, (c) +2, and (d) +3.

The overall charge of a complex ion is the sum of the charges on the metal centre and individual ligands.

Take for example, $[Fe(SCN)(H_2O)_5]^{2+}$.

Let the oxidation state of Fe be x. SCN^- has a net charge of −1. The charge on the neutral H_2O molecule is 0. The overall charge of this complex ion is +2.

Thus, we have

$$x + (-1) + 5(0) = +2$$

$$x = +3.$$

Hence, the oxidation state of Fe is +3.

Example 11.9: Identify the donor atoms, which can form dative covalent bonds, in the following ligands:

(a) CN^- **(b)** CO **(c)** SCN^- **(d)** $H_2N–CH_2–CH_2–NH_2$

(e) 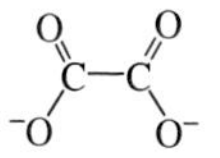 $^-O–C(=O)–C(=O)–O^-$ **(f)** 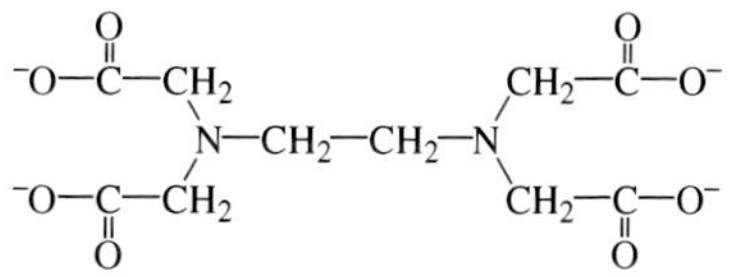 $(^-O–C(=O)–CH_2)_2N–CH_2–CH_2–N(CH_2–C(=O)–O^-)_2$

Solution:

(a) and (b) For these monodentate ligands, the donor atom is the C atom. For $^-C{\equiv}N$, the negative charge resides on the carbon atom, which makes it more electron-rich than the nitrogen atom. As for CO, the negative end of the dipole also resides on the C atom (refer to Chap. 2 on Chemical Bonding!). In addition, the C atom is less electronegative than the other bonding atom, i.e., it is more willing to donate its lone pair of electrons.

(c) For the monodentate thiocyanate ion, either the S or N atom can be considered donor atoms due to the resonance that exists within the ion:

$$S{=}C{=}N^- \longleftrightarrow {}^-S{-}C{\equiv}N.$$

(d) For the bidentate ligand ethane-1,2-diamine, the donor atoms are the two N atoms — the only atoms in the molecule with a lone pair of electrons.

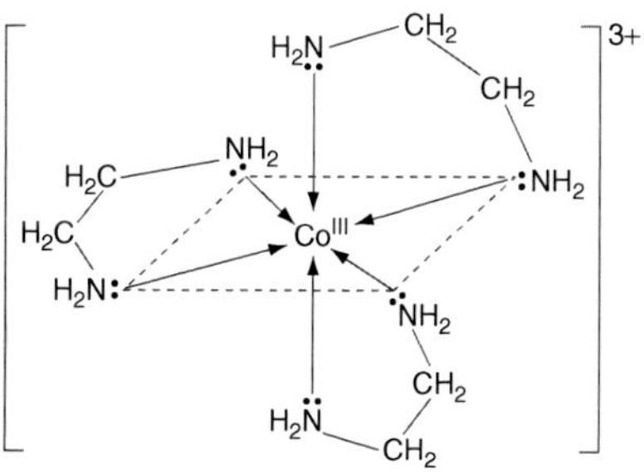

ethane-1,2-diamine (en) as a bidentate ligand in $[Co(en)_3]^{3+}$

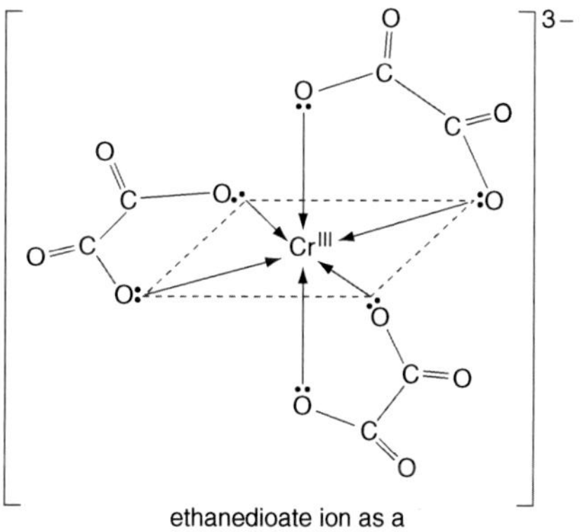

ethanedioate ion as a bidentate ligand in $[Cr(C_2O_4)_3]^{3-}$

(e) The ethanedioate ion serves as a bidentate ligand. Its donor atoms are the O atoms bearing the negative charge. These are more electron rich than the O atoms that are doubly bonded to the C atoms.

(f) The **e**thylene**d**iamine**t**etra**a**cetate ion, abbreviated as edta^{4-}, is an example of a *hexa*dentate ligand which can form six dative covalent bonds with the metal centre. Its donor atoms are the two N atoms and the four O atoms bearing the negative charge:

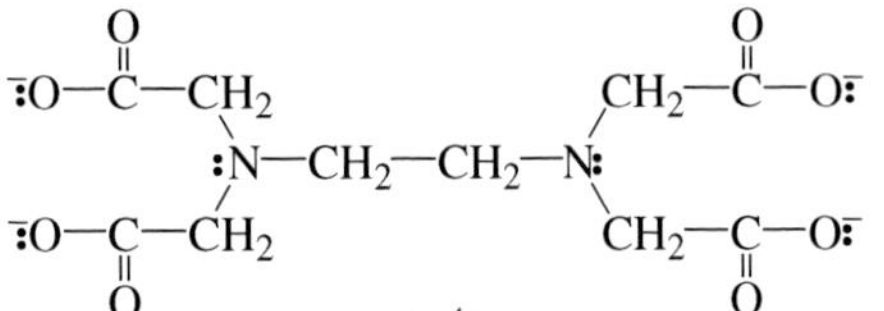

edta^{4-}

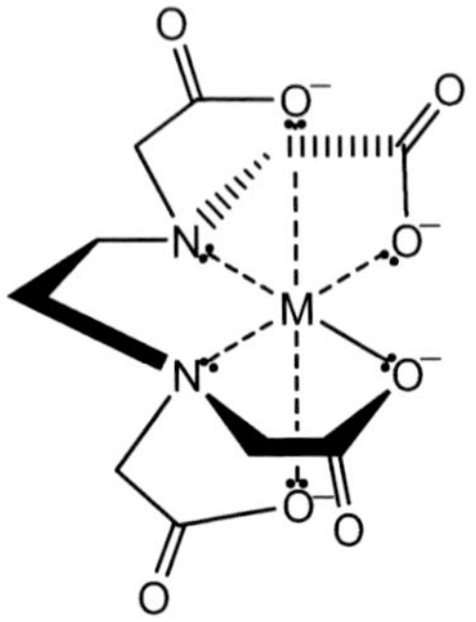

Q: Can Zn^{2+} function as a ligand since it has five pairs of electrons in its $3d$ subshell?

A: No, it cannot. Remember, in bond formation, there must be electrostatic attraction between two particles. Zn^{2+} is a positively charged species that is unlikely to use its lone pair of electrons for dative covalent bonding. In addition, if the metal centre is a cation, both species exhibit electrostatic repulsion between each other.

11.3.3.2 *Shapes of complex ions*

Ligands can be a monatomic ion or a large molecule. The difference in sizes of these ligands, coupled with the various sizes of the transition metal ions, means that the number of possible ligands that can fit around the central metal ion varies.

Nonetheless, some generalisations can be made, such as the following:

- Some transition metal ions have the same co-ordination number for various types of ligands. For instance, Ag^+ tends to form complexes with the co-ordination number of 2 and these are linear in shape, e.g., $[Ag(CN)_2]^-$ and $[Ag(NH_3)_2]^+$:

$$[H_3N: \longrightarrow Ag^I \longleftarrow :NH_3]^+ \quad [Ag(NH_3)_2]^+$$

linear

- Since H_2O is a relatively small ligand, aqua complexes are generally octahedral in shape with the co-ordination number of 6, i.e., 6 H_2O molecules surround the central metal ion, e.g., $[V(H_2O)_6]^{3+}$, $[Fe(H_2O)_6]^{3+}$ and $[Cu(H_2O)_6]^{2+}$.
- NH_3 and CN^- are also considered relatively small ligands, and complexes containing these ligands are generally octahedral in shape with the co-ordination number of 6, e.g., $[Ni(NH_3)_6]^{2+}$ and $[Fe(CN)_6]^{3-}$.

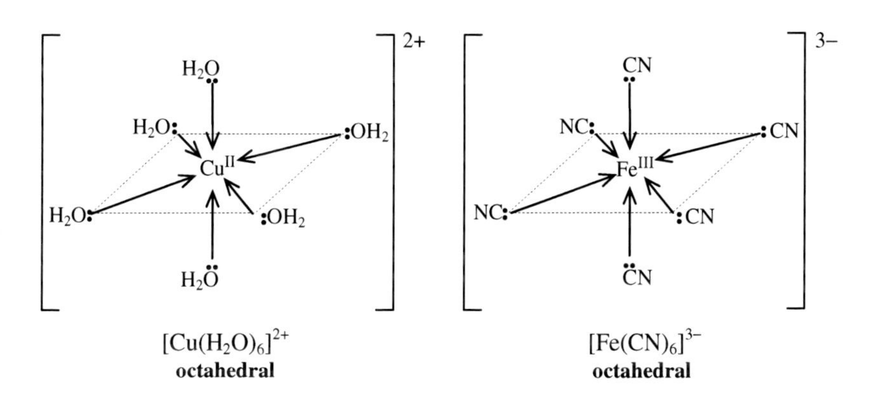

$[Cu(H_2O)_6]^{2+}$
octahedral

$[Fe(CN)_6]^{3-}$
octahedral

However, there are exceptions such as $[Cu(CN)_4]^{3-}$, which is tetrahedral in shape with the co-ordination number of 4. It is thus important to take note that the co-ordination number around the metal species is highly dependent on the nature of the metal species and the ligand involved.

- Compared to the above ligands, Cl^- is relatively larger and only four Cl^- ligands can fit around, for instance, cobalt and copper ions. For example, both $[CoCl_4]^{2-}$ and $[CuCl_4]^{2-}$ are tetrahedral.

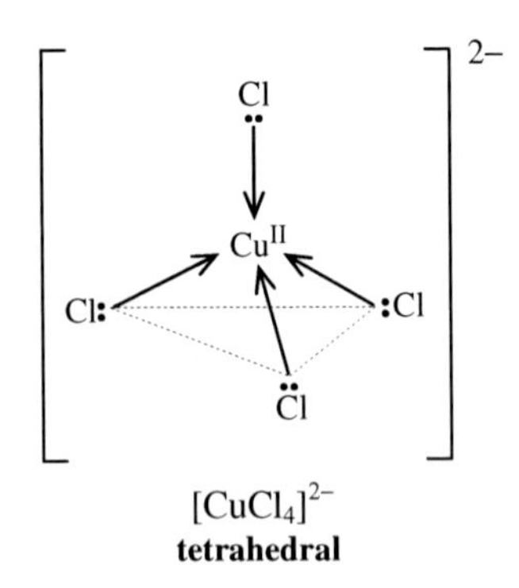

$[CuCl_4]^{2-}$
tetrahedral

11.3.3.3 *Understanding the nomenclature of complexes*

It is both important to know the chemical formula of the complex and how to name it. There is a systematic approach to naming complexes and this makes identifying them easier, but firstly, you need to learn about the rules involved.

For cationic complexes, the name of the metal is used. For example,

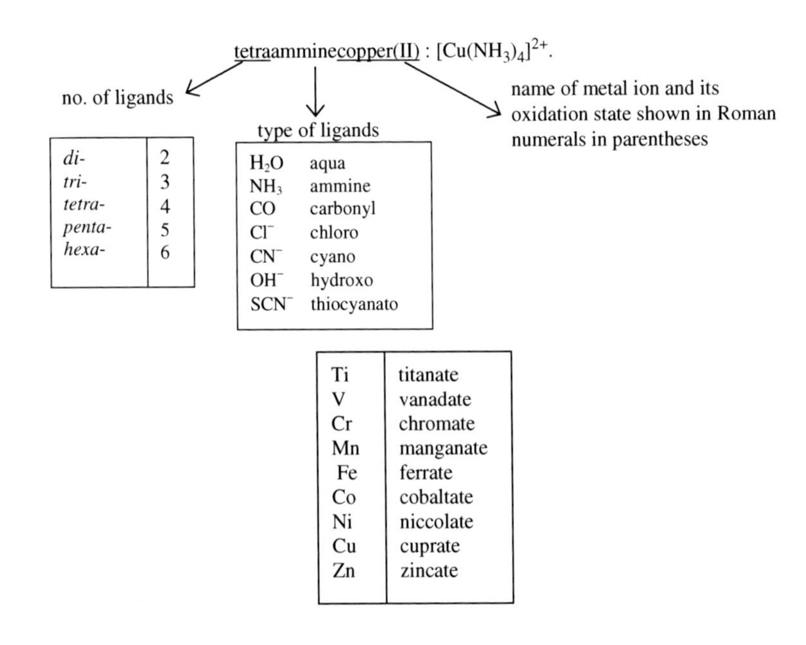

di-	2
tri-	3
tetra-	4
penta-	5
hexa-	6

H_2O	aqua
NH_3	ammine
CO	carbonyl
Cl^-	chloro
CN^-	cyano
OH^-	hydroxo
SCN^-	thiocyanato

Ti	titanate
V	vanadate
Cr	chromate
Mn	manganate
Fe	ferrate
Co	cobaltate
Ni	niccolate
Cu	cuprate
Zn	zincate

For anionic complexes, the name of the metal is modified to end in -*ate*. For example, tetracyanocuprate(I): $[Cu(CN)_4]^{3-}$.

For complexes with more than one type of ligand: the ligands are listed in alphabetical order and the number of each is stated immediately before its name. For example,

$$\underbrace{\text{pentaaqua}}_{\substack{5\,H_2O \\ \text{ligands}}}\underbrace{\text{hydroxo}}_{\substack{1\,OH^- \\ \text{ligand}}}\text{aluminium(III)}: [Al(OH)(H_2O)_5]^{2+}.$$

There are three main points to note when writing the chemical formulae of complex ions:

- The chemical formula of the complex ion is enclosed in square brackets with the charge written as a superscript.
- When there is more than one polyatomic ligand of the same kind, the chemical formula of the ligand is enclosed in curved brackets with the number of such ligands written as a subscript, e.g., 5 H_2O ligands represented in bold in the formula $[Al(OH)\mathbf{(H_2O)}_5]^{2+}$.
- Regardless of the type of complex, one always lists the chemical symbol of the metal, followed by the ligands in the order: anion and then neutral ligand. For example, $[Al(OH)(H_2O)_5]^{2+}$.

Example 11.10: Write the chemical formulae for the following metal complexes:

(a) tetracarbonylnickel(0), (b) tetraamminediaquacopper(II),
(c) hexacyanoferrate(III), and (d) potassium hexacyanoferrate(II).

Solution: (a) $Ni(CO)_4$, (b) $[Cu(NH_3)_4(H_2O)_2]^{2+}$, (c) $[Fe(CN)_6]^{3-}$, and (d) $K_4[Fe(CN)_6]$.

11.3.4 *The property of colour*

Many transition metal compounds are coloured. Common examples include the following:

CrO_4^{2-}	yellow	$[Fe(H_2O)_6]^{3+}$	yellow
$Cr_2O_7^{2-}$	orange	$[Fe(H_2O)_6]^{2+}$	pale green
$[Cr(H_2O)_6]^{3+}$	green	$[Fe(SCN)(H_2O)_5]^{2+}$	blood-red
$[Cr(H_2O)_6]^{2+}$	blue	$[CoCl_4]^{2-}$	blue
MnO_4^-	purple	$[Co(H_2O)_6]^{2+}$	pink
$[Mn(H_2O)_6]^{2+}$	faint pink	$[Cu(H_2O)_6]^{2+}$	blue

11.3.4.1 *Why are transition metal compounds coloured?*

White light is composed of the various colours that make up the visible region of the spectrum. These, in order from short to long wavelength (from around 400–700 nm), are: violet, blue, green, yellow, orange and red.

When light strikes an opaque object, some wavelengths may be absorbed and others reflected. We see an apple as "red" because it absorbs all the frequencies of visible light shining on it except for those that it reflects, which, in this case, correspond to wavelengths richer in the colour red.

Transparent objects appear coloured because they transmit a certain range of wavelengths. If they transmit all the wavelengths of white light, they will appear colourless.

An object appears white when all the incident white light is reflected. When all the wavelengths of white light are absorbed, the object will appear black.

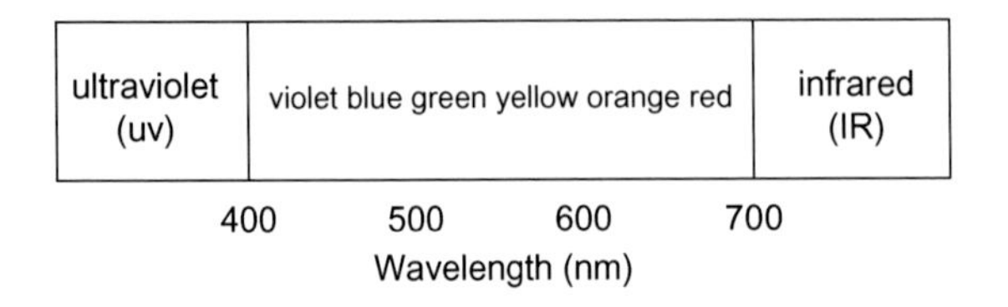

Fig. 11.8. The visible region of the spectrum.

In general, the observed colour of an object is the reflected wavelengths that are not absorbed by the object. We say that its colour is complementary to the colour of light (wavelengths) absorbed.

Going back to the example of the apple, because it absorbs green light, we see the complementary colour of green — which is red. A similar explanation is used for a green apple.

A colour wheel helps us to determine the observed colour of light that arises from the removal of certain colours from white light (i.e., those absorbed by the object). Complementary colours are those opposite each other in the colour wheel such as the pair: blue and orange. The use of mnemonics, consisting of an acronym or short phrase, can help us to remember the list of colours. For instance, the first letters of the words in the phrase "*R*ichard *O*f *Y*ork *G*ets *B*ig *V*an" correspond to the colour sequence in the colour wheel:

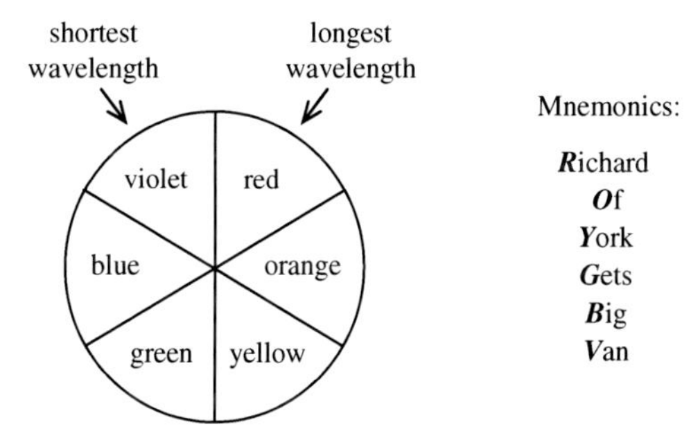

In addition, to remember the relative wavelengths of the colours, we just have to note those at both ends of the visible region of the spectrum — the colour red is of the longest wavelength and violet is of the shortest wavelength and this information can be inferred from the length of the words "Richard" and "Van", i.e., long and short, respectively.

Q: Transition elements form compounds that are coloured. A solution of V^{3+} is green in colour. What brings about the absorption of particular colours and hence the colours observed?

A: The resultant colour of the solution observed is due to the interaction between the complex species and the visible spectrum.

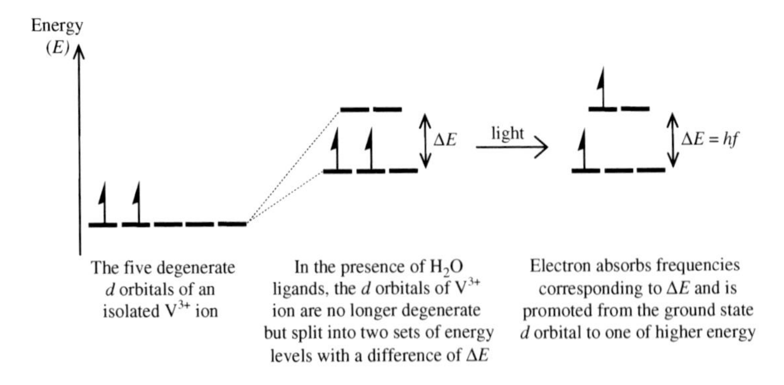

Fig. 11.9. The splitting of d orbitals by ligands and d–d transition.

In the gaseous state, all the d orbitals of the isolated V^{3+} ion have the same energy, i.e., they are degenerate.

But in aqueous solution, since V^{3+} exists as $[V(H_2O)_6]^{3+}$, there is a non-homogeneous electrostatic field surrounding the V^{3+} ion. The electrostatic field is especially strong along the x, y, and z axis (refer to Sec. 11.3.4.2 on Crystal Field Theory).

Consequently, the d orbitals experience different magnitudes of electrostatic interaction with the ligands, which causes the set of d orbitals to lose their degeneracy and split into two groups of slightly different energy levels with a difference of ΔE (see Fig. 11.9).

When visible light passes through this solution containing $[V(H_2O)_6]^{3+}$, the light energy with frequency corresponding to the energy gap, i.e., $\Delta E = hf$, is absorbed by the electron in the lower energy level. This electron transits to the higher energy level. Such electronic transition is termed ***d*–*d* transition**. To be more accurate, it should be *d*-*d** transition in which the * represents an anti-bonding orbital (refer *to Understanding Advanced Organic and Analytical Chemistry* by K. S. Chan & J. Tan for the meaning of anti-bonding orbital).

We observe the colour that is not absorbed and it is complementary to the colour that is absorbed.

Q: After the electron has absorbed the energy and been promoted to the higher energy level, does it stay there forever?

A: After the absorption of energy, the complex ion is no longer in the ground state but it is in an excited state. The excited state is not a stable state. The ion will revert back to the ground state but it does not re-emit the absorbed radiation as a photon that it absorbed originally. Instead, it loses this extra energy in the form of kinetic energy through vibration or collisions. We term such a process "relaxation".

Q: Why is a solution containing $Na^+(aq)$ colourless? Isn't it possible for an electron in the $2p$ orbital to absorb radiation and be promoted to the higher energy $3s$ orbital?

$$\begin{array}{ccc} Na^+ & & Na^+ \\ \text{(ground state)} & \longrightarrow & \text{(excited state)} \\ 1s^2 2s^2 2p^6 & & 1s^2 2s^2 2p^5 3s^1 \end{array}$$

A: Yes, it is possible for an electron in the $2p$ orbital to be promoted to the $3s$ orbital. However, because ΔE between these two energy levels is very large, the $2p$ electron needs to absorb a photon of high energy radiation, which is beyond what is available in the visible region of the spectrum. The energy required falls in the ultraviolet region. Since there is no absorption of wavelengths from the visible region of the spectrum, all the wavelengths of white light are transmitted and the solution of $Na^+(aq)$ appears colourless.

Note that in solution you also have other ions such as the anions, i.e., a solution of NaCl(aq). For the same reason as in the case of Na^+, the Cl^- ion does not absorb any of the wavelengths of white light and thus appears colourless.

Q: Why is a solution of $Zn^{2+}(aq)$ colourless even though it has electrons in the *d* orbitals?

A: This is because all the *d* orbitals are completely filled with electrons. Even if splitting of the five *d* orbitals occurs, *d*–*d* transition is not possible as the higher energy *d* orbitals are not vacant and hence cannot accept incoming electrons. As a result, none of the wavelengths corresponding to the visible spectrum are absorbed.

In conclusion, for colour to be observed as a result of *d*–*d* transition, the transition metal ion must contain a partially filled *d* subshell such that an electron from a lower energy level can be promoted to a higher energy level by absorbing the appropriate electromagnetic radiation in the visible region of the spectrum.

Can you reason why compounds containing the transition metal ions Sc^{3+}, Ti^{4+} and Cu^{+} are colourless?

11.3.4.2 *Crystal field theory*

Crystal field theory uses simple electrostatic considerations to explain how the energies of the 3*d* orbitals are affected by the presence of surrounding ligands. According to the theory, the interaction between the transition metal ion and the ligands is essentially the attraction between oppositely charged species — the cation and the negative charge (lone pair of electrons) on the donor atom of the ligand.

We will first consider the case of octahedral complexes. For an isolated transition metal ion in the gaseous phase, the five *d* orbitals are degenerate. In forming an octahedral complex, six ligands (L), which are regarded as point charges, will have to approach the cation (M) along the three axes (x-, y- and z-axes) to facilitate dative covalent bond formation in these orientations (see Fig. 11.10).

When the ligands approach the cation, there will be inter-electronic repulsion between the electrons in the *d* orbitals of the cation and the lone pair of electrons from the donor atom of the ligands. As a result, the energies of these *d* electrons increase but to different extents.

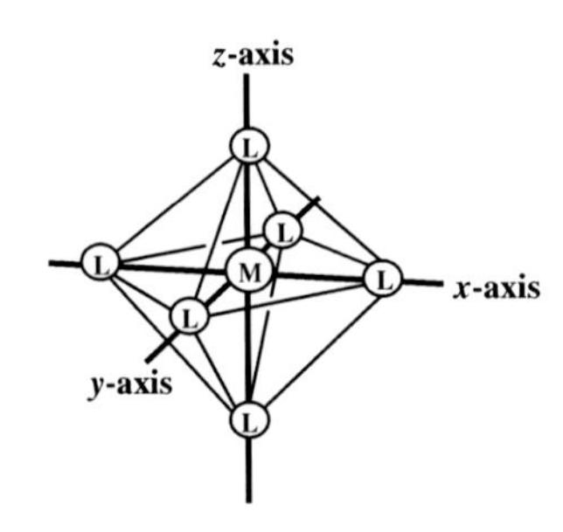

Fig. 11.10. Orientations of ligands in forming an octahedral complex.

Q. If the energies of the d electrons increase to different extents, does this mean that they experience repulsion of different magnitudes?

A. Yes, and we can explain this by considering the relative proximity of the d orbitals to the approaching ligands.

Recall the shapes and orientation of the five d orbitals? We can group these orbitals according to the distribution of maximum electron density:

- $d_{x^2-y^2}$ and d_{z^2} orbitals have maximum electron density *along* the axes;
- d_{xy}, d_{yz} and d_{xz} orbitals have maximum electron density *between* the axes.

The electrons in these two sets of orbitals will experience different extents of repulsion with the lone pair of electrons from the donor atom of the ligands. When the ligands approach along the three axes, you will expect electrons in the $d_{x^2-y^2}$, and d_{z^2} orbitals to experience greater repulsion than those in the d_{xy}, d_{yz} and d_{xz} orbitals (Fig. 11.11).

The five d orbitals are no longer degenerate. They are split into two different energy levels: the $d_{x^2-y^2}$ and d_{z^2} orbitals will have higher energy than the other three d orbitals (Fig. 11.12).

The type of ligands and numbers of d electrons that the metal centre possesses will bring about different magnitudes of the splitting, ΔE. This will in turn lead to different colours observed, e.g., $[Ni(H_2O)_6]^{2+}$(aq) is green whereas $[Ni(NH_3)_6]^{2+}$(aq) is blue. For a given central metal ion, the relative splitting effect of various ligands is presented in the spectrochemical series that is partially listed here:

$CO > CN^- > NO_2^- > NH_3 > edta^{4-} > H_2O > C_2O_4^{2-} > OH^- > F^- > Cl^- > Br^- > I^-$

stronger field ligands that cause large splitting (large ΔE) — weaker field ligands that cause small splitting (small ΔE)

The relationship between ΔE and wavelength (λ) of the radiation absorbed is given by Planck's equation:

$$\Delta E = hf = \frac{hc}{\lambda},$$

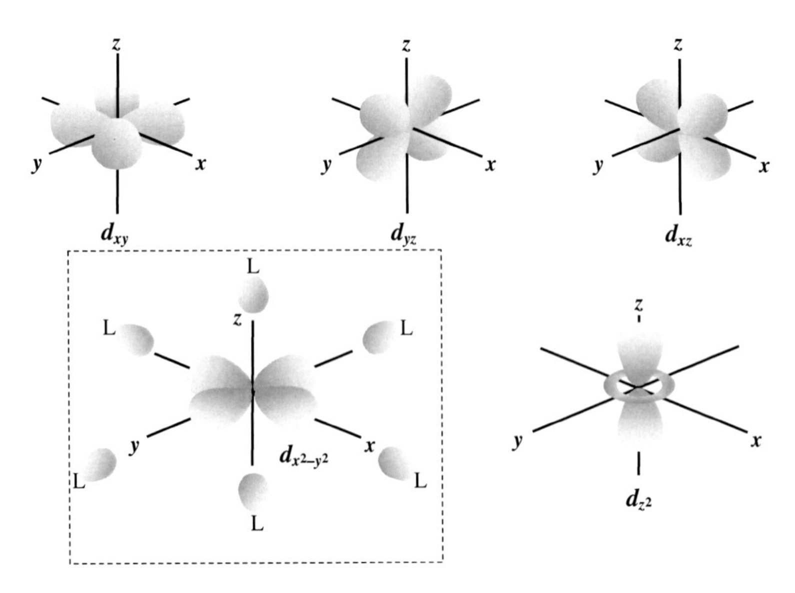

Fig. 11.11. *d* electrons repelled, to different extents, by the charges of the ligands.

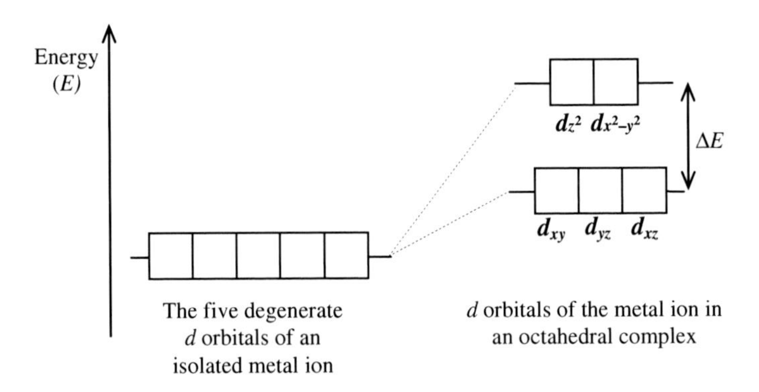

Fig. 11.12. The splitting of the *d* orbitals.

where h = Planck constant $= 6.63 \times 10^{-34}\,\mathrm{J\,s}$, f = frequency ($\mathrm{s^{-1}}$), and c = speed of light $= 3.00 \times 10^8\,\mathrm{m\,s^{-1}}$.

If we are to compare two octahedral complexes of a given metal ion, the one with the weaker field ligands will absorb light of longer wavelength (or lower frequency) since ΔE is smaller, and *vice versa.*

Can you now account for the respective colours of $[Ni(H_2O)_6]^{2+}(aq)$ and $[Ni(NH_3)_6]^{2+}(aq)$?

Crystal field theory can also be used to account for other geometries of complexes, such as tetrahedral and square planar complexes. In each of these cases, the different spatial orientation of the approaching ligands and its relative repulsion effect on the *d* orbitals will cause different splitting patterns from that seen in the case of octahedral complexes.

The size of the splitting, ΔE, determines the type of wavelengths absorbed, and hence the resultant transmission of the wavelength of the radiation affects the colour of complexes. Therefore, *d* orbital splitting, and ultimately colour, is affected by the various factors:

- nature of the transition metal ion;
- oxidation state of the metal ion, e.g., $[Fe(H_2O)_6]^{3+}$ is yellow and $[Fe(H_2O)_6]^{2+}$ is pale green;
- orientation of the ligands around the central metal ion, i.e., type of geometry;
- type of ligands.

Take note that the colours of transition elements and their compounds are not solely attributed to *d*–*d* transitions. Other possible reasons for the phenomenon of colour include the charge transfer mechanism, which accounts for the purple colour of MnO_4^-. However, these are beyond the scope of this book.

Example 11.11: Why is a solution of Al^{3+} colourless whereas an aqueous solution of Cr^{3+} is coloured?

Solution: In aqueous solutions, Al^{3+} and Cr^{3+} exists as $[Al(H_2O)_6]^{3+}$ and $[Cr(H_2O)_6]^{3+}$, respectively.

In the case of Cr^{3+}, the presence of H_2O ligands causes its five degenerate 3*d* orbitals to split into two sets of orbitals with slightly different energy levels. When white light passes through a solution of $[Cr(H_2O)_6]^{3+}$, the electrons from the lower energy level absorb certain wavelengths from the visible region of the spectrum and are promoted to the higher energy level. The colour of the complex observed is the complement of the colour absorbed.

As for Al^{3+}, it does not have any d electrons and thus d–d transition does not occur. To promote its electrons ($2p$ electrons) to vacant orbitals ($3s$ orbital) of a higher energy level requires absorption of wavelengths beyond the visible region of the spectrum. Therefore, when light passes through a solution of $[Al(H_2O)_6]^{3+}$, none of it is absorbed and as all the wavelengths of white light are transmitted, the solution of $Al^{3+}(aq)$ appears colourless.

11.3.5 *Ligand exchange reactions*

When you add excess aqueous ammonia to a solution containing $[Ni(H_2O)_6]^{2+}$ ions (e.g., nickel(II) sulfate solution), you will notice that the solution changes from its original green colour to blue. What causes the colour change?

A ligand exchange reaction has occurred, whereby a ligand in the complex ion is replaced by a different ligand. In this example, the 6 H_2O molecules are replaced by 6 NH_3 molecules to form the complex ion $[Ni(NH_3)_6]^{2+}$:

$$\underset{\text{green}}{[Ni(H_2O)_6]^{2+}}(aq) + 6NH_3(aq) \rightleftharpoons \underset{\text{blue}}{[Ni(NH_3)_6]^{2+}}(aq) + 6H_2O(l).$$

Q: How do we know if a ligand is able to replace another? Can NH_3 molecules be replaced by H_2O molecules and re-form $[Ni(H_2O)_6]^{2+}(aq)$?

A: Ligand exchange can be viewed as a competition, with different ligands vying for the same metal ion. As to which ligands will emerge the winners, we can make use of the concept of the **complexation stability constant**. In general, ligands that can form a complex of higher stability constant will replace ligands that are only capable of forming a complex of lower stability constant.

How do we calculate the complexation stability constant? The stability constant of a complex (K_{stab}) is essentially the equilibrium constant K for its formation. K_{stab} for the $[\text{Ni}(\text{NH}_3)_6]^{2+}$ complex is given as

$$K_{\text{stab}} = \frac{[\text{Ni}(\text{NH}_3)_6{}^{2+}]}{[\text{Ni}(\text{H}_2\text{O})_6{}^{2+}][\text{NH}_3]^6}\,\text{mol}^{-6}\,\text{dm}^{18}.$$

The concentration of water is taken to be a constant as the water molecules are present in large excess. Take note that the square brackets used in the equation represent the concentration of the species, including the complex ions, enclosed within them.

The calculated value for K_{stab} of $[\text{Ni}(\text{NH}_3)_6]^{2+}$ is $5 \times 10^7\,\text{mol}^{-6}\,\text{dm}^{18}$. Take note that the K_{stab} values are measured against the relative stability of the aqua complexes.

The large value of K_{stab} indicates that

- the position of the equilibrium lies to the right,
- $[\text{Ni}(\text{NH}_3)_6]^{2+}$ is a more stable complex ion than $[\text{Ni}(\text{H}_2\text{O})_6]^{2+}$,
- NH_3 is a stronger ligand than H_2O, and
- NH_3 forms a stronger dative covalent bond with the central cation than H_2O.

Since the molecules of H_2O and NH_3 are quite similar in size, there is no change in co-ordination number, i.e., it remains as 6 upon the ligand exchange. You will find this is not always the case, especially when dealing with larger ligands such as Cl^-.

The K_{stab} for some common complexes are given below. Sometimes $\log K_{\text{stab}}$ is used for easier comparison, as big numbers can be dispensed with. Basically, the greater the value of K_{stab}, the more stable the complex is.

Element	Complex Ion	K_{stab}	$\log K_{stab}$
Monodentate ligands			
Fe	$[Fe(CN)_6]^{4-}$	1×10^{24}	24
	$[Fe(CN)_6]^{3-}$	1×10^{31}	31
Ni	$[Ni(CN)_4]^{2-}$	1×10^{31}	31
	$[Ni(NH_3)_6]^{2+}$	5×10^{7}	7.7
Cu	$[Cu(CN)_4]^{3-}$	1×10^{27}	27
	$[Cu(NH_3)_4(H_2O)_2]^{2+}$	1.2×10^{13}	13.1
Bidentate ligands			
Ni	$[Ni(en)_3]^{2+}$	6×10^{18}	18.8
Cu	$[Cu(en)_2(H_2O)_2]^{2+}$	4×10^{19}	19.6
Hexadentate ligands			
Fe	$[Fe(edta)]^{2-}$	2×10^{14}	14.3
	$[Fe(edta)]^{-}$	1.3×10^{25}	25.1
Ni	$[Ni(edta)]^{2-}$	4×10^{18}	18.6
Cu	$[Cu(edta)]^{2-}$	6×10^{18}	18.8

Q: Why must $NH_3(aq)$ be added in excess? What happens when it is added in small amounts?

A: Recall that in aqueous solution the weak base NH_3 undergoes partial ionisation, producing OH^- ions:

$$NH_3(aq) + H_2O(l) \rightleftharpoons NH_4^+(aq) + OH^-(aq).$$

When a small amount of $NH_3(aq)$ is added, the $OH^-(aq)$ reacts with the $[Ni(H_2O)_6]^{2+}(aq)$ and a green precipitate of nickel(II) hydroxide is formed. Its chemical formula is $[Ni(OH)_2(H_2O)_4]$ but it is commonly written as $Ni(OH)_2$ without including the water ligands:

$$[Ni(H_2O)_6]^{2+}(aq) + 2OH^-(aq) \rightleftharpoons \underbrace{[Ni(OH)_2(H_2O)_4]}_{\substack{\text{commonly written} \\ \text{as } Ni(OH)_2}}(s) + 2H_2O(l),$$

i.e.,

$$[Ni(H_2O)_6]^{2+}(aq) + 2OH^-(aq) \rightleftharpoons \underset{\text{green ppt}}{Ni(OH)_2(s)} + 6H_2O(l). \quad (11.5)$$

Precipitation of green $Ni(OH)_2$ occurs once the ionic product exceeds its K_{sp} value. But one can actually view the formation of the precipitate as a form of ligand exchange reaction.

When a substantial amount of $NH_3(aq)$ is added, ligand exchange occurs and the more stable $[Ni(NH_3)_6]^{2+}$ ion is formed, as shown in the following equilibrium:

$$\underset{\text{green}}{[Ni(H_2O)_6]^{2+}}(aq) + 6NH_3(aq) \rightleftharpoons \underset{\text{blue}}{[Ni(NH_3)_6]^{2+}}(aq) + 6H_2O(l). \quad (11.6)$$

There are now two competing reactions as shown by Eqs. (11.5) and (11.6). Both NH_3 and OH^- are competing for the same cation, $[Ni(H_2O)_6]^{2+}$. However, when a large excess of NH_3 is added, $[NH_3]$ is high enough such that the ligand exchange reaction is favoured over precipitation, i.e., the formation of $[Ni(NH_3)_6]^{2+}(aq)$ is favoured over the formation of $Ni(OH)_2$ (see Fig. 11.13).

What actually happened? The increasing addition of $NH_3(aq)$ shifts the equilibrium position in Eq. (11.6) to the right. As the concentration of $[Ni(H_2O)_6]^{2+}(aq)$ decreases, the equilibrium position in Eq. (11.5) is shifted to the left. All of these are in accordance with Le Chatelier's Principle. What we see is the dissolving of the green precipitate of $Ni(OH)_2$ in excess $NH_3(aq)$ and the formation of a blue solution of $[Ni(NH_3)_6]^{2+}(aq)$.

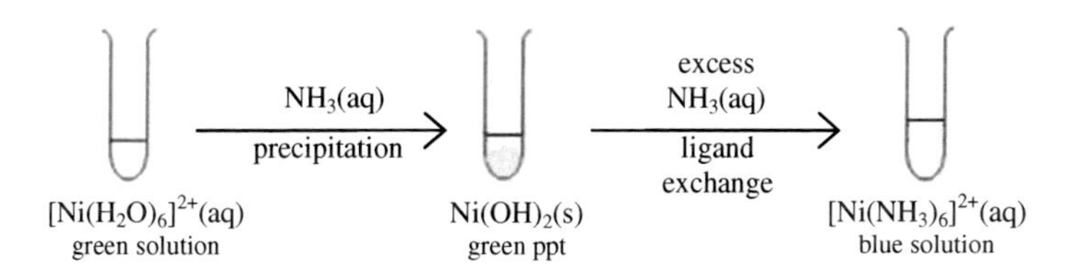

Fig. 11.13. Reaction of $Ni^{2+}(aq)$ with $NH_3(aq)$.

Example 11.12: Describe and explain what happens when gradual amounts of $NH_3(aq)$ are added, until in excess, to a solution of $CuSO_4(aq)$.

Solution: In aqueous solution, $Cu^{2+}(aq)$ exists as $[Cu(H_2O)_6]^{2+}(aq)$, which gives the solution its blue colour.

In aqueous solution, the weak base NH_3 undergoes partial ionisation, producing OH^- ions:

$$NH_3(aq) + H_2O(l) \rightleftharpoons NH_4{}^+(aq) + OH^-(aq).$$

When a small amount of $NH_3(aq)$ is added, the $OH^-(aq)$ reacts with the $[Cu(H_2O)_6]^{2+}(aq)$ and a blue precipitate of $Cu(OH)_2$ is formed:

$$\underset{\text{blue}}{[Cu(H_2O)_6]^{2+}(aq)} + 2OH^-(aq) \rightleftharpoons \underset{\text{blue ppt}}{Cu(OH)_2(s)} + 6H_2O(l). \quad (11.7)$$

Precipitation of $Cu(OH)_2$ occurs when the ionic product exceeds its K_{sp} value.

When a substantial amount of $NH_3(aq)$ has been added, ligand exchange occurs and the more stable $[Cu(NH_3)_4(H_2O)_2]^{2+}$ ion is formed as shown in the following equilibrium:

$$\underset{\text{blue}}{[Cu(H_2O)_6]^{2+}(aq)} + 4NH_3(aq) \rightleftharpoons \underset{\text{deep blue}}{[Cu(NH_3)_4(H_2O)_2]^{2+}(aq)} + 4H_2O(l). \quad (11.8)$$

There are now two competing reactions as shown by Eqs. (11.7) and (11.8). Both NH_3 and OH^- are competing for the same cation $[Cu(H_2O)_6]^{2+}$. However, when a large excess of NH_3 is added, the $[NH_3]$ is high enough such that the ligand exchange reaction is favoured over precipitation, i.e., the formation of $[Cu(NH_3)_4(H_2O)_2]^{2+}(aq)$ is favoured over the formation of $Cu(OH)_2$ (see Fig. 11.14).

The increasing addition of $NH_3(aq)$ shifts the equilibrium position in Eq. (11.8) to the right. As the concentration of $[Cu(H_2O)_6]^{2+}(aq)$ decreases, the equilibrium position in Eq. (11.7) is shifted to the left. All of these are in accordance with Le Chatelier's Principle. What we see is the dissolving of the blue precipitate of $Cu(OH)_2$

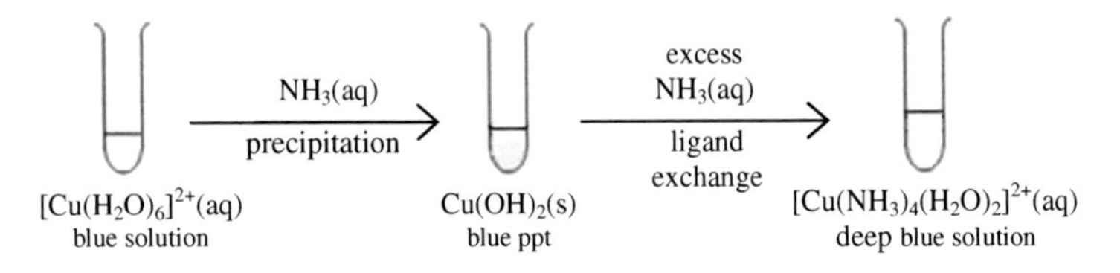

Fig. 11.14. Reaction of $Cu^{2+}(aq)$ with $NH_3(aq)$.

in excess NH_3(aq) and the formation of a deep blue solution of $[Cu(NH_3)_4(H_2O)_2]^{2+}$(aq).

- **Ligand exchange in haemoglobin**

Haemoglobin is a macro-biological molecule in the red blood cell that is responsible for transporting oxygen molecules to the other cells during respiration. The structure of haemoglobin is as follows:

- The metal centre consists of an iron(II) ion coordinated to six groups of molecules.
- Four of the co-ordination sites are taken up by nitrogen atoms from a ring system called a porphyrin which acts as a tetradentate ligand. This complex is called "haem."
- The fifth co-ordination site is taken up by the nitrogen atom of a complex protein called "globin."
- The sixth co-ordination site is reversibly bonded to an oxygen molecule. This allows haemoglobin to carry oxygen from one part of the body to another.
- The better ligands such as carbon monoxide and cyanide ions can be bonded less reversibly to the metal centre, which thus inhibits haemoglobin from carrying oxygen. This accounts for the toxic nature of such strong ligands.

Schematic representation of the haem showing the reversible uptake of the oxygen molecule:

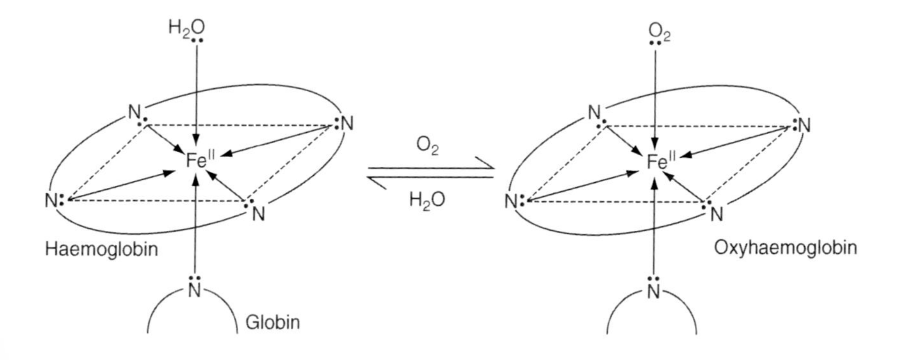

- **Effect of ligand exchange on $E^\ominus_{cell}$ values**

Two complexes with the same metal centre but different types of ligands bonded to it exhibit a differential ability to undergo reduction. For example, replacement of H_2O ligands in an aqua-complex by other ligands can cause large changes in electrode potential values.

Example 11.13:

$$[Fe(H_2O)_6]^{3+}(aq) + e^- \rightleftharpoons [Fe(H_2O)_6]^{2+}(aq), \quad E^\ominus = +0.77\,V,$$
$$[Fe(CN)_6]^{3-}(aq) + e^- \rightleftharpoons [Fe(CN)_6]^{4-}(aq), \quad E^\ominus = +0.36\,V,$$
$$\tfrac{1}{2}I_2(aq) + e^- \rightleftharpoons I^-(aq), \quad E^\ominus = +0.54\,V.$$

Replacing the H_2O ligand by the CN^- ligand causes the $E^\ominus$ value for the Fe(III)/Fe(II) system to become less positive, i.e., the Fe(III) metal centre becomes less oxidising. It is not difficult to understand this as the positively charged $[Fe(H_2O)_6]^{3+}(aq)$ is more likely to take up an electron than the negatively charged $[Fe(CN)_6]^{3-}(aq)$.

Thus, the CN^- ligand, through the formation of the $[Fe(CN)_6]^{3-}(aq)$ complex, helps to stabilise the +3 oxidation state of Fe relative to the +2 oxidation state of Fe.

The differences in the oxidising power of both the $[Fe(CN)_6]^{3-}(aq)$ and $[Fe(H_2O)_6]^{3+}(aq)$ can be demonstrated by calculating the $E^\ominus_{cell}$ values of the redox reactions as follows:

$$[Fe(H_2O)_6]^{3+}(aq) + I^-(aq) \rightarrow [Fe(H_2O)_6]^{2+}(aq) + \frac{1}{2}I_2(aq),$$
$$E^\ominus_{cell} = +0.23\,V (> 0\,V),$$
$$[Fe(CN)_6]^{3-}(aq) + I^-(aq) \rightarrow [Fe(CN)_6]^{4-}(aq) + \frac{1}{2}I_2(aq),$$
$$E^\ominus_{cell} = -0.18\,V (< 0\,V).$$

The more positive $E^\ominus_{cell}$ for the reaction between $[Fe(H_2O)_6]^{3+}(aq)$ and $I^-(aq)$ indicates that the redox reaction is more thermodynamically spontaneous under standard conditions.

Example 11.14:

$$[Co(H_2O)_6]^{3+}(aq) + e^- \rightleftharpoons [Co(H_2O)_6]^{2+}(aq), \quad E^\ominus = +1.82\,V,$$
$$[Co(NH_3)_6]^{3+}(aq) + e^- \rightleftharpoons [Co(NH_3)_6]^{2+}(aq), \quad E^\ominus = +0.10\,V,$$
$$[Co(CN)_6]^{3-}(aq) + e^- \rightleftharpoons [Co(CN)_6]^{4-}(aq), \quad E^\ominus = -0.80\,V.$$

The $E^{\ominus}$ values show that there is stabilisation of Co(III) with respect to Co(II) by the NH_3 and the CN^- ligands. Since the lone pair of electrons of NH_3 is more likely to be donated for dative covalent bonding than the lone pair of H_2O (because the effective nuclear charge of N is lower than that of the O atom), we expect the cobalt metal centre of $[Co(NH_3)_6]^{3+}(aq)$ to be more electron rich than $[Co(H_2O)_6]^{3+}(aq)$. As a result, the cobalt centre of $[Co(NH_3)_6]^{3+}(aq)$ is less likely to take up an electron. In addition, since $[Co(CN)_6]^{3-}(aq)$ is negatively charged, it is less likely to take up an electron to undergo reduction. The differential oxidising power of cobalt complexes is indicated by the following observations:

- $[Co(H_2O)_6]^{3+}$ ions are so strongly oxidising that they react with water to produce oxygen.
- $[Co(NH_3)_6]^{2+}(aq)$ and $[Co(CN)_6]^{4-}(aq)$ are so strongly reducing that they are oxidised by air.

11.3.6 *Selected reactions of some transition metals and their compounds*

- ***Chromium and its compounds***

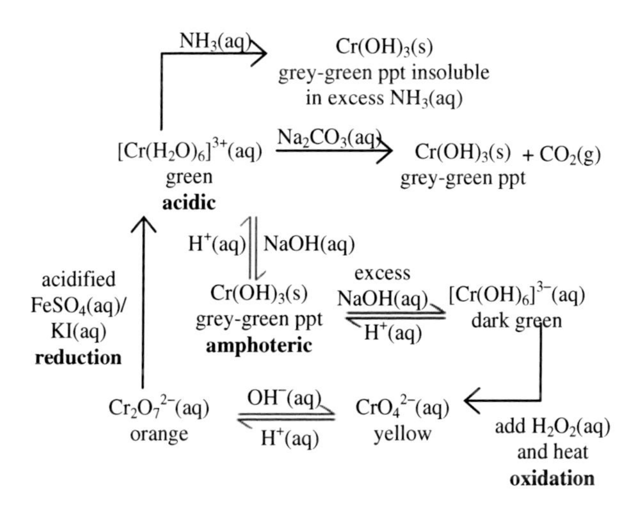

• *Manganese and its compounds*

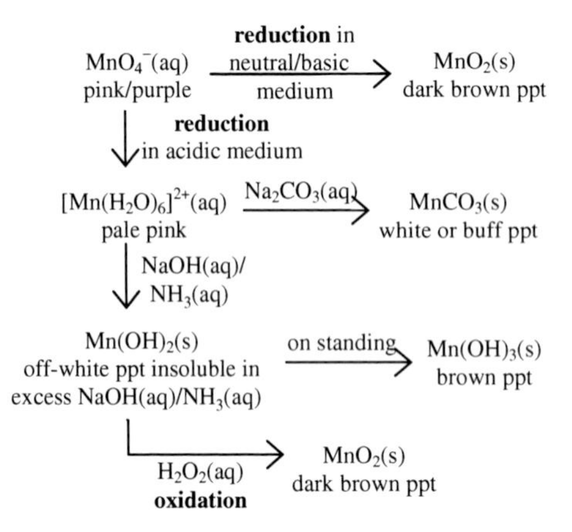

• *Iron and its compounds*

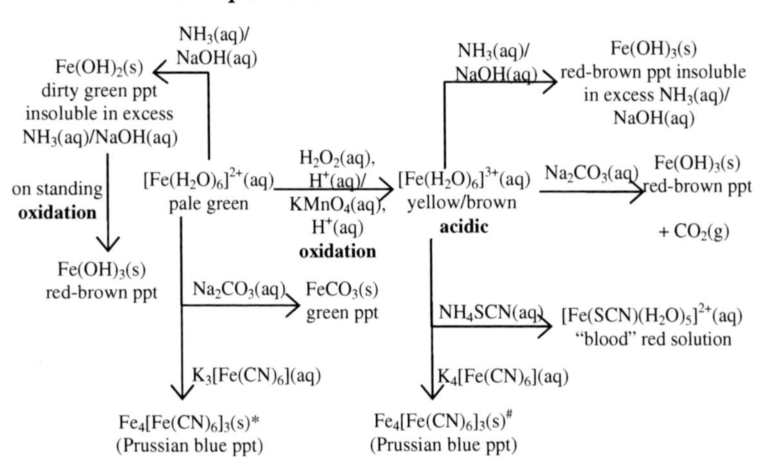

*The reaction is: $Fe^{2+}(aq) + [Fe(CN)_6]^{3-}(aq)$

$$\rightarrow Fe^{3+}(aq) + [Fe(CN)_6]^{4-}(aq),$$

$$4Fe^{3+}(aq) + 3[Fe(CN)_6]^{4-}(aq) \rightarrow Fe_4[Fe(CN)_6]_3(s).$$

#The reaction is: $4Fe^{3+}(aq) + 3[Fe(CN)_6]^{4-}(aq) \rightarrow Fe_4[Fe(CN)_6]_3(s)$

Note: $K_4[Fe(CN)_6]$ — potassium hexacyanoferrate(II)
NH_4SCN — ammonium thiocyanate
$K_3[Fe(CN)_6]$ — potassium hexacyanoferrate(III)

- ***Copper and its compounds***

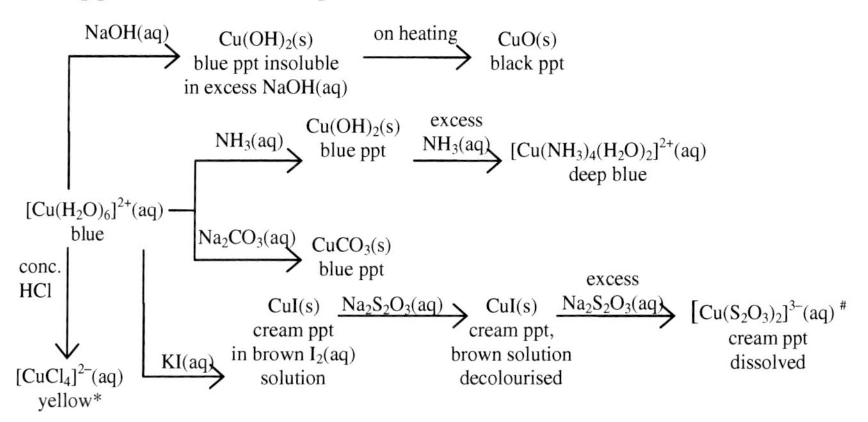

*The reaction is: $\underset{\text{blue}}{[Cu(H_2O)_6]^{2+}}(aq) + 4Cl^-(aq)$

$\rightleftharpoons \underset{\text{yellow}}{[CuCl_4]^{2-}}(aq) + 6H_2O(l).$

As this is a reversible reaction, the presence of bothblue$[Cu(H_2O)_6]^{2+}$ and yellow $[CuCl_4]^{2-}$ ions causes the resultant solution to be green in colour.

#The reaction is: $2Cu^{2+}(aq) + 4I^-(aq) \rightarrow \underset{\text{cream}}{2CuI(s)} + \underset{\text{brown}}{I_2(aq)},$

$I_2(aq) + 2S_2O_3^{\;2-}(aq) \rightarrow 2I^-(aq) + S_4O_6^{\;2-}(aq),$

$CuI(s) + 2S_2O_3^{\;2-}(aq) \rightarrow [Cu(S_2O_3)_2]^{3-}(aq) + I^-(aq).$

My Tutorial (Chapter 11)

1. (a) Discuss, in relation to their electronic structures, the variation in the number of oxidation states shown by the elements along the first transition series.

   (b) (i) What is the oxidation number of nickel in each of the following compounds?

   (A) $K_2Ni(CN)_4$,

   (B) $Ni(H_2O)_6(NO_3)_2$, and

   (C) $Ni(CO)_4$.

(ii) Suggest the likely stereochemical arrangement of ligands around the nickel atom in compounds (A) and (B).

(iii) Nickel carbonyl, (C), has a tetrahedral arrangement of the CO ligands around the nickel atom. What would you expect the stereochemical arrangement around the nickel atom to be in the compound $K_4Ni(CN)_4$? Give your reasoning.

(c) A green aqueous solution of a nickel(II) salt is converted to a blue solution containing $[Ni(NH_3)_6]^{2+}$ ions by the addition of an excess of aqueous ammonia. The green solution is then converted to a yellow solution containing $[Ni(CN)_4]^{2-}$ ions by addition of an excess of aqueous potassium cyanide. Explain why the colours of the solutions are different.

2. In addition to its widespread use as a structural material, iron is often employed as a heterogeneous catalyst for industrial reactions.

(a) What is meant by the term *heterogeneous catalysis*?

(b) Explain how iron may function as a heterogeneous catalyst in the Haber process for the synthesis of ammonia.

(c) What chemical properties of iron make it effective both as a heterogeneous and as a homogeneous catalyst?

(d) Explain, by means of an example, why iron complexes are essential to human life.

3. Replacement of water molecules by other ligands generally changes the redox potentials of transition metal ions. Use the following information in your answer to this question.

	E° (V)
$[Co(H_2O)_6]^{3+}(aq) + e^- \rightleftharpoons [Co(H_2O)_6]^{2+}(aq)$	+1.81
$\frac{1}{2}O_2(g) + 2H^+(aq) + 2e^- \rightleftharpoons H_2O(l)$	+1.23
$H^+(aq) + e^- \rightleftharpoons \frac{1}{2}H_2(g)$	0.00
$[Co(CN)_6]^{3-}(aq) + e^- \rightleftharpoons [Co(CN)_6]^{4-}(aq)$	−0.83

(a) What products are likely to be formed when cobalt(III) sulfate is dissolved in water? Give an equation for the reaction.

(b) If an aqueous solution of cobalt(II) chloride is mixed with an excess of aqueous potassium cyanide, the ion $[Co(CN)_6]^{4-}(aq)$ is formed. Explain why this mixture absorbs oxygen from the air.

(c) Suggest another reaction which might take place in the cyanide mixture if air were excluded.

4. One difference between the chemistries of calcium and manganese is that manganese can undergo disproportionation reactions whereas calcium cannot.

(a) Explain in terms of the electronic structures of calcium and manganese why disproportionation is a feature of the chemistry of many of the elements of the *d* block but none of those in the *s* block.

(b) Derive the half-equation for the reduction of manganate(VII) ions in acidic solution to MnO_2. Given that $E^\ominus$ for this reduction is +1.67 V, show that manganese(IV) oxide will not disproportionate to MnO_4^- and Mn^{2+} in acid solution.

$$MnO_2(s) + 4H^+(aq) + 2e^- \rightleftharpoons Mn^{2+}(aq) + 2H_2O(l),$$

$$E^\ominus = +1.23\,V.$$

5. (a) Give the electronic configuration of a V^{3+} ion.

(b) Vanadium is a transition element. Give two characteristic properties of such a transition element other than the ability to form coloured ions.

(c) Ammonium vanadate(V) dissolves in sulfuric acid to give a yellow solution, the colour being due to the VO_2^+ ion.

(i) What is the oxidation number of vanadium in the VO_2^+ ion?

(ii) Give the systematic name of the VO_2^+ ion.

(d) Treatment of the yellow solution from (c) with zinc causes the colour to change to green, then to blue, followed by green again, and finally violet. Give the formulae of the ions responsible for each of these colours.

(e) In the sequence of changes in (d), zinc acts as a reducing agent.

(i) State the meaning of the term *reducing agent.*
(ii) Write a half-equation showing how zinc acts as a reducing agent.
(iii) Write the half-equation for the conversion of ${VO_2}^+$ to VO^{2+} in acid solution.
(iv) Hence write the equation for the reduction of ${VO_2}^+$ to VO^{2+} by zinc.

6. Coins are made from an alloy, nickel-brass, which consists essentially of the metals copper, nickel, and zinc. A one pound coin weighing 9.50 g was completely dissolved in concentrated nitric acid, in which all three metals dissolved, to give solution A.

Dilute sodium hydroxide solution is then added carefully with stirring, until present in excess. Zinc hydroxide is amphoteric. The precipitate formed, B, was filtered off from the supernatant liquid, C. The precipitate B, was quantitatively transferred to a graduated flask of 500 cm^3 capacity. Dilute sulfuric acid was then added dropwise to dissolve the whole of precipitate B and the solution was made up to 500 cm^3 with distilled water.

25.0 cm^3 portions of this solution were pipetted into a conical flask and an excess of potassium iodide solution was added. The liberated iodine was then titrated against a sodium thiosulfate solution of concentration 0.100 mol dm^{-3}. 18.7 cm^3 of the sodium thiosulfate solution was required for a complete reaction.

(a) (i) With reference to the Data Booklet, using the appropriate half-equations, write an equation for the reaction of any one of the metals in nickel-brass with concentrated nitric acid.
(ii) What type of reaction is taking place?

(b) Identify by giving full formulas:

(i) the complex cations present in A,
(ii) the precipitates in B, and
(iii) any metal-containing anion in C.

(c) (i) Write an equation for the precipitate of any one of the metal ions in A with sodium hydroxide.
(ii) What type of reaction is occurring in c(i)?

(d) Suggest an explanation why it is necessary to add sodium hydroxide, followed by dilute sulfuric acid, before performing the titration.

(e) On addition of the potassium iodide solution, the only reaction that occurs is:

$$2Cu^{2+}(aq) + 4I^{-}(aq) \longrightarrow 2CuI(s) + I_2(aq).$$

(i) Write an equation for the reaction between sodium thiosulfate and the liberated iodine. What indicator would you use in this titration? At what stage would you add it? Give a reason for your answer.
(ii) Calculate the percentage of copper in the alloy.
(iii) Suggest why this reaction occurs in light of the $E^{\ominus}$ values from the Data Booklet.

Index